Extractive Metallurgy of
COPPER

Extractive Metallurgy of
COPPER
Sixth Edition

MARK E. SCHLESINGER
Missouri University of Science and Technology, Rolla, MO, United States

KATHRYN C. SOLE
Sole Consulting, Johannesburg, Gauteng, South Africa; University of Pretoria, Pretoria, Gauteng, South Africa

WILLIAM G. DAVENPORT
Emeritus Professor, Department of Materials Science and Engineering, University of Arizona, Tuscon, AZ, United States

GERARDO R. F. ALVEAR FLORES
Adjunct Professor, University of Queensland, Brisbane, QLD, Australia; Manager Technical Marketing, Rio Tinto Singapore Holdings Pte Ltd., Singapore

ELSEVIER

Elsevier
Radarweg 29, PO Box 211, 1000 AE Amsterdam, Netherlands
The Boulevard, Langford Lane, Kidlington, Oxford OX5 1GB, United Kingdom
50 Hampshire Street, 5th Floor, Cambridge, MA 02139, United States

Library of Congress Cataloging-in-Publication Data
A catalog record for this book is available from the Library of Congress

British Library Cataloguing-in-Publication Data
A catalogue record for this book is available from the British Library

ISBN: 978-0-12-821875-4

For information on all Elsevier publications visit our website at https://www.elsevier.com/books-and-journals

Publisher: Joseph P. Hayton
Acquisitions Editor: Susan Dennis
Editorial Project Manager: Mariana L. Kuhl
Production Project Manager: Vijayaraj Purushothaman
Cover Designer: Mark Rogers

Typeset by TNQ Technologies

CONTENTS

Preface to the sixth edition

When we began this revision in 2019, the price of copper was \$2.60/lb. As we complete it today, the price is \$4.38/lb. By the time you read these words, it may be back down to \$2.60 again! Regardless, it has been some time since there was so much interest in the availability and technology of copper supply. The long-term prospects for the industry appear to be bright, and prospectors are scouring the globe for ore, from mountaintops in the Andes to the bottom of the oceans. The lists of top-producing mines and smelters in this book demonstrate the increase in capacity of the industry from our previous edition.

Technology advances in the industry over the previous decade have been largely evolutionary rather than revolutionary. The most significant developments include the following:

- the adoption of Chinese bottom-blown and side-blown bath smelting and converting technology;
- the creation of flowsheets for handling increasingly complex ores and concentrates;
- improved technology for processing high-strength off-gases;
- higher current densities in electrorefining and electrowinning practice.

Other things have not changed. The Peirce–Smith converter is still the primary means of producing blister copper; flash furnaces are still being commissioned for concentrate smelting, and casting processes are still largely the same. Some technologies have more value than previously thought!

There has also been a change in the creative team behind this book. Matt King has decided to find other uses for his nights and weekends; in his place, we welcome to the group Dr. Gerardo Alvear Flores, currently based in Singapore. Gerardo has vast experience in the copper industry, which has taken him over his career to multiple continents and production facilities, and his experience will enhance our efforts and improve our direction.

As with previous efforts, this edition of *Extractive Metallurgy of Copper* is largely a product of the copper industry as a whole, since so many engineers and scientists volunteered their time and expertise (along with photographs and drawings) to make sure we got it right. Our guides included

Ken Armstrong (Chemetics)
Nigel Aslin (Glencore)
Martin Bakker (Glencore)
Michele Beacom (CIM)
Juan Carrasco (Glencore)

Chunlin Chen (CSIRO)
Peter Cole (deceased) (Peter Cole Metallurgical Services)
Bernadette Currie (Rio Tinto)
Julio Flores Cantillánez (Mantos Copper)
Magnus Ek (Boliden)
Jeremy Gillis (PASAR)
Cameron Harris (Canadian Engineering Associates)
David Hazelett (Hazelett Corporation)
Enrique Herrera (Southern Peru Copper)
Shawn Hinsberger (Heath and Sherwood)
Mike Hourn (Glencore)
Chris Holding (Copper Worldwide)
John Hugens
John Hugens, Jr. (Hugens Metallurgy)
Evgeni Jak (University of Queensland)
Hugo Joubert (Tenova Pyromet)
Akira Kaneda (Mitsubishi Materials)
Matt King (Worley Parsons)
Michael Marinigh (Heath and Sherwood)
Leopoldo Mariscal (Mexicana de Cobre)
Jorge Meza (Southern Peru Copper)
Yoshihiro Mine (Mitsui Mining & Smelting)
Enrique Miranda (IIMCH)
Michael Moats (Missouri University of Science and Technology)
Anthony Mukutuma (First Quantum Minerals)
Kensaku Nakamura (JX Nippon Mining and Metals)
Takeshi Nakamura (Tohoku University)
Stanko Nikolic (Glencore)
Yerzhan Ospanov (Kazakhmys)
Lauri Palmu (Metso Outotec)
John Parker (Process Ideas)
Joe Pease (Minerales)
Beatrice Pierre (Glencore)
Nikola Popovic (Glencore)
Yves Prevost (Glencore)
Venkoba Ramachandran
Pedro Reyes (Anglo American)
Carlos Risopatron (International Copper Study Group)
Joanne Roberts (Freeport McMoRan)
Tim Robinson (Metso Outotec)

Gerardo Sanchez (Codelco)
Maxsym Shevchenko (University of Queensland)
Etsuro Shibata (Tohoku University)
Denis Shishin (University of Queensland)
Alfred Spanring (RHI AG)
Andreas Specht (Aurubis)
Hitoshi Takano (Sumitomo)
Supriya Upadhyay (Rio Tinto)
Maurits Van Camp (Umicore Precious Metals Recycling)
Nicole Witaslawsky (International Copper Association)
Yutaka Yasuda (JX Nippon Metals and Mining)
Gabriel Zárate (Jetti Cobre)

CHAPTER 1

Overview

1.1 Introduction

Copper is most commonly present in the earth's crust as copper—iron—sulfide and copper sulfide minerals, primarily chalcopyrite ($CuFeS_2$) and chalcocite (CuS_2). The concentration of these minerals in a mine is low. Typical copper ores contain from 0.3% Cu (open pit mines, Fig. 1.1) to 1.7% Cu (underground mines, Fig. 1.2). The rest is noneconomic rock. Pure copper metal is mostly produced from these ores by concentration, smelting, and refining (Fig. 1.3).

Copper also occurs to a lesser extent in oxidized minerals (carbonates, oxides, hydroxysilicates, sulfates). Copper metal is usually produced from these minerals by leaching, solvent extraction, and electrowinning (Fig. 1.4). These processes are also used to treat chalcocite (Cu_2S).

Figure 1.1 Open pit Cu mine. Note the new blast holes, top right, and blasted ore to the left of them. The shovel is placing blasted ore in the truck from where it will go to processing. The water truck is suppressing dust. The front-end loader is cleaning up around the shovel. The shovel is electric. Its insulated power wire mostly lies on the surface except over the wire bridge under which all vehicles travel to and from the shovel. *(Photograph courtesy of Freeport—McMoRan Copper & Gold Inc.).*

Extractive Metallurgy of Copper, Sixth Edition
ISBN 978-0-12-821875-4, https://doi.org/10.1016/B978-0-12-821875-4.00017-1

Figure 1.2 Underground mining operation. Underground mine showing wheeled mining machine. The drills on the right end of the vehicle are drilling up into the ore body. The drilled holes are subsequently filled with explosive which, when detonated, break the ore and drop the ore pieces onto the mine floor. The ore pieces are then elevated to the mine surface where they are crushed and sent to processing. A new mine in Arizona contains ~1.5% Cu, 0.04% Mo in chalcopyrite and molybdenite minerals. ~120 kilotonnes per day of this ore will be mined ~2300 m below ground level. About one-fourth of the world's mined copper is obtained by underground mining; the remainder is by surface mining. *(Purchased from Roslyn Budd: www.buddphotography.com.au).*

A third major source of copper is scrap copper and copper alloys. Yearly production of copper from recycled used objects is 15%–20% of mine production. In addition, there is considerable remelting and refining of scrap generated during fabrication and manufacture. Total copper production in 2018 (mined and from end-of-use scrap) was 25 million tonnes.

This chapter introduces the principal processes by which pure copper is extracted from ore and scrap. It also indicates the relative industrial importance of each. Finally, it discusses worker safety—because it is so important.

1.2 Ore–rock differentiation in the mine

A mine consists of regions of economic ore and regions of noneconomic rock, e.g., quartz and feldspar (Berger et al., 2008). These regions are located by drilling, sampling, and analyzing the drill core for Cu. Where possible the noneconomic rock is not mined. When the mine can only be developed by mining intertwined ore and rock, the ore is sent to copper production and the rock goes to waste heaps. The material at the ore/rock boundary is carefully sampled and analyzed to ensure that ore is not mistakenly sent to the waste rock heaps.

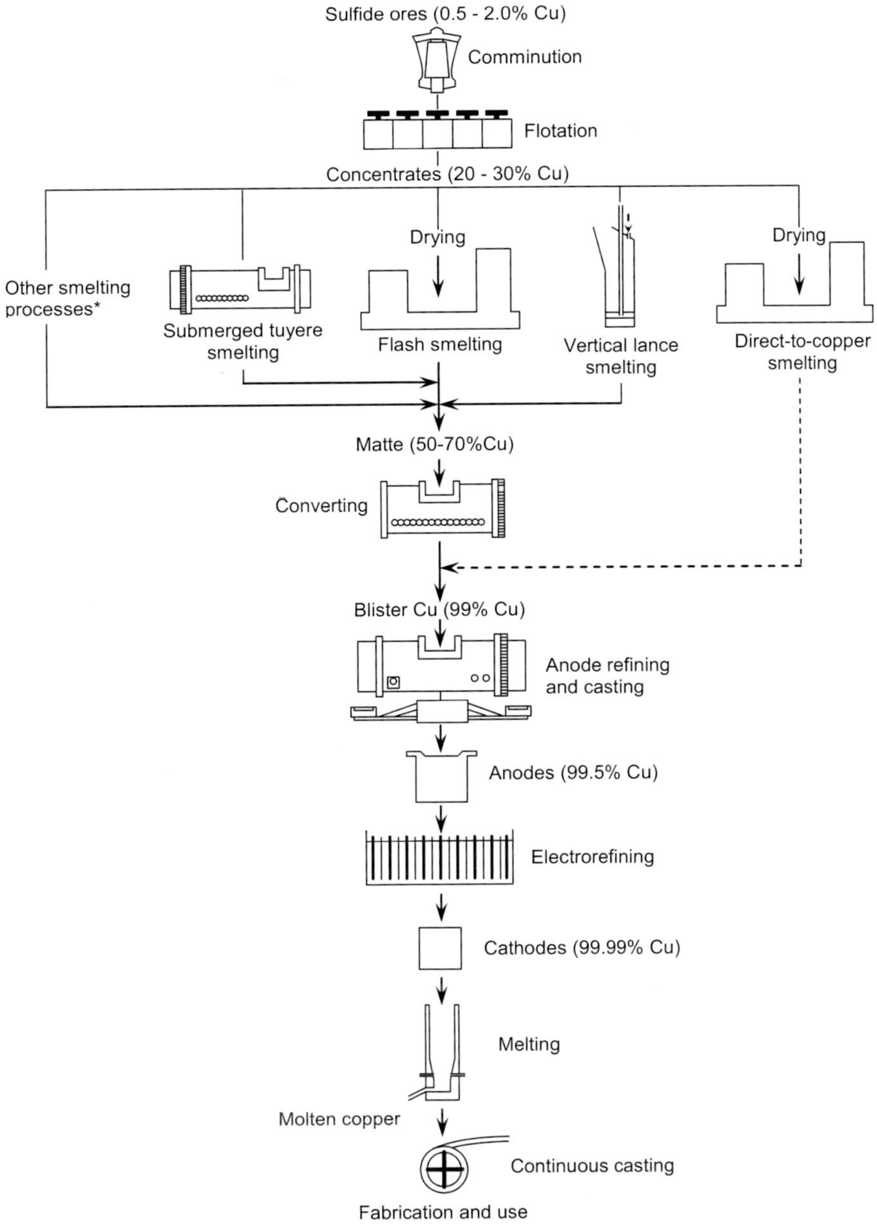

Figure 1.3 Main processes for extracting copper from sulfide ores. *Parallel lines* indicate alternative processes. *Principally Mitsubishi and Vanyukov smelting. ————and various small Chinese processes.

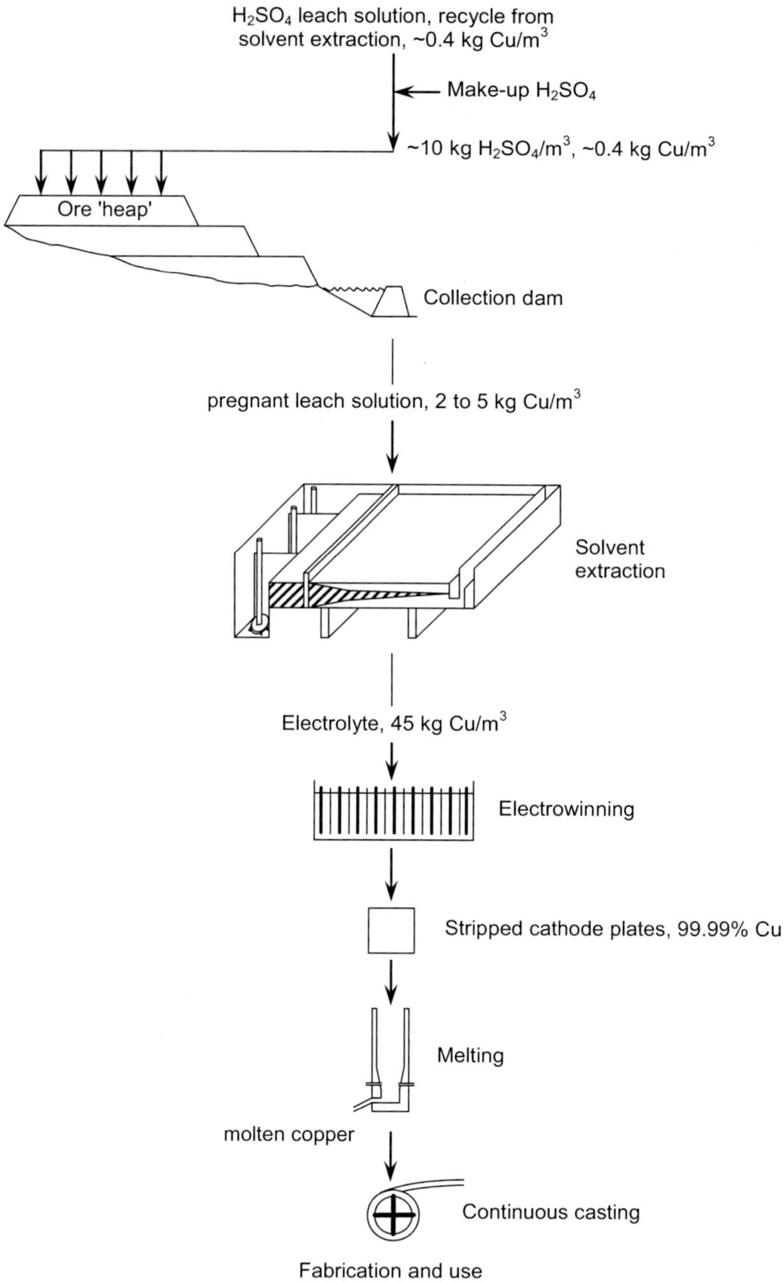

Figure 1.4 Flowsheet for leaching oxide and Cu_2S ores. The dissolved Cu is recovered by solvent extraction purification followed by electrowinning. Leaching accounts for ~20% of primary (from ore) copper production.

1.3 Extracting copper from copper–iron–sulfide ores

About 80% of the world's mined copper originates in chalcopyrite ($CuFeS_2$) ores. Chalcopyrite is extremely difficult to dissolve in aqueous solutions, so the vast majority of copper extraction from these ores is pyrometallurgical. Extraction by this approach entails

(a) isolating the ore's Cu–Fe–S and Cu–S mineral particles in a concentrate by froth flotation

(b) smelting this concentrate to molten high-Cu sulfide *matte*

(c) converting (oxidizing) this molten matte to impure molten *blister* copper

(d) fire- and electrorefining this impure copper to ultrapure copper.

Increasingly, smelting and converting are being combined, especially in China.

1.3.1 Concentration by froth flotation

Copper ores being mined in 2020 are far too dilute in Cu (0.3%–1.7%) to be smelted directly. Heating and melting their huge quantity of waste rock would require prohibitive amounts of fuel. The rock would also produce impossibly large amounts of molten oxide slag. Fortunately, the ore's Cu–Fe–S and Cu–S minerals can be isolated by physical means into high-Cu concentrate, which can then be smelted economically.

The most effective method of isolating the Cu minerals is froth flotation. This process causes small Cu–Fe–S and Cu–S mineral particles to become attached to air bubbles rising through a finely ground ore slurry (Fig. 1.5). Selectivity of flotation is created by

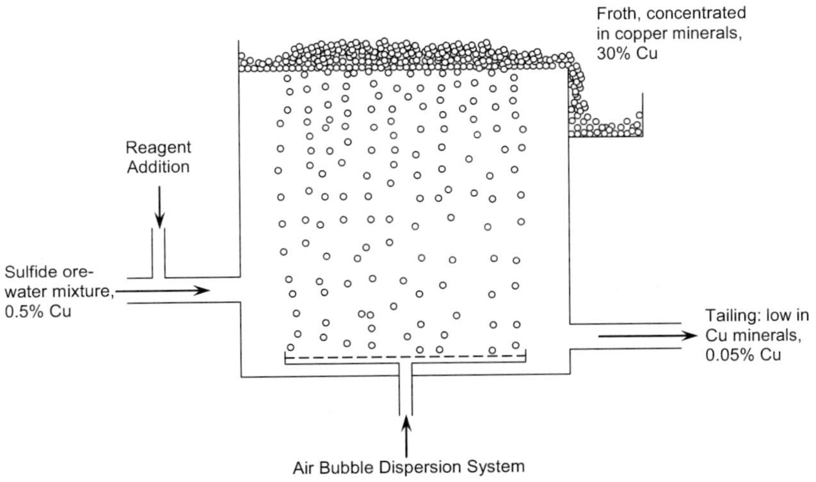

Figure 1.5 Schematic view of flotation cell. Reagents cause Cu–Fe sulfide and Cu sulfide minerals in the ore to attach to rising air bubbles, which are then collected in a short-lived froth. This froth is dewatered to become concentrate. The unfloated waste passes through several cells before being discarded as a final tailing. Many types and sizes (up to 300 m^3) of cell are used (Chapter 3).

using chemical reagents (collectors) that make the Cu minerals water-repellent while leaving waste minerals covered by water (i.e., wetted).

This water repellency causes Cu mineral particles to attach to the flotation cell's rising bubbles while the other mineral particles remain unfloated. The floated minerals overflow the cell in a froth that, when dewatered, becomes ~25% Cu *concentrate*. The low-Cu waste rock particles underflow the cell. They are dewatered and sent to tailing ponds.

Flotation is preceded by crushing and grinding the mined Cu ore to small (~100 μm diameter) particles. Its use has led to adoption of smelting processes that efficiently smelt finely ground solids.

A video of mining and mineral processing may be seen in *The Mining Process at Copper Mountain Mine*. This mine produces 30,000 tonnes of ore and 160,000 tonnes of waste rock per day.

1.3.2 Matte smelting

Matte smelting oxidizes and melts flotation concentrate in a large, hot (1250°C) furnace (Figs. 1.3 and 1.6). The objective of the smelting is to oxidize S and Fe from the Cu—Fe—S concentrate to produce an intermediate Cu-enriched molten sulfide matte. The oxidant is oxygen-enriched air. Example reactions are

$$\underset{\text{in concentrate}}{\overset{30\,^{\circ}\mathrm{C}}{2\,\mathrm{CuFeS_2(s)}}} + \underset{\text{in oxygen-enriched air}}{3.25\,\mathrm{O_2(g)}} \rightarrow \underset{\text{molten matte}}{\overset{1220\,^{\circ}\mathrm{C}}{\mathrm{Cu_2S\text{-}0.5FeS}(\ell)}} + \underset{\text{iron oxide}}{1.5\,\mathrm{FeO(s)}} + \underset{\text{in offgas}}{2.5\,\mathrm{SO_2(g)}} \tag{1.1}$$

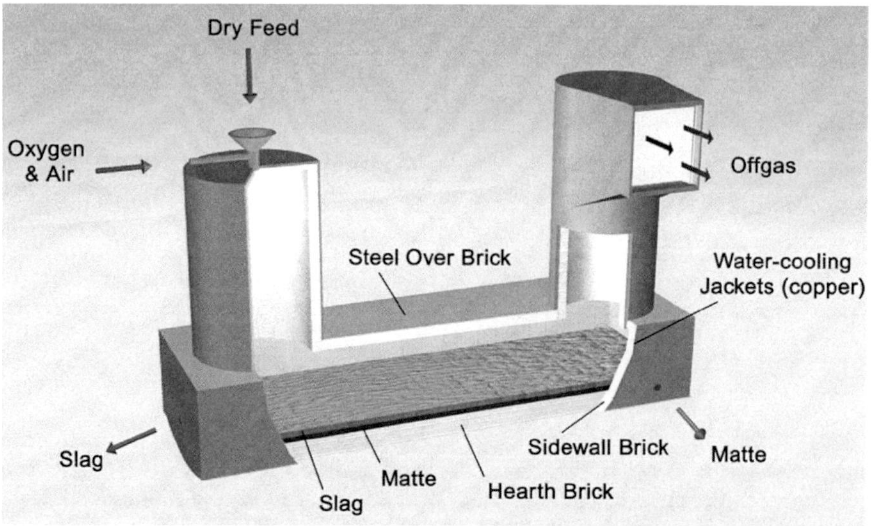

Figure 1.6 Metso Outokumpu oxygen-enriched air flash furnace. Flash furnaces are typically 20 m long and 7 m wide. They smelt 1000—3000 tonnes of concentrate per day.

Simultaneously, the iron oxide is fluxed with quartz to give a low melting point molten slag. A representative reaction is

$$\underset{\text{iron oxide}}{\overset{1220\ ^\circ\text{C}}{1.5\ \text{FeO(s)}}} + \underset{\substack{\text{in rock and}\\ \text{quartz flux}}}{\overset{30\ ^\circ\text{C}}{0.75\ \text{SiO}_2\text{(s)}}} + \text{heat} \rightarrow \underset{\text{molten slag}}{\overset{1250\ ^\circ\text{C}}{0.75\ \text{Fe}_2\text{SiO}_4(\ell)}} \tag{1.2}$$

The concentrate's *gangue* (waste rock) minerals also dissolve in the molten slag.

Reactions (1.1) and (1.2) are both exothermic; they supply most of the smelting process's heat requirement.

The products of smelting are (a) molten sulfide matte (55%—70% Cu) containing most of the Cu in the concentrate and (b) molten oxide slag with as little Cu as possible. The molten matte is subsequently converted (oxidized) in a converting furnace to form impure molten copper. The slag is treated for Cu recovery, then discarded or sold (Chapter 10).

SO_2-bearing offgas (10—60 volume% SO_2, remainder mostly N_2) is also produced. SO_2 is harmful to flora and fauna, so it must be removed before the offgas is released to the atmosphere. This is done by capturing the SO_2 as sulfuric acid (Chapter 11).

Matte smelting is mostly done in flash furnaces (Fig. 1.6). It is also done in top lance and submerged tuyere furnaces (Chapter 6). Three smelters smelt Cu-rich concentrate directly to molten copper (Chapter 10). This is unusual because single-furnace smelting with normal (25% Cu) concentrate produces too much high-Cu slag.

1.3.3 Converting

Copper converting is oxidizing the intermediate molten matte from smelting with air or oxygen-enriched air. It removes Fe and S from the matte to produce crude (99% Cu) molten copper. This molten copper is subsequently sent to fire- and electrorefining.

Most converting is done in cylindrical Peirce—Smith converters (Fig. 1.7a), but newer continuous converting processes are eating into this dominance (Chapter 8).

1.3.3.1 Peirce—Smith converting

Peirce—Smith converting entails pouring large ladles of molten smelting furnace matte (1220°C) into a cylindrical converter through a large central mouth (Fig. 1.7b). The oxidizing oxygen-enriched air *blast* is then started and the converter is rolled, forcing the blast into the molten matte through a row of tuyeres along the length of the vessel. The heat generated in the converter by Fe and S oxidation is sufficient to make the process autothermal.

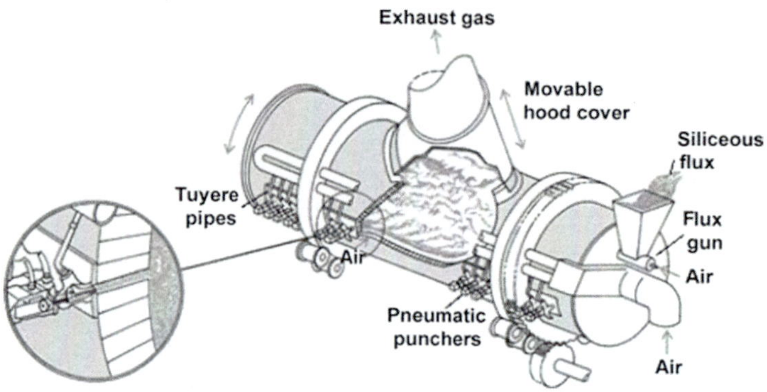

Figure 1.7a Peirce–Smith converter for producing molten "blister" copper from molten Cu–Fe–S matte. Typical production rates are 200–600 tonnes of copper per day. Oxygen-enriched air or air "blast" is blown into the matte through submerged tuyeres. Silica flux is added through the converter mouth or by air gun through an endwall. Offgas is collected by means of a hood above the converter mouth. *(From Chapter 2.1 - Copper Production in Treatise on Process Metallurgy, Volume 3: Industrial Processes, 1st Edition, Copyright Elsevier 2013 Pages 534–624).*

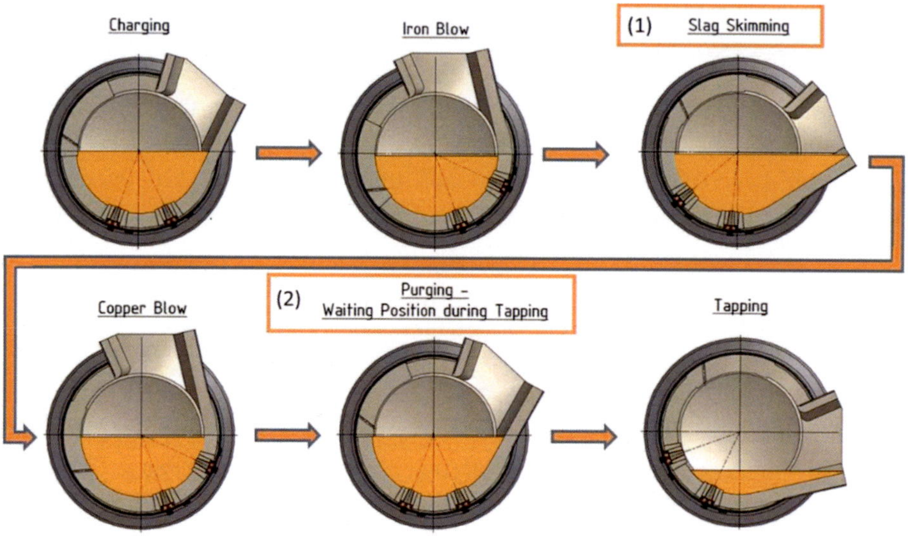

Figure 1.7b Positions of Peirce–Smith converter for charging, blowing, and skimming. SO_2 offgas escapes the system unless the hooding is tight. A converter is typically 4 or 4.5 m diameter. *(Credit: Used with permission from Mettop, https://mettop.com/api/cdn/uploads/1493364462_7v372frw.pdf Figure 4).*

The converting takes place in two sequential stages:

(a) the FeS elimination or slag-forming stage:

$$\underset{\text{in molten matte}}{\overset{1220\,^\circ\text{C}}{2\,FeS(\ell)}} + \underset{\text{in air blast}}{\overset{30\,^\circ\text{C}}{3\,O_2(g)}} + \underset{\text{in quartz flux}}{\overset{30\,^\circ\text{C}}{SiO_2(s)}} \rightarrow \underset{\text{molten slag}}{\overset{1200\,^\circ\text{C}}{Fe_2SiO_4(\ell)}} + \underset{\text{in offgas}}{\overset{1200\,^\circ\text{C}}{2\,SO_2(g)}} + \text{heat} \qquad (1.3)$$

(b) the copper-making stage:

$$\underset{\substack{\text{molten matte}}}{\overset{1200\,^\circ\text{C}}{Cu_2S(\ell)}} + \underset{\substack{\text{in air blast}}}{\overset{30\,^\circ\text{C}}{O_2(g)}} \rightarrow \underset{\substack{\text{impure molten}\\ \text{'blister' copper}}}{\overset{1200\,^\circ\text{C}}{2\,Cu(\ell)}} + \underset{\substack{\text{in offgas}}}{\overset{1200\,^\circ\text{C}}{SO_2(g)}} + \text{heat} \qquad (1.4)$$

Copper making (b) occurs only after the matte contains less than about 1% Fe, so that most of the Fe can be removed from the converter (in slag) before copper production begins. Likewise, significant oxidation of copper does not occur until the sulfur content of the copper falls below ~0.02%. Blowing is terminated near this sulfur end point. The resulting molten blister copper (1200°C) is poured out and sent to refining, whereupon the converter is refilled with molten matte and the process starts again.

Because conditions in the converter are strongly oxidizing and agitated, Peirce–Smith converter slag inevitably contains 4%–8% Cu. This Cu is recovered by settling and/or ambient temperature froth flotation. The resulting low-Cu slag is then discarded or sold (Chapters 10 and 20).

SO_2, 8–12 vol.% in the converter offgas, is a byproduct of both converting reactions. It is combined with smelting furnace gas and captured as sulfuric acid. There is, however, some leakage of $SO_2(g)$ into the atmosphere during charging and pouring. This problem is encouraging development of continuous converting processes. The most prominent of these is *flash converting* (Chapter 8).

1.3.4 Direct-to-copper smelting

Smelting and converting are sequential steps in oxidizing Cu–Fe–S concentrates to metallic copper. It would seem natural that these two steps should be combined to produce *blister* copper directly in one furnace. It would also seem natural that this should be done continuously rather than batchwise.

In 2020, *blister* copper is made in a single furnace at only three places—Olympic Dam, Australia; Glogow, Poland; and Chingola, Zambia—all using a flash furnace. The strong oxidizing conditions in a direct-to-copper furnace produce a slag with 14%–24% oxidized Cu. The expense of reducing this dissolved Cu back to copper metal has so far restricted the process to low-Fe concentrates, which produce little slag.

Continuous smelting/converting, even in more than one furnace, has energy, SO_2 collection, and cost advantages. Metso Outotec flash, Mitsubishi lance, and Noranda submerged-tuyere smelting/converting all use this approach, described in Chapter 6 through Chapter 8.

1.3.5 Fire refining and electrorefining of blister copper

The *blister* copper from the above processing is electrochemically refined to high-purity cathode copper. This final copper contains less than 20 ppm undesirable impurities. It is suitable for electrical and almost all other uses.

Electrorefining requires strong, thin, flat anodes to interleave with cathodes in an electrorefining cell, Fig. 1.8. These anodes are produced by (a) sequential air oxidation then natural gas reduction of blister copper then (b) casting the resulting *fire-refined* copper in open, anode-shape molds (Figs. 12.2 and 12.3.).

Copper electrorefining entails

(a) electrochemically dissolving copper from fire-refined anodes into $CuSO_4-H_2SO_4-H_2O$ electrolyte

(b) electrochemically plating pure copper (without the anode impurities) from the electrolyte onto stainless-steel cathodes by passing DC electrical current through the electrolyte between anodes and cathodes.

Copper is deposited on the cathodes for 7–14 days. The cathodes are then removed from the cell. The copper is stripped, washed, and (a) sold or (b) melted and cast into useful products.

The electrolyte is an aqueous solution of H_2SO_4 (150–200 kg/m^3) and $CuSO_4$ (40–50 kg Cu/m^3). In use, it also contains impurities and trace amounts of chloride and organic *addition agents*, Section 14.5.1.

Figure 1.8 Casting of anodes for electrorefining. *(Photo courtesy of Miguel Palacios.).*

Many anode impurities are insoluble in this electrolyte (Au, Pb, Pt metals, Sn). They do not interfere with the electrorefining. They are collected as solid *slimes* and treated for Cu and byproduct recovery, Chapter 20.

Other impurities, such as As, Bi, Fe, Ni, and Sb, are partially or fully soluble. Fortunately, they do not plate with the copper at the low voltage of the electrorefining cell (~0.3 V). They must, however, be kept from accumulating in the electrolyte to avoid physical contamination of the cathode copper. This is done by continuously bleeding part of the electrolyte through a purification circuit.

An older form of the process is to plate cathode copper onto thin pure copper sheets. In this case the entire cathode is lifted from the cell, washed, and sent to users.

1.4 Hydrometallurgical extraction of copper

About 80% of copper from ore is produced by flotation, smelting, and refining. The remaining 20% is produced hydrometallurgically. Hydrometallurgical extraction entails (Fig. 1.3)

(a) sulfuric acid leaching of Cu from broken or crushed ore to produce impure Cu-bearing aqueous solution

(b) transfer of Cu from the resulting weak, impure solution to pure, high-Cu-strength electrolyte via solvent extraction

(c) electrodepositing pure cathode copper sheets from this pure electrolyte

The ores most commonly treated this way are (a) naturally oxidized Cu minerals, including carbonates, hydroxy-silicates, sulfates, and (b) chalcocite, Cu_2S.

The leaching is mostly done by dripping dilute sulfuric acid (~10 kg H_2SO_4 per m^3 of solution) onto a layer of broken or crushed ore (0.3%–2.3% Cu) and allowing the acid to trickle through to collection ponds, Figs. 1.4 and 1.5. Several months of leaching are required for efficient Cu extraction from the ore.

After this first extraction another layer of ore is placed on top of the leached layer and the process is repeated and so on.

Oxidized minerals are dissolved by the sulfuric acid by reactions like

$$\underset{\text{in ore}}{CuO(s)} + \underset{\text{in 30 °C sulfuric acid}}{H_2SO_4(\ell)} \rightarrow \underbrace{Cu^{2+}(aq) + SO_4^{2-}(aq) + H_2O(\ell)}_{\text{aqueous pregnant leach solution}} \quad (1.5)$$

Sulfide minerals, on the other hand, require oxidation, schematically:

$$\underset{\text{in ore}}{Cu_2S(s)} + \underset{\text{in air}}{2.5O_2(g)} + \underset{\text{in sulfuric acid}}{H_2SO_4(\ell)} \xrightarrow[\text{30 °C}]{\text{bacterial enzyme catalyst}} \underbrace{2Cu^{2+}(aq) + 2SO_4^{2-}(aq) + H_2O(\ell)}_{\text{pregnant leach solution}} \quad (1.6)$$

As indicated, sulfide leaching is greatly speeded up by bacterial action (Chapter 15).

Leaching is occasionally applied to Cu-bearing flotation tailings, mine wastes, old mines, and fractured ore bodies. However, leaching of ore heaps is far and away the largest hydrometallurgical Cu extraction technique.

1.4.1 Solvent extraction

The solutions from heap leaching contain ~3 g/L of Cu and ~3 g/L of H_2SO_4 plus dissolved impurities, such as Fe and Mn. These solutions are too dilute in Cu and too impure for direct electrodeposition of pure copper metal. The Cu must be transferred to pure, high-Cu electrolyte to make pure copper.

The transfer is achieved by

(a) extracting Cu from an impure leach solution into a Cu-specific liquid organic extractant

(b) separating by gravity the Cu-loaded extractant from the Cu-depleted leach solution

(c) stripping Cu from the loaded extractant into 180 g/L H_2SO_4 electrolyte.

Extraction and stripping are carried out in large mixer-settlers, Fig. 1.9.

The solvent extraction process is represented by the reaction:

$$\underset{\text{in pregnant leach solution}}{Cu^{2+}(aq) + SO_4^{2-}} + \underset{\text{in organic solvent}}{2RH} \xrightarrow{30\,°C} \underset{\text{in organic solvent}}{R_2Cu} + \underset{\substack{\text{in aqueous solution,}\\ \text{recycle to leach}}}{2H^+ + SO_4^{2-}} \tag{1.7}$$

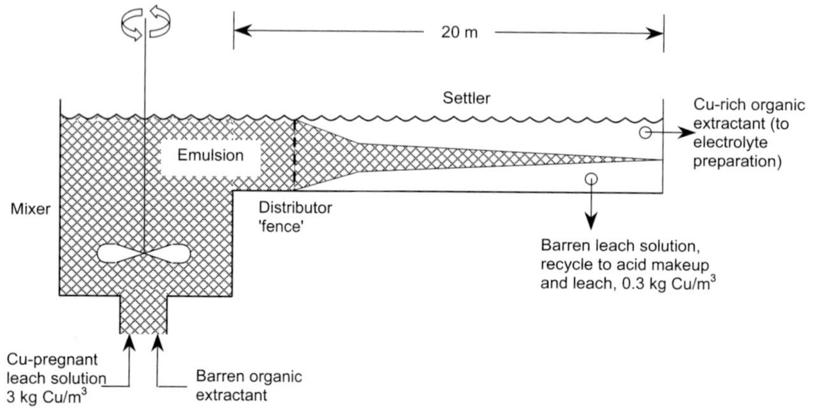

Figure 1.9 Schematic view of solvent extraction mixer-settler for extracting Cu from pregnant leach solution into organic extractant. The Cu-loaded organic phase goes forward to another mixer/setter ("stripper") where Cu is stripped from the organic into pure, strongly acidic, high-Cu electrolyte for electrowinning. The process is continuous.

This shows that a low-acid (i.e., low H^+) aqueous phase causes the organic extractant to load with Cu (as R_2Cu). It also shows that a strong acid solution causes the organic to unload (i.e., *strip*).

Thus, when organic extractant is contacted with weak acid pregnant leach solution [step (a) above], Cu is loaded into the organic phase. Then when the organic phase is subsequently put into contact with high (~180 g/L) H_2SO_4 electrolyte [step (c) above], the Cu is stripped from the organic into the electrolyte at high Cu^{2+} strength, suitable for electrowinning.

The extractants complex with considerable Cu but almost no impurities. They produce electrolytes that are strong in Cu (~50 g/L) but dilute in impurities.

1.4.2 Electrowinning

The Cu in the above electrolytes is mostly recovered by electrodepositing pure metallic copper onto stainless-steel cathodes, Fig. 1.10. Electrowinning is similar to electrorefining except that the anode is inert, almost always 99% Pb alloy.

The cathode reaction is

$$\underset{\text{in sulfate electrolyte}}{Cu^{2+}(aq)} + \underset{\substack{\text{electrons from}\\ \text{external DC power supply}}}{2e^-} \xrightarrow{60\,^\circ C} \underset{\substack{\text{pure metal deposit}\\ \text{on stainless steel cathode}}}{Cu(s)} \tag{1.8}$$

Figure 1.10 Plates of electrowon copper on stainless-steel cathodes after removal from an electrowinning cell. These cathodes are ~1 m wide and 1.3 m deep. They are carefully washed to remove electrolyte. The copper plates are then stripped off in automatic machines and sent to market. They typically contain <20 parts per million impurities. Note the polymer side strips—they prevent copper from plating around the edges. *(Photograph courtesy of Freeport-McMoRan Copper & Gold Inc).*

The anode reaction is

$$\underset{\text{in electrolyte}}{H_2O(\ell)} \xrightarrow{60\,^\circ C} \underset{\substack{\text{evolved on}\\ \text{inert anode}}}{0.5\,O_2(g)} + \underset{\substack{\text{formed in electrolyte,}\\ \text{recycle to solvent}\\ \text{extraction}}}{2\,H^+(aq)} + \underset{\substack{\text{electrons to}\\ \text{external DC}\\ \text{power supply}}}{2\,e^-} \quad (1.9)$$

As with electrorefining, the copper is plated on the cathode by passing DC current through the electrolyte between anodes and cathodes. About 2 V is required.

Pure metallic copper (less than 20 ppm undesirable impurities) is plated on the cathode. Oxygen gas is evolved at the anode.

A recent development is the use of precious metal-coated titanium cathodes, which lowers electrowinning energy consumption.

1.5 Melting and casting cathode copper

The first steps in making products from electrorefined and electrowon copper are melting and casting, Fig. 1.11. The melting is almost always done in vertical shaft furnaces, Fig. 19.2. The copper cathode sheets are charged to the top of the furnace. They descend into the furnace and are melted by ascending hot combustion gases. Low-sulfur fuels prevent sulfur pickup. Reducing flames prevent excessive oxygen pickup.

The molten copper is mostly cast in continuous casting machines from where it goes to rolling or extrusion then manufacturing. An especially important combination is continuous bar casting/rod rolling, Chapter 19.

1.5.1 Types of copper product

The copper described above is Grade 1 copper. It contains ~0.025% oxygen and the cathode impurities. It is the most common type of primary copper product.

A second type is oxygen-free copper (<5 ppm O). It is used for highly demanding applications, such as high-end audio equipment. It is made from cathode copper by electric induction melting with added pure graphite for removing O as CO(g).

About 35% of copper is used in alloy form as brasses, bronzes, etc. Much of the copper for these alloys comes from recycle scrap.

1.6 Recycle of copper and copper alloy scrap

Copper production from recycling copper and copper alloy end-of-life objects (*old* scrap) amounts to 15%–20% of the above-described copper-from-mine production (International Copper Study Group, 2020).

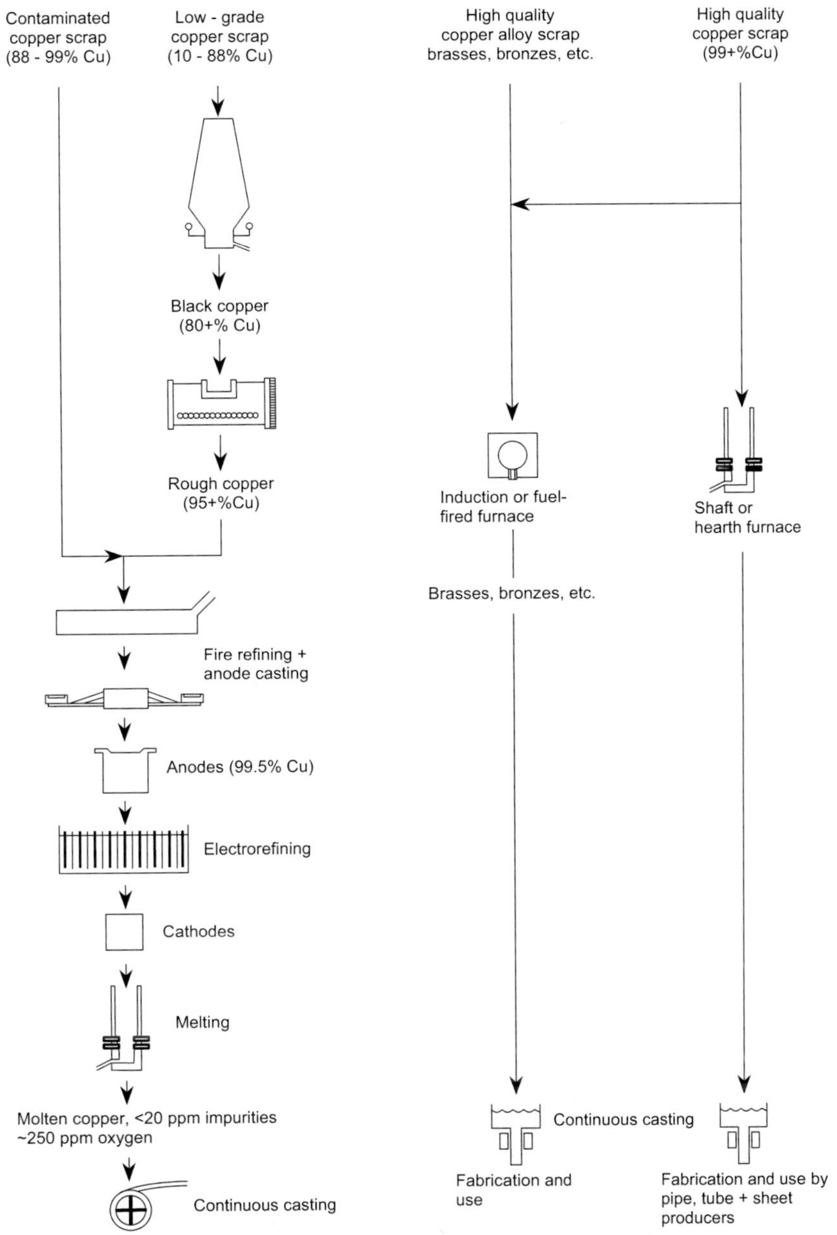

Figure 1.11 Flowsheet of processes for recovering copper and copper alloys from scrap. Low-grade scrap is usually smelted in shaft furnaces but other furnaces (e.g., electric) are also used.

Processing and manufacturing also produce copper and alloy scrap, called *new scrap*. Most of this scrap is easily recovered by internal recycle.

Production of copper from scrap (a) requires considerably less energy than mining and processing of copper ore; (b) minimizes carbon emissions; (c) avoids mine, concentrator, leach, and smelter wastes; and (d) helps to ensure the availability of copper for future generations.

The treatment given to copper scrap depends on its purity, Fig. 1.10. The lowest grade scrap is smelted and refined like concentrate in primary or secondary (scrap) smelters/refineries. Higher-grade scrap is fire-refined then electrorefined. The highest-grade scrap (mainly manufacturing waste) is often melted and cast without refining. Its copper is used for nonelectrical products, including tube, sheet, and alloys.

Alloy scrap (brass, bronze) is melted and cast as alloy. There is no advantage to separating it into pure metals. Some slagging is done during melting to remove dirt and other contaminants. The product alloy composition is adjusted by combining alloy and individual metal scrap as needed.

1.7 Safety

Of paramount concern in copper mining—manufacturing operations is safety (Mathis & Galloway, 2016). A safe working environment is fostered by

(a) setting an accident-free workplace as a primary goal
(b) close attention to safety by management, even in tough economic times
(c) inclusion of safety considerations during design and construction
(d) thorough worker safety training with regular updates
(e) mandatory wearing of safety apparel, such as steel-toed boots and hard hats
(f) effective lighting in all work areas, especially underground
(g) thorough maintenance and hazard identification/elimination
(h) thorough investigation of any accident or near-miss and resolution of any indicated problems
(i) "stay safe" discussions before every work shift
(j) personal responsibility for safety
(k) three-point contact when ascending or descending stairs and ladders.

Detailed safety concerns in various segments of copper making (e.g., smelting) are presented in later chapters. Safety excellence is detailed in *STEPS to Safety Culture Excellence* (Mathis & Galloway, 2013).

1.8 Environment

Production of pure metallic copper from mineral deposits (mines) involves many steps, including mining, concentration, smelting, leaching, and electrorefining or electrowinning, as well as land and water transportation.

Mining raises dust during blasting and trucking. It also alters the landscape by removing material from the surface and dumping waste rock in piles around the mine.

Concentration raises dust during crushing and produces small-particle tailings waste during concentration. Most concentrators minimize water use and contamination by recycling as much water as possible from the tailing ponds.

Smelting produces SO_2-bearing offgas. Fortunately, processes have been developed to capture this $SO_2(g)$ and make it into sulfuric acid, itself a valuable product. Smelters also produce solidified oxide slag which, in some locations, is a valuable abrasive and cement-making raw material.

Heap leaching alters the landscape by placing ore on the earth's surface and acid leaching it in the open. Extreme caution must be taken to prevent the acid from reaching the water table.

Electrorefining and electrowinning are contained in buildings, so their impact on the environment is not large. Generation of sulfuric acid mist above the electrolytic cells can adversely affect the working environment. Covering the cells and providing effective extraction fans ameliorates this problem. Well-designed workplace clothing and equipment are also essential. An important example is filtered breathing equipment for employees. These issues are discussed in depth in their specific chapters.

Beyond legal requirements, an important goal of the copper industry is to produce pure copper while minimizing deleterious effects on the environment. In 2020, this is a top goal of copper management and employees. Post-mine-life landscape rehabilitation is an important part of attaining this goal. Several videos describe rehabilitation (Hanson, 2016; Munsey, 2015; Teck Resources Limited, 2016).

1.9 Summary

About 80% of the world's primary copper is produced by concentration, smelting, and refining of sulfide ores. The other 20% is produced by heap leaching, solvent extraction, and electrowinning of oxidized and chalcocite ores. Secondary copper production from recycled end-of-use objects is 15%–20% of primary copper production (ICSG, 2018).

Electrochemical processing is always used to produce high-purity copper: electrorefining in the case of pyrometallurgical extraction and electrowinning in the case of hydrometallurgical extraction. The principal final copper product is Grade 1 copper ($\sim$250 ppm oxygen and <20 ppm unwanted impurities). It is suitable for nearly all uses.

The trend in copper extraction is toward processes that do not harm the environment and consume less energy and water. This has led to

(a) energy- and pollution-efficient oxygen-enriched smelting
(b) efficient recovery of $SO_2(g)$ from smelter offgas as sulfuric acid
(c) leaching–solvent extraction–electrowinning flowsheets
(d) increased water recycle
(e) increased recycle of end-of-use scrap
(f) reclamation of end-of-life mines and hydrometallurgical leach sites.

References

Berger, B. R., Ayuso, R. A., Wynn, J. S., & Seal, R. R. (2008). *Model of porphyry copper deposits. USGS open file report 2008-1321.* https://doi.org/10.3133/sir20105070B> Accessed 31.05.20

Boldt, J. R., & Queneau, P. (1967). *The winning of nickel.* Toronto: Longmans Canada Ltd.

Hanson, A. (2016). *Digging in to save the earth — mine reclamation* (Accessed 31 May 2020) https://www.youtube.com/watch?v=UHfMsgEJXXo.

International Copper Study Group. (2018). *The world copper factbook 2018* (Accessed 01 November 2018) www.icsg.org.

Mathis, T. L., & Galloway, S. M. (2016). *Steps to safety culture excellence from ProAct Safety* (Accessed 31 May 2020) https://proactsafety.com/solutions/consulting/steps.

Munsey, Tim (2015). *Mine reclamation in Virginia* (Accessed 31 May 2020) https://www.youtube.com/watch?v=_b_-E7Wt988&t=23s.

Teck Resources Limited. (2016). *Teck: Our approach to mine reclamation (2016)* (Accessed 31 May 2020) https://www.youtube.com/watch?v=RP9zCiftIWk.

Suggested reading

Copper. (2016). In *Proceedings of the conference (2016).* Tokyo: MMIJ.

Extraction. (2018). In *Proceedings of the first global conference on extractive metallurgy (2018).* Warrendale, PA: TMS.

John, D. A. (2010). *Porphyry copper deposit model. USGS scientific investigations report 2010-5070-B* (Accessed 31 May 2020) https://pubs.usgs.gov/sir/2010/5070/b/pdf/SIR10-5070B.pdf.

Proceedings of the 58th conference of metallurgists hosting the international copper conference 2019. (2019). Montreal: CIM.

Further reading

Briggs, D. F. (2014). *What is a porphyry copper deposit?* (Accessed 31 May 2020) https://arizonageologicalsoc.org/resources/Documents/Short%20Papers%20by%20Members/2014/Mining%20and%20You%20-%201-11-2014.pdf.

Moats, M., Wang, S., Filzwieser, A., Davenport, W., & Robinson, T. (2016). Survey of copper electrorefining operations. In *Proceedings of copper 2016* (pp. 1914–1923). Tokyo: MMIJ.

Moats, M., Wang, S., Filzwieser, A., Davenport, W., & Robinson, T. (2019). Survey of copper electrorefining operations. In *Proceedings of the 58th conference of metallurgists hosting the international copper conference 2019.* Montreal: CIM (Article 594940).

Rio Tinto mine of the future.(2014) (Accessed 31 May 2020) https://www.youtube.com/watch?v=9L_EbGLItyY.

CHAPTER 2

Production and use

Metallic copper occasionally occurs in nature. For this reason, it has been known to man since about 10,000 B.C. (Copper Development Association, 2020). Its early uses were in jewelry, utensils, tools, and weapons. Although its use gradually increased over the years, it was only in the 20th century, with the global adoption of electricity, that copper usage really expanded. Copper is essential to economic and technological development, so substantial growth in its applications has continued in the 21st century, driven by the rapid industrialization of China and other emerging-market economies (Fig. 2.1).

This chapter discusses production and use of copper around the world. It gives production, use and price statistics, and identifies and locates the world's largest copper-producing plants.

2.1 Properties and uses of copper

Copper is an excellent conductor of electricity and heat. It also resists corrosion. It is easily fabricated into wire, pipe, and other forms, and is easily joined by soldering and brazing. It is also malleable, ductile, and tough (not brittle). All of these properties are widely exploited (Table 2.1).

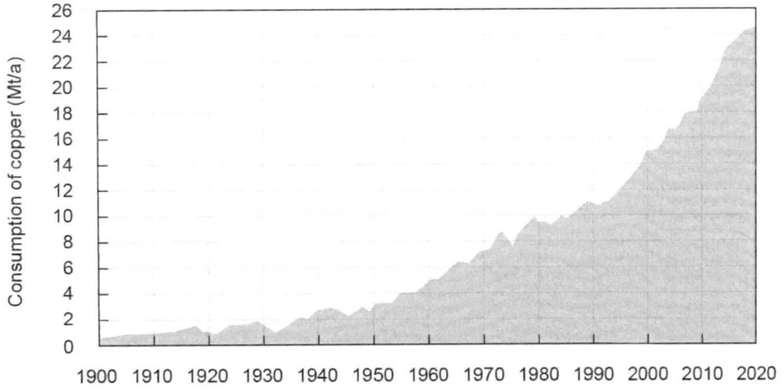

Figure 2.1 World use of refined copper, which has grown from less than 500,000 t in 1900 to more than 24.5 million tons in 2018, representing a compound annual growth rate of 3.4% (International Copper Study Group, 2019).

Extractive Metallurgy of Copper, Sixth Edition
ISBN 978-0-12-821875-4, https://doi.org/10.1016/B978-0-12-821875-4.00015-8

Table 2.1 Properties of copper (Copper Development Association, 2020).

Exploited property	% of use
Electrical conductivity	65
Corrosion resistance	16
Thermal conductivity	11
Mechanical and structural	6
Aesthetics	2

Copper is readily alloyed with zinc, tin, nickel, and other metals to improve its castability, strength, and corrosion resistance. Brass (Cu—Zn alloy) is an especially useful alloy that has a low solidification temperature and high fluidity, allowing efficient casting into intricate shapes. Beryllium—copper is heat treatable, is as hard and strong as steel, and can be used for several applications (Materion, 2020). Alloys account for about 25% of copper usage.

Among its significant beneficial properties, copper

- is easily drawn into extremely fine wire;
- is easily extruded into thin pipe and other shapes;
- is easily machined and threaded;
- is readily cast into intricate shapes, especially in alloy form;
- is antibacterial (useful in medical piping and other equipment);
- is nonmagnetic;
- has an attractive sheen and color;
- exhibits low metal-against-metal friction (in alloy form);
- has catalytic properties; and
- does not oxidize during use, so has a long service life and can be recycled without loss.

Fig. 2.2 shows the distribution of copper usage in various industries (International Copper Study Group, 2019). Its major use is in conduction of electricity, ranging from small domestic installations to high-voltage heavy-current applications. It is widely used in generators, motors, transformers, and in renewable energy. Its other major use today is in communication technologies, such as high-speed data transmission, area networks, cellphones, personal computers, and other electronic equipment. Copper can be used in construction for building wire, plumbing, and heating and refrigeration, as well as for its aesthetic appeal. Use of copper-based building materials in regions with high seismic activity can limit earthquake damage owing to its ability to absorb energy.

Its durability, machinability, and ability to be cast with high precision makes copper suitable for the manufacture of industrial equipment, such as gears, bearings, and turbine blades, and for use in the transport industry, including aircraft, high-speed rail, and vehicles. Its excellent heat-transfer properties find use in heat exchangers and pressure vessels. Copper and its alloys are widely used in marine environments, owing to good corrosion

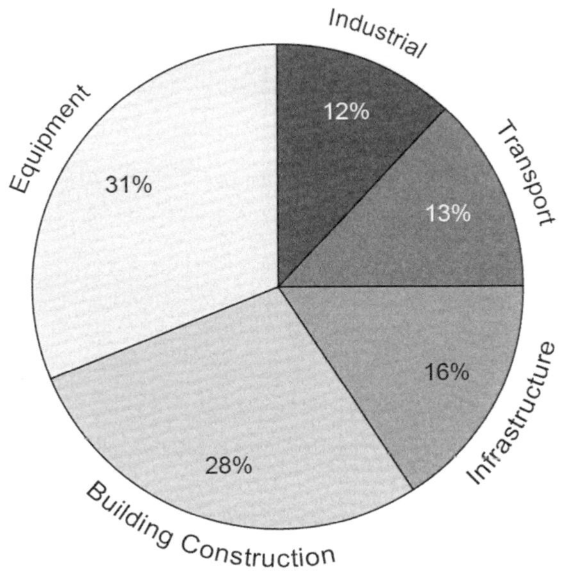

Figure 2.2 Major uses of copper by main industry (International Copper Study Group, 2019).

resistance. Copper also exhibits antimicrobial properties, which make it suitable for selected medical applications and for high-use surfaces, such as door knobs.

2.2 Global copper production

Fig. 2.3 shows global copper mine production. Mine output was almost 21 million tonnes in 2019 (International Copper Study Group, 2020), most of which was in the

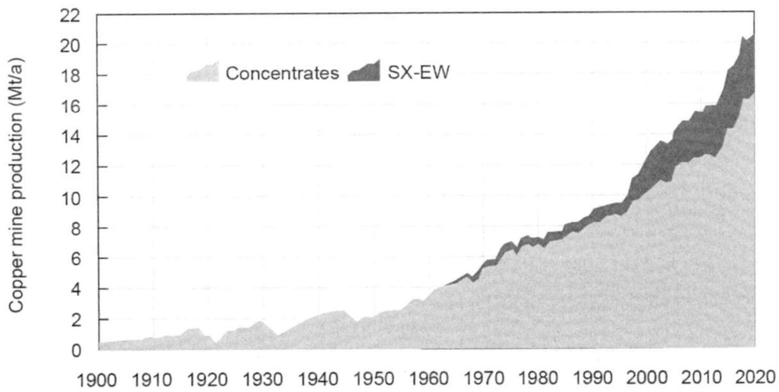

Figure 2.3 World copper mine production from 1900 to 2018. Hydrometallurgical production of copper only started to become significant in the 1970s, and today represents about 20% of global production (International Copper Study Group, 2019).

Table 2.2 Major world producers of copper in 2019, kt of contained copper (International Copper Study Group, 2020).

Country	% of total	Mine production (kt Cu)
Chile	34.8	5787
Peru	14.8	2455
China	10.1	1684
United States	7.7	1284
Democratic Republic of Congo	7.7	1283
Australia	5.6	929
Zambia	4.8	797
Russian Federation	4.8	791
Mexico	4.6	769
Kazakhstan	4.2	696
Canada	3.4	573
Poland	2.4	399
Brazil	2.2	362
Indonesia	2.2	361

form of concentrates (see Chapter 3); about 20% of mine production uses hydrometallurgical processing involving solvent extraction and electrowinning (SX—EW) of copper cathode (Chapters 15 and 16).

The Andes mountain region of South America (Chile and Peru) is the world's largest source of copper, accounting for about one-third of global output (Table 2.2). China and the United States of America are major producers, followed by the African Copper belt (Democratic Republic of Congo (DRC) and Zambia). The remaining production is scattered around the world. The world's 20 largest mines are listed in Table 2.3 and plotted in Fig. 2.4.

2.3 Copper minerals, mines, and cut-off grades

Table 2.4 lists the main copper minerals. These minerals occur at low concentrations in ores. The remainder of the ore is mostly oxide rock, such as quartz and potassium feldspar, with a small amount of waste sulfide minerals, such as pyrite.

A copper mine consists of regions of (i) ore, containing economic quantities of Cu minerals, and (ii) waste rock, containing low, uneconomic quantities of Cu minerals. These regions are located by drilling, sampling, and analyzing drill core for Cu. Where possible, the waste rock is not mined. Where the ore and rock are intertwined, they must be mined together and separated later, as described in Chapters 3 and 4.

Mined Cu ore typically contains 0.5% Cu (open-pit mines) and 2% Cu (underground mines). Many Cu ores also contain small amounts of valuable minor metals, such as gold,

Table 2.3 The world's 20 highest production Cu mines (International Copper Study Group, 2019). These account for about 40% of total world mined copper production.

	Mine	Country	Production (kt/a contained Cu)	Products
1	Escondida	Chile	1400	Concentrate and SXEW
2	Collahuasi	Chile	570	Concentrate and SXEW
3	Buenavista del Cobre	Mexico	525	Concentrate and SXEW
4	Morenci	United States	520	Concentrate and SXEW
5	Cerro Verde III	Peru	500	Concentrate
6	Antamina	Peru	450	Concentrate
7	Norilsk Polar Division	Russia	450	Concentrate
8	Las Bambas	Peru	430	Concentrate
9	El Teniente	Chile	422	Concentrate and SXEW
10	Chuquicamata	Chile	390	Concentrate and SXEW
11	Los Bronces	Chile	390	Concentrate and SXEW
12	Los Pelambres	Chile	370	Concentrate
13	Kansanshi	Zambia	340	Concentrate and SXEW
14	Radomiro Tomić	Chile	330	Concentrate and SXEW
15	Grasberg	Indonesia	300	Concentrate
16	Kamoto	DRC	300	SXEW
17	Bingham Canyon	United States	280	Concentrate
18	Toquepala	Peru	265	Concentrate and SXEW
19	Sentinel	Zambia	250	Concentrate
20	Olympic Dam	Australia	225	Concentrate and SXEW

DRC, Democratic Republic of Congo.

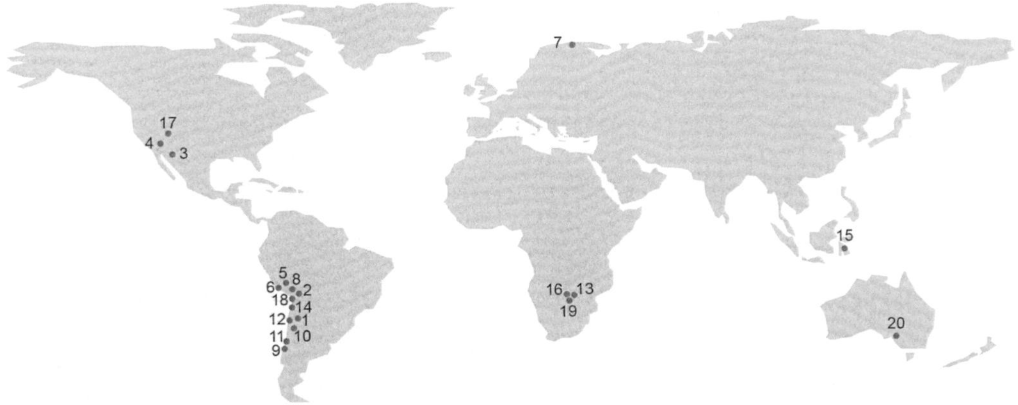

Figure 2.4 Map showing the location of the world's largest copper mines. The numbers refer to the data in Table 2.3. Notice the density of mines in the Chilean and Peruvian Andes mountains. Most are porphyry copper deposits (International Copper Study Group, 2019).

Table 2.4 Main commercial copper minerals. Chalcopyrite is by far the largest source of copper. Sulfide minerals are treated by pyrometallurgical processing (Fig. 1.1); carbonates, chlorides, oxides, silicates, and sulfates are treated by hydrometallurgical processing (Fig. 1.2), although secondary sulfides, e.g. chalcocite are increasingly being in this manner (see Chapter 14).

Type	Common minerals	Chemical formula	Theoretical % Cu
Primary sulfides			
Hypogene sulfides	Chalcopyrite	$CuFeS_2$	34.6
	Bornite	Cu_5FeS_4	63.3
Secondary minerals			
Supergene sulfides	Chalcocite	Cu_2S	79.9
	Covellite	CuS	66.5
	Digenite	$Cu_{1.8}S$	78.1
Native copper	Metal	Cu	100
Carbonates	Malachite	$CuCO_3 \cdot Cu(OH)_2$	57.5
	Azurite	$(CuCO_3)_2 \cdot Cu(OH)_2$	55.3
Hydroxysilicates	Chrysocolla	$CuO \cdot SiO_2 \cdot 2H_2O$	36.2
Oxides	Cuprite	Cu_2O	88.8
	Tenorite	CuO	79.9
Sulfates	Antlerite	$CuSO_4 \cdot 2Cu(OH)_2$	53.7
	Brochantite	$CuSO_4 \cdot 3Cu(OH)_2$	56.2

silver, molybdenum, rhenium, and cobalt, which can be profitably recovered during processing. Custom smelters can profit from the recovery of nickel, selenium, tin, zinc, and lead that is present in copper concentrates.

The average grade of ore that is extracted from any given mine is determined by the *cut-off grade* (% Cu). This value determines the split between materials classified as *ore* from that considered *waste*. Material with a copper content of less than the cut-off grade cannot be profitably treated for copper recovery and is considered waste. It is removed to large waste dumps.

Cut-off grade depends on the current copper selling price and the mining and extraction costs. If, for example, the price of copper rises and costs are constant, it may become profitable to treat lower-grade material. This means that cut-off grade (and average ore grade) will decrease. Lower copper prices and increased costs have the opposite effect. Higher levels of gold, silver, and molybdenum extraction decrease Cu cut-off grade because these elements provide additional income per tonne of ore.

Average world Cu ore grades (%Cu) are falling at about 25% per decade as copper-rich mines become depleted and newly located deposits contain less copper (Calvo et al., 2016). As shown in Fig. 2.5, it is now rare to find a large copper deposit averaging more than 1% or 2% Cu; furthermore, the mineralogy of new orebodies is becoming more

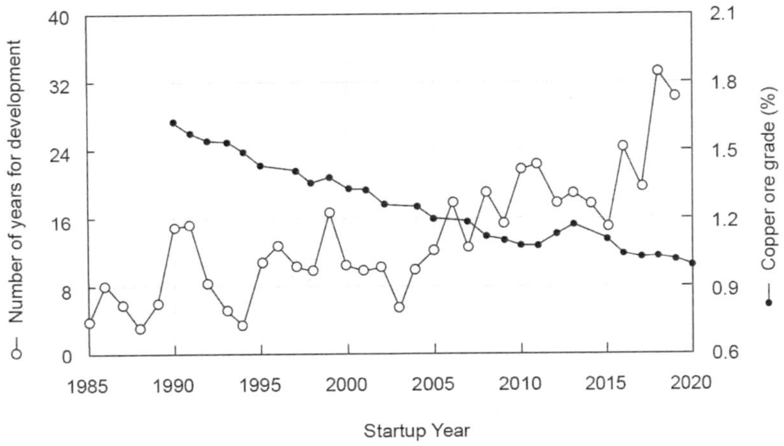

Figure 2.5 Trend of declining ore grade and increasing time to commercialization of copper deposits, owing to increasingly complex mineralogy of exploited copper ores. *(Adapted from Winton Group. (2017), Copper-bottomed booms and busts. Retrieved from https://www.winton.com/longer-view/copper-bottomed-booms-and-busts. Accessed 19 September 2020, with data from Brook Hunt.)*

complex and the time required from discovery of the deposit to commissioning of a mine has substantially increased (Winton Group, 2017).

2.4 Locations of processing plants

2.4.1 Smelters

The first stage of copper extraction from sulfide ores is production of high-grade (~25% Cu) particulate concentrate from low-grade (~0.5%–2% Cu) ore (Chapters 3 and 4). This is always done at the mine site to avoid costly long-distance transport of the ore's waste rock.

The resulting concentrate is usually transported by truck or train from the mine to a regional port, then by ship to coastal or riverside smelters around the world, located mainly in China, Japan, and Europe. China is by far the biggest importer, processing ~7000 kt/a of Cu-in-concentrate. Chile is by far the largest concentrate exporter, shipping out ~3000 kt/a of contained Cu.

The world's largest smelters are listed in Table 2.5 and their locations are shown in Fig. 2.6. In 2018, China accounted for about 40% of world smelter production, followed by Chile (10%) and Japan (7%).

2.4.2 Electrorefineries

The main product of smelting is copper anodes, which are treated in an electrorefinery to produce pure high-grade (<25 ppm impurities) copper cathode. Copper electrorefineries are usually built near the smelter that supplies them with anodes or in a convenient central

Table 2.5 The world's 20 largest Cu smelters in descending order of smelter capacity (International Copper Study Group, 2019). They account for about 60% of the world's primary copper production.

	Smelter	Country	Capacity (kt/a Cu)	Smelting process
1	Guixi	China	900	Metso Outotec flash
2	Birla	India	500	Metso Outotec flash, Mitsubishi
3	Chuquicamata	Chile	450	Metso Outotec flash + Teniente converter
4	Jinchuan (Fangchenggang)	China	450	Ausmelt® TSL Process – Metso Outotec + reverberatory
5	Hamburg	Germany	450	Metso Outotec flash, Contimelt, electric
6	Besshi/Ehime (Toyo)	Japan	450	Metso Outotec flash
7	Saganoseki/Ooita	Japan	450	Metso Outotec flash
8	El Teniente (Caletones)	Chile	400	Reverberatory + Teniente converter
9	Jinchuan	China	400	Reverberatory + Kaldo converter
10	Jinchuan	China	400	Metso Outotec flash
11	Jinlong (Tongdu)	China	400	Metso Outotec flash
12	Xiangguang	China	400	Metso Outotec flash
13	Sterlite (Tuticorin)	India	400	Isasmelt
14	Norilsk (Mikelevy, Medny)	Russia	400	Reverberatory, electric, Vanyukov
15	Pirdop	Bulgaria	360	Metso Outotec flash
16	Ilo	Peru	360	ISASMELT™
17	Onahama/Fukushima	Japan	354	Mitsubishi + Reverberatory
18	Naoshima/Kagawa	Japan	342	Mitsubishi
19	Isabel/Leyte	Philippines	330	Metso Outotec flash
20	Huelva	Spain	320	Metso Outotec flash

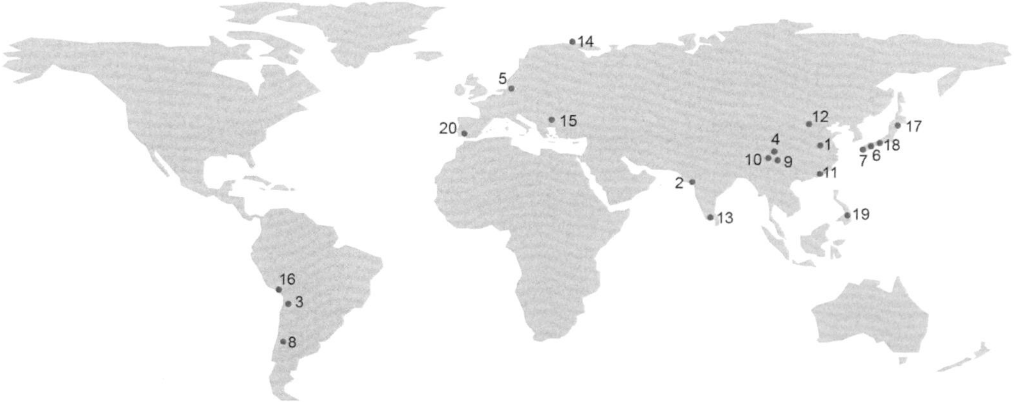

Figure 2.6 Map showing the locations of the world's 20 largest copper smelters. The numbers refer to the data in Table 2.4. Notice the large concentrations in China and Japan. These smelters mostly import Cu concentrate from which they make Cu anodes.

Table 2.6 The world's 20 largest Cu refineries in descending order of copper production capacity (International Copper Study Group, 2019). They account for about 60% of the world's refined copper production.

	Refinery	Country	Capacity (kt/a Cu)
1	Guixi	China	1000
2	Jinchuan	China	650
3	Daye/Hubei	China	600
4	Yunnan	China	500
5	Birla	India	500
6	Jinchuan	China	450
7	Toyo/Niihama (Besshi)	Japan	450
8	Amarillo	United States	450
9	Chuquicamata	Chile	450
10	Onsan I	Korea	440
11	Hamburg	Germany	416
12	El Paso	United States	415
13	Las Ventanas	Chile	410
14	Jinguan	China	400
15	Jinlong (Tongdu)	China	400
16	Xiangguang	China	400
17	Shandong Fangyuan	China	400
18	Sterlite	India	400
19	Pyshma	Russia	390
20	Canadian Copper Refinery	Canada	370

location close to several smelters. There is also a small amount of world trade in anodes (International Copper Study Group, 2017), which favors coastal and riverside electrorefineries. The world's largest electrorefineries are listed in Table 2.6 and shown in Fig. 2.7.

2.4.3 Hydrometallurgical plants

Leach—SX—EW operations are always located next to their mines. This is because ores that are suitable for leaching are low in copper grade, and hence uneconomic to transport. The world's largest copper leach—SX—EW operations are listed in Table 2.7 and their locations are shown in Fig. 2.8. This technology dominates in Chile and the African Copper Belt.

2.5 Price of copper

Fig. 2.9 shows the price of copper from 1958 to 2020. The dollar price increased steadily until 2002 then rose rapidly, reaching record high values in 2011. The post-2005 high prices were due to the huge demand for copper by China during its rapid industrialization and infrastructure growth. The large swings in price since then are mostly the result of short-term variations in Chinese demand, with some due to arbitrary U.S. tariffs.

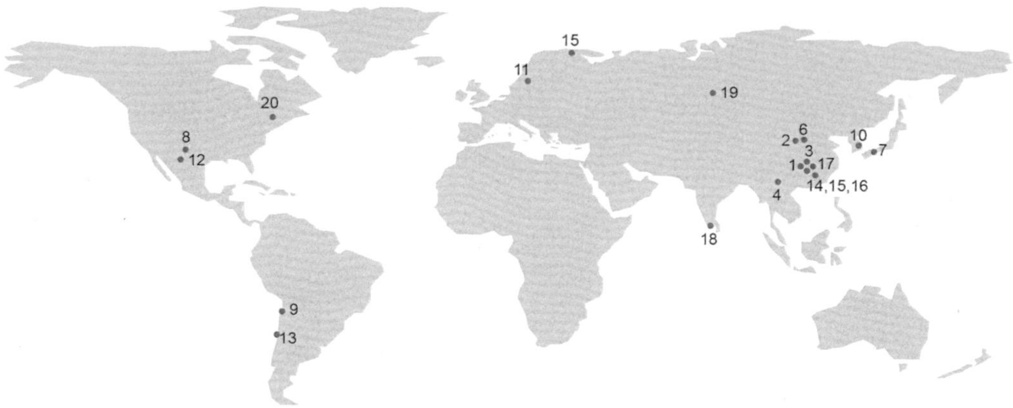

Figure 2.7 Map showing the locations of the world's 20 largest copper electrorefineries. The numbers refer to the data in Table 2.5. The large number of electrorefineries in China is due to its rapid growth of electrical infrastructure and other industrial and consumer expansions.

Table 2.7 The world's 20 largest Cu solvent extraction–electrowinning plants in descending size (International Copper Study Group, 2017, 2018). These plants account for almost 70% of the world's electrowon copper production.

	SXEW Refinery	Country	Capacity (kt/a Cu)
1	Morenci	United States	350
2	Escondida	Chile	270
3	Radomiro Tomić	Chile	220
4	Mutanda	DRC	220
5	Tenke Fungurume	DRC	214
6	Buenavista del Cobre	Mexico	175
7	Kamoto	DRC	165
8	Spence	Chile	160
9	El Abra	Chile	150
10	Konkola	Zambia	130
11	Dikuluwe	DRC	125
12	Gabriella Mistral	Chile	120
13	Safford	United States	110
14	Zaldivar	Chile	100
15	Sepon	Laos	90
16	Antucoya	Chile	85
17	Chuquicamata	Chile	80
18	Kinsevere	DRC	80
19	Lomas Bayas	Chile	75
20	Centinela	Chile	70

DRC, Democratic Republic of Congo.

2.6 Future outlook

China, which accounts for more than half of world copper consumption, is currently developing several megacities and high-speed rail networks that will require large amounts of copper (Winton Group, 2017). The increasing role of renewable energy

Figure 2.8 Map showing the locations of the world's 20 largest copper solvent extraction–electrowinning plants. Most of these are in Chile and the African Copper Belt. The numbers refer to the data in Table 2.6.

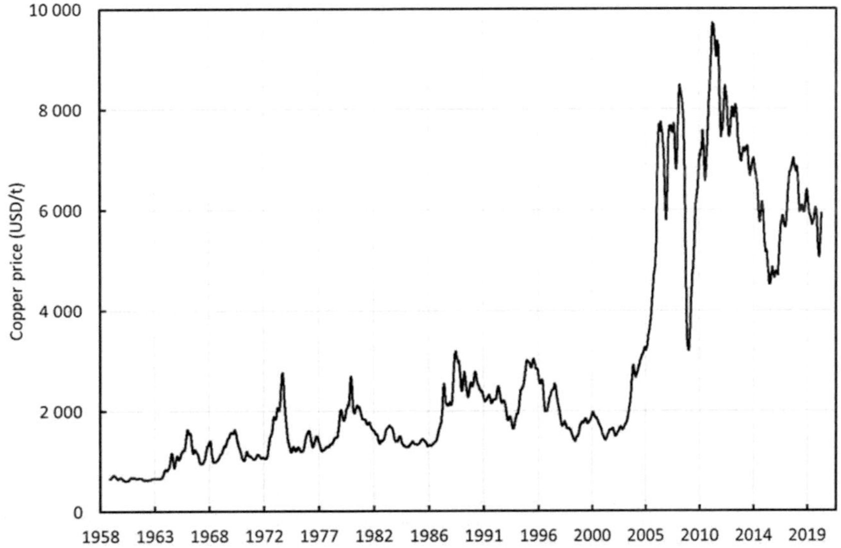

Figure 2.9 London Metal Exchange cash price for Grade A copper since 1958. The price climbed rapidly from 2002 owing to the rapid industrialization of China. Specifications for Grade A copper are given in Chapter 13. *(Data from Macrotrends. (2020). Copper prices — 45 year historical chart. Retrieved from https://www.macrotrends.net/1476/copper-prices-historical-chart-data. Accessed 19 September 2020.)*

will also boost consumption because wind turbines and solar panels require larger amounts of copper, per unit of energy generated, than traditional coal and gas plants. Electric and hybrid cars also use considerably more copper than conventional vehicles (International Copper Study Group, 2019).

2.7 Summary

Copper is produced around the world. About 40% is mined in the western mountain region of South America. Mined ore grades (%Cu) have fallen over the years as today's relatively high Cu grade deposits are mined out and newly discovered deposits contain smaller amounts of Cu.

Concentrators and leach—solvent extraction—electrowinning plants are located near their mines. Smelters and refineries are increasingly located near sea ports so that they can receive concentrates from mines around the world.

Copper's most exploited property is its high electrical conductivity, in conjunction with its excellent corrosion resistance, formability, and joinability. Its high thermal conductivity and corrosion resistance are also exploited in many heat transfer applications. Copper is readily alloyed with other metals (mostly zinc) to make complex, strong, corrosion-resistant castings.

Worldwide, about 21 million tonnes of copper are mined each year. Currently, about one-third of copper is recycled at the end of its useful life (Statistica, 2020) (Chapter 18), which adds an additional 5 million tonnes into the market.

References

Calvo, G., Mudd, G., Valero, A., & Valero, A. (2016). Decreasing ore grades in global metallic mining: A theoretical issue or a global reality? *Resources, 5*(4), 36—49.

Copper Development Association. (2020). *60 centuries of copper.* https://www.copper.org/education/history/60centuries/ Accessed 19.09.20.

International Copper Study Group. (2017). *Directory of copper mines and plants up to 2020.* Lisbon: International Copper Study Group.

International Copper Study Group. (2018). *ICSG 2018 statistical yearbook.* Lisbon: International Copper Study Group.

International Copper Study Group. (2019). *The world copper factbook 2019.* Lisbon: International Copper Study Group.

International Copper Study Group. (2020). *Copper bulletin July 2020* (Monthly publication No. 7, Vol. 27). Lisbon: ICSG.

Macrotrends. (2020). *Copper prices — 45 year historical chart.* https://www.macrotrends.net/1476/copper-prices-historical-chart-data. Accessed 19.09.20.

Materion. (2020). *Advanced copper beryllium alloys — reliability in demanding applications.* https://materion.com/products/highperformance-alloys/copper-beryllium-alloys. Accessed 19.09.20.

Statistica. (2020). *Global recycling input rate for copper scrap from 2004 to 2014.* https://www.statista.com/statistics/280982/global-recycling-rate-for-copper-scrap/. Accessed 19.09.20.

Winton Group. (2017). *Copper-bottomed booms and busts.* https://www.winton.com/longer-view/copper-bottomed-booms-and-busts. Accessed 19.09.20.

CHAPTER 3

Production of high copper concentrates—comminution and flotation (Johnson et al., 2019)

The grade of copper ores is typically too low (0.5%—2% Cu) for economic direct smelting. Heating and melting the huge quantity of largely worthless rock would require too much energy and too much furnace capacity. For this reason, all ores destined for pyrometallurgical processing are physically concentrated before smelting. The product of this step is *concentrate*, which contains ~30% Cu (virtually all as sulfide minerals).

Ores destined for hydrometallurgical processing are rarely concentrated. Cu is usually extracted from these ores by direct leaching of crushed or milled ore (Chapter 14).

This chapter describes the production of high-grade concentrate from low-grade ore. Processing of sulfide minerals is emphasized because these minerals account for virtually all Cu concentration.

3.1 Concentration flowsheet

Concentration of Cu ores consists of isolating the copper—containing minerals from the rest of the ore. It entails the following:

(a) blasting, crushing, and grinding the ore to a size where the Cu mineral grains are liberated from the non-Cu mineral grains, known as *comminution*, and

(b) physically separating the liberated Cu minerals from non-Cu minerals by *froth flotation* to generate a Cu-rich concentrate and Cu-barren *tailings* Chapter 4.

Fig. 1.2 shows these processes in relation to the overall copper-making flowsheet. Figs. 3.1 and 3.11 describe them in detail.

Copper concentrators typically treat 10,000—150,000 tonnes of ore per day, depending on the production rate of their mines (ICSG, 2010).

Extractive Metallurgy of Copper, Sixth Edition
ISBN 978-0-12-821875-4, https://doi.org/10.1016/B978-0-12-821875-4.00011-0

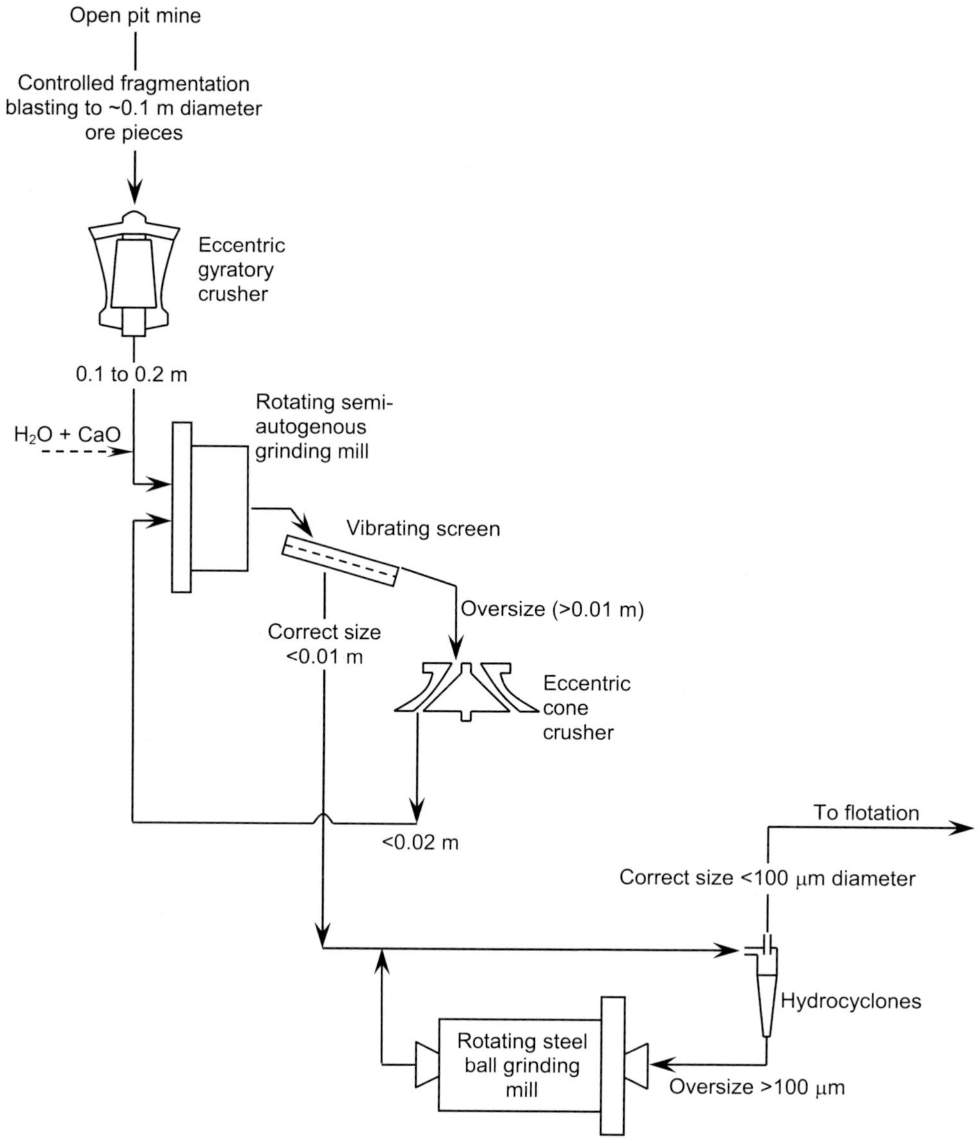

Figure 3.1 Flowsheet for preparing small flotation feed particles from as-mined ore fragments. One stage of crushing and two stages of grinding are shown. The crushing is open-circuit, i.e., there is no recycle loop. The two grinding circuits are closed circuit, i.e., oversize material is recycled for recrushing or regrinding to specified particle sizes. *(International Copper Study Group (ICSG) World Copper Factbook, IWCC and ICA.)*

3.2 The comminution process

To liberate the copper-containing minerals from the gangue materials in the ore (Chapter 2) and enable them to be collected by flotation into a concentrate, the ore should be finely ground.

Comminution is performed in three stages:

(a) *breaking* the ore by explosions in the mine (blasting);
(b) *crushing* large ore pieces by compression in gyratory or roll crushers; and
(c) wet *grinding* of the crushed ore in rotating tumbling mills, where abrasion, impact, and compression all contribute to breaking the ore.

The final fineness of grind is mainly determined by the number of times an ore particle passes through the grinding mills. Separate crushing and grinding steps are necessary because it is not possible to break massive run-of-mine ore pieces while at the same time controlling the fineness of grind that is necessary for the flotation step.

3.2.1 Crushing

Crushing is mostly done in the mine, mostly using *gyratory crushers* (Bearman, 2019, Fig. 3.2). This permits ore to be transported out of an open-pit mine by conveyor. It also permits easy hoisting of ore out of an underground mine. The crushed ore is stored in a coarse-ore stockpile, from which it is sent by conveyor to a grinding mill. The ore from crushing is then sent for grinding.

3.2.2 Grinding

3.2.2.1 Grind size and liberation of copper minerals

To isolate the copper-containing minerals into a concentrate, the ore should be finely enough ground to liberate the Cu mineral grains from the non-Cu mineral grains. The extent of grinding required to do this is determined by the size of mineral grains in the ore. Laboratory-scale flotation tests on materials of different particle sizes are usually required to determine the grind size that is required to liberate the copper minerals.

Fig. 3.3A shows the effect of grind size on recovery of Cu into the concentrate, while Fig. 3.3B shows the corresponding Cu concentration in the tailings. There is an optimum grind size for maximum recovery of Cu to the concentrate in the subsequent flotation step. Too large a grind size causes some Cu mineral grains to remain combined with or *occluded* by non-Cu mineral grains, preventing their flotation; too fine a grind size causes the rising bubbles in the flotation process to push the very tiny Cu-containing particles aside, preventing their contact with the bubble and reducing their recovery to the concentrate. In addition, very fine rock and Fe-sulfide mineral particles may agglomerate with very fine Cu-sulfide particles, preventing them from floating.

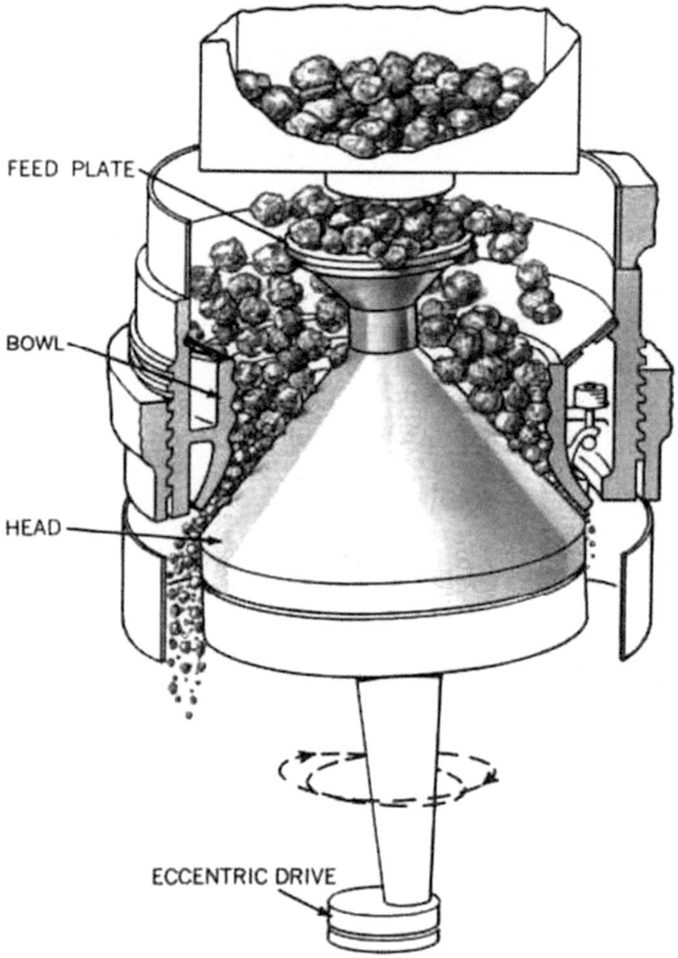

Figure 3.2 Gyratory crusher for crushing as-mined ore to ~0.2 m diameter pieces. The crushing is done by compression of ore pieces between the eccentrically rotating spindle and the fixed crusher wall. The crushing surface on the spindle can be up to 3 m high, 1.5 m diameter. Crushing rates are 10,000–150,000 tonnes of ore per day. *(International Copper Study Group (ICSG) World Copper Factbook, IWCC and ICA.)*

Liberation of mineral grains from each other generally requires grinding to ~100 μm diameter particle size. Flotation begins to be adversely affected when particles less than ~10 μm are produced.

Grinding requires considerable electrical energy (Table 3.1). This is another reason to avoid overgrinding.

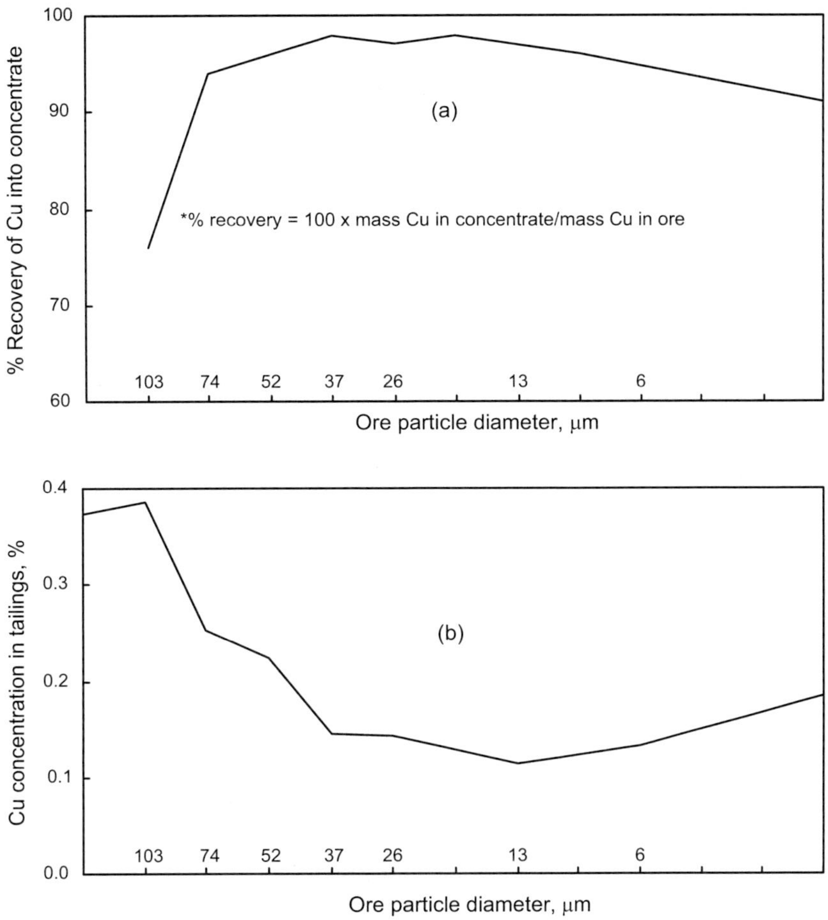

Figure 3.3 Effect of grind particle size on (A) Cu recovery to concentrate and (B) % Cu in tailings. The presence of an optimum is shown (Taggart, 1954). *(International Copper Study Group (ICSG) World Copper Factbook, IWCC and ICA.)*

3.2.2.2 Grinding equipment

Grinding produces ore particles of sufficient fineness for Cu mineral recovery by flotation. The most common grinding mills are *semiautogenous* (*SAG*) and *autogenous* (*AG*) mills and cone mills (Fig. 3.4; Giblett, 2019; Jones & Pena, 1999). A grinding circuit usually consists of one SAG or AG mill and one or two cone mills. Grinding is continuous and fully integrated with the subsequent flotation operation. Grinding is always done wet, with mixtures of ~70 mass% solids in water.

3.2.2.3 Autogenous and semiautogenous mills

The crusher product is ground in an SAG or AG mill (Giblett, 2019). AG mills crush the ore without the need for iron or steel grinding media. They are used when the ore is hard

Table 3.1a Industrial crushing and grinding data for three copper concentrators. They all treat ore from large open-pit mines. Flotation details are given in Table 3.5.

Concentrator	Candelaria, Chile	El Soldado, Chile	Los Bronces, Chile
Ore treated per year, tonnes	25,000,000 (2001)	7,700,000 (2010)	21,000,000 (2010)
Ore grade, %Cu	0.9–1.0	0.63	1.055
Crushing		Chalcopyrite	Chalcopyrite
Primary gyratory Crusher	1	1	1
Diameter × height, m	1.52 × 2.26	1.1 × 1.65	1.4 × 1.9
Power rating, kW	522	300	430
Rotation speed, RPM			6.5
Product size, m	0.1–0.13	0.2	0.2
Energy consumption, kWh per tonne of ore	0.3 (estimate)	0.2	0.25 (0.18 m diameter product
Secondary crushers	No	Yes, 4	No
First stage grinding			
Mill type	Semiautogenous	Semiautogenous	Semiautogenous
Number of mills	2	1	2
Diameter × length, m	11 × 4.6	10.4 × 5.2	8.5 × 4.3
Power rating each Mill, kWh	12,000	11,300	5590
Rotation speed, RPM	9.4–9.8	10.2	11.3
Vol. % "steel" in mill	12–15	15	13%–15%
Ball size, initial	12.5 cm	12.7	12.7
Ball consumption, kg per tonne of ore	0.3	0.32	0.35
Feed	70% ore, 30% H_2O	80% ore 20% water	70% ore 30% water
Product size	80% < 140 μm	200 μm	3 cm
Oversize treatment	22% ore recycle through two 525 kW crushers		Oversize to pebble crushers
Energy consumption, kWh per tonne of ore	7.82		6.5
Second stage grinding			
Mill type	Ball mills	Rod + ball mills	Ball mills
Number of mills	4	4 rod, 7 ball	3
Diameter × length, m	6 × 9	3.6 × 2.5	7 × 11
Power rating each Mill, kW	5600	Rod 350 Ball 550	10,800
Rotation speed, RPM		28	12
Vol. % "steel" in mill		47	35
Feed		70% ore, 30% w	65% ore, 35% water
Product size			80% < 212 μm

Table 3.1a Industrial crushing and grinding data for three copper concentrators. They all treat ore from large open-pit mines. Flotation details are given in Table 3.5.—cont'd

Concentrator	Candelaria, Chile	El Soldado, Chile	Los Bronces, Chile
Energy consumption, kWh per tonne of ore	7 (estimate)	~8	7.2
Hydrocyclones	14 Krebs (0.5 m diameter)	14 (~0.6 m diameter)	3 (Krebs + Cavex) (0.66 m diameter)
Particle size monitor		Yes PS1200	Yes PS1200

Table 3.1b Industrial crushing and grinding data for three copper concentrators, 2010. Flotation details are given in Table 3.5.

Concentrator	Africa, open pit	Africa, u-ground	Mantos Blancos, Chile
Ore treated per year, tonnes	4,000,000	900,000	4,500,000
Ore grade, %Cu	0.15	0.45	1.13
Crushing	Chalcocite, chalcopyrite	Chalcopyrite	Chalcocite
Primary gyratory Crusher	1 jaw crusher	1 jaw crusher	1
Diameter × height, m	4.2 × 6.5	8.5 × 11	1.1 × 1.65
Power rating, kW	250	132	300
Rotation speed, RPM			
Product size, m	80% < 0.25	80% < 0.25	0.15
Energy consumption, kWh per tonne of ore	5.4	4.5	0.4
Secondary crushers	None	Yes, jaw crusher	Yes, 4
First stage grinding			
Mill type	Semiautogenous	Semiautogenous	Ball mills
Number of mills	1	1	2
Diameter × length, m	8.53 × 4.35	6.1 × 2.75	3.8 × 4.6
Power rating each Mill, kWh	6000	1300	1120
Rotation speed, RPM	14	12	21.6
Vol. % "steel" in mill	22%	25%	38–40
Ball size, initial, cm	12.5	12.5	
Ball consumption, kg per tonne of ore	0.41	0.9	
Feed	85% ore, 15% water	85% ore, 15% water	84% ore, 16% water
Product size	80% < 800 μm	80% < 600 μm	75% < 212 μm
Oversize treatment	Pebble crushers, SAG	Recycle to SAG	
Energy consumption, kWh per tonne of ore	8.5	6.5	13.5

Continued

Table 3.1b Industrial crushing and grinding data for three copper concentrators, 2010. Flotation details are given in Table 3.5.—cont'd

Concentrator	Africa, open pit	Africa, u-ground	Mantos Blancos, Chile
Second stage grinding			
Mill type	Ball mill	Ball mill	Ball mill
Number of mills	1	1	1
Diameter × length, m	6.1 × 9.05	4.27 × 7.01	3.5 × 5.2
Power rating each Mill, kW	6000	1860	940
Rotation speed, RPM	18	16	23
Vol. % "steel" in mill	35%	32	38–40
Feed	85% ore, 15% water	85% ore, 15% water	85% ore, 15% water
Product size	80% < 125 μm	80% < 100 μm	75% < 212 μm
Energy consumption, kWh per tonne of ore	16.8	15	20
Hydrocyclones	8 (1.1 m dia)	8 (0.25 m dia)	5 (0.5 m diameter
Particle size monitor	Outotec PSI	Outotec PSI	

Table 3.1c Industrial crushing and grinding data for three copper concentrators, 2010. They all treat ore from large open-pit mines. Flotation details are given in Table 3.5.

Concentrator	Cerro Verde Peru	Ray, Arizona	Sierrita, U.S.A.
Ore treated per year, tonnes	3,900,000	10,000,000	34,000,000
Ore grade, %Cu	0.6 (chalcopyrite)	0.453 in Chalcopyrite and Chalcocite	0.23% Cu in cpy 0.03 Mo in MoS_2
Crushing			
Primary gyratory Crusher	1	1	2
Diameter × height, m	1.5 × 2.87	1.5 × 2.3	1.5 × 2.3
Power rating, kW	750		600
Rotation speed, RPM			
Product size, m	0.125	0.152trun -1	80% < 0.1
Energy consumption, kWh per tonne of ore	5		0.08
Secondary crushers	Yes	No	4
First stage grinding			
Mill type	Ball mills	Semiautogenous	Ball mills

Table 3.1c Industrial crushing and grinding data for three copper concentrators, 2010. They all treat ore from large open-pit mines. Flotation details are given in Table 3.5.—cont'd

Concentrator	Cerro Verde Peru	Ray, Arizona	Sierrita, U.S.A.
Number of mills	4	1	16
Diameter × length, m	7.3 × 11	10.4 × 5.2	5 × 5.8
Power rating each Mill, kWh	13,000	12,500	3000
Rotation speed, RPM	12.2	8.3−10.8	13.8
Vol. % "steel" in mill	38	13	39
Ball size, initial, cm	6.35	14	10, 7.5
Ball consumption, kg per tonne of ore	0.67	0.15	0.39
Feed	75% ore/25% water	83% ore/17% H_2O	98% ore, 2% H_2O
Product size	80% < 150 μm	6.35 cm	80% < 350
Oversize treatment		Cone crushers	
Energy consumption, kWh per tonne of ore	10	12−18	8.4
Second stage grinding			None
Mill type		Ball mills	
Number of mills		2	
Diameter × length, m		5.5 × 9.34	
Power rating each Mill, kW		4500	
Rotation speed, RPM		13	
Vol. % "steel" in mill		35−40	
Feed		85% ore/15% H_2O	
Product size		80% < 147 μm	
Energy consumption, kWh per tonne of ore		8	
Hydrocyclones	9 per ball mill (0.84 m diameter)	16 (1.5 m diameter)	2 per ball mill (0.84 m dia.)
Particle size monitor	Yes	No	No

enough for the tumbling ore to grind itself. In SAG milling, ~0.15 m^3 of 13 cm diameter iron or steel balls are added into the mill per 0.85 m^3 of ore (i.e., 15 vol.% "steel") to assist grinding. SAG mills are much more common.

The mill product is usually passed over a large vibrating screen to separate oversize *pebbles* from ore particles of the correct size. The correct size material is sent forward

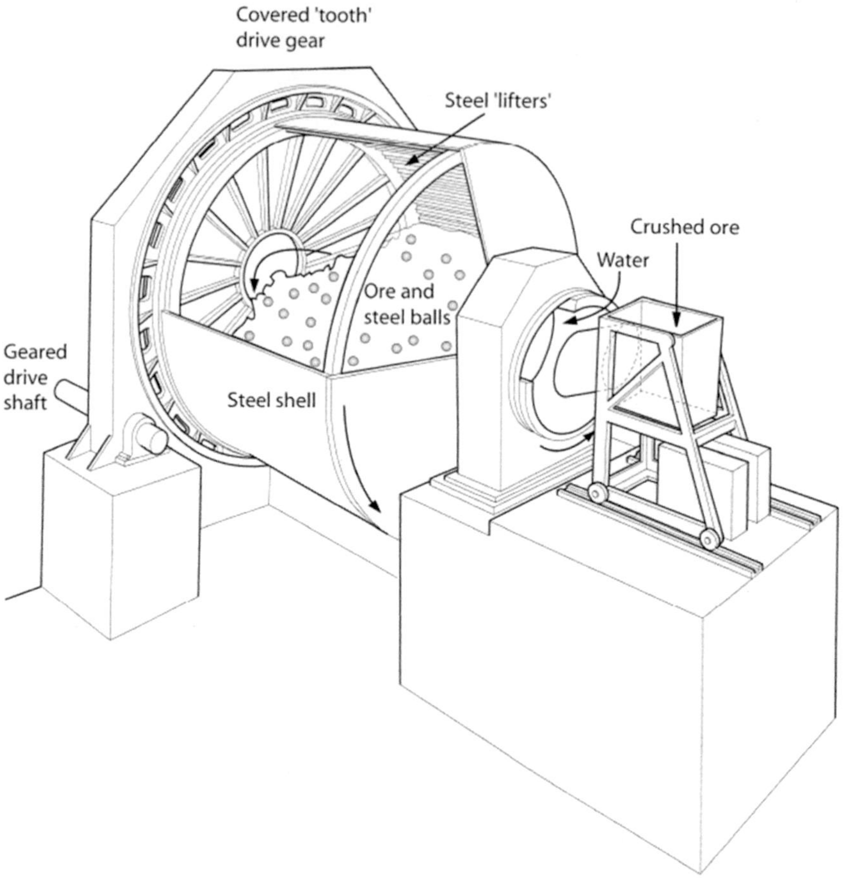

Figure 3.4 Semiautogenous grinding mill. *(International Copper Study Group (ICSG) World Copper Factbook, IWCC and ICA.)*

to a eccentric cone mill for final grinding. The oversize pebbles are recycled through a small ball mill for final reduction. This procedure maximizes ore throughput and minimizes electrical energy consumption.

3.2.2.4 Ball mills (Giblett, 2019)

The ball mill accepts the oversize from hydrocyclone sizing of the SAG mill product. Ball mills give a controlled final grind and produce flotation feed of a uniform size. Ball mills tumble iron or steel balls with the ore. The balls are initially 5–10 cm diameter but gradually wear away as grinding of the ore proceeds. The feed to ball mills (dry basis) is typically 75 vol.% ore and 25% steel.

The ball mill is operated in closed circuit with a particle-size measurement device and size-control cyclones. The cyclones send correct-size material on to flotation and direct oversize material back to the ball mill for further grinding.

3.2.2.5 HPGR

Fig. 3.5 sketches a high pressure grinding (HPGR) roll crusher. This equipment is used extensively in the cement, diamond, and iron ore industries. First deployed in the copper industry in 2006 at the Cerro Verde concentrator in Peru, it has since been adopted by several copper producers as a replacement for the ball mills used to grind oversize product from the SAG mill (Fig. 3.1; Burchardt et al., 2010). It is increasingly used as a replacement for the SAG mill itself (Daniel et al., 2019).

Its perceived advantages are 15%–25% lower electrical energy consumption per tonne of ore and its ability to fracture ore at the mineral grain boundaries (although this is probably more advantageous for leaching than for froth flotation). Its major disadvantage appears to be its difficulty in handling wet, sticky ores and high clay ores and a 10%–25% higher initial capital cost (Daniel et al., 2019). HPGR mills are lower capacity than SAG mills, so operating multiple units in parallel is required in a production environment.

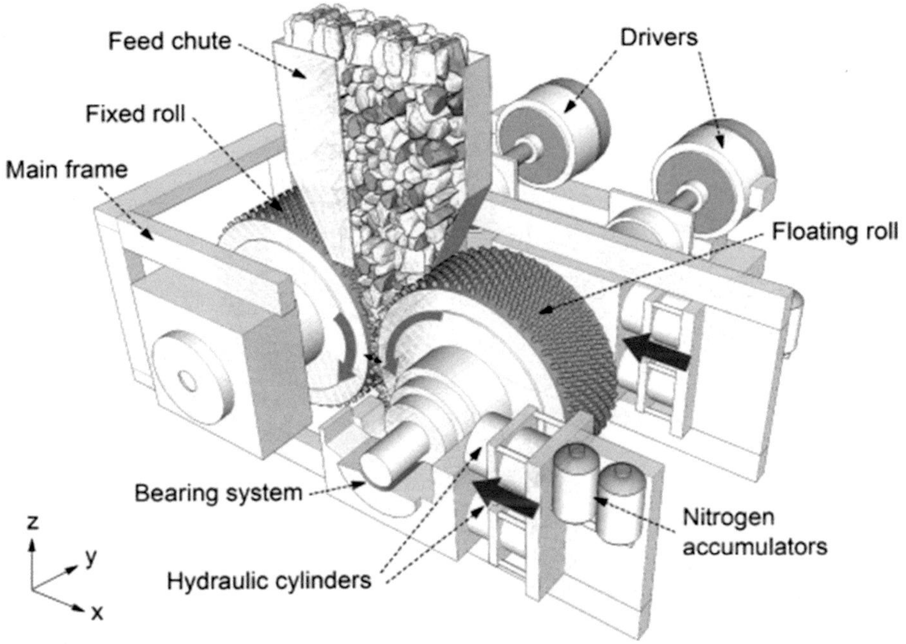

Figure 3.5 Sketch of high pressure roll crusher. Ore is force-fed between the two rotating, hard-surfaced rolls. Constant crushing pressure is provided by hydraulic pistons pressing on the movable left roll housing. Representative details are as follows:

2.4 m roll diameter × 1.65 m length
21 rotations per minute
2500–3000 tonnes/h ore feed rate
<0.05 m feed diameter
<0.0055 m product diameter
2 × 25 kW power requirement
130–160 bar pressure on moveable roll housing *(Credit: Barrios & Tavares, 2016.)*

3.3 Particle size control of flotation feed

A critical step in grinding is ensuring that its product particles are fine enough for efficient flotation. Coarser particles must be isolated and returned for further grinding.

Size control is universally done by *hydrocyclones* (Fig. 3.6; Flintoff & Knorr, 2019). A hydrocyclone uses the principle that, under the influence of a force field, large ore particles in a water–ore mixture (slurry) tend to move faster than small ore particles.

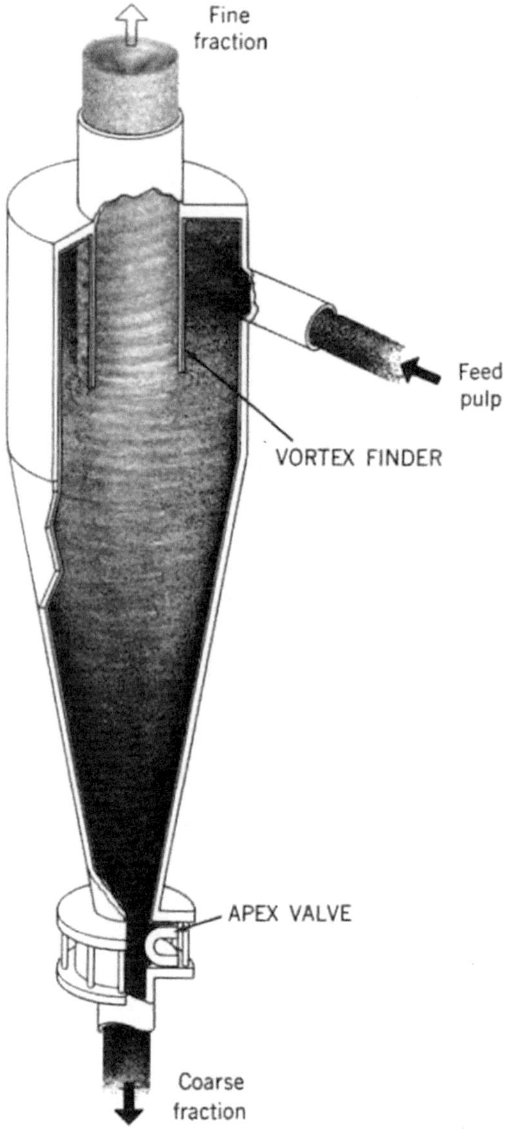

Figure 3.6 Cutaway view of hydrocyclone showing (A) tangential input of water–ore particle feed and (B) separation of the feed into fine particle and coarse particle fractions. *(Reproduced from Boldt, J. R., & Queneau, P. (1967).* The winning of nickel. *Toronto: Longmans Canada Ltd. courtesy Vale Limited.)*

This principle is put into practice by pumping the grinding mill discharge into a hydrocyclone at high speed, 5—10 m/s. The slurry enters tangentially (Fig. 3.6), giving it a rotational motion inside the cyclone. This creates a centrifugal force which accelerates ore particles toward the cyclone wall. The water content of the slurry, typically ~60% by mass, is adjusted so that (a) the oversize particles are able to reach the wall, where they are dragged out by water flow along the wall and through the apex of the cyclone, and (b) the correct (small) size particles do not have time to reach the wall before they are flushed out with the main slurry flow through the vortex finder.

The principal control parameter for the hydrocyclone is the water content of the incoming slurry. An increase in the water content of the slurry gives less hindered movement of particles and allows a greater fraction of the input particles to reach the wall and pass through the apex. This increases the fraction of particles being recycled for regrinding and ultimately leads to a more finely ground final product. A decrease in water content has the opposite effect.

3.3.1 Instrumentation and control

Grinding circuits are extensively instrumented and closely controlled (Fig. 3.7, Table 3.2). The objectives of the control are to

(a) produce particles of appropriate size for efficient flotation recovery of Cu minerals;
(b) produce these particles at a rapid rate;
(c) produce these particles with a minimum consumption of energy.

The most common control strategy is to ensure that the sizes of particles in the final grinding product are within predetermined limits, as sensed by an on-stream particle-size analyzer (Outotec, 2016; O'Keefe et al., 2014, pp. 694—703), and then optimize production rate and energy consumption while maintaining this correct size. Fig. 3.7 and the following describe one such control system.

3.3.1.1 Particle-size control

The particle-size control loop in Fig. 3.7 controls the particle size of the grinding product by automatically adjusting the rate of water addition to the hydrocyclone feed sump. If, for example, the flotation feed contains too many large particles, an electronic signal from the particle-size analyzer (S) automatically activates water valves to increase the water content of the hydrocyclone feed. This increases the fraction of the ore being recycled to the ball mills and gives a finer grind.

Conversely, too fine a flotation feed automatically cuts back on the rate of water addition to the hydrocyclone feed sump. This decreases ore recycle to the grinding mills, thereby increasing the particle size of the flotation feed. It also permits a more rapid initial feed to the ball mills and minimizes grinding energy consumption.

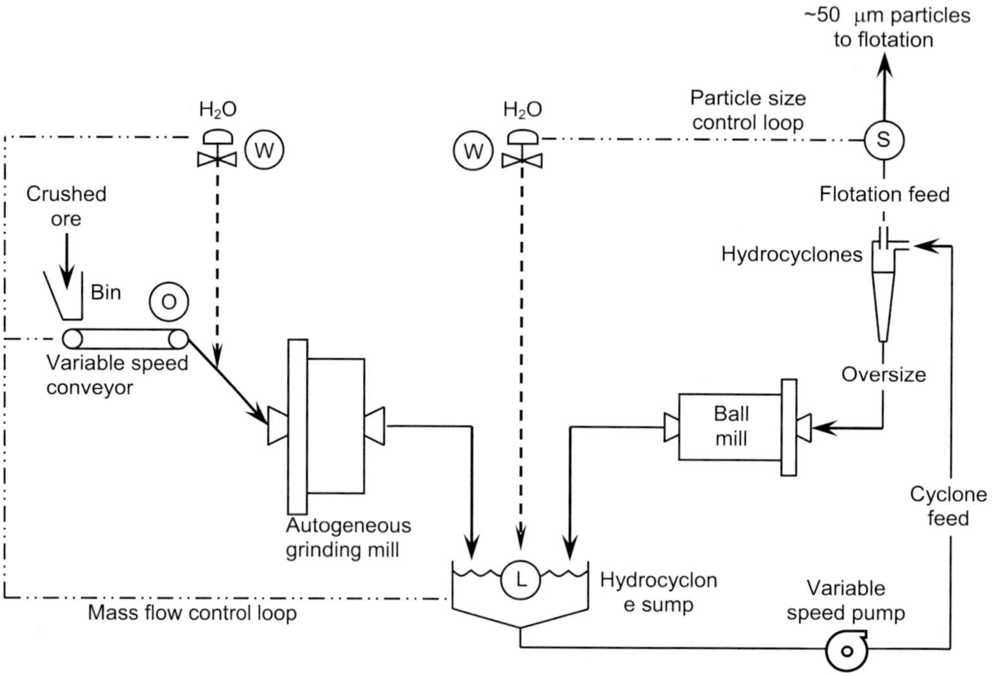

Figure 3.7 Control system for grinding mill circuit (ore flow, water flow; electronic control signals). The objective is to quickly and cheaply produce correct size ore particles for efficient Cu mineral flotation. The *circled* symbols refer to the sensing devices in Table 3.2.

3.3.1.2 Ore throughput control

The second control loop in Fig. 3.7 gives maximum ore throughput rate without overloading the ball mill. Overloading might become a problem if, for example, the ball mill receives tough, large particles which require extensive grinding to achieve the small particle size needed by flotation.

The simplest mass flow control scheme is to use the hydrocyclone sump slurry level to adjust the ore feed rate to the grinding plant. If, for example, slurry level sensor (L) detects that the slurry level is rising (due to tougher ore and more hydrocyclone recycle), it automatically slows the input ore feed conveyor. This decreases flow rates throughout the plant and stabilizes ball mill loading and sump level.

Detection of a falling sump level, on the other hand, automatically increases ore feed rate to the grinding plant, to a prescribed rate or to the maximum capacity of another part of the concentrator, such as the flotation circuit.

There is, of course, a time delay (5—10 min) before the change in ore feed rate is felt in the hydrocyclone feed sump. The size of the sump must be large enough to accommodate further buildup (or draw-down) of slurry during this delay.

Table 3.2 Sensing and control devices for Fig. 3.8 grinding control circuit.

Sensing instruments	Symbol in Fig. 3.8	Purpose	Type of device	Use in automatic control system
Ore input rate weightometer	O	Senses feed rate of ore into grinding circuit	Load cells, conveyor speed	Controls ore feed rate
Water flow gages	W	Sense water addition rates	Rotameters	Control water-to-ore ratio in grinding mill feed
On-stream particle size analyzer	S	Senses a critical particle size parameter (e.g., percent minus 70 μm) on the basis of calibration curves for the specific ore	Measures diffraction of laser beam by particles in an automatically taken slurry sample (Outotec, 2010)	Controls water addition rate to hydrocyclone feed (which controls recycle and the size of the final grinding circuit product)
Hydrocyclone feed sump level indicator	L	Senses changes of slurry level in sump	Bubble pressure tubes; electric contact probes; ultrasonic echoes; nuclear beam	Controls rate of ore input into grinding circuit (prevents overloading of ball mills)

3.3.2 Automated mineralogical analysis

An important modern tool for optimizing concentrate production is quantitative scanning electron microscopy which, for any assemblage of ore particles, (a) indicates the mineral grain makeup of every particle in the assemblage, showing liberated, partially liberated, and unliberated copper mineral grains, and (b) suggests whether more or less grinding will give improved Cu recovery to the concentrate (Aylmore, 2019; Zahn et al., 2007). Fig. 3.8 shows an example concentrate assemblage. This can be done routinely or specifically when Cu recovery to concentrate unexpectedly decreases. Automated mineralogical analysis can also be useful in the initial design of a new concentrator by indicating the minerals and mineral grain associations of ore samples from around the entire mineral deposit.

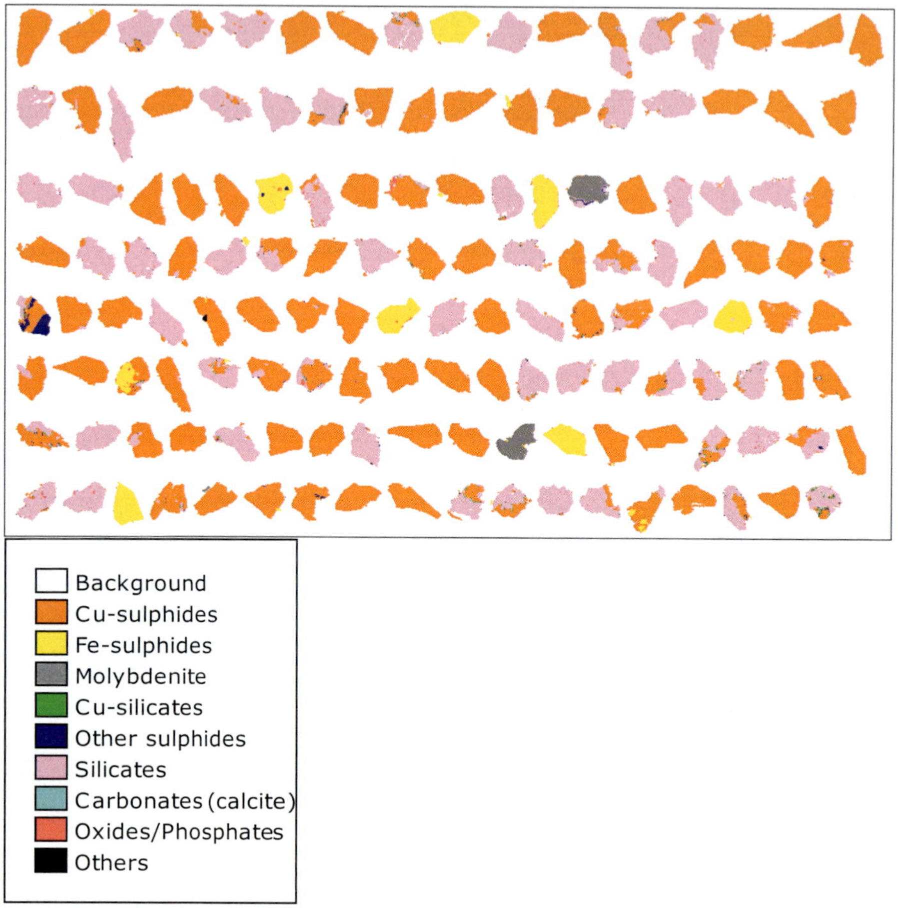

Figure 3.8 Scanning electron microscope image of Cu sulfide concentrate particle assemblage. The key is shown at left. The bottom right particle is ~70 μm across. As expected, most of the particles are Cu sulfide minerals. However, the concentrate contains many silicate particles, which indicates that concentrate cleaning could be improved. Fe sulfides have also floated so that depression could also be improved. Further grinding might liberate a few more Cu sulfide particles, but for the most part grind size is about optimum. *(Courtesy of Anglo American: Research.)*

3.4 Froth flotation fundamentals

The indispensable tool for Cu concentrate production is *froth flotation* (Fuerstenau et al., 2007). This technique is used to upgrade an ore to a concentrate by selectively floating copper-containing minerals away from noncopper minerals. The principles of froth flotation are as follows:

(a) Sulfide minerals are normally wetted by water (*hydrophilic*), but they can be conditioned with reagents (known as *collectors*) that cause them to become water repellent (*hydrophobic*).

(b) Cu sulfide minerals can selectively be made hydrophobic by their interaction with the collector, leaving other minerals wetted.

(c) Collisions between ~1 mm diameter rising air bubbles and the now water-repellent Cu minerals result in attachment of the Cu mineral particles to the bubbles (Fig. 3.9). The Cu minerals are *floated* to the surface of the slurry by the air bubbles.

(d) The still wetted noncopper mineral particles do not attach to the rising bubbles and remain in the slurry.

Industrially, the process entails (a) conditioning a water–ore mixture (*slurry*) to make the Cu minerals water repellent while leaving the non-Cu minerals hydrophilic and (b) passing a dispersed stream of small bubbles up through the slurry. These procedures cause the Cu mineral particles to attach to the rising bubbles which carry them to the top of the flotation cell (Fig. 3.10). The other minerals are left behind. They depart the cell through an underflow system. These are mostly nonsulfide *gangue* with a small amount of pyrite.

The last step in the flotation process is creating a strong but short-lived froth by adding *frother* when the bubbles reach the surface of the slurry. This froth prevents bursting of the bubbles and release of the Cu mineral particles back into the slurry. The froth overflows the flotation cell, often with the assistance of paddles (Fig. 3.10), and into a trough. There it collapses and flows into a collection tank.

Figure 3.9 Photograph of water-repellent mineral particles attached to rising bubbles. The input bubbles in industrial flotation are ~1 mm diameter, which is about the size of the largest bubble in this photograph. *(Photograph by H. Rush Spedden.)*

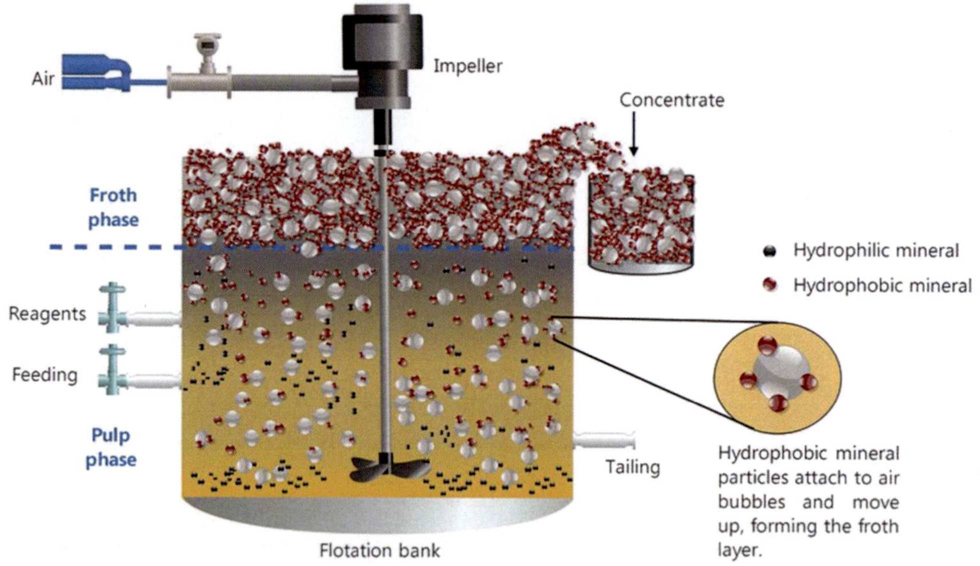

Figure 3.10 Cutaway view of cubic mechanical flotation cell. The methods of producing bubbles and gathering froth are shown. Mechanical flotation cells in recent design Cu concentrator are up to 300 m^3 cylindrical tanks. They operate continuously. *(Credit: Li et al., 2019, Figure 2.)*

A sequence of flotation cells is designed to optimize Cu recovery and the Cu grade in the concentrate (Fig. 3.11). The froth from the last set of flotation cells is, after water removal, the *Cu concentrate*.

3.5 Flotation chemicals (Nagaraj et al., 2019; Woodcock et al., 2007)

3.5.1 Collectors

The reagents (collectors) which create the water-repellent surfaces on sulfide minerals are heteropolar molecules. They have a polar (charged) end and a nonpolar (hydrocarbon) end. They attach their polar (charged) end to the mineral surface (which is itself polar), leaving the nonpolar hydrocarbon end extended outward (Fig. 3.12). It is this orientation that imparts the water-repellent character to the conditioned mineral surfaces.

A commonly used collector is xanthate, e.g.,

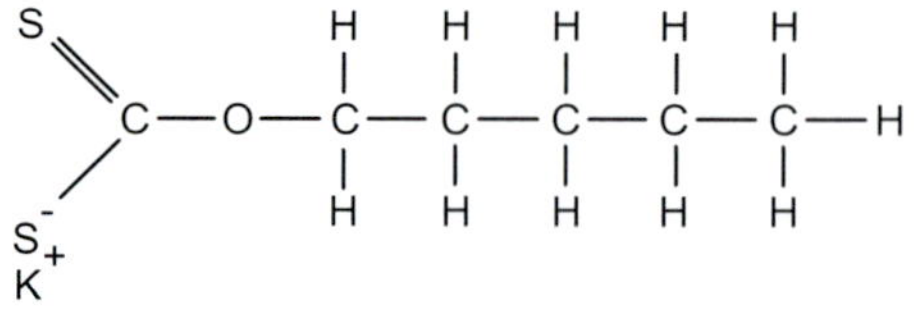

(potassium amyl xanthate).

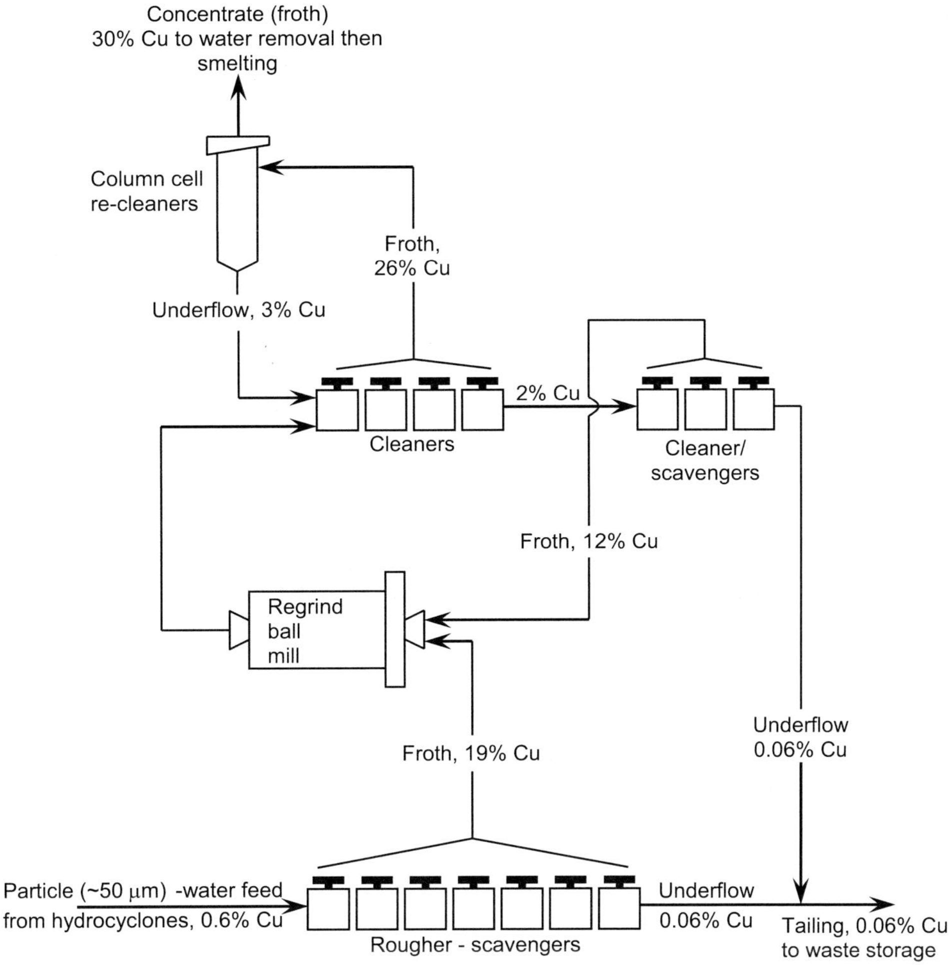

Figure 3.11 Flowsheet for floating Cu sulfide concentrate from gangue minerals and Fe sulfides. Residence times in each sector (e.g., rougher–scavenger cells) are 10–20 min.

Other sulfur-containing molecules, particularly dithiophosphates, mercaptans, and thiocarbamates, are also used. Commercial collectors are often blends of several of these reagents. About 10–50 g collector is required per tonne of ore feed.

3.5.2 Selectivity in flotation

The simplest froth flotation separation is that of sulfide minerals from waste oxide minerals, such as andesite, granodiorite, and granite. It uses collectors which, when dissolved in a water–ore slurry, preferentially attach themselves to sulfides. These collectors mostly have a sulfur group at the polar end, which attaches to sulfide minerals but ignores oxides.

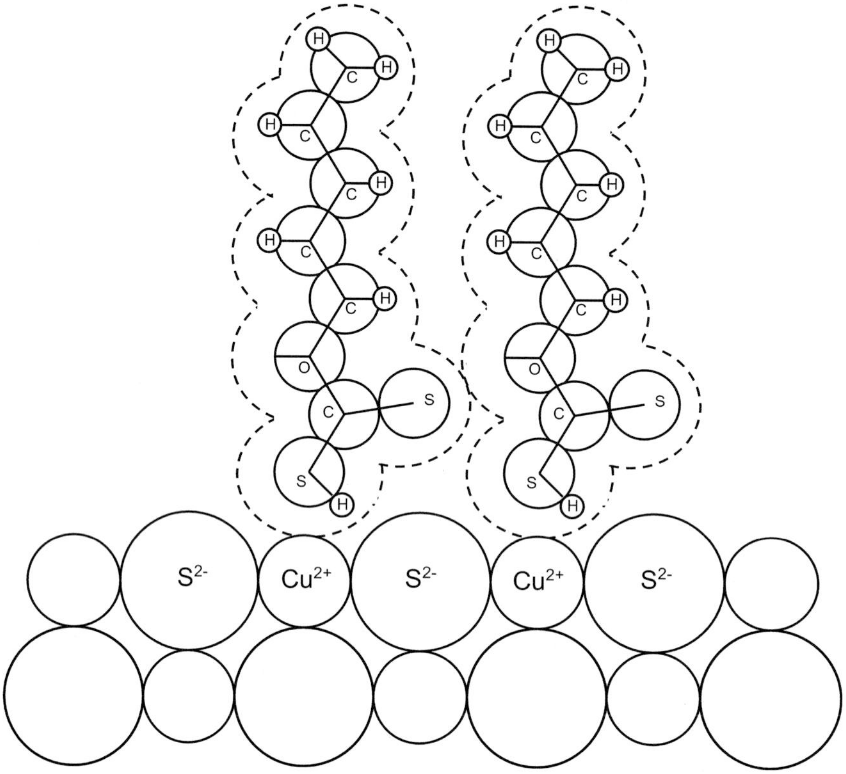

Figure 3.12 Sketch of attachment of amyl xanthate ions to covellite. There is a hydrogen atom hidden behind each carbon of the hydrocarbon chain. *(After Hagihara, H. (1952). Mono- and multiplayer adsorption of aqueous xanthate on galena surfaces.* Journal of Physical Chemistry, *56, 616–621.)*

3.5.3 Differential flotation modifiers

Separating sulfide minerals, such as chalcopyrite from pyrite, is somewhat more complex. It relies on modifying the surfaces of non-Cu sulfides so that the collector does not attach to them while still attaching to Cu sulfides.

The most common *modifier* is the OH^- (hydroxyl) ion. Its concentration is varied by adjusting the basicity of the slurry with burnt lime (CaO), occasionally with sodium carbonate (Na_2CO_3). The effect is demonstrated in Fig. 3.13, which shows how chalcopyrite, galena, and pyrite can be floated from each other. Each line on the graph marks the boundary between float and nonfloat conditions for the specific mineral—the mineral floats to the left of its curve, but not to the right.

This graph shows that:

(a) up to pH 5 (acid slurry), $CuFeS_2$, PbS, and FeS_2 all float;
(b) between pH 5 and pH 7.5 (neutral slurry), $CuFeS_2$ and PbS float, while FeS_2 is depressed;
(c) between pH 7.5 and pH 10.5 (basic slurry), only $CuFeS_2$ floats.

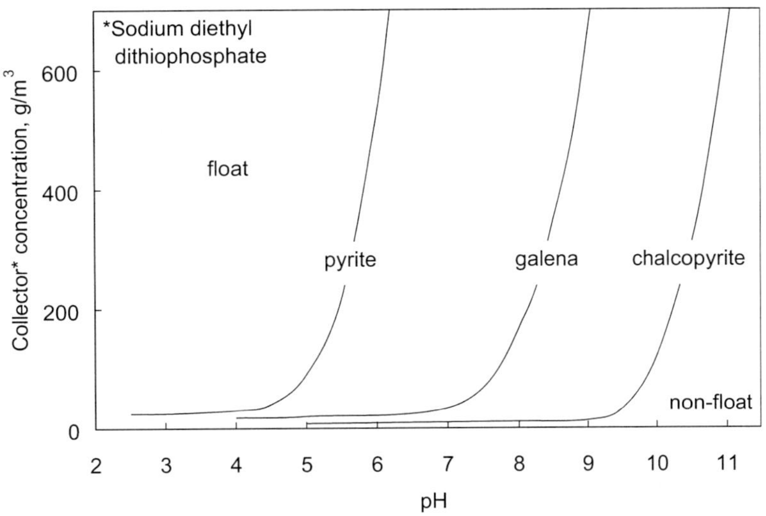

Figure 3.13 Effects of collector concentration and pH on the floatability of pyrite, galena, and chalcopyrite. Each line marks the boundary between float and nonfloat conditions for the specific mineral (Wark & Cox, 1934). Precise float/nonfloat boundary positions depend on collector, mineral, and water compositions.

A bulk Pb—Cu sulfide concentrate could therefore be produced by flotation at pH 6.5. The Pb and Cu sulfides could then be separated at pH 9, which occurs after addition of more CaO.

The modifying effect of OH^- is due to its competition with collector anions, such as xanthates, for a place on the mineral surface. OH^- ions are, for example, selectively adsorbed on pyrite. This prevents appreciable xanthate adsorption on the pyrite, selectively *depressing* it. However, too many OH^- ions will also depress chalcopyrite, so excess CaO must be avoided.

3.5.4 Frothers

Collectors and modifiers give selective flotation of Cu minerals from non-Cu minerals. *Frothers* create the strong but short-lived froth which holds the floated Cu minerals at the top of the cell. Frothers produce a froth which is strong enough in the flotation cell to support the floated Cu minerals, but breaks down quickly once the froth and its attached minerals overflow the cell.

Thousands of compounds have been developed as frothers over the years. Most are either short chain aliphatic alcohols (linear, branched, or MIBC) or polyglycols (polyethylene or polypropylene). Once the standard choice, pine oil has become much less popular in recent years. Most commercial frothers are mixtures, and formulation changes are frequent.

Frothers stabilize the froth by absorbing their OH^- polar end in water, while their branched (nonpolar) chains form a cross-linked network in air. The froth should not be long lived, so the branched chain hydrocarbon tails should not be too long. The addition of frothers also decreases bubble size, which provides more relative surface to attach mineral particles to.

3.6 Flotation of Cu ores

Selective flotation of Cu sulfide minerals (chalcopyrite, chalcocite, and bornite) from oxide rock and Fe sulfide minerals (mainly pyrite) is usually done with xanthate, dithiophosphate, mercaptan, or thionocarbamate collectors; burnt lime (CaO) for pH (OH^- ion) control; and mixed alcohol frothers. A representative flowsheet, industrial data, and example reagents are shown in Fig. 3.11 and Tables 3.3–3.5.

The flowsheet in Fig. 3.11 shows four sets of flotation cells:

(a) the *rougher–scavenger cells* float the incoming ground–ore slurry under conditions which give efficient recovery of Cu to the froth with a reasonable froth grade (15%–20% Cu);

(b) the *cleaner cells* in which flotation of non-Cu minerals in reground concentrate from the rougher–scavenger cells is depressed with CaO to produce a higher grade Cu concentrate;

(c) the *recleaner cells* maximize concentrate grade (%Cu) by giving rock and Fe sulfide minerals a final depression under quiescent conditions;

(d) the *cleaner–scavenger cells* scavenge the last traces of Cu to the froth from the cleaner tails by the addition of more collector before the tails are discarded.

The froths from the rougher–scavengers and cleaner–scavengers are ground before being sent to the cleaners (Fig. 3.11). This releases previously unliberated Cu mineral grains.

The rougher–scavenger and cleaner–scavenger cells are designed to maximize Cu recovery to concentrate. The cleaner and recleaner cells maximize concentrate grade.

Table 3.3 Cu contents of ores around the world. Industrial recoveries to concentrate are also shown.

Country	Mine	%Cu in ore	Recovery of Cu to concentrate, %
Chile	Candelaria	0.54	91
Indonesia	Grasberg open pit	0.9	84
Peru	Cerro Verde	0.4	86
Poland	Lower Silesia	1.64	~90
USA	Bagdad	0.36	85
USA	Bingham Canyon	0.6	84
USA	Sierrita	0.25	82

Table 3.4 %Cu in concentrate feeds at smelters around the world (Kapusta, 2004). The feeds are sometimes combinations of several concentrates, but they give an idea of Cu concentrate grades around the world. Pure chalcopyrite contains 35.4% Cu.

Country	Smelter location	%Cu in concentrate
Australia	Mount Isa	25–27
Brazil	Caraiba	32
Canada	Noranda	22
Chile	Caletones (Teniente)	31
China	Guixi	28
Finland	Harjavalta	28
Germany	Hamburg	30–33
India	Tuticorin	30–32
Indonesia	Gresik	30
Iran	Khatoon Abad	27
Japan	Toyo (Niihama)	32
Korea	Onsan	32.5
Mexico	La Caridad	26
Peru	Ilo	28
Philippines	Isabel (Leyte)	25–32
Poland	Lower Silesia	23
Russia	Norisk	26
South Africa	Palabora	30–35
Spain	Huelva	31
Sweden	Ronnskar	30
U.S.A.	Kennecott	26.5
Zambia	Mopani	28–42

Table 3.5a Industrial flotation data from three copper concentrators. All three treat ore from large open-pit mines. The equivalent crushing/grinding data are given in Table 3.1.

Concentrator	Candelaria, Chile	El Soldado, Chile	Los Bronces, Chile
Ore treated per year, tonnes	25,000,000 (2001)	7,700,000 (2010)	20,500,000 (2010)
Concentrate, tonnes/year		67,000	676,000
Ore grade, %Cu	0.9–1.0	0.63	
Sulfide		0.54	1.055
"Oxide"		0.09	0.049
Concentrate grade, % Cu	29–30	22.5	28.4
Tailings grade, % Cu		0.18	0.133
Cu recovery to concentrate, %	95% Cu, 82% Au 87% Mo		

Continued

Table 3.5a Industrial flotation data from three copper concentrators. All three treat ore from large open-pit mines. The equivalent crushing/grinding data are given in Table 3.1.—cont'd

Concentrator	Candelaria, Chile	El Soldado, Chile	Los Bronces, Chile
Rougher—scavenger flotation			
Feed	Cyclone overflow	Cyclone overflow	Grinding product
Number of cells	24	32	30
Volume of each cell, m^3	85 and 128	42.5	3 × 127 + 27 × 85
Cell type	Eimco	Wemco	Wemco
Mass% solids in feed		38	30
pH	10.4	10.3	10.3
Collector	SF 323	K dithiophosphate	Dithiophosphate and polysulfate
kg/tonne of ore		0.19	0.035
Oily collector, kg/t of ore		no	0.012
Frother	MIBC	Dowfroth 250: MIBC 1:5	5 Nalfloat: 1 Dowfoth 250
kg/tonne of ore		0.033	0.022
CaO, kg/tonne of ore	0.7	0.65	0.45
Residence time, minutes		33	25—30
Cleaner flotation			
Feed	Rougher concentrate ground in 4.27 × 6.7 m ball mill	Rougher—scavenger concentrate	Reground rougher and scavenger concentrate
Number of cells		1 square + 1 round columns	5
Volume of each cell, m^3		8.8 m^2 square, 16 m^2 round	224 c/u
Cell type		Column	Column
Mass% solids in feed		16	16
pH		10.5	11.3
Reagents		CaO	No
residence time, minutes			15.8
Cleaner—scavenger flotation			No
Feed		Cleaner tails	Cleaner tails
Number of cells		11 Wemco	21
Volume of each cell, m^3		10 × 28.3 + 1 × 127	42.5
Cell type		Mechanical	Wemco
Mass% solids in feed		18	16
pH		10—10.5	11.3
Reagents		No	No
Residence time, minutes			15

Circuits such as that in Fig. 3.11 give ~90% recovery of Cu sulfide minerals and ~30% Cu concentrate grade (Tables 3.4 and 3.5).

3.7 Flotation cells

Fig. 3.10 shows a *mechanical* flotation cell (Nelson & Lelinski, 2019). Air bubbles are introduced into the slurry through a rotating agitator at the bottom of the cell. The agitator shears the air into the 1 mm diameter bubbles needed for Cu mineral attachment. It also disperses the bubbles across the cell.

3.7.1 Column cells

Many new Cu flotation plants use column cells for cleaner flotation (Tables 3.5 and 3.6; Figs. 3.14 and 3.15). These cells provide separate zones for particle–bubble attachment and for draining of nonattached low-Cu particles from the froth (Mankosa et al., 2019).

Table 3.5b Industrial flotation data from three copper concentrators, 2010. The equivalent crushing/grinding data are given in Table 3.1.

Concentrator	Africa open pit	Africa under ground	Mantos Blancos Chile, open pit
Ore treated per year, tonnes	4,000,000	900,000	4,500,000
Concentrate, tonnes/year	35,000	21,000	125,000
Ore grade, %Cu	0.21	0.45	1.13
Sulfide	0.15	Chalcopyrite	0.92
"Oxide"	0.06	–	0.21
Concentrate grade, % Cu	25	17	35.2
Tailings grade, % Cu	0.018	0.04	0.12
Cu recovery to concentrate, %	80	85	89
Rougher–scavenger flotation			
Feed	Cyclone overflow	Cyclone overflow	Cyclone overflow
Number of cells	5	7	11
Volume of each cell, m^3	150	40	43
Cell type	OK150 tank	OK40 tank	Wemco
Mass% solids in feed	30	36	42
pH	9.0	7.9	9.5
Collector	K amyl xanthate Aerophine 3418A	K amyl xanthate Aerophine 3418A	sodium di-isobutyl dithiophosphate
kg/tonne of ore	0.12 g PAX, 0.08g Aerophine	0.1g PAX, 0.03g Aerophine	0.042

Continued

Table 3.5b Industrial flotation data from three copper concentrators, 2010. The equivalent crushing/grinding data are given in Table 3.1.—cont'd

Concentrator	Africa open pit	Africa under ground	Mantos Blancos Chile, open pit
oily collector, kg/t of ore	no	no	no
Frother	MIBC	Interfroth 64	MB78-MIBC
kg/tonne of ore	0.06 g	0.06 g	0.02
CaO, kg/tonne of ore	0.5	2	0.808
Residence time, minutes	25	35	5
Cleaner flotation			
Feed	Various reground concentrates 80% < 30 μm	Various reground concentrates 80% < 20 μm	Rougher, scavenger and sand flotation concentrates
Number of cells	7	6	1
Volume of each cell, m^3	8	17	86
Cell type	OK8 trough cell	Mechanical	Column
Mass% solids in feed	14	20	15
pH	11.5	11.5	11
Reagents	CaO, 3418A, CMC (depressant)	CaO, 3418A, CMC (depressant)	None
Residence time, minutes	9	12	16.5
Cleaner–scavenger flotation	None		
Feed		Cleaner tails	Cleaner tails
Number of cells		4	6
Volume of each cell, m^3		17	3 × 28 + 3 × 14
Cell type		OK 16 mechanical	Wemco
Mass% solids in feed		16	18
pH		11.5	11.5
Reagents		Aerophine 3418A	0.005 kg/tonne
Residence time, minutes		10	16.5
	Recleaner float	**Releaner Float**	**Sand flotation**
Feed	Cleaner concentrate	Cleaner concentrate	Various tailings
Number of cells	5	Various	20
Cell volume, m^3	4.25	Various	14
Cell type	OK3HG trough cell	Various	Wemco
Mass% solids in feed	10	24	42
pH	11	11	11.5
Reagents	CaO, Aerophine 3418A, MIBC	CaO, 3418A, Inter-froth, CMC	NaHS (4%) + CytecAero 3753
Residence time, minutes	14	16	19

Table 3.5c Industrial flotation data from three copper concentrators, 2010. All three treat ore from large open-pit mines. The equivalent crushing/grinding data are given in Table 3.1.

Concentrator	Cerro Verde, Peru	Sierrita, U.S.A.	Ray, U.S.A.
Ore treated per year, tonnes	39,000,000	37,000,000	10,000,000
Concentrate, tonnes/year	750,000	325,000	150,000
Ore grade, %Cu		0.26	0.453
Sulfide	0.6	0.23 chalcopyrite and chalcocite	chalcopyrite and chalcocite
"Oxide"	0.02	0.03	0.025
Concentrate grade, % Cu	25	26	28
Tailings grade, % Cu	0.075	0.041	0.035
Cu recovery to concentrate, %	88	84.3	90
Rougher–scavenger flotation			
Feed	Cyclone overflow	Cyclone overflow	Cyclone overflow
Number of cells	40	380	15
Volume of each cell, m^3	160	2.9; 14.2; 29.3	9.4
Cell type	Wemco	Denver and Wemco	Wemco
Mass% solids in feed	30	35	35
pH	11.2	~11.8	10.5
Collector	Thionocarbamate Na isopropyl xanthate	Nalco 9740 + K amyl xanthate	Xanthate
kg/tonne of ore	0.01	0.01 N and 0.005 K	0.015
Oily collector, kg/t of ore	0.012	Chevron Phillips MCO, 0.05	0.012
Frother	Molycop H 75	Nalco 9873/Ore Prep X-133 80/20	Pine oil
kg/tonne of ore	0.02	0.018	0.015
CaO, kg/tonne of ore	1.26	1.02	0.92
Residence time, minutes	18	8	27
Cleaner flotation			
Feed	Reground cleaner-scavenger and rougher–scavenger froth and column	Rougher froth	Reground rougher–scavenger and cleaner–scavenger froth

Continued

Table 3.5c Industrial flotation data from three copper concentrators, 2010. All three treat ore from large open-pit mines. The equivalent crushing/grinding data are given in Table 3.1.—cont'd

Concentrator	Cerro Verde, Peru	Sierrita, U.S.A.	Ray, U.S.A.
	tail, 80% < 101 μm		
Number of cells	6	12	6
Volume of each cell, m^3	160	8.5	42.5
Cell type	Wemco	Wemco	Denver Agitair
Mass% solids in feed	19	20	14
pH	12.2		11
Reagents	CaO, Molyflow	none	CaO, collector, frother
Residence time, minutes	20	6.2	12
Cleaner–scavenger flotation			
Feed	Cleaner tails	Cleaner + recleaner tails	Cleaner tails
Number of cells	4	6	4
Volume of each cell, m^3	160	14.2	42.5
Cell type	Wemco	Wemco	Denver Agitair
Mass% solids in feed	19	16	11.3
pH	12.2		11.5
Reagents	CaO, Molyflow	none	CaO, xanthate
Residence time, minutes	23	4.2	8
Recleaner flotation			No
Feed	Rougher and cleaner froths	Ground cleaner froth 80% < 44 μm	
Number of cells	4	12	
Cell volume, m^3	236	54.5	
Cell type	Minovex column cell	Column	
Mass% solids in feed	27	20	
pH	12.3		
Reagents	CaO, Molyflow	None	
Residence time, minutes	36		

Column cells provide a long vertical particle/bubble contact zone and a well-controlled froth-draining zone. They are an excellent tool for maximizing %Cu in the final concentrate.

Table 3.6 Column flotation cells around the world. Virtually all are used for cleaning gangue from the final concentrate (thereby increasing Cu grade).

Country	Mine	Column cell use	Column diameter × height, m	Number
Australia	Cobar	Cu cleaner	2.3 × ?	1
Brazil	Sossego	Cu cleaners	4.3 × 14	6
Canada	Duck Pond	Cu cleaners	1.2 × 10	2
Chile	Escondida	Cu cleaners	4 × 4 × 13.9	8
China	Dashan copper	Cu cleaners	4.3 × 12	7
Indonesia	Batu Hijau	Cu cleaners	4.3 × 14	2
Mexico	Capstone mining	Cu cleaner	3.1 × 7.6	1
Mongolia	Oyu Tolgoi	Cu cleaners	5.5 × 16	4
Peru	Toquepala	Cu	2.4 × 13.3	3
Philippines	Philex Mining	Cu cleaner	3.1 × 12	1
USA	Bingham Canyon	Cu cleaners,	4.3 × 9.6	4
		scavengers	4.3 × 9.6	4
Venezuela	Auriferas Brisas	Cu cleaners	3.1 × 9	2
Zambia	Nchanga	Cu cleaners	4 × ?	6
		Cu cleaners	2.5 × ?	2

3.8 Flotation process control

Modern flotation plants are equipped with sensors and automatic control systems (Shang et al., 2009). The principal objectives of the control are to maximize Cu recovery, concentrate grade (% Cu), and ore throughput rate. The principal variables sensed are as follows (Table 3.7):

(a) ore particle size after grinding and regrinding (Cirulis & Poplaswski, 2019);
(b) % Cu, % solids, pH, and mass flowrate of the process streams (especially the input and output streams);
(c) froth height in the flotation cells.

Impeller speeds and air input rates in the flotation cells are also often sensed. The adjustments made on the basis of the sensor readings are as follows:

(a) water flowrates into hydrocyclone feed sumps to control grinding recycle, and thus flotation feed (ore) particle size;
(b) flotation reagent (collector, frother, and depressant) and water addition rates throughout the flotation plant;
(c) slurry level in the flotation cells, by adjusting the underflow valves in each cell.

3.8.1 Continuous chemical analysis of process streams

Of particular importance in flotation control is continuous measurement of Cu concentration in the solids of the process streams. This is typically done by X-ray fluorescence analysis, which may be accomplished by either sending samples to a central X-ray analysis

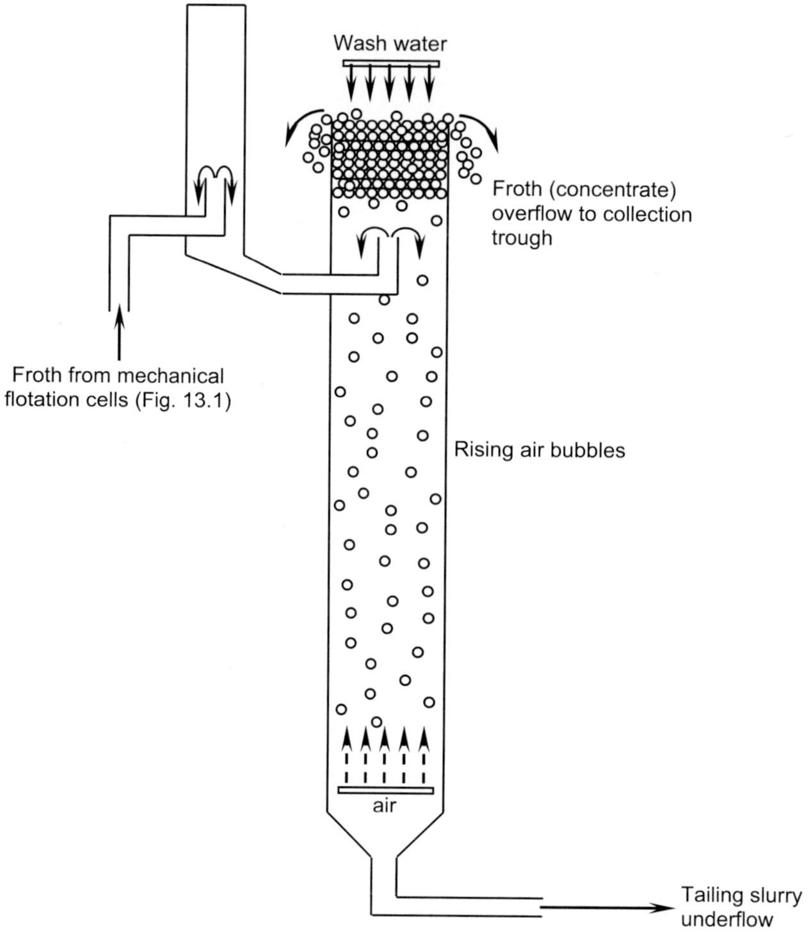

Figure 3.14 Sketch of column flotation cell. Air bubbles carry Cu sulfide particles up to the top of the column where they overflow in concentrate froth. Water spraying on top of the froth layer washes weakly held pyrite and gangue from the froth back into the slurry and eventually out in the underflow tailings slurry. Industrial column cell dimensions are given in Table 3.6.

unit or by using small X-ray units in the process streams themselves. The analyses are often done by fixed-crystal wavelength—dispersive spectrometry (Aylmore, 2019; Xore, 2014).

These analyses are used to monitor and optimize plant performance by automatically controlling (for example) reagent addition rates, grind size, and flotation cell operation. In modern plants, the control is done by a supervisory computer (Barrette et al., 2009). Advanced mineralogical techniques are also being more commonly employed to improve the performance and control of flotation circuits (Celik et al., 2011).

Figure 3.15 Industrial column flotation cells. Their sizes around the world are given in Table 3.6.

Table 3.7 Sensors and their use in automatic flotation control and optimization.

Sensing instrument	Purpose	Type of device	Use in automatic control
Process stream particle size analyzer	senses particle sizes after grind and regrind mills	laser beam diffraction	controls water addition rates to hydrocyclone feed (which controls particle recycle to ball mills and final grind size)
In-stream X-ray chemical analyzer	senses Cu content of solids in process streams	X-ray fluorescence analysis of automatically taken process stream samples	controls collector, frother, depressant, and water addition rates throughout the plant. Adjusts valves in flotation cells to vary froth thickness.
Slurry level sensor	indicates slurry surface location	float level, hydrostatic pressure, conductivity	adjusts underflow valves to maintain prescribed froth layer thicknesses
Slurry mass flow gage and % solids in slurry gage	determines mass and volumetric flow rates of process streams	magnetic induction, Doppler effects, ultrasonic energy loss	measures flows in flotation circuit, permits optimization of recycle streams

3.8.2 Machine vision systems

Several large concentrators have installed machine vision systems on their flotation cells (Outotec, 2018). These systems determine froth quality and froth overflow rate. They are used to automatically adjust grind size, reagent addition rates, and froth depth, to maximize Cu recovery in their rougher and scavenger cells and Cu grade in their cleaner and recleaner cells (Massinaei, 2015; Supomo et al., 2008).

Freeport's Grasberg concentrator installed 172 machine vision systems. These systems and new process strategies increased Cu-to-concentrate recovery by several percent (Supomo et al., 2008).

3.9 Flotation product processing

3.9.1 Thickening and dewatering

The product from flotation contains 75—80 wt-% water, most of which must be removed before the concentrate can be transported and smelted. Most of this dewatering is performed by settling in large quiescent *thickeners* (Schoenbrunn et al., 2019). The solids settle under the influence of gravity to the bottom of the thickener from where they are scraped to a central discharge by a slowly rotating rake (Parsapour et al., 2019). Faster settling is encouraged by adding small quantities of organic flocculants, such as polyacrylamides (Ferrera et al., 2009), to the input slurry. These cause the fine particles to flocculate, which results in faster settling velocities.

The underflows from the thickeners still contain 30%—40% water. This is reduced to 10%—15% in rotary vacuum filters and then further reduced to 8%—12% in pressure filters or ceramic disc vacuum filters (Meadows, 2019; Outotec, 2020a,b). The water content of the concentrate when it is shipped (approximately 8%) is a compromise between the cost of shipping water and the need to minimize concentrate losses as dust during transport (Garshasb & Saulters, 2019).

3.9.2 Tailings

Flotation tailings account for ~98% of the ore fed to the concentrator. They are stored in large tailings dams near the mine property. Water is reclaimed from the dams and recycled to the concentrator (ICMM, 2012).

Most concentrators are zero water discharge plants, which means that no water can be discharged from the plant but must be internally recycled. This minimizes water consumption and avoids mixing concentrator effluents with the surrounding water table. The pH of the tailings water is close to that required for rougher—scavenger flotation, so this recycle minimizes CaO consumption. Cu flotation tailings typically contain 0.02%—0.15% Cu on a dry basis (Table 4.3).

3.10 Other flotation separations

Copper flotation consists mainly of separating Cu sulfide minerals from nonsulfide rock and Fe-sulfide minerals. Many Cu deposits also contain molybdenite. Others contain pentlandite ($(Ni,Fe)_9S_8$), sphalerite (ZnS), or galena (PbS). These can all be separated from Cu minerals by selective flotation. Molybdenite flotation is discussed in Chapter 20. The flotation of sphalerite, galena, and Ni and Cu oxides is discussed by Biswas and Davenport (1994). Pentlandite flotation is described by Crundwell et al. (2011).

3.10.1 Gold flotation

Au is present in many Cu sulfide ores, from just traces to ~1 g/t Au (10^{-4} wt.-% Au). Au is mostly present in Cu sulfide ores as metal and Ag—Au alloy (electrum) grains. Some may also be present as tellurides. These grains of Au and Au alloy may be liberated particles, attached to other minerals, encapsulated in sulfide mineral particles, or encapsulated in oxide particles. They seem to be preferentially associated with bornite and pyrite (Agorhom et al., 2015).

The Au-containing mineral grains in Cu—Au ores may be smaller than their companion copper sulfide mineral grains, so higher Au recovery requires finer grinding. Optimum grind size is best determined by controlled in-plant testing.

Fortunately, all of these gold mineral types float efficiently under the same conditions as copper sulfide minerals, using xanthate, dithiophosphate, and thionocarbamate collectors and conventional frothers such as methyl isobutyl carbinol (Agorhom et al., 2015; Woodcock et al., 2007). Generally speaking, liberated Au metal, Au alloy, and Au mineral particles float with chalcopyrite and other sulfide minerals and report to the bulk concentrate (Zahn et al., 2007). Au recoveries to the concentrate of 80%—85% are typical. However, efforts to depress pyrite flotation to reduce sulfur levels in the concentrate can also reduce gold recovery.

Kendrick et al. (2003) suggest that Au recovery to concentrate can be increased ~10% by decreasing the pH of the ore slurry from 10.7 to 9.5. This also appreciably lowers CaO consumption. Special collectors have also been developed for copper ores with significant gold content. These collectors are a blend of monothiophosphates and dithiophosphates, and may be worth considering for Cu—Au concentrators.

3.11 Summary

Copper ores typically contain 0.5%—2% Cu. They must be concentrated to ~30% Cu before smelting. The universal concentration technique is froth flotation of finely ground ore particles.

The feed to froth flotation is produced by comminution, which includes the following:

(a) blasting ore fragments from the mine walls (~0.1 m diameter);
(b) crushing these fragments in eccentric and roll crushers (~0.01 m diameter);
(c) grinding the crushed ore in rotating tumbling mills (<100 μm diameter).

The resulting finely ground particles are then sent continuously to froth flotation.

Froth flotation entails attaching fine Cu sulfide mineral particles to bubbles and "floating" them out of a water–ore mixture. The flotation is made selective by using reagents which make the Cu sulfide minerals water repellant (hydrophobic), while leaving the other minerals wetted (hydrophilic).

Typical Cu sulfide recoveries to concentrate are 85%–90%. Typical concentrate grades are 30% Cu (higher with chalcocite, bornite, and native copper mineralization). Column flotation cleaner cells are particularly effective at giving a high Cu grade in the concentrate.

Modern concentrators are automatically controlled to give maximum Cu recovery, maximum %Cu in the concentrate, and maximum ore throughput rate at minimum cost. Expert control systems help optimize the performance of flotation plants. On-stream particle size and on-stream X-ray fluorescence analyses are key components of this automatic control.

References

Agorhom, E. A., Lem, J. P., Skinner, W., & Zanin, M. (2015). Challenges and opportunity in the recovery/rejection of trace elements in copper flotation – a review. *Minerals Engineering, 78*, 45–57.

Aylmore, M. G. (2019). Automated mineralogy. In R. C. Dunne, S. K. Kawatra, & C. Y. Young (Eds.), *SME minerals processing and extractive metallurgy handbook* (pp. 43–68). Littleton, CO: SME.

Barette, R., Bom, A., Taylor, A., & Lawson, V. (2009). Implementation of expert control systems at Vale Inco Limited's Clarabelle Mill. In H. Shang, L. Ryan, & S. Kennedy (Eds.), *Process control: Applications in mining and metallurgical plants* (pp. 25–37). Montreal, Canada: CIM.

Barrios, G.K.P., & Tavares, L.M. (2016). A preliminary model of high pressure roll grinding using the discrete element method and multi-body dynamics coupling. International Journal of Mineral Processing, *156*, 32–42. https://www.sciencedirect.com/science/article/pii/S0301751616301442.

Bearman, R. (2019). Gyratory and roll crushers. In R. C. Dunne, S. K. Kawatra, & C. Y. Young (Eds.), *SME minerals processing and extractive metallurgy handbook* (pp. 391–416). Littleton, CO: SME.

Biswas, A. K., & Davenport, W. G. (1994). *Extractive metallurgy of copper* (3rd ed.). New York: Elsevier Science Press.

Boldt, J. R., & Queneau, P. (1967). *The winning of nickel.* Toronto: Longmans Canada Ltd.

Burchardt, E., Patzelt, N., Knecht, J., & Klymowsky, R. (2010). HPGRs in copper ore comminution – a technology broke barriers. In *Copper 2010, Volume 7: Plenary, mineral processing, recycling, posters, GDMB* (pp. 2621–2635). Germany: Clausthal–Zellerfeld.

Celik, I. B., Can, N. M., & Sherazadishvili, J. (2011). Influence of process mineralogy on improving metallurgical performance of a flotation plant. *Mineral Processing and Extractive Metallurgy Review, 32*(1), 30–46.

Cirulis, D., & Poplaswski, J. (2019). Novel technology provides for on-line measurement of particle size in individual cyclones. In *Copper 2019.* Montreal: CIM. Paper 594127.

Crundwell, F., Moats, M., Ramachandran, R., Robinson, T., & Davenport, W. G. (2011). *Extractive metallurgy of nickel, cobalt and Pt group metals.* Oxford: Elsevier.

Daniel, M., Klein, B., & Wang, C. (2019). High—pressure grinding roll technology. In R. C. Dunne, S. K. Kawatra, & C. Y. Young (Eds.), *SME minerals processing and extractive metallurgy handbook* (pp. 444—464). Littleton, CO: SME.

Ferrera, V., Arinaitwe, E., & Pawlik, M. (2009). A role of flocculant conformation in the flocculation process. In C. O. Gomez, J. E. Nesset, S. R. Rao, & Montreal CIM (Eds.), *Advances in mineral processing science and technology: Proceedings of the 7th UBC—MCGill—UA international symposium on fundamentals of mineral processing* (pp. 397—408). Canada.

Flintoff, B., & Knorr, B. (2019). Cyclones. In R. C. Dunne, S. K. Kawatra, & C. Y. Young (Eds.), *SME minerals processing and extractive metallurgy handbook* (pp. 595—610). Littleton, CO: SME.

Fuerstenau, M. C., Jameson, G., & Yoon, R.-H. (Eds.). (2007). *Froth flotation, A century of innovation*. Littleton, CO, USA: SME.

Garshasb, M., & Saulters, R. (2019). Physical and mineralogical impact of copper concentrate on Transportable Moisture Limit (TML). In *Copper 2019*. Montreal: CIM. Paper 599011.

Giblett, A. (2019). Grinding technologies. In R. C. Dunne, S. K. Kawatra, & C. Y. Young (Eds.), *SME minerals processing and extractive metallurgy handbook* (pp. 517—528). Littleton, CO: SME.

Hagihara, H. (1952). Mono- and multiplayer adsorption of aqueous xanthate on galena surfaces. *Journal of Physical Chemistry, 56*, 616—621.

ICMM. (2012). *Water management in mining: A selection of case studies*. www.icmm.com/website/publications/pdfs/water/Water-management-in-mining_case-studies-1.pdf. (Accessed 16 July 2020).

ICSG. (2010). *Directory of copper mines and plants 2008 to 2013*. Lisbon: International Copper Study Group.

Johnson, A., Meadows, D. G., & Capuccitti, F. (2019). Copper mineral processing. In R. C. Dunne, S. K. Kawatra, & C. Y. Young (Eds.), *SME minerals processing and extractive metallurgy handbook* (pp. 1615—1641). Littleton, CO: SME.

Jones, S. M., & Pena, R. F. (1999). Milling for the millennium. In B. A. Hancock, & M. R. L. Pon (Eds.), *Copper 99—cobre 99, Vol. II: Mineral processing/environment, health and safety* (pp. 191—204). Warrendale, PA: TMS.

Kapusta, J. P. T. (2004). JOM world nonferrous smelters survey, part I: Copper. *JOM, 56*(7), 21—27.

Kendrick, M., Baum, W., Thompson, P., Wilkie, G., & Gottlieb, P. (2003). The use of the QemSCAN automated mineral analyzer at the Candelaria concentrator. In C. O. Gomez, & C. A. Barahona (Eds.), *Copper—cobre 2003, Vol. III: Mineral processing* (pp. 415—430). Montreal: CIM.

Li, Z., Huang, M., Gui, W., & Jiang, Z. (2019). Data-driven adaptive optimal control for flotation processes with delayed feedback and disturbance. *IEEE Access, 7*, 163138—163149. https://doi.org/10.1109/ACCESS.2019.2952396

Mankosa, M. J., Kohmuench, J. N., & Luttrell, G. H. (2019). Column flotation. In R. C. Dunne, S. K. Kawatra, & C. Y. Young (Eds.), *SME minerals processing and extractive metallurgy handbook* (pp. 913—930). Littleton, CO: SME.

Massinaei, M. (2015). Estimation of metallurgical parameters of flotation process from froth visual features. *International Journal of Mining and Geological Engineering, 49*, 75—81.

Meadows, D. G. (2019). Filtration. In R. C. Dunne, S. K. Kawatra, & C. Y. Young (Eds.), *SME minerals processing and extractive metallurgy handbook* (pp. 1081—1106). Littleton, CO: SME.

Nagaraj, D. R., Farinato, R. S., & Arinaitwe, E.,I. (2019). Flotation chemicals and chemistry. In R. C. Dunne, S. K. Kawatra, & C. Y. Young (Eds.), *SME minerals processing and extractive metallurgy handbook* (pp. 965—1010). Littleton, CO: SME.

Nelson, M. G., & Lelinski, D. (2019). Mechanical flotation. In R. C. Dunne, S. K. Kawatra, & C. Y. Young (Eds.), *SME minerals processing and extractive metallurgy handbook* (pp. 891—912). Littleton, CO: SME.

O'Keefe, C., Rothman, P., Maron, R., Newton, D., Mercuri, J., Cirulis, D., & Holdsworth, M. (2014). Grind circuit optimization at Rio Tinto Kennecott using real-time measurement of individual hydrocyclone overflow stream particle size enabled by novel CYCLONEtrac^{SM} technology. In *Comminution 2014*. London: MEI. www.cidra.com/default/files/document_library/BI0614-CiDRA-Comminution-2014-Ver201440321-v3-Color-1.pdf. (Accessed 18 July 2020).

Outotec. (2016). *Advanced grinding circuit control using online analyzer systems*. https://www.outotec.com/products-and-services/newsletters/minerva/minerva-issue-3–2016/advanced-grinding-circuit-control-using-online-analyzer-systems/. (Accessed 18 July 2020).

Outotec. (2018). *Outotec Frothsense: Real—time froth analysis.* https://www.outotec.com/products-and-services/technologies/analyzers-and-automation/frothsense-sensor-system/. (Accessed 20 July 2020).

Outotec. (2020a). *Outotec Larox PF® pressure filter.* https://www.outotec.com/products-and-services/technologies/filtration/larox-pf-pressure-filter/. (Accessed 14 July 2020).

Outotec. (2020b). *Outotec Larox PF® pressure filter. Outotec (2020) Outotec Larox CC® ceramic filter.* https://www.outotec.com/products-and-services/technologies/filtration/larox-cc-ceramic-filter/. (Accessed 14 July 2020).

Parsapour, G., Molhem, F., Montazari, H., Arghavani, E., Binaz, G., & Banisi, S. (2019). Performance improvement of the Sarcheshmeh Copper Complex Cu—Mo thickener using a vane type feedwell. *International Journal of Mining and Geological Engineering, 53*, 79—82.

Schoenbrunn, F., Laros, T., Henriksson, B., & Arbuthnot, I. (2019). Sedimentation equipment. In R. C. Dunne, S. K. Kawatra, & C. Y. Young (Eds.), *SME minerals processing and extractive metallurgy handbook* (pp. 1029—1040). Littleton, CO: SME.

Shang, H., Ryan, L., & Kennedy, S. (Eds.). (2009). *Process control applications in mining and metallurgical plants.* Montreal, Canada: CIM.

Supomo, A., Yap, E., Zheng, X., Banini, G., Mosher, J., & Partanen, A. (2008). PTFreeport Indonesia's mass—pull control strategy for rougher flotation. *Minerals Engineering, 21*, 808—816.

Taggart, A. F. (1954). *Handbook of mineral dressing.* New York: John Wiley and Sons.

Wark, I. W., & Cox, A. G. (1934). Principles of flotation, III, an experimental study of influence of cyanide, alkalis and copper sulfate on effect of sulfur—bearing collectors and mineral surfaces. *AIME Transactions, 112*, 288.

Woodcock, J. T., Sparrow, G. J., Bruckard, W. J., Johnson, N. W., & Dunne, R. (2007). Plant practice: Sulfide minerals and precious metals. In M. C. Fuerstenau, G. Jameson, & R.-H. Yoon (Eds.), *Froth flotation, A century of innovation* (pp. 781—817). Littleton, CO, USA: SME.

Xore. (2014). *Xore customer case Boliden, Aitik mine.* https://www.xore.se/boliden-aitik-mine/. (Accessed 20 July 2020).

Zahn, R. P., Baum, W., Magnuson, R., Jensen, D., & Thompson, P. (2007). QEMSCAN characterization of selected ore types for plant optimization. In R. del Villar, J. E. Nesset, C. O. Gomez, & A. W. Stradling (Eds.), *Copper 2007, volume II: Mineral processing* (pp. 147—156). Montreal: CIM.

Outotec. (2010). *Grinding control solutions* Accessed 01 February 2011.

Suggested reading

Fiscor, S. (2010). Design considerations for a tailings disposal system. *Engineering and Mining Journal.* https://www.e-mj.com/features/design-considerations-for-a-tailings-disposal-system/. (Accessed 16 July 2020).

Williams, R. (2019). *When to use a high—density thickener.* https://www.mclanahan.com/blog/when-to-use-a-high-density-thickener. (Accessed 16 July 2020).

Wills, B. A., & Napier—Munn, T. J. (2006). *Mineral processing technology* (7th ed.). Oxford: Elsevier.

CHAPTER 4

Pyrometallurgical processing of copper concentrates

4.1 Fundamental thermodynamic aspects associated with pyrometallurgical copper processing

Before introducing basic chemical reactions involved in the pyrometallurgical processing of copper concentrates, it is necessary to understand some basic thermochemical concepts.

Pyrometallurgical processing of metals in the presence of a combined oxygen and sulfur atmosphere involves the following two reactions:

$$2M + O_2 = 2\,MO \tag{4.1}$$

$$2M + S_2 = 2\,MS \tag{4.2}$$

For reactions (4.1) and (4.2), their respective equilibrium constants can be represented by Eqs. (4.3) and (4.4), respectively.

$$\text{Log } K_{(4.1)} = 2\log(a_{MO}/a_M) - \log pO_2 \tag{4.3}$$

$$\text{Log } K_{(4.2)} = 2\log(a_{MS}/a_M) - \log pS_2 \tag{4.4}$$

In these expressions, K_i, P, and a represent the equilibrium constant for each reaction and the partial pressures and activities of elements and compounds. Combination of reactions (4.1) and (4.2) to generate Eq. (4.5) indicates how oxides and sulfides will interact under an oxygen—sulfur environment.

$$2\,MS + O_2 = 2\,MO + S_2 \tag{4.5}$$

$$\log K_{(4.5)} = 2\log(a_{MO}/a_{MS}) + \log pS_2 - \log pO_2 \tag{4.6}$$

where Eq. (4.6) indicates the relationship between the activities of the metal oxide and sulfide for given oxygen and sulfur partial pressures at a defined temperature. In this expression, the equilibrium constant K is indirectly calculated from the relationship between the free energy at equilibrium for reaction (4.5), assuming pure compounds (unitary activity), from the following expression:

$$\Delta G^\circ = -2.303\,RT\log K_{(4.5)} \tag{4.7}$$

In expression 4.7, R is the ideal gas constant, T the absolute temperature, and K the equilibrium constant for reaction 4.5.

Extractive Metallurgy of Copper, Sixth Edition
ISBN 978-0-12-821875-4, https://doi.org/10.1016/B978-0-12-821875-4.00004-3

An additional relationship required to understand the behavior between oxides and sulfides is given by the relationship between oxygen and sulfur, represented by reaction (4.8):

$$1/2\ S_2(g) + O_2(g) = SO_2(g) \tag{4.8}$$

$$\log K_{(4.8)} = \log pSO_2 - 1/2 \log pS_2 - \log pO_2 \tag{4.9}$$

Expressions (4.6) and (4.9) can be represented in an oxygen–sulfur potential (or predominance) diagram, as shown in Fig. 4.1. The figure indicates that iron oxide is more stable than copper oxide at low oxygen potentials, with both metal sulfides stable at similar sulfur potentials. The diagram also indicates that it is thermodynamically feasible to oxidize iron sulfide and increase the copper concentration in a sulfide phase. This principle is represented by the following chemical reaction and its equilibrium constant:

$$FeS(l) + 1/2O_2(g) = FeO(l) + 1/2\ S_2(g) \tag{4.10}$$

$$\log K_{(4.10)} = \log(a_{FeO}/a_{FeS}) + 1/2(\log pS_2 - \log pO_2) \tag{4.11}$$

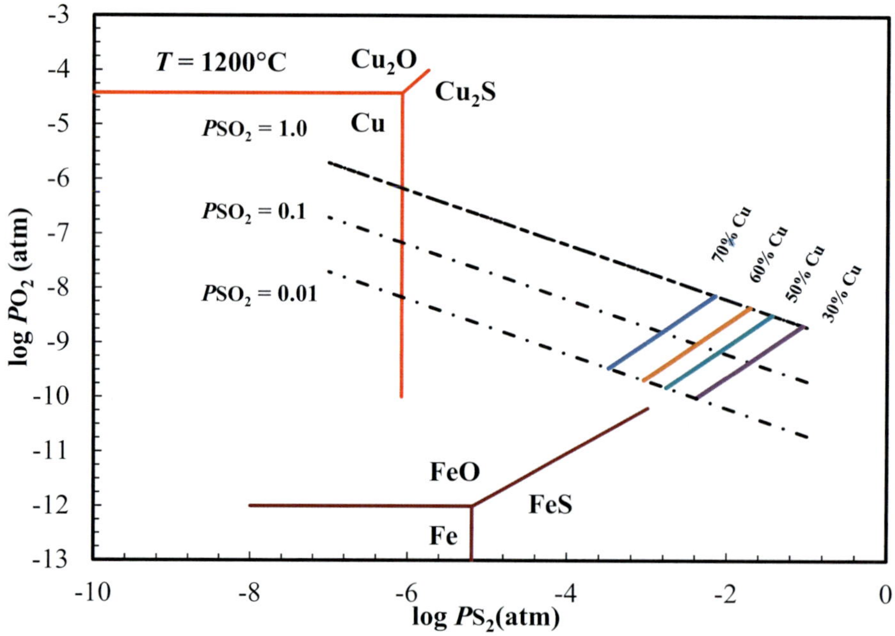

Figure 4.1 Calculated predominance diagram for Cu–S–O and Fe–S–O at 1200°C. Calculations done using public thermodynamic information.

Eq. (4.11) is represented by the iso-concentration copper lines in Fig. 4.1, indicating the equilibrium between slag and copper matte phases for a given sulfur and oxygen potential in the graph.

Fig. 4.1 also shows the sulfur dioxide isobars for 0.01, 0.1 and 1.0 atm. The area between both isobars represents the conditions in which copper smelting and converting processes are conducted, depending on the following:

- Oxygen enrichment defined for the process (from air to 100% oxygen)
- Moisture of the materials fed to the process (bath smelting up to 10%)

The operating isobar for each operation will be defined based on type of feed (dry or feed), amount of fuel required to support the reactions, availability of industrial oxygen, and type of concentrate to be treated. However, most of the technologies will take advantage of the intensive use of oxygen. This will then translate into higher sulfur dioxide partial pressure for the generated process gases.

4.2 The Yazawa diagram and pyrometallurgical copper processing

The oxygen—sulfur potential diagram concept shown in Fig. 4.1 was used by Yazawa (1979) to describe, from the metallurgical thermodynamics point of view, aspects involved in the pyrometallurgical processing of copper bearing materials, using thermodynamic and phase equilibria information available at that time. This representation can be considered a sort of navigation tool to understand the theoretical restrictions of copper pyrometallurgical processes, and behavior and potential interaction of copper with iron and minor elements. For a deeper understanding on how different phase equilibria data were combined to construct the Yazawa diagram, the reader should refer to the recommended reading section of this chapter.

Fig. 4.2 shows an updated version of the Yazawa diagram, prepared with support of Pyrosearch—University of Queensland, using the FactSage public database (Jak et al., 2016). Fig. 4.2 also shows the original boundary lines calculated by Yazawa (1979). From Fig. 4.2, we can identify the main "metallurgical road" for copper processing, A → B → C (C′) → D → E and B → B′, as follows:

- Copper smelting: Following the path A → B copper concentrates are gradually combusted (oxidized) to a given matte grade by controlled oxygen addition (specific oxygen per unit of fed material). Through this process, iron is oxidized, sulfur is combusted to SO_2, and slag is formed by adding silica as a flux.
- Slag—Matte separation: Following path B → B′, matte and slag are given proper conditions to allow settling of matte droplets and recovery of copper.
- Converting:
  - o Batch Process—Slag Blowing following path B → C: Copper matte is oxidized to finalize the controlled oxidation of iron. The formed slag has a significant amount of copper and must be treated to recover the contained copper units (flotation or recycling to smelting unit).

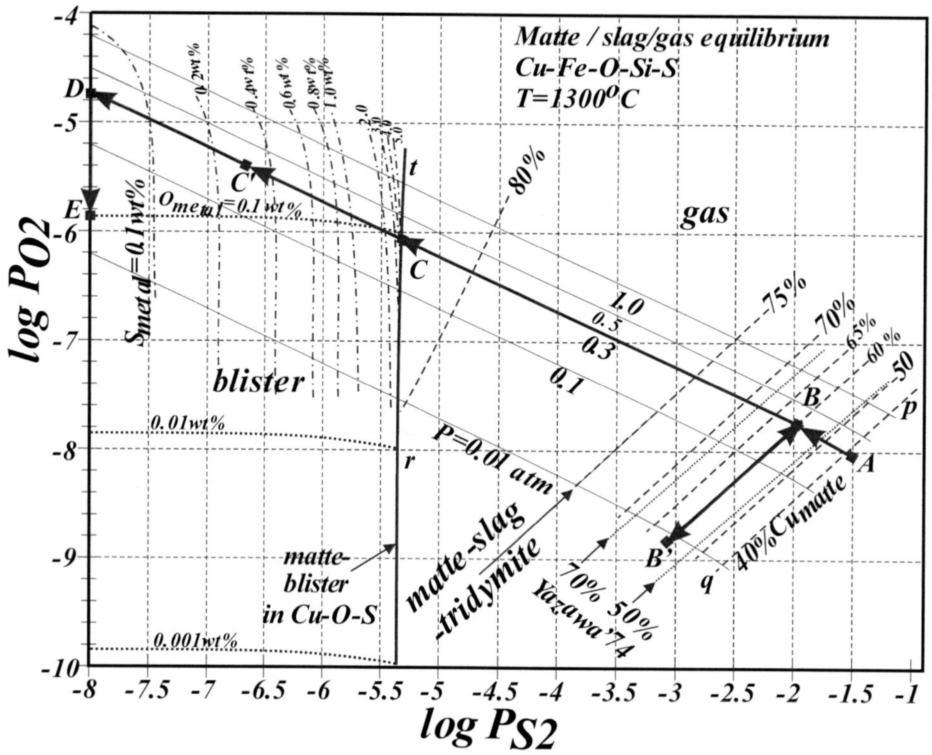

Figure 4.2 Predominance diagram describing copper making process at 1300°C based on calculations using FactSage public data base and original Yazawa diagram. *(Figure prepared with support of M. Shevchenko, D. Shishin, and E. Jak, Pyrosearch—University of Queensland, Australia.)*

- o Batch Process—Copper Blowing following path C → D: After removal of slag generated in the previous step, the high-grade matte is converted to blister copper, by removing most of the remaining sulfur (over 20% of the high-grade copper matte is sulfur).
- o Continuous Converting following the path B (close to the 70% Cu matte grade) → C′: Copper matte is directly converted to blister, avoiding an intermediate step as in the batch process. Slag is processed or recycled to the smelting step to recover the contained copper units.

- Direct to Blister process: For special concentrates, normally low in iron, it is possible to produce blister copper in a single step. This follows the path A → C′ (Location of C′ will depend on the final S content of blister of the respective process).
- Refining:
  - o Desulfurization from an intermediate location between C′ and D to D, sulfur is removed by addition of oxygen.
  - o Reduction from D to E, excess oxygen is removed to about 1000 ppm to allow casting of copper into anodes.

4.3 Smelting: the first processing step

As shown in Fig. 4.2, production of blister copper involves a sequential movement across the oxygen—sulfur potential diagram, with controlled increase of the oxygen potential which reduces the sulfur potential. This sequence must be followed to

- Achieve a controlled release of heat generated by the combustion of copper concentrates,
- Control copper losses and the amount of slag.

From a theoretical point of view, copper concentrates could be directly combusted (point A) (oxidized) to produce metallic Cu and ferrous oxide (point D), as follows:

$$CuFeS_2(s) + 5/2\ O_2(g) \rightarrow Cu^{\circ}(l) + FeO(s) + 2\ SO_2(g) \quad (4.12)$$

$$Cu_2S(s) + O_2(g) \rightarrow 2Cu^{\circ}(l) + SO_2(g) \quad (4.13)$$

$$FeS_2(s) + 5/2\ O_2(g) \rightarrow FeO(s) + 2\ SO_2(g) \quad (4.14)$$

Additional gangue material contained in the copper concentrates, such as alumina, magnesia, and calcia, will join the iron oxide and with the addition of a proper fluxing material (silica) will form slag.

Reactions 4.12—4.14 are exothermic, meaning that they generate heat that must be released from the reactor without comprising its integrity and the safety of people. In addition, the large jump in oxygen potential from point A to D also involves the oxidation of copper sulfide as follows:

$$Cu_2S(s) + 3/2O_2(g) \rightarrow Cu_2O(l) + SO_2(g) \quad (4.15)$$

Therefore, the outcome of direct processing of chalcopyrite-based copper concentrate will result in a large amount of slag, due to the oxidation of iron (normally 25%—20% of Fe in these type of concentrates), and large copper losses in the form of copper oxide. These outcomes make this process unattractive from the commercial point of view as primary copper recovery will be low and an additional process step will be required to recover the contained copper in the slag.

Therefore, the art of smelting consists of defining an optimum matte grade to control the heat release for a given reactor, the amount of generated slag, and the oxidation of copper, in which reaction (4.15) can manage the resulfidization of copper oxide, as follows:

$$FeS(l, \text{in matte}) + Cu_2O(l, \text{in slag}) \rightarrow Cu_2S(l, \text{in matte}) + FeO(l, \text{in slag}) \quad (4.16)$$

$$K_{(4.15)} = (a_{Cu_2S} \cdot a_{FeO})/(a_{FeS} \cdot a_{Cu_2O}) \quad (4.17)$$

The expression for the equilibrium constant of this reaction is shown in Eq. (4.17). This reaction has a strong tendency to shift to the right side of the equation with an equilibrium constant of $2.78 \cdot 10^4$ at 1200°C, meaning that any oxidation of copper will be buffered by iron sulfide. In other words, if enough iron sulfide is present in the molten phase, copper will be resulfidized. The general smelting principle represented by reaction (4.15) will control the overall reaction of combustion of the copper concentrate expressed by reaction (4.17).

$$\underset{\text{Feed}}{CuFeS_2(s)} + \underset{\text{Combustion Agent}}{O_2(g)} + \underset{\text{Flux}}{SiO_2(s)} \rightarrow \underset{\text{Matte}}{Cu\text{–}Fe\text{–}S(l)} + \underset{\text{Slag}}{FeO \cdot SiO_2\,(l)} + \underset{\text{Gas}}{SO_2(g)} \quad (4.18)$$

Reaction (4.18) shows that the combustion of a chalcopyrite concentrate will form three phases: matte, slag, and gas. A third element on the reactant side is the fluxing agent, required to allow proper separation of the produced molten product, an oxysulfide mixture that with the addition of a solvent (fluxing agent) in the proper proportion allows the formation of a sulfide-rich phase (copper matte) and an oxide-rich phase (liquid slag). The gas phase will contain the combusted sulfur in the form of sulfur dioxide.

The extraction process (matte—slag separation) represented by reaction 4.18 can be illustrated with the help of the SiO_2—FeO—FeS phase diagram, shown in Fig. 4.3. The left edge of the diagram represents a solution consisting only of FeS and FeO. In silica-free melts with FeS concentrations above ~31 mass%, a single oxysulfide liquid is formed. However, when silica is added, a liquid-state miscibility gap appears. This gap becomes larger as more silica is added.

Lines *a*, *b*, *c*, and *d* represent the equilibrium compositions of the two liquids. The sulfide-rich melt is known as *matte*. The oxide-rich melt is known as *slag*.

The process of separation between matte and slag, represented by reaction Eq. 4.16, is known as *matte smelting*. It accomplishes the partial removal of Fe needed to make the final copper making stage successful.

4.3.1 Slag phase: $FeO—Fe_2O_3—SiO_2$ system

From a value point, the slag phase is a residue from the smelting stage. However, from the metallurgical point of view, it has a predominant role to play to ensure proper refining of the metal containing phase. The chemical and physical properties of the slag are therefore of paramount importance for an efficient metallurgical process.

Slag is a solution of molten oxides. These oxides include FeO from Fe oxidation, SiO_2 from flux and concentrate, and other oxide impurities from concentrate. Oxides commonly found in slags include ferrous oxide (FeO), ferric oxide (Fe_2O_3), silica (SO_2), alumina (Al_2O_3), calcia (CaO), and magnesia (MgO). As Fig. 4.3 shows, small amounts of sulfides can also be dissolved in $FeO—SiO_2$ slags. Small amounts of calcia and alumina in slags decrease this sulfide solubility, Table 4.1.

The molecular structure of molten slag is described by dividing its oxides into three groups: acidic, basic, and neutral. The best-known acidic oxides are silica and alumina.

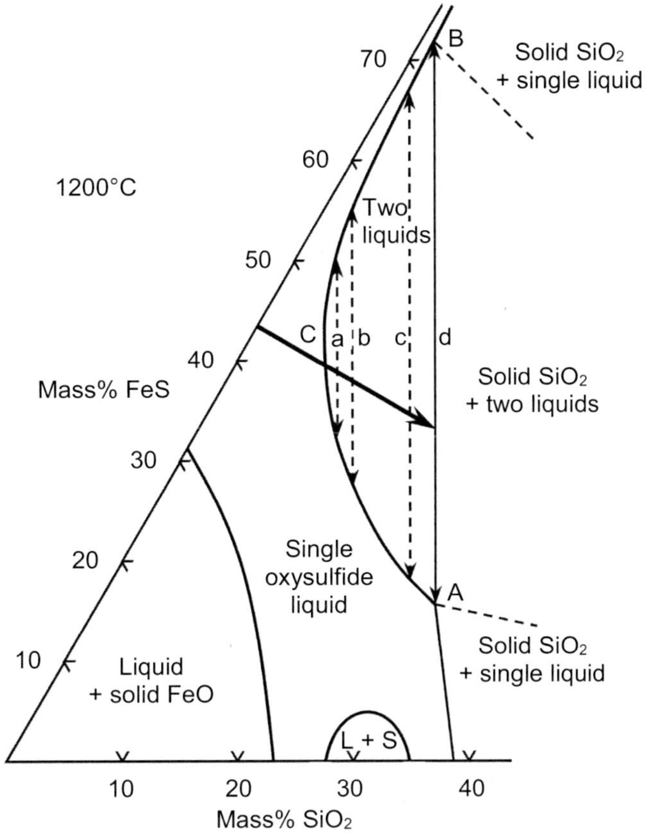

Figure 4.3 Simplified partial phase diagram for the Fe—O—S—SiO_2 system showing liquid–liquid (slag-matte) immiscibility caused by SiO_2. The *heavy arrow* shows that adding SiO_2 to an oxysulfide liquid causes it to split into FeS-rich matte and FeS-lean slag. The compositions of points A and B (SiO_2 saturation) and the behavior of Cu are detailed in Table 4.1. (Yazawa & Kameda, 1953).

When these oxides melt, they polymerize, forming long polyions such as those shown in Fig. 4.4. These polyions give acidic slags high viscosities, making them difficult to work with. Acidic slags also have low solubilities for other acidic oxides. This can cause difficulty in copper making because impurities which form acidic oxides (e.g., As_2O_3, Bi_2O_3, Sb_2O_3) will not be removed in slag, i.e., they will remain in matte or copper.

Adding basic oxides such as calcia and magnesia to acidic slags breaks the polyions into smaller structural units. As a result, basic slags have low viscosities and high solubilities for acidic oxides. Up to a certain limit, adding basic oxides also lowers the melting point of a slag. Copper making slags generally contain small amounts of basic oxides.

Neutral oxides such as FeO and Cu_2O react less strongly with polyions in a molten slag. Nevertheless, they have much the same effect. FeO and Cu_2O have low melting

Table 4.1 Compositions of immiscible liquids in the SiO_2-saturated Fe—O—S system, 1200°C (Yazawa & Kameda, 1953).

		Composition (mass%)					
System	**Phase**	**FeO**	**FeS**	**SiO_2**	**CaO**	**Al_2O_3**	**Cu_2S**
$FeS—FeO—SiO_2$	"A" slag	54.82	17.90	27.28			
	"B" matte	27.42	72.42	0.16			
$FeS—FeO—SiO_2 + CaO$	Slag	46.72	8.84	37.80	6.64		
	Matte	28.46	69.39	2.15			
$FeS—FeO—SiO_2 + Al_2O_3$	Slag	50.05	7.66	36.35		5.94	
	Matte	27.54	72.15	0.31			
$Cu_2S—FeS—FeO—SiO_2$	Slag	57.73	7.59	33.83			0.85
	Matte	14.92	54.69	0.25			30.14

Points A (slag) and B (matte) correspond to A and B in Fig. 4.3. Added Cu_2S (bottom data set) widens the miscibility gap. The Cu_2S reports almost entirely to the matte phase.

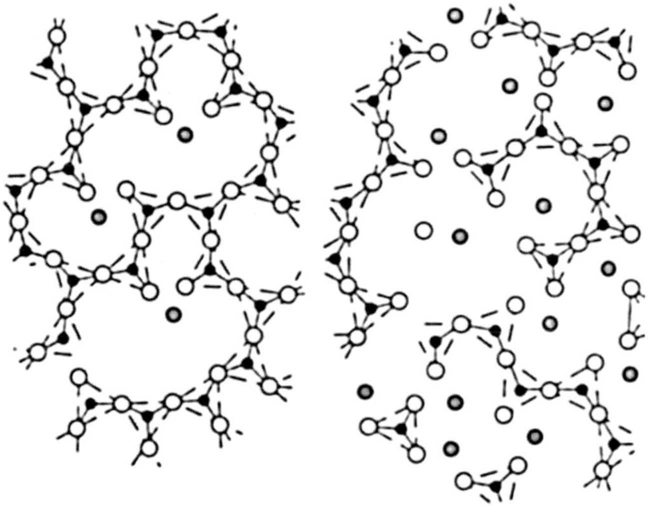

Figure 4.4 Impact of basic oxides on the structure of silica polyions in molten slags. Adding basic oxides like CaO and MgO breaks up the polyions, reducing the melting point and viscosity of the slag. • = Si; ¡ = O; £ = Ca^{++} or Mg^{++}.

points, so they tend to lower a slag's melting point and viscosity. The slags produced in industrial matte smelting consist primarily of FeO, Fe_2O_3, and SO_2, with small amounts of Al_2O_3, CaO, and MgO, Table 4.2.

Fig. 4.5 shows the composition limits for the liquid region in the $FeO—Fe_2O_3—SiO_2$ system at 1200 and 1250°C. Along the top line, the slag is saturated with solid silica.

Table 4.2 Compositions of industrial concentrates, fluxes, mattes, slags, and dusts for various matte-smelting processes.

Smelter/Process	Concentrate					Flux			Matte			Slag			
	Cu	Fe	S	SiO_2	H_2O	SiO_2	CaO	Fe	Cu	Fe	S	Cu	SiO_2	Total Fe	CaO
La Caridad Metso Outotec flash	26	26	35	6	8	90	0.1	0.4	65	10	21	2–3	31	43	
Caraiba Metso Outotec flash	32	25	33	7	8	97	2	0	62	12	22	2.3	31	42	
Huelva Metso Outotec flash	31	23	31	7	7–8	94–97	0.2	0.9	62	13	20	1.5	28	42	2
Hayden Inco flash	28	25	31	6	10	92	0.4	1.3	59	14	23	1.2	34	36	2
Potrerillos Teniente	30	25	32	4	8	86	0.2	0.5	71	5	21	8	25	38	
Horne Noranda	22	30	30	7	5–10	75	2	2	70	3	20	6.0	30	36	1
Mount Isa Isasmelt™	25–27	24–26	27–29	11–13	7	87	0.3	1	60	15	23	2.6	33	41	1.5
Zhongtiaoshan Metso Outotec Ausmelt®	23–26	26–29	29–32	5–13	7–12				60	15	23	0.7	32	36	5
Glogow 2 Metso Outotec flash direct-to-copper	28	3.3	11	17		92–95		1–2	>95		0.5	14.3	31	6.4	14
Gresik Mitsubishi	30	24	29	10		90	0	<2	68	8	21	0.7	34	36	5
Naoshima Mitsubishi	34	21	24	17	12	85			68	8	21	0.7	36	40	5
SUMZ Vanyukov	15-–20	35	33	5	8–10	90	1.5	2.0	45	30	25	0.7	32	30	3
Pasar Electric	25–32	22–30	25–33	6–10	10	>95	<1.5	<2	55	24	23	<0.7	36	38	2
Palabora Reverberatory	30–35	20–30	20–30	3–5	6–7	60–80	0–15	0–15	40–50	25–30	20–27	0.5–0.8	30–40	30–40	0–10

Information extracted from (Ramachandran et al., 2003).

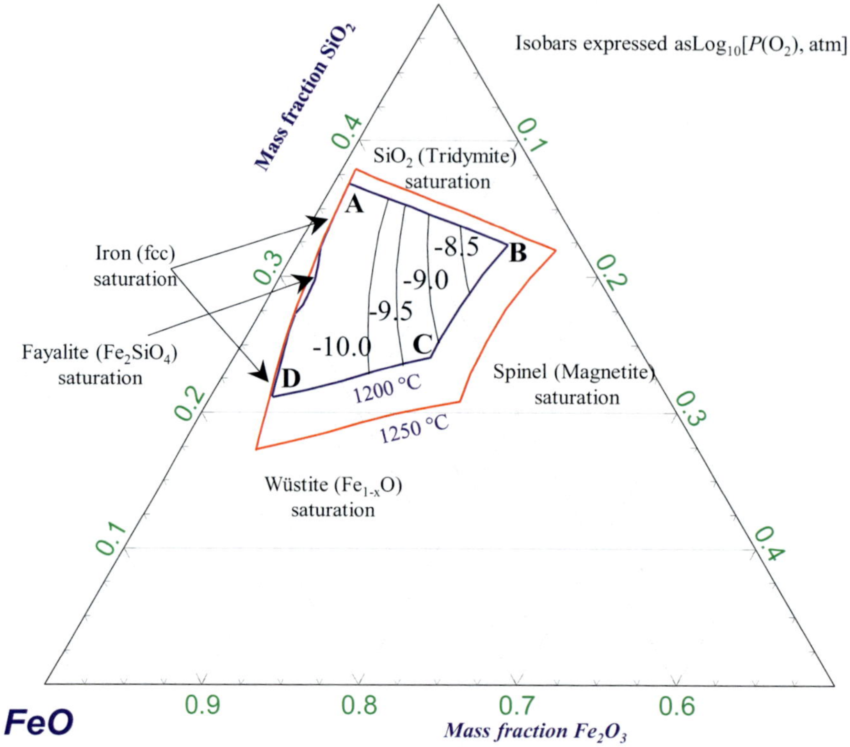

Figure 4.5 Liquidus surface in the FeO—Fe_2O_3—SiO_2 system at 1200 and 1250°C (Muan, 1955), recalculated using FactSage 7.3 using model parameters from Hidayat et al. (2017). *(Figure prepared with support of D. Shishin and E. Jak, Pyrosearch—University of Queensland, Australia.)*

Along the bottom boundary line, the slag is saturated with solid FeO. The boundary at right marks the compositions at which dissolved FeO and Fe_2O_3 react to form solid magnetite, according to reaction (4.19):

$$FeO(s) + Fe_2O_3(s) \rightarrow Fe_3O_4(s) \tag{4.19}$$

Most of the smelters will operate close to magnetite saturation conditions, represented by the line BC and control the precipitation of magnetite by the addition of silica. From the thermodynamic point of view, this will translate in lowering the activity of FeO in the slag by the formation of fayalite as follows:

$$FeO(s) + SiO_2(s) \rightarrow FeO - SiO_2(l,\ slag) \tag{4.20}$$

However, again there is a trade-off. Flux costs money and the energy required to heat and melt; it also costs more as more silica flux is used and the operational window to have a liquid slag is also narrowed, as shown in Fig. 4.6. To have a suitable slag with the right physical and chemical properties, the metallurgist will control the Fe/SiO_2 ratio to ensure that the slag is within the operational window suitable for the process. Fig. 4.6 shows the

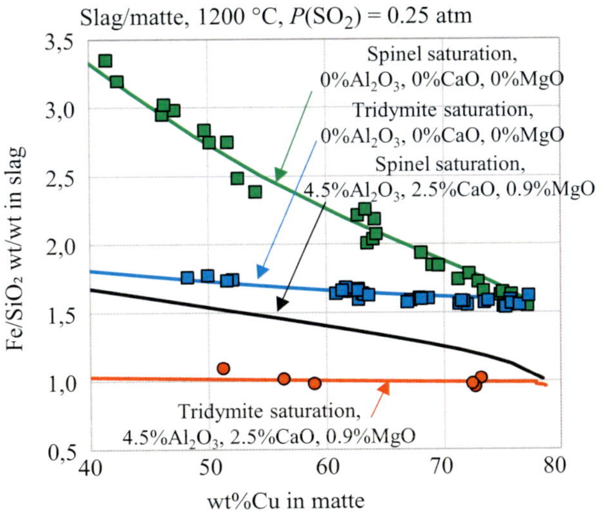

Figure 4.6 Impact of minor oxides in the Fe/SiO_2 ratio as function of copper matte grade for the Iron silicate slag system at 1200°C and $p_{SO2} = 0.25$ atm. Diagram calculated using FactSage 7.3 and model parameters from (Shishin et al., 2019; Shishin, Decterov, & Jak, 2018). *(Figure prepared with support of D. Shishin and E. Jak, Pyrosearch—University of Queensland, Australia.)*

relationship between matte grade and Fe/SiO_2 ratio at 1200°C and a sulfur dioxide partial pressure of 0.25 atm. The liquid operational window for the Fe/SiO_2 ratio is strongly influenced by the presence of minor oxides for a given matte grade.

In addition, viscosities of smelting slags increase as the silica concentration increases. This makes slag handling more difficult and reduces the rate at which matte particles settle through the slag layer. If the matte particles settle too slowly, they will remain entrained in the slag when it is tapped. This increases Cu losses. As a result, the correct levels of FeO and SiO_2 in the slag require another balancing act.

Extensive oxidation and lower smelting temperatures encourage the formation of Fe_2O_3 in the slag and further formation of magnetite. Avoiding and/or proactively controlling these conditions minimizes magnetite precipitation and potential foaming of the molten material from the reactor.

Along the left-hand boundary, the slag is saturated either with metallic iron or solid fayalite (Fe_2SiO_4). Under the oxidizing conditions of industrial copper smelting, this never occurs. Table 4.2 lists the compositions of some smelter slags, including their Cu content. Controlling the amount of Cu dissolved in smelting slag is an important part of smelter strategy, as demonstrated by Chapter 10.

In addition to the chemical and compositional properties of the slag, physical properties are also relevant to ensure a proper metal extraction process. Among these properties, the slag viscosity is one of the most relevant in the copper smelting process. A slag with adequate viscosity will help to control copper losses due to lower entrainment and in bath smelting applications (Chapter 10) to help protect refractory materials in the reactor.

Many measurements have been made of the viscosities of molten slags. These have been used to develop a model which calculates viscosities as a function of temperature and composition (Utigard & Warczok, 1995). The model relies on calculation of a viscosity ratio (*VR*). *VR* is the ratio of *A* (an equivalent mass% in the slag of acidic oxides) to *B* (an equivalent mass% of basic oxides):

$$VR = A/B \tag{4.21}$$

$$A = (\%SiO_2) + 1.5(\%Cr_2O_3) + 1.2(\%ZrO_2) + 1.8(\%Al_2O_3) \tag{4.22}$$

$$\begin{aligned} B = {} & 1.2(\%FeO) + 0.5(\%Fe_2O_3 + \%PbO) + 0.8(\%MgO) + 0.7(\%CaO) \\ & + 2.3(\%Na_2O + \%K_2O) + 0.7(\%Cu_2O) + 1.6(\%CaF_2) \end{aligned} \tag{4.23}$$

Utigard and Warczok related *VR* to viscosity by regression analysis against their existing database, obtaining the following:

$$\log \mu(\mathrm{kg/m-s}) = -0.49 - 5.1\sqrt{VR} + \frac{-3660 + 12\,080\sqrt{VR}}{\mathrm{T(K)}} \tag{4.24}$$

Since the work of Utigard and Warczok, several efforts have been conducted to measure viscosities and develop models to calculate them. Among these efforts, CSIRO (Commonwealth Scientific Industrial Research Organization, Australia) developed a thermodynamic computational package, MPE (Multiphase Equlibrium), for ferrous and nonferrous smelting applications (Zhang, 2002). The software allows prediction of equilibrium compositions and viscosities and solidus content in slags. Fig. 4.7 shows the application of the CSIRO MPE package to calculate the effect of Fe/SiO_2 ratio under copper smelting conditions at 1200 and 1250°C. The secondary vertical axis shows the % solids in the slag only at 1200°C, as the slag is fully liquid at 1250°C.

4.3.2 Calcium ferrite and olivine slags systems

The ferrous silicate melts discussed in the previous section have been the basis for the slags used in most primary copper smelting and converting technologies. However, ferrous silicate slags have limitations. The most important is the need to control magnetite formation (Eq. 4.19). In bath smelting or converting, this is a problem because the magnetite crust that forms during the initial reaction can act as a process barrier. The second is the relatively small window of compositions where the slag is molten at smelting temperatures (Fig. 4.8). Ferrous silicate slags are relatively acidic, which makes it more difficult to use them to remove impurities such as As, Bi, and Sb that form acidic oxides.

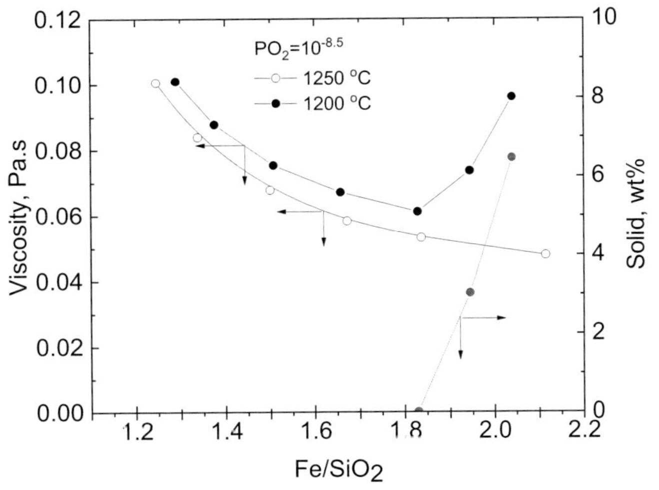

Figure 4.7 MPE predictions for the effect of Fe/SiO_2 ration on the slag viscosity at 1200 and 1250°C. The secondary vertical axis shows the percentage of solid in the slag at 1200°C. *(Figure courtesy of Active Integrated Matter Future Science Platform—Processing Mineral Resources CSIRO, Australia.)*

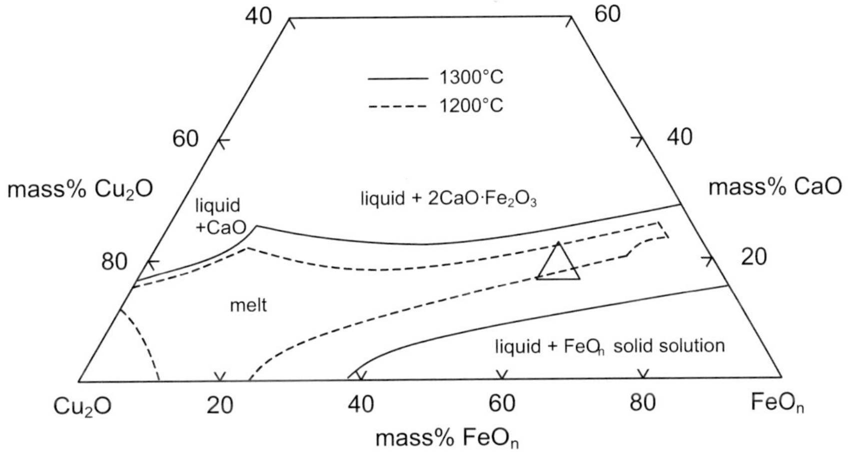

Figure 4.8 Liquidus lines at 1200 and 1300°C for calcium ferrite ($CaO—Cu_2O—FeO_n$) slags in air, 1 atm pressure (Goto & Hayashi, 1998). The triangle represents Mitsubishi converting furnace slag if all its Cu exist as Cu_2O. Slag compositions inside the solid line are fully liquid at 1300°C. Slag compositions inside the dashed lines are fully liquid at 1200°C.

The development of the Mitsubishi converter (Chapter 8) was accompanied by the adoption of CaO-based (rather than SiO_2 based) slag, better known as *calcium ferrite* (Jahanshahi & Sun, 2003). Early in the development of the process, it was found that blowing O_2-rich blast onto the surface of fayalite slag produced a crust of solid magnetite.

This made further converting impossible. CaO, on the other hand, reacts with magnetite, molten Cu, and O_2 to form a molten $Cu_2O-CaO-Fe_3O_4$ slag (Fig. 4.9). The slag typically contains 14%–16% Cu, 40%–55% Fe (mostly Fe^{3+}), and 15%–20% CaO.

Calcium ferrite slag has a low viscosity (0.025–0.04 kg/m-s), and specific gravities are 3.1–3.8, less than that of comparable fayalite slags (Vadász et al., 2001). Surface tensions are 0.58–0.60 N/m. This improves settling rates after smelting. The CaO in the slag lowers the activity of Fe_2O_3, and this prevents magnetite formation. The activity coefficient of Cu_2O is also higher in calcium ferrite slags. This means that the solubility of Cu_2O is reduced, which means lower copper losses. Calcium ferrite slags are more basic than fayalite slags, which lower the activity coefficient of impurity oxides such as As_2O_3 and Sb_2O_3. The result is enhancing removal of these impurities. Calcium ferrite slags are also less likely to engage in foaming.

The advantages of calcium ferrite slag have led to its adoption for bath converting furnaces (Chapter 9). However, there are drawbacks to the use of this slag, especially in smelting. Fig. 4.9 shows the $CaO-FeO_x-SiO_2$ phase diagram at 1300°C (Tsukihashi & Kimura, 2000). As it shows, the solubility of silica in $CaO-FeO_x$ melts is low. This makes smelting of concentrates with significant levels of silica difficult (Vartiainen et al., 2003). The other significant difficulty with the use of calcium ferrite slags is their aggressive behavior toward the mag-chrome ($MgO-Cr_2O_3$) refractories used in copper

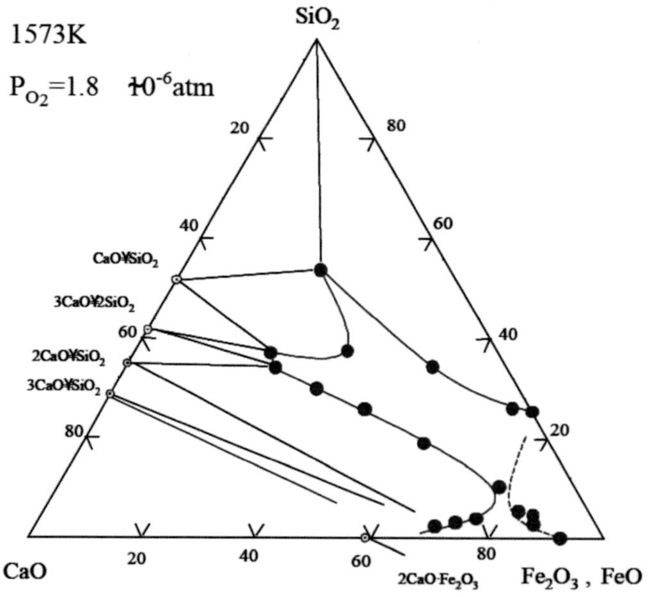

Figure 4.9 Liquidus line for the $CaO-SiO_2-(FeO_x)$ system at 1300°C (Tsukihashi & Kimura, 2000). The minimal solubility of silica in calcium ferrite slags is apparent, as is the wider solubility range of olivine (ferrous calcium silicate) slags.

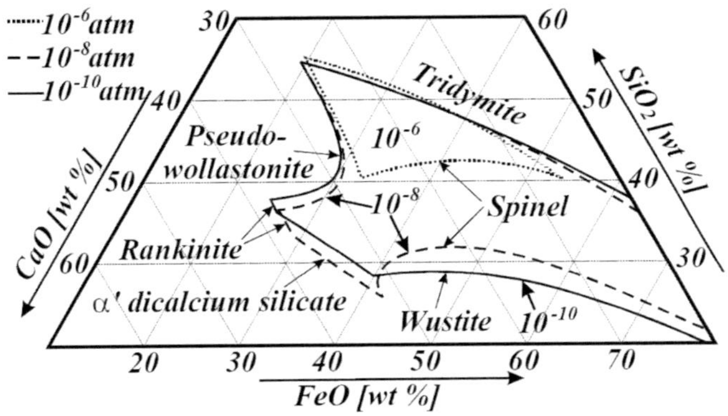

Figure 4.10 Liquidus isotherms in the $CaO-FeO_x-SiO_2$ system at 1250°C and varying oxygen potential (Jak et al., 2000).

smelting and converting furnaces (Fahey et al., 2004). The frequent lining repair and replacement that occurs as a result increases downtime and expense. As a result, extensive research has been conducted over the past decade on a more suitable type of slag. Fig. 4.10 shows the liquid region at 1250°C for slags in the $CaO-FeO_x-SiO_2$ system at different oxygen potentials (Jak et al., 2000). Slags in this region have been called *ferrous calcium silicate (FCS)* or *olivine* ($(Fe,Ca)_2SiO_4$) slags. Their properties represent a compromise between those of calcium ferrite and fayalite slags. As a result, they are a possible alternative to fayalite slags for copper smelting. They have also been chosen for the Metso Outotec Ausmelt® C3 converting process (Wood et al., 2016).

4.3.3 Matte (Sundström et al., 2008)

As Fig. 4.3 shows, immiscibility between matte and fayalite slag increases with increasing silica content (Yazawa, 1956). A high sulfur/iron ratio also increases the completeness of separation as do calcia and alumina, Table 4.1.

There is some silica and oxygen solubility in matte, but Li and Rankin (1994) demonstrated that increasing Cu_2S in matte decreases these solubilities "dramatically." As a result, the typical industrial matte contains only about 1% oxygen.

Mattes do not consist of polyions like those in slags. They appear instead to be best represented as molten salts (Shimpo et al., 1986). Their specific gravity is higher than that of slags and so they form the bottom layer in smelting furnaces. As Fig. 4.11 shows, their melting points are lower than the 1200°C of most slags.

Their viscosities are low as well: ~0.003 kg/m·s versus 0.2–1 kg/m-s for typical slags. Nevertheless, smelting furnaces are operated at about 1250°C, to ensure molten slag and superheated matte. This ensures that the matte and slag stay molten during tapping and transfer.

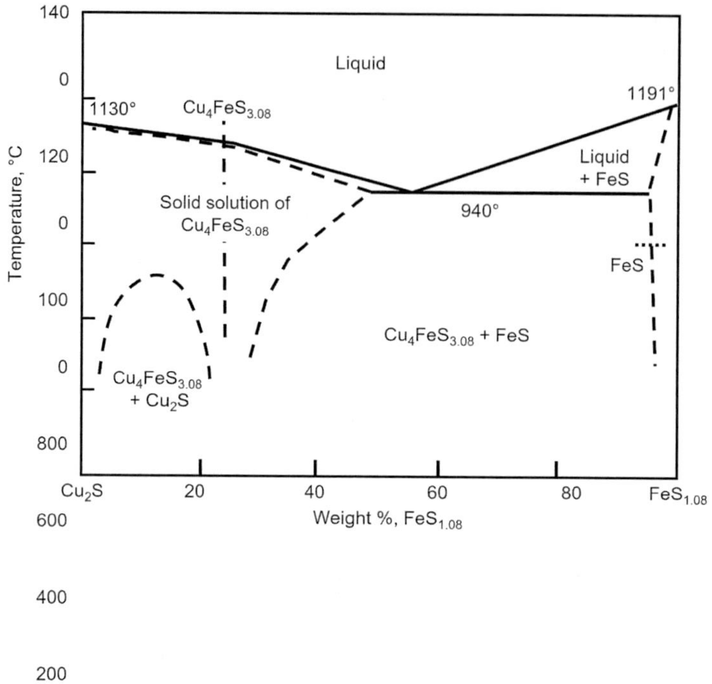

Figure 4.11 Cu_2S—FeS phase diagram redrawn from Schlegel & Schuller, 1952. Actual matte melting temperatures are lower than the liquidus line temperature due to impurities in the matte.

The surface tension of Cu_2S—FeS mattes ranges from 0.33 to 0.45 N/m, increasing with Cu_2S content. Temperature has little effect (Kucharski et al., 1994).

Specific gravity of matte ranges linearly from 3.9 for pure FeS to 5.2 for pure Cu_2S. It decreases slightly with increasing temperature. Multiplying these specific gravities by the kinematic viscosities measured by Nikiforov et al. (1976) yields viscosities of about 0.003 kg/m-s for pure Cu_2S at 1250°C, falling to about 0.002 kg/m-s for mattes with 35% FeS. The value then rises rapidly with increasing FeS. It decreases slowly with increasing temperature.

Measurements of interfacial tension between molten mattes and slags were reviewed by Nakamura and Toguri (1991). Interfacial tension increases from near zero in low-Cu mattes to about 0.20 N/m for high-Cu mattes (−70 mass% Cu_2S). Matte-specific electrical conductances are 200—1000 Ω-cm (Liu et al., 1980; Pound et al., 1955).

4.3.4 Off-gas

The off-gas from smelting contains SO_2 generated by the smelting reactions, N_2 from the air used for oxidizing the concentrate, and small amounts of CO_2, H_2O and volatilized impurity compounds. The strength of the off-gas is usually 10–60 vol.-% SO_2. The strength depends on the type of O_2-containing gas used for smelting, the amount of air allowed to leak into the furnace, and the grade of matte produced. The concentration of SO_2 (vol.-%SO_2) in smelter off-gases has risen in recent years. This is due to increased use of oxygen in smelting, which reduces the amounts of nitrogen and hydrocarbon combustion gases passing through the furnace.

Smelter off-gases may also contain substantial levels of dust (up to 0.3 kg/Nm^3). This dust comes from (a) small particles of unreacted concentrate or flux, (b) droplets of matte/slag that did not settle into the slag layer in the furnace, and (c) volatilized elements in the concentrate such as arsenic, antimony, bismuth, and lead, which have either solidified as the gas cools or reacted to form nonvolatile compounds. The dust generally contains 20–40 mass% Cu, making it potentially valuable. It is nearly always recycled to the smelting furnace, but it may be treated hydrometallurgically to recover Cu and remove deleterious impurities from the smelting circuit (Chapter 21).

4.4 The copper converting process

During smelting, formation of magnetite (reaction 4.18) has been controlled by addition of silica (4.19) and activity of FeS has gradually decreased. The molten products of the smelting process are then separated, normally in a settling process within the smelting reactor or in a separate vessel. Matte then is ready to continue its copper making journey.

From the thermodynamic point of view, using the Yazawa predominance diagram, copper matte must be transformed to blister copper by further reducing the sulfur potential and increasing the oxygen potential. This process is represented in Fig. 4.12 by the black arrow across the path A → B → C → D for batch converting, as follows:

- Path A → B: Slag Blowing:
- Path B → C: Copper blowing:

In continuous converting, on the other hand, the process will follow the path A′ → C′ for a continuous converting process. Location of points A′ and C′ in the predominance diagram will depend on the smelting practice and type of continuous converting process.

4.4.1 Reactions involved in batch converting

During the slag blowing stage, copper content in matte is increased by the elimination of remaining FeS in the copper matte and formation of slags:

$$FeS(l) + 3/2O_2(g) \rightarrow FeO(s) + SO_2(g) \tag{4.25}$$

$$FeO(s) + SiO_2(s) \rightarrow FeO - SiO_2(l,\ \mathit{slag}) \tag{4.20}$$

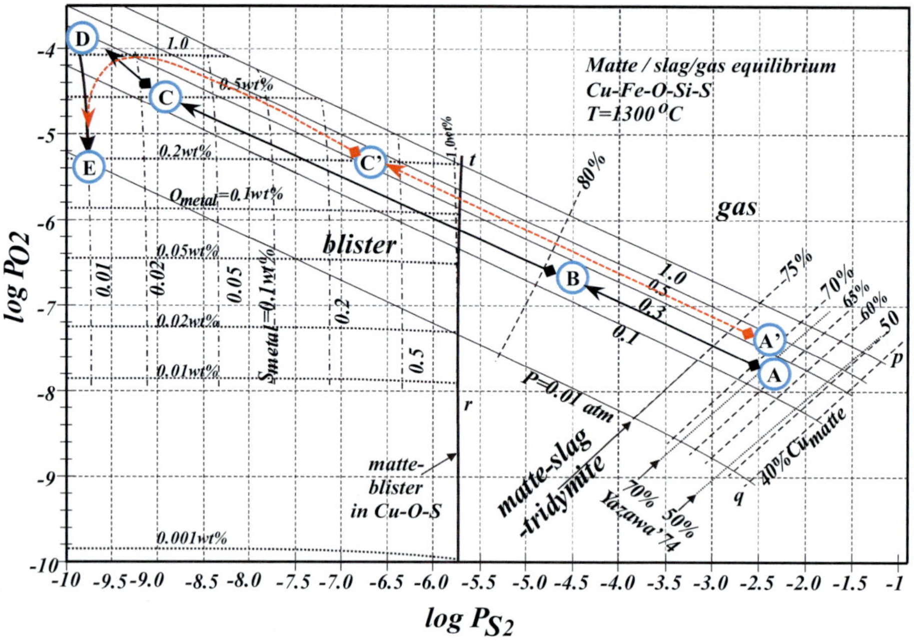

Figure 4.12 Predominance diagram describing copper converting and refining at 1300°C based on calculations using FactSage public data base and original Yazawa diagram. *(Figure prepared with support of Pyrosearch—University of Queensland, Australia.)*

$$2\ FeS(l) + 7/2O_2(g) \rightarrow Fe_2O_3(s) + 2\ SO_2(g) \quad (4.26)$$

$$FeO(s) + Fe_2O_3(s) \rightarrow Fe_3O_4(s) \quad (4.19)$$

$$6\ FeO(s) + O_2(g) \rightarrow 2\ Fe_3O_4(s) \quad (4.27)$$

Copper oxide formation will also increase as the activity of FeS decreases, and there is not enough FeS to drive the equilibrium of reaction 4.15 to the formation of Cu_2S:

$$FeS\ (l,\ in\ matte) + Cu_2O\ (l,\ in\ slag) \rightarrow Cu_2S\ (l,\ in\ matte) + FeO\ (l,\ in\ slag) \quad (4.16)$$

At a given point of the process, slag is skimmed (point B), and matte is charged once again, resulting in an increase of the sulfur potential and therefore a reduction in copper matte grade (backwards in path AB), and the oxidation process continues. This process is repeated several times until there is enough high-grade matte (point B) to continue with the converting of copper sulfide.

Copper blowing will eliminate a small amount of the remaining iron (~2% Fe in matte) and gradually remove the remaining sulfur from Cu_2S, as follows:

$$Cu_2S(s)+O_2(g)\rightarrow 2\ Cu^{\circ}(l)+SO_2(g) \quad (4.13)$$

$$2Cu_2O(l) + Cu_2S(g)\rightarrow 6Cu(l) + SO_2(g) \quad (4.28)$$

The copper blows will end when reaction 4.28 cannot proceed any more as Cu_2S has completely converted. The end point represented by Point D in Fig. 4.12 corresponds to sulfur levels between 300 and 500 ppm.

4.4.2 Reactions involved in continuous converting

Continuous copper converting chemistry varies depending on the technology applied to convert copper matte. In general terms, these processes can be classified into the following categories:

- Continuous converting with two molten phases: Copper sulfide is fully converted to copper, with the remaining sulfur dissolved in the copper. Molten phases are slag and copper
- Continuous converting with three molten phases: Blister copper is in equilibrium with copper sulfide (white metal) and slag.

The above classification can be explained with the help of the Cu–S phase diagram shown in Fig. 4.13. Let us consider a continuous converting process at 1200°C. The endpoint of the converting process is defined by the specific amount of oxygen provided for the converting reactions. If the amount of oxygen is below the amount required to completely convert Cu_2S to Cu, represented by Point c in the phase diagram, the generated blister copper will be in equilibrium with Cu_2S at a composition of 80% copper, represented by Point a, in the figure. Therefore, the blister copper will be saturated with Cu_2S. The minimum amount of sulfur required to achieve this condition is about 1% at 1200°C. On the other hand, if enough oxygen is supplied to convert Cu_2S, the generated blister copper will be in the region represented by the range of compositions betweenc and d, and only two molten phases will be in equilibrium.

From the chemistry point of view, chemical reactions will depend on the type of process, the choice of flash or bath converting, and the slag chemistry involved (calcium ferrite or olivine slags). A more detailed explanation is given in the continuous converting chapter.

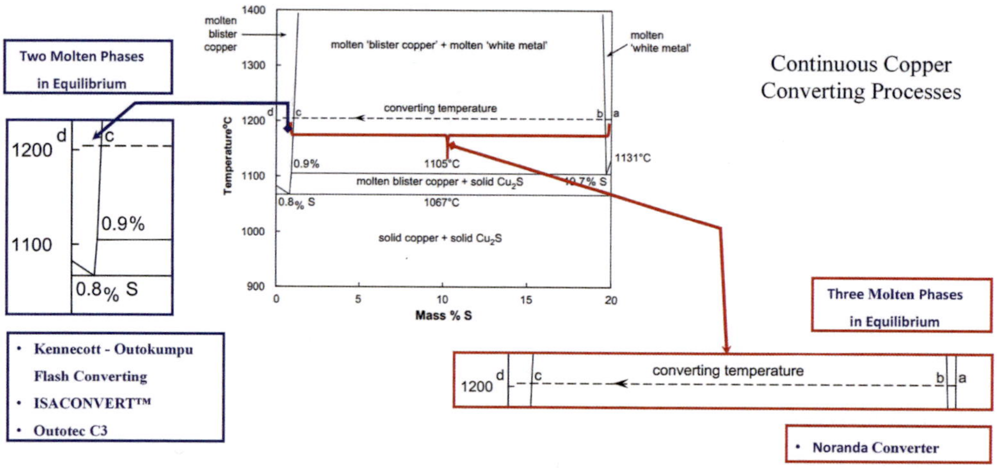

Figure 4.13 Graphical representation of continuous copper converting processes, in terms on molten phases in equilibrium. Two molten phases in equilibrium (zone d—c) correspond to blister copper in equilibrium with molten slag. Three molten phases in equilibrium (zone b—c) correspond to molten copper saturated with sulfur, molten white metal (Cu_2S) and slag. Figure adapted from Figure 8.4 (Sharma, R. C., & Chang, Y. A. (1980)).

4.5 The refining process

The last step in the copper making process is the refining of blister, represented by the path C → D → E for batch converting and C′→ E for continuous converting. The process consists of two steps:

- Oxidation to remove the remaining sulfur
- Reduction aimed to reduce oxygen

The main chemical reactions involved are as follows:

Oxidation:

$$O_2(g) = 2\underline{O}(\text{in copper}) \tag{4.29}$$

$$\underline{S}(\text{in copper}) + 2\,\underline{O}(\text{in copper}) = SO_2(g) \tag{4.30}$$

Reduction:

$$C(s) + \underline{O}(\text{in copper}) = CO(g) \tag{4.31}$$

$$CO(g) + \underline{O}(\text{in copper}) = CO_2(g) \tag{4.32}$$

$$H_2(g) + \underline{O}(\text{in copper}) = H_2O(g) \tag{4.33}$$

A detailed explanation of this process is given in Chapter 12.

4.6 Minor elements

Copper concentrates, given their nature, also contain minor valuable and deleterious elements. These elements have been present in relatively small concentrations compared with copper, sulfur, iron, and gangue materials. Typical minor metals are gold and silver (precious metals), nickel, lead, cobalt, bismuth, zinc, arsenic, antimony, bismuth, selenium, and tellurium. In addition, halides and radioactive elements such as uranium also report in copper concentrates.

However, as copper consumption has dramatically increased in recent years (Alvear Flores et al., 2018), more complex copper concentrates have been processed in copper smelters, gradually increasing the concentration of minor elements. Elements of concern include arsenic, antimony, and bismuth due to their nature and detrimental impact during the electrorefining process (Moats et al., 2016).

A great effort has been done to understand the behavior of these elements under copper making conditions (Chaubal et al., 1989; Chen et al., 2010; Itagaki & Yazawa, 1982; Nagamori & Chaubal, 1982a,b; Takeda et al., 1983). Behavior of minor elements in the overall copper making process will depend on their specific mineralogy, physical and chemical properties, and affinity with the respective formed phases in smelting, converting, and refining.

Process engineers need to understand the behavior of each one of these elements to

- Optimize recovery of valuable metals that will contribute to the smelting business.
- Optimize the removal of deleterious elements that will affect the copper cathode properties.
- Define suitable slag compositions to optimize removal of certain elements.
- Define individual minor element capacities for each smelting technology.
- Define optimum outlets for minor elements.

The recovery of minor valuable elements is normally achieved via the following:

- Leaching of metallurgical dusts generated in the copper smelting and converting processes,
- Concentration and crystallization of electrolyte,
- Recovery from copper slimes from the electrorefining process.

4.6.1 Deportment of minor elements

Once minor elements are under given oxygen and sulfur potentials, they will distribute among the condensed and gas phases, depending on their relative affinity for each phase. Minor element deportment to each phase will be defined by the following aspects, among others:

- Matte grade (and therefore oxygen and sulfur potential),

- Stability of each element as oxide, sulfide, or metallic form,
- Partial pressure of each formed minor element species,
- Process temperature,
- Total off-gas volume,
- Slag capacity for each minor element,
- Activity coefficient of each minor element species in their respective molten phase,
- Degree of agitation of the reactor.

Several authors have developed thermodynamic models to explain the deportment of minor elements during smelting, converting, and refining Alvear Flores et al., 1999; Chaubal et al., 1989; Chen & Jahanshahi, 2010; Itagaki et al., 2002; Nagamori & Chaubal, 1982a, 1982b; Tan & Zhang, 1997; Kim and Sohn, 1996).

Thermodynamic packages have also incorporated existing information to predict minor element behavior. Fig. 4.14 shows the predictions for As distribution based on the MPE CSIRO package.

Minor element deportment between phases varies depending on process conditions and applied technologies. Table 4.3 compares some key minor element distributions during smelting for the flash furnace (Moats et al., 2016), TSL—ISASMELT, Dongying Fangyuan, and Mitsubishi processes. Technology providers normally will argue about their specific technology advantages for efficient removal of minor elements to the gas and slag phases. However, process engineers must be careful when using this information, as operating conditions under which these data were collected are sometimes not directly comparable.

In addition, smelting process conditions and, subsequently, deportment behavior must be revised in light with increasing environmental restrictions. This is especially applicable to smelting slags, if they are to be valorized for other applications (Klaffenbach et al., 2018).

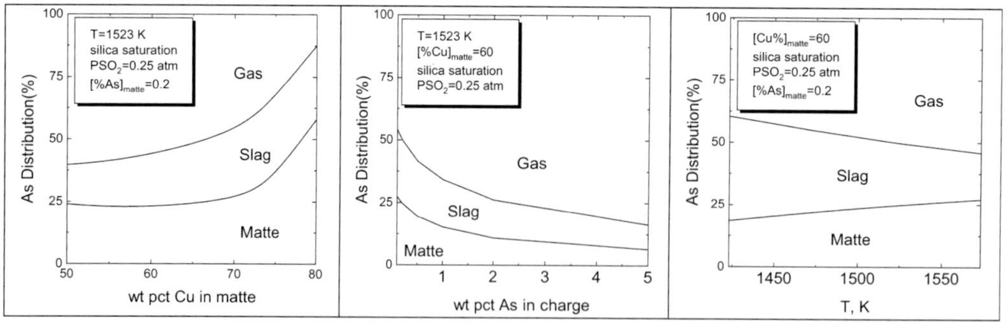

Figure 4.14 Calculated distribution of As among the gas, silica-saturated iron silicate slag, and copper matte phases during continuous smelting of charge matte (40 pct Cu_2S, 59.8 pct FeS, and 0.2 pct As) at 1523 K (1250 C) with $PSO_2 = 0.25$ at, using MPE. Figure prepared with support of Dr. Chunlin Chen, CSIRO.

Table 4.3 Compositions of As, Sb, Bi, Pb, and Zn distribution for the Flash smelting, TSL ISASMELT, Dongying Fangyuan SLS Smelting, and Mitsubishi processes (Liao et al., 2016; Ajima et al., 1995; Moats et al., 2016; Surapunt & Hasegawa, 2003; Wang et al., 2017).

Process	As	Sb	Bi	Pb	Zn
Flash smelting	Several process conditions				
Matte	15–40	60–70	30–75	45–80	30–50
Slag	5–25	5–35	5–30	15–20	50–60
Gas	35–80	5–25	15–65	5–40	5–15
TSL (ISASMELT™)	Process conditions: Cu matte grade: 60%–65%, temperature: 1200°C, oxygen enrichment: 60%–65%				
Matte	6–10	20–40	14–25	60–65	25–30
Slag	5–25	30–40	1–2	15–20	60–70
Gas	35–80	17–45	75–85	15–25	5–10
Dongying Fangyuan SLS furnace	Process conditions: Cu matte grade: 75%, temperature: 1200°C, oxygen enrichment 73%				
Matte	5.9	12.3	19.1	55.6	17.8
Slag	12.1	71.1	11.4	24.9	64.9
Gas	82.0	16.6	69.5	19.5	17.4
Mitsubishi	Process conditions: Cu matte grade: 65%, temperature: 1230°C, oxygen enrichment 73%				
Matte	8–12	31–33	38–45	33–40	15
Slag	12–18	53–57	5–12	12–14	70
Gas	70–80	10–15	50–60	45–55	15

4.6.2 Recovery of minor elements

The focus for the removal of deleterious elements is to define appropriate operating conditions to achieve the expected anode specifications required by the electrorefining process. On the other hand, for valuable elements, process engineers are expected to define suitable conditions to maximize their recovery. Both objectives although seemingly contradictory have in common the need for secondary processes to recover these elements, followed by disposal or sale. These objectives are normally achieved via the following:

- Leaching of metallurgical dusts generated by copper smelting and converting processes
- Removal of impurities from the electrolyte
- Neutralization of acid plant effluents
- Recovery from copper slimes from the electrorefining process.

4.6.2.1 Dust leaching

Used for removal and stabilization of deleterious elements and recovery of copper and precious metals. This option is an alternative for smelters that requires to bleed impurities as their main outlets so that their levels in slag and anode copper are below standard limits. Industrial examples for this alternative are the Ecometales dust leaching plant in Calama, Chile, and the Glencore Mount Isa Mine smelting dust leaching plant in Queensland, Australia.

4.6.2.2 Removal of impurities from the electrolyte

Integrated to the copper electrorefining process, some amount of electrolyte must be purified to maintain the electrolyte solution under the required standard to achieve efficient copper electrodeposition. This process is carried out in three main steps:

- Copper electrowinning,
- Electrowinning of As, Sb, and Bi,
- Evaporation and crystallization of nickel

These processes are discussed in the copper electrorefining chapter.

4.6.2.3 Neutralization of effluents of the acid plant

As minor elements are volatilized from the smelting and converting processes, some of the formed species cannot condense and are carried out to the gas cleaning section of the acid plant and removed into a liquid effluent that must be neutralized.

4.6.2.4 Removal and recovery from copper anode slimes

Precious metals and other valuable elements such as selenium, tellurium, and tin precipitate to the bottom of the electrolytic cells during the electro refining processes. Tank house sections are periodically cleaned, and slimes are removed from each cell and filtered. The recovered slimes are processed in a dedicated precious metal plant and valuable elements are recovered via a series of pyrometallurgical and hydrometallurgical processes.

4.7 Summary

The pyrometallurgical production of copper from sulfide concentrates is a journey involving the gradual oxidation of copper concentrates in sequential steps. These steps are smelting, converting, and refining. In some special cases, copper can be produced in a single step, when iron content is low enough to produce a reasonably low amount of slag. The overall copper making process can be explained by using the oxygen—sulfur predominance diagrams. This tool is a navigation chart to understand the overall copper making process.

A proper understanding of the influence of the sulfur and oxygen potential is needed to define proper operating points by selecting appropriate matte grade for the smelting process, adequate slag compositions to ensure suitable slag quality, and controlled copper losses.

References

Ajima, S., Hayashi, I., Hasegawa, & Nozaki, M. (1995). The distribution of minor elements at Naoshima. In T. Lehner (Ed.), *Co-products and minor elements in non-ferrous smelting* (pp. 13–26). Warrendale, PA: TMS.

Alvear Flores, G. R. F., Löbbus, M., Springub, B., Fallah-Mehrjardi, A., & Tappe, A. (2018). Sulfide smelting: Thirty–five years of continuous efforts to find new value adding solutions. In B. Davis, et al. (Eds.), *Extraction 2018* (pp. 39–56). Warrendale, PA: TMS.

Alvear Flores, G. R. F., Tanaka, A., Fukuyama, H., Fujisawa, T., & Yamauchi, C. (1999). Application of Na_2CO_3 flux to the removal of impurity elements from molten copper. *Shigen to Sozai, 115*(2), 89–96.

Chaubal, P. C., Sohn, H. Y., George, D. B., & Bailey, L. K. (1989). Mathematical modeling of minor–element behavior in flash smelting of copper concentrates and flash converting of copper mattes. *Metallurgical and Materials Transactions B, 20B*, 39–51.

Chen, C., & Jahanshahi, S. (2010). Thermodynamics of arsenic in $FeOx-CaO-SiO_2$ slags. *Metallurgical and Materials Transactions B*, 41B1166–1174.

Chen, C., Zang, L., & Jahanshahi, S. (2010). Thermodynamic modeling of arsenic in copper smelting processes. *Metallurgical and Materials Transactions B, 41B*, 1175–1185.

Fahey, N. P., Swinbourne, D. R., Yan, S., & Osborne, J. M. (2004). The solubility of Cr_2O_3 in calcium ferrite slags at 1573 K. *Metallurgical and Materials Transactions B, 35B*, 197–202.

Goto, M., Oshima, E., & Hayashi, M. (1998). Control aspects of the Mitsubishi continuous process. *JOM, 50*(4), 60–64.

Hidayat, T., Shishin, D., Decterov, S. A., & Jak, E. (2017). Experimental study and thermodynamic Re-evaluation of the $FeO-Fe_2O_3-SiO_2$ system. *Journal of Phase Equilibria and Diffusion, 38*, 477–492.

Itagaki, K., Sohn, H. Y., & Pérez–Tello, M. (2002). Basic principles of sulfide smelting and converting with oxygen–rich gas. In R. L. Stephens, & H. Y. Sohn (Eds.), *Sulfide smelting* (pp. 15–40). Warrendale, PA: TMS.

Itagaki, K., & Yazawa, A. (1982). Thermodynamic evaluation of distribution of arsenic in copper smelting. *Transactions of the Japan Institute of Metals, 23*, 759–767.

Jahanshahi, S., & Sun, S. (2003). Some aspects of calcium ferrite slags. In F. Kongoli, C. Yamauchi, & H. Y. Sohn (Eds.), *Materials processing fundamentals and new technologies: Vol. 1. Yazawa international symposium* (pp. 227–244). Warrendale, PA: TMS.

Jak, E., Hidayat, T., Shinshin, D., Fallah Mehrjardi, A., Chen, J., Decterov, S., & Hayes, P. (2016). Integrated experimental and thermodynamic modelling research methodology for metallurgical slags with examples in the copper production field. In R. G. Reddy, P. Chaubal, C. Pistorius, & U. Pal (Eds.), *Advances in molten slags, fluxes, and salts (MOLTEN16)* (pp. 947–959). Warrendale, PA: TMS.

Jak, E., Zhao, B., & Hayes, P. (2000). Phase equilibria in the system $FeO-Fe_2O_3-Al_2O_3-CaO-SiO_2$, with applications to non-ferrous smelting slags. In S. Seetharaman, & D. Sichen (Eds.), *6th Int. Conf. Slags, fluxes, molten salts* (pp. 147–152). Stockholm: KTH.

Kim, H. G., & Sohn, H. Y. (1996). Thermodynamic modelling of minor-element behaviour in in-bath copper smelting and converting with calcium ferrite slag. *Transactions of the Institution of Mining and Metallurgy, 105*, 151–163.

Klaffenbach, E., Alvear Flores, G. R. F., Guo, M., & Blanpain, B. (2018). Thermodynamic consideration of copper matte smelting conditions with respect to minor element removal and slag valorization options. In B. Davis, et al. (Eds.), *Extraction 2018* (pp. 491–504). Warrendale, PA: TMS.

Kucharski, M., Ip, S. W., & Toguri, J. M. (1994). The surface tension and density of Cu_2S, FeS, Ni_3S_2 and their mixtures. *Canadian Metallurgy Quarterly, 33*, 197–203.

Li, H., & Rankin, J. W. (1994). Thermodynamics and phase relations of the $Fe-O-S-SiO_2$ (sat) system at 1200 °C and the effect of copper. *Metallurgical and Materials Transactions B, 25B*, 79–89.

Liao, L., Qinmeng, W., Qinghua, T., & Xueyi, G. (2016). Multiphase equilibrium modeling study on the oxygen bottom blowing copper smelting (SKS) process. In *Proceedings of copper 2016* (pp. 1252–1264). Tokyo: MMIJ.

Liu, C., Chang, M., & He, A. (1980). Specific conductance of Cu_2S, Ni_3S_2 and commercial matte. *Chinese Nonferrous Metals, 32*, 76–78.

Moats, M., Aslin, A., Pranowo, A., & Alvear Flores, G. R. F. (2016). Behaviour and benefits of arsenic in copper electrorefining. *Comercio e Industria de la Madera J*, 7(3), 151–157.

Muan, A. (1955). Phase equilibria in the system $FeO-Fe_2O_3-SiO_2$. *Transactions of the AIME, 205*, 965–976.

Nagamori, M., & Chaubal, P. C. (1982a). Thermodynamics of copper matte converting: Part III. Steady-state volatilization of Au, Ag, Pb, Zn, Ni, Se, Te, Bi, Sb, and as from slag, matte, and metallic copper. *Metallurgical and Materials Transactions B, 13B*, 319–329.

Nagamori, M., & Chaubal, P. C. (1982b). Thermodynamics of copper matte converting: Part IV. *A priori* predictions of the behavior of Au, Ag, Pb, Zn, Ni, Se, Te, Bi, Sb, and as in the noranda process reactor. *Metallurgical and Materials Transactions B, 13B*, 331–338.

Nakamura, T., & Toguri, J. M. (1991). Interfacial phenomena in copper smelting processes. In C. Díaz, C. Landolt, A. A. Luraschi, & C. J. Newman (Eds.), *Pyrometallurgy of copper: vol. IV. Copper 91–Cobre 91* (pp. 537–551). New York: Pergamon Press.

Nikiforov, L. V., Nagiev, V. A., & Grabehak, V. P. (1976). Viscosity of sulfide melts. *Inorganic Materials, 12*, 985–988.

Pound, G. M., Derge, G., & Osuch, G. (1995). Electrical conductance in molten Cu–Fe–sulphide mattes. *Transactions of the AIME, 204*, 481–484.

Ramachandran, V., Díaz, C., Eltringham, T., Jiang, C. Y., Lehner, T., Mackey, P. J., Newman, C. J., & Tarasov, A. (2003). Primary copper production — a survey of operating world copper smelters. In C. Díaz, J. Kapusta, & C. Newman (Eds.), *Pyrometallurgy of copper (Book 1), the Herman Schwarze symposium on copper pyrometallurgy: Vol. IV. Copper 2003–Cobre 2003* (pp. 3–106). Montreal: CIM.

Schlegel, H., & Schuller, A. (1952). Das zustandsbild kupfer–eisen–schwefel. *Zeitschrift fur Metallkunde, 43*, 421–428.

Shimpo, R., Goto, S., Ogawa, O., & Asakura, I. (1986). A study on the equilibrium between copper matte and slag. *Canadian Metallurgy Quarterly, 25*, 113–121.

Shishin, D., Decterov, S. A., & Jak, E. (2018). Thermodynamic assessment of slag–matte–metal equilibria in the Cu–Fe–O–S–Si system. *Journal of Phase Equilibria and Diffusion, 39*, 456–475.

Shishin, D., Hidayat, T., Fallah Mehrjardi, A., Hayes, P. C., Decterov, S. A., & Jak, E. (2019). Integrated experimental and thermodynamic modelling study of the effects of Al_2O_3, CaO and MgO on slag–matte equilibria in the Cu–Fe–O–S–Si–(Al, Ca, Mg) system. *Journal of Phase Equilibria and Diffusion, 40*, 445–461.

Sundström, A. W., Eksteen, J. J., & Georgalli, G. A. (2008). A review of the physical properties of base metal mattes. *Journal of the South African Institute of Mining and Metallurgy, 108*, 431–448.

Surapunt, S., & Hasegawa, N. (2003). Distribution Behaviour of Arsenic, Antimony and Bismuth in the Smelting Stage of the Mitsubishi Process. *Yazawa International Symposium, 1*, 375–681.

Takeda, Y., Ishiwata, S., & Yazawa, A. (1983). Distribution equilibria of minor elements between liquid copper and calcium ferrite slag. *Transactions of the Japan Institute of Metals, 24*, 518–528.

Tan, P., & Zhang, C. (1997). Computer model of copper smelting process and distribution behaviors of accessory elements. *Journal of Central South University Technology, 4*, 36–41. https://doi.org/10.1007/s11771-997-0027-y.

Tsukihashi, F., & Kimura, H. (2000). Phase diagram for the $CaO-SiO_2-FeO_x$ system at low oxygen partial pressure. In S. Seetharaman, & D. Sichen (Eds.), *6th int. conf. slags, fluxes, molten salts (article 22)*. Stockholm: KTH.

Utigard, T. A., & Warczok, A. (1995a). Density and viscosity of copper/nickel sulphide smelting and converting slags. In W. J. Chen, C. Diaz, A. Luraschi, & P. J. Mackey (Eds.), *Pyrometallurgy of copper: Vol. IV. Copper 95–Cobre 95 Proceedings of the third international conference* (pp. 423–437). Montreal: CIM.

Vadász, P., Tomášek, K., & Havlik, M. (2001). Physical properties of $FeO-Fe_2O_3-SiO_2-CaO$ melt systems. *Archives of Metallurgy, 46*, 279–291.

Vartiainen, A., Kojo, I. V., & Acuña, R. C. (2003). Ferrous calcium slags in direct-to-blister flash smelting. In F. Kongoli, C. Yamauchi, & H. Y. Sohn (Eds.), *Materials processing fundamentals and new technologies: Vol. 1. Yazawa international symposium* (pp. 277–290). Warrendale, PA: TMS.

Wang, Q., Guo, X., Tian, Q., Jiang, T., Chen, M., & Zhao, B. (2017). Effects of Matte Grade on the Distribution of Minor Elements (Pb, Zn, As, Sb, and Bi) in the Bottom Blown Copper Smelting Process. *Metals, 7*, 502. https://doi.org/10.3390/met7110502.

Wood, J., & Hughes, S. (2016). Future development opportunities for the Outotec® Ausmelt process. In *Proceedings of copper 2016* (pp. 361–372). Tokyo: MMIJ.

Yazawa, A. (1956). Copper Smelting. V. mutual solution between matte and slag produced in the Cu_2S-FeS-FeO-SiO_2 system. *Journal of the Mining Institute of Japan, 72*, 305–311.

Yazawa, A. (1979). Thermodynamic evaluations of extractive metallurgical processes. *Metallurgical and Materials Transactions B, 10B*, 307–321.

Yazawa, A., & Kameda, A. (1953). Copper smelting. I. Partial liquidus diagram for $FeS-FeO-SiO_2$ system. *Technology Reports of the Tohoku University, 18*, 40–58.

Zhang, L., Jahanshahi, S., Sun, S., Chen, C., Bourke, B., Wright, S., & Somerville, M. (2002). CSIRO's multiphase reaction models and their industrial applications. *JOM, 54*(11), 51–56.

Suggested reading

Mackey, P. J. (1982). The physical chemistry of copper smelting slags – a review. *Canadian Metallurgy Quarterly, 21*, 221–260.

CHAPTER 5

Theory to practice: pyrometallurgical industrial processes

5.1 General considerations

As indicated in Chapter 4, pyrometallurgical processing of copper concentrates is based on the continuous management of the sulfur and oxygen potentials. This is done in sequential stages to control the heat released from the chemical reactions, and the overall copper recovery.

The sulfur—oxygen potential management principle is the corner stone used by technologists for the development and design of pyrometallurgical processes. This metallurgical principle is applied in conjunction with core technology principles to define appropriate metallurgical conditions under which a defined technology could be applied to the processing a given feedstock.

Selection of operating parameters will be driven by design considerations associated with aspects such as the following:

- Concentrate mineralogy and particle size distribution for each of the present species.
- Igniting characteristics of the copper concentrate to be processed.
- Composition and nature of gangue materials: Relevant to define suitable slag composition, and overall heat load that the process can handle.
- Matte grade: relevant to define overall smelting-converting heat release for each stage and first-pass copper recovery.
- Operating temperature: Some technologies will choose to operate with superheated slag or under the slag liquidus temperature. This strategy will impact the reactor design and expected furnace campaign life, depending one use of intensive reactor cooling.
- Oxygen enrichment: A relevant element to determine the overall reactor heat balance, off-gas handling requirements, and volatilization of minor elements.
- Fe/SiO_2 ratio: To define slag generation and first copper pass recovery.
- Furnace heat management: Define the degree of cooling required by the reactor and indirectly the overall furnace campaign.
- Moisture content: Required in some cases to minimize feed carryover and heat balance of the process (bath smelting). However, increases the off-gas volume handling and affects the water balance in the gas cleaning plant.

The above-mentioned parameters are part of general design criteria used by technologist to engineer pyrometallurgical process and operating parameters to be controlled by

Extractive Metallurgy of Copper, Sixth Edition
ISBN 978-0-12-821875-4, https://doi.org/10.1016/B978-0-12-821875-4.00006-7

process engineers. The next chapters will show how these parameters are applied in the design and operation of the flash and bath smelting technologies, currently under use in pyrometallurgical processing.

5.2 Technology evolution since 1970

The first edition of *Extractive Metallurgy of Copper* was published during a pivotal time in the development of the copper smelting industry. The world was facing an oil crisis that transformed the industry, gradually replacing reverberatory furnaces into flash smelting and bath smelting technologies, increasing the sulfur capture, and reducing the consumption of fuel. Many smelters, operating old fashion high oil consumption technologies (reverberatory furnace technology), were forced to shut down as their operating costs were not competitive at that time (Mackey et al., 2014). Energy consumption was in comparison 60% higher than current average of modern bath and flash smelting technologies (Coursol et al., 2015).

At the time that the first edition of *Extractive Metallurgy of Copper* was published, new processes, such as the Noranda, El Teniente, and the Mitsubishi, were in their initial stage of development, while the flash smelting was the main technology option available to modernize smelters, with 22 new flash smelters licensed by 1976 (Mackey et al., 2014). Since then, the copper industry has seen several technological changes aimed to support to growing demand for copper (three times increase from 1976 to 2018). Technologies such as the Noranda, El Teniente, Top Submerged Lance (ISASMELT™ and Ausmelt® (now Metso-Outotec-Ausmelt®), Mitsubishi Process, Vanyukov, and most recently the full spectrum of new Chinese technology developments are now treating large tonnages of copper concentrate in a single furnace.

During the last 40 years, the copper industry has witnessed dramatic technology changes to support the growing copper demand. The first golden age of development from 1970 to 1985, with the development of bath smelting technologies, has been followed by the gradual modernization of smelters, driven by increasing copper demand, cost reduction, and increasing environmental regulations. Then, since 1995, the copper industry has witnessed the implementation of brown and greenfield projects, and the increasing of smelters average copper processing capacities from below 100,000 tonnes per year to levels of 400,000 tonnes per year (Risopatron, 2020). Fig. 5.1 shows a comparison of smelter capacities between 1980 and 2013.

Fig. 5.2 identifies some technological milestones over the last 40 years in the copper smelting industry. Some selected milestones have been the following:

- 1973 – Development of the Noranda Reactor: Originally aimed to produce blister copper, this bath smelting technology allowed the production of high matte grade (Tarassoff, 1984).

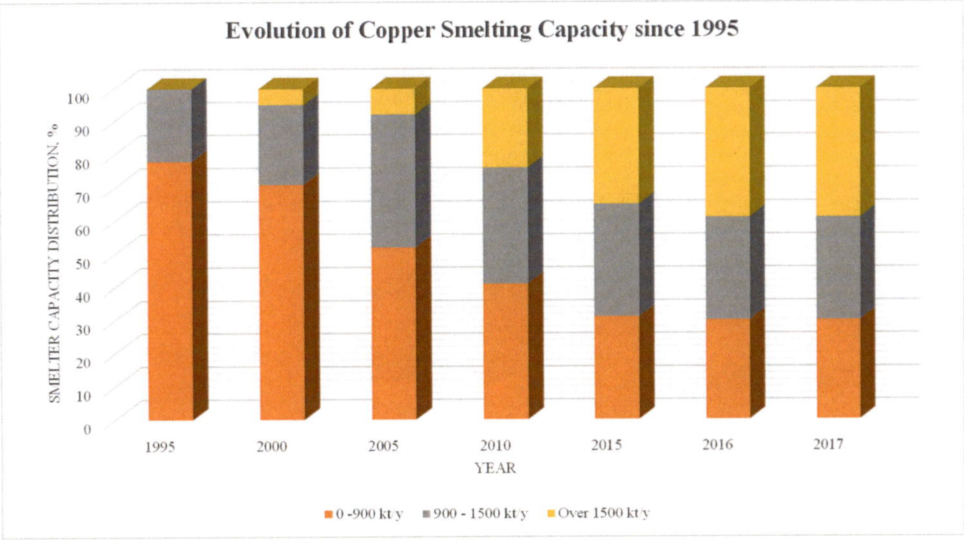

Figure 5.1 Comparison of smelter capacities between 1995 and 2013 (Risopatron, 2020).

- 1974 – Mitsubishi process: First commercial installation in Japan commissioned at Naoshima, with an initial capacity of 48,000 tonnes per year (Goto et al., 1998).
- 1977 – SiroSmelt lance industrial tests in Mount Isa Smelter: Developed by CSIRO-Australia, the technology was industrially developed by Mount Isa Mines (ISASMELT™) and Metso-Outotec-Ausmelt® (Previously Ausmelt) and applied to the processing of base metals (Arthur & Hunt, 2005; Floyd, 2005).
- 1977 – EL Teniente Reactor: Following the path of the Noranda process, the first industrial installation commenced operations in Chile (Campos & Torres, 1993).
- 1977 – Vanyukov process commercialized at Norilsk smelter (Fyodorov et al., 1992).
- 1978 – KGHM Direct to Blister: Developed by Outokumpu (Metso Outotec) and applied by KGHM to the processing of copper concentrates to blister at Glogow smelter (Czernecki, 1999).
- 1983 – Inco Flash Furnace Hayden Smelter commissioning: First application of the INCO flash process to copper concentrates (Marczeski & Aldrich, 1986).
- 1992 – ISASMELT industrial scale applications commissioned at Cyprus Miami, Arizona, and Mount Isa, Australia (Bhappu, Larson, & Tunis, 1993).
- 1995 – Kennecott-Outokumpu Flash Converting plant starts operating, providing a new option for continuous converting of copper (Newman, 1998).
- 1995 – Transformation of the Hoboken flowsheet and development of the Umicore Precious Metals Complex Metallurgy Process using copper and lead metallurgy (Vanbellen & Chintinne, 2006).

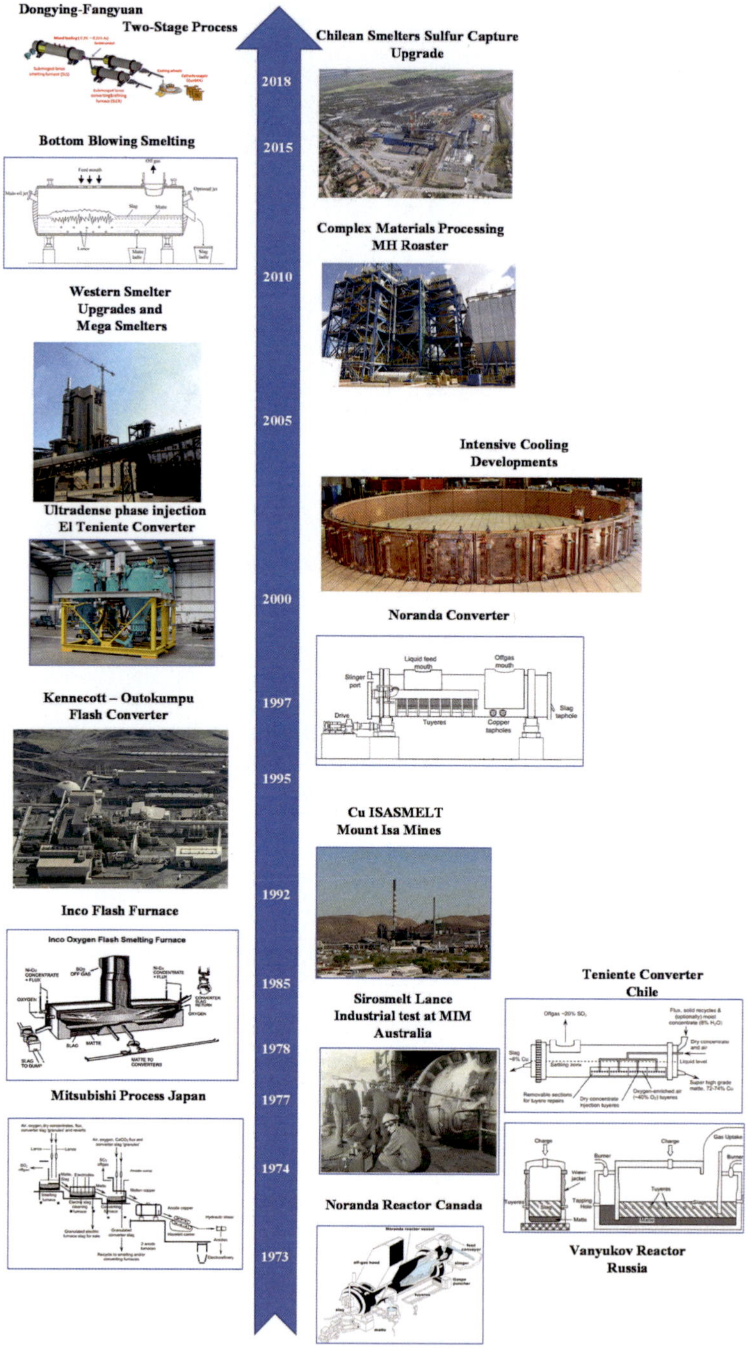

Figure 5.2 Evolution of main smelting and converting technologies and selected milestone over the last 40 years.

- 1996 onward — Large-scale smelting in single reactors with productions over 3000 tpd of dry feed (Hirai, 1999).
- 1997 — Noranda Converting process first commercial installation starts operation at Horne smelter, processing matte from the Noranda Reactor (Prévost et al., 1999).
- 2000 onward — Large capacity smelters start to be built across the world (Mackey, 2014).
- 2002 — Commissioning of the Kayser Recycling System by Huettenwerke Kayser now Aurubis (Ayhan, 2001).
- 2006 — African copper belt renaissance with modernization of Mufulira smelter in 2006; and new smelters of Nchanga in 2008, Chambishi in 2009, and Kansanshi in 2014 (ICSG, 2017).
- 2007 — Second flash converter installed in Xiangguang Copper Smelter at Yanggu (Kojo, Lahtinen, & Miettinen, 2009).
- 2008 onward — Chinese smelting and converting technology developments: Bottom blowing smelting and converting, side blowing smelting and suspension smelting (Guo, Zhang, Wang, & Yuan, 2016).
- 2013 — Ministro Hales Roaster commissioning (Holmström, Andersson, Carrasco, & Deutsch, 2013).
- 2015 — Dongying Fangyuan SLS-SLCR two-stage process started up (Chi et al., 2016).
- 2018 — Implementation of new environmental regulations for sulfur emission in Chile and the United States of America (Arizona).

This technological evolution has been also driven by the increasing demand for high environmental performance standards required by local authorities to copper smelters, aimed to increase sulfur capture from main off-gas streams, secondary emissions from converter aisles, and management of deleterious elements such as arsenic, lead, and halogens elements, among others.

Over the last 20 years, several smelters were required to update their technologies to meet these new standards. Table 5.1 shows a selection of modernization projects executed in the last 20 years.

In terms of technology, Fig. 5.3 compares the melting technology distribution since 1995 (Risopatron, 2020). Flash smelting continues to be the dominant technology; however, bath smelting technologies have grown in acceptance since 1995.

By 2020, there are not many copper smelters left behind to transform their respective operational configurations from old technology to modern technologies and several Chinese smelters are now in operation with a combination of Flash Furnace, TSL furnaces, Bottom Blowing Smelting, Side Blowing Smelting, among others. Table 5.2 lists a comparison between main smelting and converting technologies in 1995 and 2019.

Table 5.1 Selected smelter modernization projects executed between 2000 and 2020 (excludes greenfield projects).

Project	Year	Country	Scope	References
Pirdop Smelter, Aurubis AG	1997–2018	Bulgaria	• Several modernization and capacity increase projects since 1997 • Environmental performance enhancement projects	Homepage of Bulgarian Association of the Metallurgical Industry (2020)
Ilo Smelter Southern Peru Copper Ilo Smelter	2006	Peru	• Flowsheet modification, replacing El Teniente Converter and Rev. furnaces by ISASMELT™ technology and two rotary holding furnaces • New acid plant • Anode and casting system	Homepage Fluor Corporation (2020)
Mopani Copper Smelter	2006	Zambia	• Flowsheet modification, replacing electric furnace by ISASMELT™ • Acid plant • Converters and anode furnace	Homepage ISASMELT (2020)
Expansion and Modernization Project, Mexicana de Cobre La Caridad Smelter	2006–08	Mexico	• Gas handling project for El Teniente and Pierce-Smith Converters • Pneumatic concentrate injection project. • Dust leaching project.	Homepage Southern Peru Copper Corporation (2007)

Table 5.1 Selected smelter modernization projects executed between 2000 and 2020 (excludes greenfield projects).—cont'd

Project	Year	Country	Scope	References
Karabash Smelter	2006	Russia	• Flowsheet modernization using Metso-Outotec-Ausmelt® technology	Homepage Yokogawa (2020)
Kazzinc, Ust Kamenogorsk Metallurgical Complex	2010	Kazakhstan	• New copper smelter for complex materials using ISASMELT™ technology, Peirce–Smith Converters, Anode Furnace and IsaKidd technology. • Integrated with Pb and Zn plant	Homepage Kazzinc (2020)
Daye Nonferrous Metals Co.	2010	China	• Metso-Outotec-Ausmelt® furnace • Electric furnace • New double contact acid plant	ICSG Data base (2017) Wan et al. (2013)
Sarcheshmeh Smelter complex, National Iranian Copper Industries Co.	2010–14	Iran	• Flowsheet modernization using Flash smelting technology	Homepage Mobina (2020)
Tsumeb Smelter	2012–18	Namibia	• Several environmental upgrades	Homepage SLG Global Environmental Solutions (2017)
Pasar Copper Smelter	2015	Philippines	• Flash furnace upgrade, increasing plant throughout	Homepage Campbell & Associates (2020)
Zijin Bor Copper	2015	Serbia	• Flowsheet modernization	

Continued

Table 5.1 Selected smelter modernization projects executed between 2000 and 2020 (excludes greenfield projects).—cont'd

Project	Year	Country	Scope	References
			replacing old reverberatory furnace by a Flash furnace	Homepage SeenNews (2020)
Asarco Hayden Smelter	2016	USA	• Replacement of 5 small Pierce–Smith Converters for three larger ones • Hooding and off-gas capture improvements	Homepage Gas Cleaning Technologies (2020)
Freeport Miami Smelter	2016	USA	• Environmental upgrade to capture converter aisle fugitive emissions	Homepage Arizona Department of Environmental Quality (2020)
Glogow I, KGHM	2013–17	Poland	• Replacement of Blast furnace for Outotec Direct to Bister	ICSG Data base (2017)
Altonorte	2017	Chile	• Environmental upgrades to meet new local regulations	(Ramírez, 2019)
Chagres, AngloAmerican	2017	Chile	• Environmental upgrades to meet new local regulations	(Ramírez, 2019)
Las Ventanas Smelter, Codelco	2017	Chile	• Environmental upgrades to meet new local regulations	(Ramírez, 2019)
Caletones Smelter, Codelco	2019	Chile	• Environmental upgrades to meet new local regulations	(Ramírez, 2019)
Potrerillos Smelter, Codelco	2019	Chile	• Environmental upgrades to meet new local regulations	(Ramírez, 2019)
Chuquicamata Smelter, Codelco	2019	Chile	• Environmental upgrades to meet new local regulations	(Ramírez, 2019)
HVL Paipote Smelter, Enami	2019	Chile	• Environmental upgrades to meet new local regulations	(Ramírez, 2019)

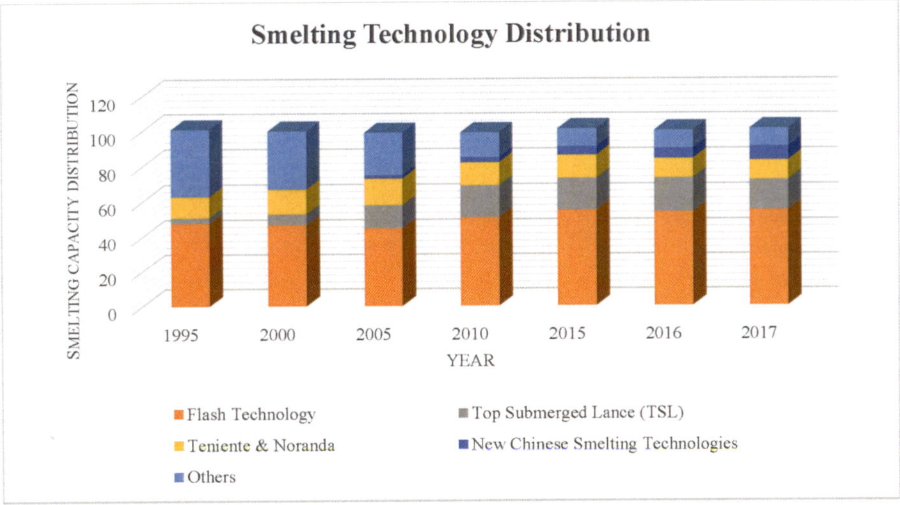

Figure 5.3 Smelting technologies distribution since 1995. TSL defined as ISASMELT and Metso-Outoted-Ausmelt®; New Chinese Technologies such as side blowing, bottom blowing smelting; and others such as electric furnace, Mitsubishi (Risopatron, 2020).

5.3 Copper making technology classification

For purpose of proper metallurgical and engineering design understanding, copper making technologies are classified as follows:

- Flash smelting and converting technologies
- Bath smelting and converting technologies
- Continuous copper making technologies

Table 5.3 describes main features of copper making technologies and current applications for each category.

For flash and cyclone smelting technologies, the main principle is the ignition of dried feed with enriched air or pure oxygen at the outlet of a burner, followed by the free fall and combustion of the particles into a settler to allow reduction of overoxidized particles and separation of phase. Fundamental principles and detailed technology discussions for the Flash and Cyclone smelting and converting technologies are presented in Chapter 6.

For bath smelting and converting technologies, the main principle is the reaction between feed (concentrate or matte) and oxygen (air or oxygen-enriched air) into the copper matte/blister/slag molten phases, depending on the process and technology. The main chemical reactions and more details discussions are presented for the following bath smelting technologies in Chapter 7:

- Noranda, El Teniente, and Vanyukov Technologies.
- Bottom Blowing and Side Blowing Technologies.
- Top Submerged Lance technology (ISASMELT,™ Metso Outotec Ausmelt®Met).

Table 5.2 Comparison of smelting and converting technologies used in copper smelting and converting between 1983 and 2020.

Smelting processes in 1983	Converting processes in 1983
Proven • Noranda • El Teniente • Oxy-fuel Reverberatory • Electric Furnace • Flash Outotec • Flash Inco Under Development • Foam smelting • Sisrosmelt • QS Process • TBRC • Contop • Kivcet • Sprinkle • Flame Cyclone	Proven • Peirce–Smith Converting • Hoboken Converting Under Development • Continuous converting of high-grade matte • Kennecott Flash Converting Direct smelting • Noranda Process • Outokumpu Direct to Blister Process
Continuous copper making processes in 1983	
• Mitsubishi Process	
Smelting Processes in 2020	**Converting Processes in 2020**
Proven • Noranda • El Teniente • Metso-Outotec Ausmelt® • ISASMELT™ • Vanyukov • Bottom Blowing Smelting • Side Blowing • Submerged Lance Smelting • Metso-Outotec Flash Smelting • Inco Flash Smelting • Cyclone smelting	Proven • Peirce–Smith Converting • Hoboken Converting • Mitsubishi Converting • Metso-Outotec Ausmelt® Converting • Submerged Lance Converting-Refining Under Development • ISACONVERT™ (first plant installed, 2019)
Continuous copper making processes in 2020	
Proven • Metso-Outotec Direct to Blister Process • Mitsubishi Process • Dongying Fangyuan two-stage process	

Table 5.3 Classification and main features of copper smelting and converting, and continuous copper making technologies.

Category	Main principle	Technologies currently applied
Flash Smelting	• Ignition at the discharge of a burner of dried concentrate with enriched air or pure oxygen to allow combustion of the involved material and freefall into a settler to allow separation between slag, matte, and gas phase. • Secondary reduction reactions allow the resulfurization of overoxidized copper into the matte phase. • Process heat is provided by the controlled combustion of the feed.	Metso Outotec Flash Smelting INCO Flash Smelting Cyclone Smelting
Flash Converting	• Ignition at the discharge of a burner of dried copper matte with enriched air or pure oxygen to allow combustion of the involved material and freefall into a settler to allow separation between slag, copper, and gas phase. • Secondary reduction reactions allow the resulfurization of overoxidized copper into the copper phase. • Process heat is provided by the controlled combustion of the feed.	Kennecott-Metso Outotec Flash Converter Cyclone Converter
Bath Smelting	• Oxygen-enriched air/oxygen reacts with the feed in the molten phase (matte, slag). • Feed can be injected dry or wet from the top or side of the furnace. • Process is in general continuously fed and batch tapped.	Noranda El Teniente Bottom Blowing Smelting Ausmelt® TSL ISASMELT™ TSL Vanyukov Side-blowing
Bath Converting	• Oxygen-enriched air/oxygen reacts with the molten or dry matte in the molten phase (matte, slag). • Gradual oxidation of iron and sulfur until achieving defined quality (sulfur in blister). • Continuous process, control to a single sulfur—oxygen potential (operating point) to achieve a defined blister quality (varies depending on used technology).	Noranda Converter ISACONVERT™Metso Outotec Ausmelt® C3 Process Bottom-Blowing Converting Peirce—Smith Hoboken

Continued

Table 5.3 Classification and main features of copper smelting and converting, and continuous copper making technologies.—cont'd

Category	Main principle	Technologies currently applied
Continuous copper making processes	One-stage process: Direct to blister • All applications currently using flash smelting/converting principles. • Ignition at the discharge of a burner of dried copper concentrate with enriched air or pure oxygen to allow combustion of the involved material and freefall into a settler to allow separation between slag, copper, and gas phase. • Secondary reduction reactions allow the resulfurization of overoxidized copper into the copper phase. • Process heat is provided by the controlled combustion of the feed. • Product quality indirectly controlled by content of copper in the slag. • Final product blister copper. • Requires slag reduction to recover oxidized copper. Two-stage process (Dongying Fangyuan) • Based on a cascade of smelting and combined converting–refining processes. • Use combination of a single continuous smelting furnace and two batch converting–refining furnaces. • Converting–refining furnace is filled with matte from smelting furnace to start the gradual oxidation of the batch. • Second converting–refining vessel is filled up, while first one is processing. • Enriched air–Natural gas flows are injected via tuyeres (lances) to achieve required level of oxygen sulfur potential. Flows are controlled to adjust oxygen and sulfur potential. • Final product anode copper, Three-stage process • Based on a cascade arrangement of smelting–slag cleaning–converting furnaces	Metso Outotec Direct to Blister Process Dongying Fangyuan Process Mitsubishi process

Table 5.3 Classification and main features of copper smelting and converting, and continuous copper making technologies.—cont'd

Category	Main principle	Technologies currently applied
	• Molten phases are kept under constant volume and compositions in each furnace • Feeding materials injected into the matte phase in the smelting furnace which kept a thin layer of slag within the furnace • Copper matte flows into the converting furnace reacting with copper oxide from the slag to produce blister • Enriched air is added to the converting furnace to allow further oxidation and regeneration of copper oxide as required.	

Continuous or direct production of copper has been also focusing of development in the copper making industry. The following applications are based on the number of stages required to produce copper:

Single stage direct to blister production: Based in flash smelting/converting principles. Dry feed is ignited with oxygen-enriched air at the outlet of a burner, reduction of overoxidized materials in a settler, and subsequent separation of gas and molten phases (blister copper from slag). Slag is further reduced in an electric furnace allowing recovery of copper. Copper is further refined in anode furnaces. The principles and applications of this technology are discussed in Chapter 9.

Two stages anode copper production: Newly developed technology by Dongying Fangyuan in China. This combines continuous smelting of copper concentrate into high matte grade in the smelting reactor and continuous matte tapping into a converting—refining reactor. Slag from the smelting reactor is batch tapped. The converting—refining process is gradually done in a single reactor by managing the oxygen—sulfur potential with addition of air—oxygen—natural gas, as required. The process is continuous from the chemical reaction point of view but batch from the tapping point of view. Oxygen-enriched air is added via bottom tuyeres (lances) into the reactors. Two converting—refining reactors give the required buffer capacity required by the smelting reactor. Copper is cast into anodes and sent to the electrorefining process.

Three stages blister copper production: The Mitsubishi process is a cascade arrangement of three reactors for smelting, slag cleaning, and converting (Goto et al., 1998). Feeding materials are injected into the matte phase in the smelting furnace which kept a thin layer of slag within the furnace. Matte and slag flow into the slag cleaning furnace where matte and slag are separated. Copper matte flows into the converting furnace where is further oxidized into blister copper. Oxygen-enriched air is added by lances into the molten matte (smelting) and copper (converting furnace). Further details of the Mitsubishi process are discussed in Chapter 9. In addition, Chinese technologists have developed a three-stage copper making process using bottom blowing technology as core principle (Yongcheng et al., 2019).

5.4 Evolution to large-scale smelting

Fig. 5.4 shows the evolution of the instantaneous smelting rate for the Flash smelting, Noranda-Altonorte-El Teniente, TSL ISASMELT and Ausmelt-Outotec, and the newly adopted Chinese Dongying Fangyuan smelting process (Alvear et al., 2010; Cui et al., 2016; Hunt & Vries, 2019; Mackey, 2014). The steep curves, in the last decade, are

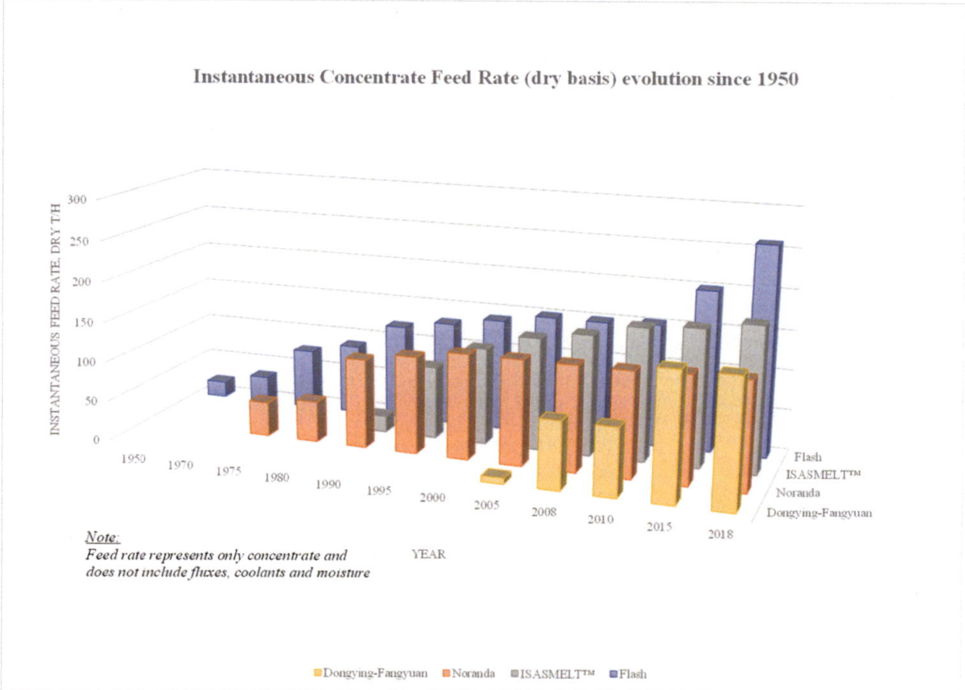

Figure 5.4 Instantaneous concentrate smelting rate comparison between Flash smelting, Noranda/EL Teniente/Alto Norte, Dongying Fangyuan SLS reactor, and TSL (ISASMELT and Outotec-Ausmelt). Data collected trom the following sources: *(Alvear, F., Gerardo, R. F., Arthur, P. S., & Partington, P. J. (2010). Feasibility to profitability with copper ISASMELT™. In:* Copper 2010, volume 2, pyrometallurgy, GDMB, Clausthal-Zellerfeld, Germany, *615–630; Cui, Z., Wang, Z., Wang H., & Wei, C. (2016). Two-step copper smelting process at dongying fangyuan. In: Copper 2016 pyrometallurgy, MMIJ, Paper PY16-4; Mackey, P.J. (2014). Evolution of the large copper smelter — 1800s to 2013. In: P. J. Mackey, E. J. Grimsey, R. T. Jones, G.A. Brooks (Eds.),* Celebrating the megascale, *Springer, Cham. https://doi.org/10.1007/978-3-319-48234-7_2.)*

demonstrations of the efforts made to reduce operating cost, increase productivity, improve environmental performance, and remain competitive in a low margin business.

Since the beginning of the 21st century, the copper smelting technology business has witnessed the consolidation of large-scale smelting. As shown in Fig. 5.1, the joint proportion of smelters between 900–1500 and over 1500 kt/y increased from a combined total of 29%–70% from year 2000 to year 2017, with most of the western plants situated in the range of 900–1500 kt/y. Table 5.4 shows a selected list of large-scale

Table 5.4 Selected large-scale smelter plants (excluding China), using a single smelting reactor (the list excludes smelters where the plant capacity is achieved with more than one smelting reactor) (ICSG Data base, 2017; Kapusta & Watt, 2019).

Technology Configuration	Example	Capacity kt Cu/y (2003)	Capacity kt Cu/y (2019)
Flash smelting Peirce–Smith converting	• Saganoseki Smelter, Japan • Aurubis Hamburg, Germany • Atlantic Copper, Huelva, Spain • Aurubis Pirdop, Bulgaria • Toyo Smelter, Japan	473 450 320 210 260	450 416 320 360 450
Flash smelting Flash converting	• Garfield Kennecott Smelter, Salt Lake City, USA	272	320
Direct to blister	• Glogow II, Poland • Nchanga, Zambia • Olympic Dam	150 — 235	310 310 250
ISASMELT Peirce–Smith converting	• Ilo SPCC, Ilo, Peru • Tuticorin Smelter, Sterlite Industries, India • Kansanshi Smelter, Zambia • Mount Isa Copper Smelter, Australia	316 180 – 232	360 400 300 300
Mitsubishi Continuous Process	• Naoshima Smelter, Japan • PT Gresik Smelter	222 246	342 300
Reverberatory-Mitsubishi	• Onahama Smelter, Japan	221	354
Altonorte reactor	• Altonorte Smelter, La Negra, Chile	290	310

western smelting plants with smelting capacities over 300 ktpa achieved with one smelting reactor (Concentrate throughput equivalent to over 1.2 mtpa) (ICSG Data base, 2017).

Most of the plants listed in Table 5.4 were upgraded and have increased their capacity since year 2000 and their instantaneous smelting throughput.

High-intensity smelting has been facilitated by technical developments aimed to enhance the performance of plants in aspects such as the following:

- Ultradense phase injection in bath smelting furnaces to increase instantaneous feed rate at higher availabilities and equipment reliability (Coleman and Money, 2008; Coleman, 2016)
- Intensive cooling for reactors to increase furnace campaign and throughput (Konefal et al., 2016; Joubert et al., 2016)
- Improvements in drilling and tapping operations to improve overall plant availability (Musuka et al., 2019)
- Development of more efficient and environmentally friendly rotary dryers (Tuominen & Berg, 2019)
- Conveying and feeding systems (Firky et al., 2019; Tajla et al., 2019)
- Improvements in flash smelter burners to reduce dust generation and improve combustion quality (Chen et al., 2016)
- Further development in boiler performance by using special coatings to increase tube life (Martinez, 2013)
- On-line monitoring systems for refractory thickness (Sadri et al., 2016; Vukovic et al., 2016)
- Furnace integrity and Refractory material improvements to increase furnace campaign length (Degel et al., 2016; Degel, Joubert, Filwiezer, Filwiezer, & Hanel, 2019; Wallner et al., 2016)
- Advanced process controls strategies based on plant-wide controls systems and digitalization (Jansson et al., 2016; Dyussekenov et al., 2019)
- Operational diagnostics, On-line process control to improve furnace campaign with online monitoring (Walters et al., 2016)
- Applications of on-line sensors to improve system reliability (Stokreef et al., 2019)
- Use of data science to process valuable information (Jansson et al., 2016)
- Proactive maintenance (Moreno-Leiva, Haas, Junne, & Kracht, 2019)
- Health and safety strategies to improve operational standards (Degel et al., 2019; Gregorczyk & Gobka, 2019; Kapobe et al., 2019; Thakurdin & Gonzales, 2019).

The above-mentioned aspects have been widely reported, mainly in the Copper Conference Series organized by TMS, SME, METSOC, IIMCH, MMIJ, and GDMB, in 2010 (Hamburg), 2013 (Santiago), Kobe (2016), and 2019 (Vancouver).

5.5 Chinese technology developments since 2000

China has become the largest smelting country in the world; as in the last decade, an increasing effort has been done by Chinese technologists to adapt existing processes and/or create new ones to meet their country needs. Table 5.5 shows a selected list of brownfield and greenfield projects implemented in China in the last 15 years using with different technology configurations (Tang, 2018; ISCG, 2019; Tang, 2019; Zhang, 2018). Flash technologies are still the most relevant technology package selected for large-scale smelting. However, for intermediate size smelting, bath smelting technologies seem to be predominant.

The main Chinese technology innovations in the last 20 years have been focused on the following aspects:

- Development and implementation of the cyclone smelting technology
- Development and implementation of the bottom blowing smelting and converting technology
- Dongying Fangyuan two-stage copper making process and
- Development and implementation of the side blowing smelting (Wang, 2019)

In addition to their own developments, Chinese technologists have also developed large-scale smelting using Flash smelting and Converting and Top Submerged Lance technologies (ISASMELT and Ausmelt-Outotec).

Bing Tan and Supin Yao (Tang, 2019) described in detail the main copper pyrometallurgical technological developments in China (2019). They highlight the following aspects:

- **Cyclone smelting technology**: Chinese improvements of the original Flash Smelting technology, developing a new flash burner that allowed a reduction in dust generation and copper losses with an increasing availability
- **Bottom blowing smelting (BBS):** Based on the QSL and Noranda-El Teniente processes, this innovation injects high pressure—enriched air into the bath, allowing the production of high copper matte grade at a relatively low copper content when compared with Noranda and Teniente technologies (Baojun reference)
- **Continuous Copper Smelting and Converting (BBRs-BBRc):** Aimed to small to medium-size smelters, this technology applies the BBS principle to the converting of copper matte by continuous transferring of copper matte into the converting reactor
- **Two-stage copper smelting process (SLS and SLCR):** With its origins on the SKS (BBS) process, the two-stage smelting process combines the smelting of copper concentrate in the SLS furnace with the converting and refining to produce anode quality copper in the SLCR furnace.

- **Side blowing smelting technology:** Originally developed to improve performance of reverberatory, blast, and Vanyukov furnaces. The newest version corresponds to the Jifeng side blowing furnace technology. Oxygen-enriched air is blown into the slag close to matte layer for smelting, and tuyeres on top of the furnace are used to supply secondary air.
- **Side-blown Smelting + Top-blown multilance continuous converting technology:** Matte flows continuously from the side blowing smelting furnace into the converting furnaces where top blowing lances provide the required oxygen for the converting process.

Selected process parameters applied in Chinese smelting and converting technologies are shown in Table 5.6. A more detailed descriptions of each of these technologies are presented in the respective bath smelting, flash smelting, and converting chapters.

Table 5.5 Selected smelting process configurations used in China for brownfield and greenfield projects in the last 15 years (ICSG, 2017; Tang, 2019; Zhang, 2018).

Technology configuration	Example	Capacity kt Cu/y
Flash smelting – Peirce–Smith converting	• Guixi Copper Smelter • Jinlong Copper Smelter • Baiyin Nonferrous Group	490 305 200
Flash smelting – Flash converting	• Jinguan Copper Smelter • Fangchenggang Copper Smelter • CHALCO Southeast Copper	400 400 400 400
Cyclone smelting – Cyclone converting	• Xiangguang Copper Smelter	400
ISASMELT – Peirce–Smith converting	• Yunnan Copper	200
Ausmelt-Outotec smelting – Peirce–Smith converting	• Daye Copper Smelter	300
Ausmelt-Outotec smelting –converting	• Yuanqu Smelter of Zhongtiaoshan Group • Yun Tin Group Smelter	60 100
Side blow Baiyin smelting – Peirce –Smith converting	• Baiyin Nonferrous Group	200
Jinfeng side blowing smelting – Peirce–Smith converting	• Zhejiang Heding Copper Smelter	300
Jinfeng side blowing smelting – Continuous top blowing converting	• Nanguo Copper Smelter • Yunnan Copper Chifeng Smelter	300 400
Bottom Blowing (Smelting –Converting)	• Lingbao Copper Smelter	120
Submerged Lance Smelting (SLS) – Submerged Lance Converting and Refining (SLCR)	• Dongying Fangyuan Copper Smelter	300
Bottom blown smelting – Flash converting	• Sinogold Copper Smelter • Zhongyuan Smelter	300 400

Table 5.6 Selected process parameters applied in Chinese smelting and converting technologies (Guo, et al., 2016; Tang, 2019; Wang, 2015; Yongcheng et al., 2019).

	Technology Configuration					
Parameter	**FSF + PSC**	**CS + CC**	**TSL + PSC**	**SB + CTBC**	**SKS + BCC**	**SLS + SLCR**
Typical smelter	**Guixi II**	**Yanggu**	**Daye**	**Chifeng**	**SDIC Jincheng**	**Dongying**
Concentrate throughput, t/a	1,250,000	1,660,000	1,410,000	680,000	680,000	1,510,000
Cu in feed, %	23–25	23–25	23–24	18–22	15–20	24–25
Feed rate, tph	172	260	200	83	110–130	210
Oxygen enrichment	80–85	80–90	60–65	85	70–73	70–75
Matte grade, %Cu	60	70	57	75	70–72	72–74
Cu in smelting slag, %Cu	1.5	0.9–1.2	0.8–1.0	2.0	1.5–2.5	2.8

5.6 Summary

Pyrometallurgical processing of copper concentrates to produce pyrorefined copper is based on the continuous management of the sulfur and oxygen potentials. The oxygen–sulfur potential management is done in sequential stages to control heat released from the chemical reactions and the overall copper recovery. This principle has been the corner stone applied by technologists to conceptualize and develop pyrometallurgical copper making technologies.

Developments over the last 40 years have been focused on increase smelting capacity in single reactors increasing from the range of 80 tph to over 200 tph. This increase has been supported by technology developments in several aspects such as reactor design, sulfur capture, peripherical equipment, plant reliability, and process control.

Since year 2000, the copper business has witnessed an explosive demand for copper consumption, a global shift of the smelting business to China and the development and commissioning of mega smelters across the globe with productions to the level of 400,000 tpy of copper.

The last 10 years have been also crucial with the incorporation of innovative smelting technologies in China.

References

Alvear, F., Gerardo, R. F., Arthur, P. S., & Partington, P. J. (2010). Feasibility to profitability with copper ISASMELT™. In *Copper 2010, volume 2, pyrometallurgy, GDMB, Clausthal-Zellerfeld, Germany* (pp. 615–630).

Arthur, P. S., & Simon, S. P. (2005). ISASMELT^{TM} — 25 years of continuous evolution (2005). In M. Nilmani, & W. J. Rankin (Eds.), *Sustainable developments in metals processing: Proceedings of the John Floyd international symposium, Melbourne, vic., July 3—6* (pp. 73—94).

Ayhan, M. (2001). *The new technology for HK processing of copper* (pp. 197—207). Paper Presented at the GDMB Copper Group.

Bhappu, R. R., Larson, K. H., & Tunis, R. D. (1993). Cyprus miami mining corporation smelter modernization project summary and status. In G. Warren (Ed.), *EPD Congress 1994* (pp. 555—570). Ihe Minerals, Metals & Materials Society.

Campos, R., & Torres, L. (1993). Caletones smelter two decades of technological improvements. In C. A. Landolt (Ed.), *The Paul E. Queneau international symposium extractive metallurgy of copper, Nickel and Cobalt volume II: Copper and Nickel smelter operations* (pp. 1441—1460). Warrendale, PA: TMS.

Chen, S., Ranasunghe, J., West, R. C., White, M., Jastrzebski, M., & Taylor, W. (2016). Sustainable improvement of concentrate burner efficiency in the flash smelting furnace at olympic dam. In *Copper 2016 pyrometallurgy*. Kobe: MMIJ. Paper PY4-5.

Coleman, M. (2016). Solutions for pyrometallurgical process plants for copper 2016, the ninth international copper conference. In *Copper 2016 pyrometallurgy*. Kobe: MMIJ. Paper PY4-5.

Coleman, M., & Money, G. (2008). Increasing capacity and productivity in the metals market through pneumatic conveying and process injection technologies. In *Lead and zinc 2008* (pp. 1—16). The South African Institute of Mining and Metallurgy.

Coursol, P., Mackey, P. J., Kapusta, J. P. T., et al. (2015). Energy consumption in copper smelting: A new Asian horse in the race. *Journal of the Minerals Metals & Materials Society, 67*, 1066—1074. https://doi.org/10.1007/s11837-015-1380-1

Cui, Z., Wang, Z., Wang, H., & Wei, C. (2016). Two-step copper smelting process at dongying fangyuan. In *Copper 2016 pyrometallurgy*. MMIJ. Paper PY16-4.

Czernecki, J. (1999). Copper metallurgy at the KGHM polska miedz S.A. - present state and perspectives. In D. B. George, W. J. Pete) Chen, P. J. Mackey, & A. J. Weddick (Eds.), *Copper 99 - Cobre 99 international conference, volume V smelting operations and advances* (pp. 189—203). The Minerals, Metals & Materials Society.

Degel, R., Filwiezer, A., Hanel, M., Krassnig, H. J., Lux, T., & Bergs, A. (2016). ILTEC technology — new pathways towards safe and effective cooling. In *E proceedings copper 2016 pyrometallurgy*. Kobe: MMIJ. Paper PY11.2.

Degel, R., Joubert, H., Filwiezer, A., Filwiezer, I., & Hanel, M. B. (2019). Innovative solutions in copper production lines. In *E-proceedings of copper 2019 pyrometallurgy*. Paper 599293. ISBN978-1-926872-44-5.

Dyussekenov, N., Rajith, J., Hanschar, L., Mascrenhas, P., & deVries, D. (2019). Peirce-smith converters: automation, design, operation, and environmental improvements at Kansanshi smelter. In *E-proceedings of copper 2019 pyrometallurgy*, ISBN 978-1-926872-44-5. Paper 593106.

Firky, A. M., Sundana, D., Kununose, K., & Unno, Y. (2019). Modification of feeding system for the smelting furnace and other recent improvements at Gresik copper smelter. In *E-proceedings of copper 2019 pyrometallurgy*, ISBN 978-1-926872-44-5. Paper 594775.

Floyd, J. M. (2005). Converting an idea into a worldwide business commercializing smelting technology. *Metallurgical and Materials Transactions B, 36*, 557—575. https://doi.org/10.1007/s11663-005-0047-7

Fyodorov, A. N., Komkov, A. A., & Sorokin, M. L. (October 1992). The use of the Vanyukov process for the smelting of various charges. In Kalgoorlie Ausimm (Ed.), *Proceedings of extractive metallurgy of gold and base metals* (pp. 26—28).

Goto, M., Oshima, E., & Hayashi, M. (1998). Control aspects of the mitsubishi continuous process. *Journal of the Minerals Metals & Materials Society, 50*, 60—64. https://doi.org/10.1007/s11837-998-0271-0

Gregorczyk, R., & Gabka, T. (2019). Improvements of elements of safety culture in KGHM Glogow copper smelter. In *E-proceedings copper 2019 pyrometallurgy*. Vancouver: CIMM, ISBN 978-1-926872-44-5. Paper 591052.

Guo, X., Zhang, Y., Wang, Q., & Yuan, Z. (2016). Advanced copper smelting technologies used to quadruple china copper production between 2000 and 2015. In *copper 2016 pyrometallurgy*. Kobe: MMIJ. Paper PY1-2.

Hirai, Y. (1999). Recent operation of the flash smelting furnace at Saganoseki smelter. In D. B. George, (Pete), W. J. Chen, P. J. Mackey, & A. J. Weddick (Eds.), *Copper 99 - Cobre 99 international conference, volume V smelting operations and advances* (pp. 671–684). Warrendale: TMS.

Homepage Arizona Department of Environmental Quality.(2020). https://azdeq.gov/miami-sulfur-dioxide-so2-nonattainment-area. (Accessed 11 August 2020).

Holmström, A., Andersson, M., Carrasco, S., & Deutsch, D. (2013). Development of Partial Roasting Technology for Arsenic Containing Copper Concentrates. In Copper 2013 Pyrometallurgy, Volume III, Nickolas Themelis Symposium on Pyrometallurgy and Process Engineering (pp. 25–38). Santiago: IIMCH.

Homepage of BAMI. (2020). *Bulgarian association of the metallurgical industry.* https://bami.bg/media/docs/20180227%20_aurubis%20bulgaria_general%20presentation.pdf. (Accessed 11 August 2020).

Homepage of Fluor Corporation. (2020). *Ilo smelter modernization (2020).* https://www.fluor.com/projects/ilo-smelter-modernization-peru. (Accessed 11 August 2020).

Homepage of gas cleaning and energy solutions website.(2020). https://gcteng.com/gas-cleaning/. (Accessed 11 August 2020).

Homepage of ISASMELT.(2020). https://www.isasmelt.com/en/download/TechnicalPapersIsasmelt/XTpaper_Mopani_MSUPyromet05.pdf. (Accessed 11 August 2020).

Homepage of Kazzinc.(2020). http://www.kazzinc.com/en/Ust-Kamenogorsk_Metallurgical_Complex. (Accessed 11 August 2020).

Homepage of Mobina. (2020). *Sarcheshmesh smelter project.* http://www.miteceng.com/SarcheshmehComplexFlashSmelterProject.html. (Accessed 11 August 2020).

Homepage of Seenews.(2020). https://seenews.com/news/serbias-rtb-bor-to-fire-up-new-flash-smelting-furnace-on-march-5-466302. (Accessed 11 August 2020).

Homepage of SLR, global environmental solutions. (2017). *Consolidated environmental management plan for the Sumeb smelter.* https://s21.q4cdn.com/589145389/files/doc_documents/EN/Appendix-K-Consolidated-EMP.pdf. (Accessed 11 August 2020).

Homepage of Southern Peru Copper Corporation. (2007). *Annual report.* http://www.southernperu.com/ENG/invrel/2007/AnnualReport/m2007i.pdf. (Accessed 11 August 2020).

Homepage of Yokogawa Electric Corporation.(2020). https://www.yokogawa.com/library/resources/references/centum-cs-3000-for-russian-copper-smelter/. (Accessed 11 August 2020).

Hompage I.A. Cambell Associates.(2020). https://campbell.com.ph/project/pasar-plant-expansion/. (Accessed 11 August 2020).

Hunt, S., & de Vries, D. (2019). Kansanshi copper smelter: The first four years of operation. In *E-proceedings of the 58th annual conference of metallurgists (COM) hosting the 10th international copper conference 2019*, ISBN 978-1-926872-44-5. Paper 593109.

International Copper Study Group. (2017). *Smelter data base.* Portugal: Lisbon. https://www.icsg.org/.php/component/jdownloads/finish/166-other/1575-icsg-directory-of-mines-and-plants-2011-03?Itemid=0.

Jansson, B. P., Lahtinen, M., Heinone, H., Jafs, M., & Fagerlund, K. (2016). The future of smelting is digital – outotec solutions. In *Copper 2016 pyrometallurgy.* Kobe: MMIJ. Paper PY11.2.

Joubert, H., Nikolic, S., Baker, M. L., & McDougall, I. (2016). Tenova pyromet-cooled copper furnace elements. In *Copper 2016 pyrometallurgy.* Kobe: MMIJ. Paper PY14.2.

Kapole, J., Lemmer, A., Milupi, M., Phiri, R., & Mazala, C. (2019). Continuous training – insurance for sustainable operations. In *E-proceedings copper 2019 pyrometallurgy.* CIMM, Vancouver, ISBN 978-1-926872-44-5. Paper 593054.

Kapusta, J., & Watt, L. (2019). The 2019 copper smelting survey. In *e-proceedings copper 2019 pyrometallurgy.* Vancouver: CIMM, ISBN -978-1-926872-44-5. Paper 592732.

Kojo, I., Lahtinen, M., & Miettinen, E. (2009). Flash converting - Sustainable technology now and in the future. In Joel Kapusta, & Tony Warner (Eds.), *International Peirce-Smith Converting Centennial.* TMS, 2009.

Konefal, R., Gawlowski, T., Gabka, T., & Rozycki, A. (2016). Development process of inhouse copper cooling blocks manufacturing for the purposes of flash smelting and electric furnaces based on 30+ year history of the casting plant. In *Copper 2016 pyrometallurgy.* Kobe: MMIJ. Paper PY4-4.

Mackey, P. J. (2014). Evolution of the large copper smelter — 1800s to 2013. In P. J. Mackey, E. J. Grimsey, R. T. Jones, & G. A. Brooks (Eds.), *Celebrating the megascale*. Cham: Springer. https://doi.org/10.1007/978-3-319-48234-7_2

Mackey, Phillip J., Alvear, F., & Gerardo, R. F. (2014). Developments in copper smelting: A discussion of changes in copper technology and science as seen through the lens of professor davenport's texts. In *Com 2014 - conference of metallurgists proceedings*, ISBN 978-1-926872-24-7.

Marczeski, W. D., & Aldrich, T. L. (1986). *Retrofitting hayden plant to flash smelting*. Warrendale, FA: TMS. Paper A86-65.

Martinez, A. (2013). Use of waste heat boilers to save resources. In *Copper 2013 pyrometallurgy, volume III, Nickolas Themelis symposium on pyrometallurgy and process engineering* (pp. 25—38). Santiago: IIMCH.

Moreno-Leiva, S., Haas, J., Junne, T., & Kracht, W. (2019). Strategic scheduling of preventive maintenance in copper processing as a flexibility option for highly renewable energy systems. In *e-proceedings copper 019 pyrometallurgy*. CIMM, Vancouver, ISBN 978-1-926872-44-5. Paper 595223.

Musuka, F., Jenkins, D., & Keyes, A. (2019). Risk management approach to taphole integrity at olympic dam copper smelter. In *E-proceedings copper 2019 pyrometallurgy*. Vancouver: CIMM, ISBN 978-1-926872-44-5. Paper 593054.

Newman, C. J. (1998). Kennecott Utah copper smelter modernization. In L. A. Asteljoki, & R. L. Stephens (Eds.), *Sulfide smelting '98: Current and future practices* (pp. 205—215). The Minerals, Metals &: Materials Society.

Prévost, Y., Lapointe, R., Levac, C. A., & Beaudoin, D. (1999). First year of operation of the Noranda continuous converter. In D. B. George, (Pete), W. J. Chen, P. J. Mackey, & A. J. Weddick (Eds.), *Proceedings of copper 99 - Cobre 99 international conference, volume V smelting operations and advances* (pp. 269—282). The Minerals, Metals & Materials Society.

Ramírez, J. P. (2019). ¿Cuál es la situación de las fundiciones en Chile?. In *minería chilena*. https://www.mch.cl/reportajes/la-situacion-las-fundiciones-chile/#. (Accessed 11 August 2020).

Risopatron, Carlos (2020). *Global industrial copper use: Re-shoring and raw materials constraints 2020—2023*. Lisbon, Portugal: International Copper Study Group.

Sadri, A., Ying, G. W. L., & Loveless, M. (2016). Monitoring a flash smelting furnace hearth by a non-destructive testing (NDT) method. In *Copper 2016 pyrometallurgy*. Kobe: MMIJ. Paper PY5.1.

Stokreef, A., Lebel, T., Alvear Flores, G. R. F., Kadiyski, M., Loock, P., Barnes, J., & Davis, B. (2019). Monitoring the burner emission spectrum in a commercial flash furnace using a novel optical probe. In *E-proceedings copper 2019 pyrometallurgy*. Vancouver: CIMM, ISBN 978-1-926872-44-5. Paper 598094.

Tajla, J. J., Jahila, R., & Lehtila, E. (2019). Challenges in pneumatic conveying of concentrate and flue dust at copper smelters. In *E-proceedings copper 2019 pyrometallurgy*. Vancouver: CIMM, ISBN 978-1-926872-44-5. Paper 593054.

Tang, B. (2018). *The developments and selection of copper smelting technology in china*. Santiago, Chile: Expomin 2018. https://congreso.expomin.cl/descargas/presentaciones/8_FURE_jueves_26_am/4_Tang_Bin_NERIN_(modulo_I).pdf. (Accessed 5 August 2020).

Tang, B. (2019). Application and developments of modern copper metallurgical technologies in China's copper industry. In *E-proceedings copper 2019 pyrometallurgy*. Vancouver: CIMM, ISBN 978-1-926872-44-5. Paper 588809.

Tarassoff, P. (1984). Process R & D—the Noranda process. *Metallurgical and Materials Transactions B, 15*, 411—432. https://doi.org/10.1007/BF02657372

Thakurding, S., & Gonzales, T. (2019). A structured approach to conducting site risk reviews. In *E-proceedings copper 2019 pyrometallurgy*. Vancouver: CIMM, ISBN 978-1-926872-44-5. Paper 594933.

Tuominen, J., & Berg, C. G. (2019). Maximizing energy efficiency in smelter feed srying. In *E-proceedings copper 2019 pyrometallurgy*. Vancouver: CIMM, ISBN 978-1-926872-44-5. Paper 593054.

Vanbellen, F., & Chintinne, M. (2006). The precious art of metals recycling. In F. Kongoli, & R. G. Reddy (Eds.), *Proceedings of Sohn international symposium, volume 7, industrial practices* (pp. 43—52). TMS (The Minerals, Metals & Materials Society).

Vukovic, G., Zivanovic, B., Drew, B., & Gamweger, K. (2016). Continuous on-line temperature measurement system. In *Copper 2016 pyrometallurgy*. Kobe: MMIJ. Paper PY24-1.

Wallner, S., Ruhs, S., Filwiezer, I., Hanel, B. N., & Filwiezer, I. (2016). Furnace integrity — mettop's holistic approach to face todays challenges. In *Copper 2016 pyrometallurgy*. Kobe: MMIJ. Paper PY11.2.

Walters, G., Cramer, M., Veenstra, R., Gonzales, T., Sterling, G., & Jastrzebski, M. (2016). Countering declining copper grades through operational diagnostics and innovation. In *Copper 2016 pyrometallurgy*. Kobe: MMIJ. Paper PY2.5.

Wang, W. (2019). China nerin's side blow furnace technology and its application. In *E-proceedings copper 2019 pyrometallurgy*. Vancouver: CIMM, ISBN 978-1-926872-44-5. Paper 547428.

Wan, J., Luo, Y., Chen, B., Swayn, G. P., Wood, J. T., & Creedy, S. J. (2013). Design and commissioning of an Outotec® Ausmelt copper smelter for daye non-ferrous metals company ltd. In *Copper 2013 pyrometallurgy, volume III, Nickolas Themelis symposium on pyrometallurgy and process engineering* (pp. 357—365). Santiago: IIMCH.

Yongcheng, Z., Bing, L., & Kai, L. (2019). Engineering and production practice of doible bottom blowing continuous copper smelting process with "three connected furnaces" arrangment. In *E-proceedings copper 2019 pyrometallurgy*. Vancouver: CIMM, ISBN 978-1-926872-44-5. Paper 592917.

Zhang, X. (2018). Investment decisions and implementation of copper smelting projects in China. In *XV Congreso international expomin, Santiago, Chile* (Accessed 6 August 2020) https://congreso.expomin.cl/descargas/presentaciones/8_FURE_jueves_26_am/3_ZHANG_XU_ENFI_(modulo_I).pdf. (Accessed 11 August 2020).

Suggested readings

Alvear Flores, G. R. F., Löbbus, M., Springub, B., Fallah-Mehrjardi, A., & Tappe, A. (2018). Sulfide smelting: Thirty-five years of continuous efforts to find new value adding solutions. In B. Davis, et al. (Eds.), *Extraction 2018* (pp. 39—56). Warrendale, PA: TMS. https://doi.org/10.1007/978-3-319-95022-8_3

Arthur, P. S., & Simon, S. P. (2005). ISASMELT^{TM} — 25 years of continuous evolution. In M. Nilmani, & W. J. Rankin (Eds.), *Sustainable developments in metals processing: Proceedings of the John Floyd international symposium, melbourne, vic., July 3—6, 2005.*

CHAPTER 6

Flash smelting (Davenport et al., 2001)

Flash smelting accounts for around half of all Cu matte smelting. It entails blowing oxygen, air, dried Cu—Fe—S concentrate, silica flux, and recycle materials into a 1250 °C hearth furnace. Once in the hot furnace, the sulfide mineral particles of the concentrate (e.g., $CuFeS_2$) react rapidly with the O_2 of the blast. This results in controlled oxidation of the concentrates Fe and S, a large evolution of heat and melting of the solids.

The process is continuous. When extensive oxygen enrichment of the blast is practiced, it is nearly autothermal. It is perfectly matched to smelting fine particulate concentrates (~100 μm) produced by froth flotation.

The products of flash smelting are the following:

1. molten Cu—Fe—S matte, ~65% Cu, considerably richer in Cu than the input concentrate
2. molten iron—silicate slag containing 1 or 2% Cu
3. hot dust—laden offgas containing 30—70 vol.-% SO_2.

Three flash furnaces produce molten copper directly from concentrate (Chapter 5). Two other Metso Outotec flash furnaces produce molten copper from solidified/ground matte. This is flash converting (Chapter 8).

The goals of flash smelting are to produce the following:

1. molten matte at constant composition and temperature for feeding to converters
2. slag which, when treated for Cu recovery, contains only a tiny fraction of the Cu input to the flash furnace
3. offgas strong enough in SO_2 for its efficient capture as sulfuric acid.

There are two types of flash smelting: the Metso Outotec process (~44 furnaces in operation), and the Inco process (2 furnaces in operation). The Outotec process was formerly known as the Outokumpu process, and may occasionally be referred to in this chapter by the previous name.

6.1 Metso Outotec flash furnace

Fig. 6.1 shows a typical 2010 design Metso Outotec flash furnace. It is 26 m long, 8 m wide, and 2 m high (all dimensions inside the refractories). It has a 7 m diameter, 8 m high reaction shaft and a 5 m diameter, 10 m high off-gas uptake. It has one concentrate burner and is designed to smelt up to 4500 tonnes of concentrate per day. It has four matte tapholes and four slag tapholes.

Outotec flash furnaces vary considerably in size and shape (Tables 6.1 and 6.2). However, they all have the following five main features:

Extractive Metallurgy of Copper, Sixth Edition
ISBN 978-0-12-821875-4, https://doi.org/10.1016/B978-0-12-821875-4.00022-5

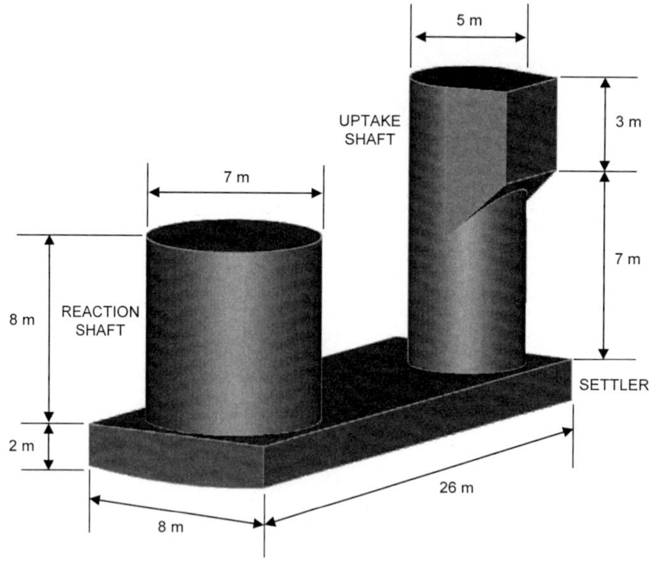

Figure 6.1 Schematic of a year 2010 Metso Outotec flash furnace. This furnace is designed to smelt up to 4500 tonnes of concentrate per day. A concentrate burner is shown in Fig. 6.2. It sits atop the reaction shaft.

1. concentrate burners (usually one, but a few have four) which combine dry particulate feed with O_2-bearing blast and blow them downward into the furnace
2. a reaction shaft where most of the reaction between O_2 and Cu—Fe—S feed particles takes place
3. a settler where molten matte and slag droplets collect and form separate layers
4. water-cooled copper block tapholes for removing molten matte and slag
5. an uptake for removing hot SO_2-bearing offgas.

6.1.1 Construction details (Fagerlund et al., 2010)

The interior of a Metso Outotec flash furnace consists of high-purity direct-bonded magnesia—chrome bricks. The bricks are backed by water-cooled copper cooling jackets on the walls and by sheet steel elsewhere. Reaction shaft and uptake refractory is backed by water-cooled copper cooling jackets or by sheet steel, cooled with water on the outside.

The furnace rests on a 2-cm thick steel plate on steel-reinforced concrete pillars. The bottom of the furnace is air cooled by forced convection. Much of the furnace structure is in operating condition after 11 years of use (Janney et al., 2007). Slag line bricks may have eroded, but the furnace can usually continue to operate without them. This is because magnetite-rich slag deposits on cool regions of the furnace walls.

Table 6.1 Dimensions and production details of four Metso Outotec flash furnaces.

Smelter	PASAR Leyte, Philippines	Aurubis Hamburg, Germany	Anglo American Chagres, Chile	JX metals smelting Saganoseki, Japan
Startup date	**1983**	**1972**		**1973**
Size, inside brick, m				
Hearth: w × l × h	7.5 × 20.2 × 2.6	6 × 20 × 3	5.8 × 17.9 × 1.9	8.0 × 21.7 × 2.2
Reaction shaft				
Diameter	6.5	6	4.5	6.2
Height above settler roof	6.5	7.5	5.3	5.9
Gas uptake				
Diameter	2.5	4 × 8	2.7	Rectangular cross-section opening size w3 × h4
Height above roof	9.4	10		6.3
Slag layer thickness	0.3–0.6	0.7–0.9	0.4	0.5
Matte layer thickness	0.65–0.85	0.2–0.5	0.6	0.7
Active slag tapholes	2	2	2	2
Active matte tapholes	4	4	5	6
Concentrate burners	1	1	1	1
Feed details tonnes/day				
New concentrate (dry)	2000 (23%–29% Cu)	3300 (29% Cu)	2710	4000–4400
Silica flux	100	350–450	20	450–800
Oxygen	440		340	
Recycle flash furnace dust	95–120	270	80	180–250
Converter dust	20–30	15	No	
Slag concentrate	75	No	No	
Reverts		150 (ladle sculls, slimes, various dusts)	30	
Other	Coal 36 t/d	430 molten converter slag	-	

Continued

Table 6.1 Dimensions and production details of four Metso Outotec flash furnaces.—cont'd

Smelter	PASAR Leyte, Philippines	Aurubis Hamburg, Germany	Anglo American Chagres, Chile	JX metals smelting Saganoseki, Japan
Startup date	**1983**	**1972**		**1973**
Blast details				
Temperature, °C	200	Ambient	30	Ambient
Volume% O_2	45–52	50–65	55	70–80
Flowrate, 1000 Nm^3/h	36	45	11.5	35–45
Production details				
Matte, tonnes/day	1080 (54–56%Cu)	1500 (63% Cu)	630	1650 (63% Cu)
Slag, tonnes/day	800–900	2000	800	2700
Slag, mass% Cu	0.5–0.75	1.5	1.5	0.8
Slag mass% SiO_2/mass% Fe	0.87–1	0.82		0.96
Cu recovery, flash slag	Electrodes and coal in flash furnace	Electric furnace		Electric furnace
Cu recovery, converter slag	Solidify/flotation	Recycle to flash furnace		Solidify/flotation
Offgas, 1000 kNm^3/h	36.6	50–68	35	35–45
Vol. % SO_2, leaving furnace	28–30	30–45	40	50–60 (calculated)
Dust production, tonnes/day	48–60 flash furnace 28–30 dryer dust	270	80	250–400
Matte/slag/offgas temperature,°C	1220/1240/750	1220/1250/1350	1280/1300/1350	1260/1260/1260
Fuel inputs, kg/h				
Hydrocarbon fuel burnt In reaction shaft	0–0.2 m^3/h bunker C oil, coal added to solid feed	None	None	None
Hydrocarbon fuel In settler burners	None, but 1.2–1.5 MW electric power	None	400	None

Table 6.2 Dimensions and production details of three Metso Outotec flash furnaces.

Smelter	Grupo Mexico La Caridad Nacozari, Mexico	Sumitomo Toyo Toyo, Japan	Mitsui Tamaño Smelter Tamano, Japan
Startup date	**1986 (upgraded 2013)**	**1971**	**1972**
Size, inside brick, m			
Hearth: w × l × h	7.1 × 24.6 × 2.2	6.7 × 19.9 × 3.3	7.0 × 19.75 × 2.5
Reaction shaft			
Diameter	6.0	6.1	6
Height above settler roof	6.3	6.4	6.9
Gas uptake			
Diameter	5.1	4	2.5 × 4.4
Height above roof	10.7	7.5	8.4
layer thickness	0.2–0.55	0.2–0.4	0.4
Matte layer thickness	0.2–0.44	0.5–0.7	0.8
Active slag tapholes	2	2	2
Active matte tapholes	7	4	4
Concentrate burners	1	1	1
Feed details tonnes/day			
New concentrate (dry)	2040 (20%–25% Cu)	3830 (26.3% Cu)	2204
Silica flux	273	691	323
Oxygen	486	1037	
Recycle flash furnace dust	120–170	151	174
Converter dust	0	51	28
Slag concentrate	0	0	94
Reverts	85 (ladle sculls)	14	131
Other	20 matte from lead refining 12 Ag–Pb concentrate 6 dust and effluent plant residue	106	19

Continued

Table 6.2 Dimensions and production details of three Metso Outotec flash furnaces.—cont'd

Smelter	**Grupo Mexico La Caridad Nacozari, Mexico**	**Sumitomo Toyo Toyo, Japan**	**Mitsui Tamaño Smelter Tamano, Japan**
Startup date	**1986 (upgraded 2013)**	**1971**	**1972**
Blast details			
Temperature, °C	200–220	215	145
Volume% O_2	35–45	84	80
Flowrate, 1000 Nm^3/h	35–45	38	27.1
Production details			
Matte, tonnes/day	650 -700 (63%–65% Cu)	1753 (62% Cu)	1013
Slag, tonnes/day	1470	2973	1395
Slag, mass% Cu	1.5–2.5	0.86	0.7
Slag mass% SiO_2/mass% Fe	0.78	1.13	0.87
Cu recovery, flash slag	Electric furnace	Electric furnace with coal	Electric furnace
Cu recovery, converter slag	Recycled to electric furnace	Solidify/mill plant/flash furnace	Flotation/solidification
Offgas, 1000 Nm^3/h	40–85	37	60–70
Vol. % SO_2, leaving furnace	32–35.6	67	20–25
Dust production, tonnes/day	120–170	151	174
Matte/slag/off-gas temperature,°C	1235/1325/1340	1230/1255/1300	1233/1275/1200
Fuel inputs, kg/h			
Hydrocarbon fuel burnt In reaction shaft	60 kg/h natural gas	439 oil	500–600 kg/h fine coal with feed
Hydrocarbon fuel In settler burners	998 kg/h natural gas	-	None

6.1.2 Cooling jackets

Recent design cooling jackets are solid copper with Monel 400 (Ni–30%Cu) alloy tube embedded inside. The tube is bent into many turns to maximize heat transfer from the solid copper to water flowing in the Monel tube (McKenna, Voermann, et al., 2010). The hot face of the cooling jacket is cast in a waffle shape. This provides a jagged face for refractory retention and magnetite slag deposition. Jackets are typically 0.75 × 0.75 × 0.1 m thick with 0.03 m diameter, 0.004 m wall Monel tube. The use of these jackets over increasing areas of the furnace is a key to increasing campaign life and reducing air infiltration to the furnace (Jin et al., 2016; Romppanen et al., 2019). It is now possible to cover almost all of the furnace.

A problem with the use of water in cooling jackets is the possibility of leaks into the furnace (Jin et al., 2016). These leaks are a safety risk, due to the potential for steam explosions (Kennedy et al., 2015). As a result, several alternative cooling fluids are being considered as a replacement, both for furnace and taphole cooling elements. New systems created by Metso Outotec can detect water leaks and minimize water usage (Bjöklund et al., 2013).

6.1.3 Concentrate burner (Fig. 6.2)

Dry concentrate and O_2-rich blast are combined in the furnace reaction shaft by blowing them through a concentrate burner. Dry flux, recycle dust, and crushed reverts (matte and slag inadvertently frozen during transport around the smelter) are also added through the burner. A year 2010 concentrate burner consists of the following:

1. an annulus through which O_2-rich blast is blown into the reaction shaft
2. a central pipe through which concentrate falls into the reaction shaft
3. a distributor cone at the burner tip, which blows air horizontally through the descending solid feed.

Special attention is paid to uniform distribution of blast and solid feed throughout the reaction shaft. It is achieved by introducing blast and solids vertically and uniformly into quadrants around the burner and by blowing the solids outwards with central jet distributor air.

Continued development of the flash furnace over the past decade has focused on improving burner operations (Romppanen et al., 2019). The use of computational fluid dynamics to predict the impact of changing conditions has been especially important (Jastrzebski et al., 2013; Jokilaakso & Taskinen, 2019; Otaka et al., 2016). The goals of burner improvements include increasing capacity, decreasing dust generation, and maximizing oxygen efficiency. Important factors include the levels of oxygen enrichment, combustion gas velocity, feed injection angle, and burner aperture size.

A variation of this burner design is the Sumitomo-type concentrate burner which uses an oxygen–fuel burner located at the central jet distributor to accelerate the ignition of the concentrate particles (Nagai et al., 2013; Otaka et al., 2016). Experiments have shown that blowing some of the air from the side reduces dust generation by about one fourth, although at some cost to oxygen efficiency.

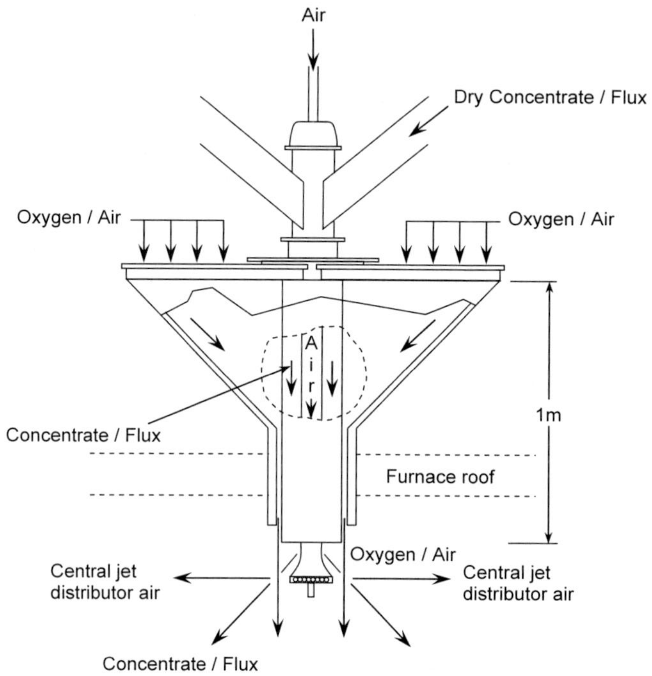

Figure 6.2 Central jet distributor Metao Outotec concentrate burner. The main goal of the burner is to create a uniform concentrate-blast suspension 360 degrees around the burner. This type of burner can smelt up to 200 tonnes of feed per hour. Its feed consists mainly of dry Cu–Fe–S concentrate, ~100 μm, silica flux, ~1 mm, recycle dust, and recycle crushed reverts, ~1 mm.

A significant new development in burner design has been the invention of a "pulse cyclone" burner to replace the sleeve-type style burner designed by Outokumpu (Zhou, 2013). This replaces the central feed nozzle with tangential introduction of concentrate to the burner head using the combustion "air" (70%–90% O_2), which causes a swirling cyclonic motion, rotating at 100–200 m/s. A small amount of oxygen is pulsed through the center of the cyclone. This changes the nature of the cones of swirling particles exiting the burner, The result is called "rotation–suspension" smelting, which reduces the dust generation, allows processing of up to 400 tonnes per hour of concentrate, and apparently eliminates the need for fuel combustion. This burner is used so far only in China, but there may be more like these in the future.

6.1.4 Supplementary hydrocarbon fuel burners

All Metso Outotec flash furnaces are equipped with hydrocarbon fuel burners atop the reaction shaft and through the settler walls and roof. Shaft-top burners keep the process in thermal balance. Settler burners eliminate cool zones in the furnace. They are also used to adjust slag temperature.

6.1.5 Matte and slag tapholes

Molten matte and slag are tapped through single-hole water-cooled copper *chill blocks* embedded in the furnace walls. The holes are typically 60–80 mm diameter. They are plugged with moist fireclay which is solidified by the heat of the furnace when the clay is pushed into the hole. They are opened by chipping out the clay and by melting it out with steel oxygen lances. Matte is tapped via copper or refractory-lined steel launders into cast steel ladles for transport to converting. Matte is also granulated to ~0.5 mm granules in a water torrent when Outotec flash converting is used instead of Peirce–Smith converting (Chapter 8). Slag is tapped down water-cooled copper launders into (a) an electric settling furnace for Cu settling and recovery or (b) ladles for truck haulage to Cu recovery by slow cooling/grinding/flotation.

Both withdrawals are only partial. Reservoirs of matte and slag (~0.5 m deep each) are maintained in the furnace. Tapping of matte is continuously rotated around its tapholes. This washes out solid buildups on the furnace floor by providing matte flow over the entire hearth.

Tapping is one of the more dangerous aspects of flash furnace operation, especially when oxygen lances are used to open the taphole (Romppanen et al., 2019). As furnace capacities grow, continuous tapping may be more feasible in the future (Jokilaakso & Taskinen, 2019).

6.2 Peripheral equipment

The Metso Outotec flash furnace is surrounded by the following:

1. concentrate blending equipment
2. solids feed dryer
3. flash furnace feed bins and feed system
4. oxygen plant
5. blast preheater (optional)
6. waste heat boiler (Chapter 11)
7. dust recovery and recycle system
8. gas cleaning system (Chapter 11)
9. sulfuric acid plant (Chapter 11)
10. secondary gas collection and treatment system (Chapter 11)
11. Cu-from-slag recovery system (Chapter 10)

6.2.1 Concentrate blending system

Most flash furnaces smelt several concentrates plus small amounts of miscellaneous materials, such as precipitate Cu. They also smelt recycle dust, sludges, slag flotation concentrate, and reverts. These materials are blended to give constant composition feed to the flash furnace. Constant composition feed is the surest way to ensure smooth flash furnace operation and continuous attainment of target compositions and temperatures.

Two techniques are used:

1. bin-onto-belt blending by which individual feed materials are dropped from holding bins at controlled rates onto a moving conveyor belt
2. bedding, where layers of individual feed materials are placed on long (occasionally circular) shaped piles, then reclaimed as vertical slices of blend (Urbaniak et al., 2019).

The blended feed is sent to a dryer. Flux may be included in the blending or added just before the dryer.

6.2.2 Solids feed dryer

Concentrate and flux are always dried prior to flash smelting to ensure even flow through the concentrate burner. Steam and rotary dryers are used (Chen & Mansikkaviita, 2006; Hoshi et al., 2010; Talja et al., 2010). The water contents of moist and dry feed are typically 6%–14% and 0.2 mass% H_2O.

Rotary dryers evaporate water by passing hot gas from natural gas or oil combustion through the moist feed. The temperature of the drying gas is kept below ~500°C (by adding nitrogen, recycle combustion gas, or air) to avoid spontaneous oxidation of the concentrate. Some smelters recycle anode furnace offgas (Chapter 11) or warm air from acid plant SO_3 coolers to their rotary dryers to decrease fuel consumption. In addition, steam from the flash furnace waste heat boiler can also be used to indirectly preheat rotary kiln air to decrease fuel consumption (Hoshi et al., 2010).

Steam dryers rotate hot, steam-heated stainless steel coils through the moist feed. Modern designs have a drum that rotates in the same direction as the steam coils to minimize tube wear and improve water evaporation efficiency.

Steam drying has the advantages of the following:

1. efficient use of flash furnace waste heat boiler steam
2. little SO_2, dust, and off-gas evolution because hydrocarbon combustion is not used
3. low risk of concentrate ignition because steam drying is done at a lower temperature (~200°C) than combustion gas drying (~500°C)
4. approximately 14% less energy consumption than a similar capacity rotary dryer (Chen & Mansikkaviita, 2006)
5. steam drying is being adopted widely in new and existing Outotec flash smelters with capacities in a single dryer up to 200 t/h wet feed (Talja et al., 2010)

6.2.3 Bin and feed system

Dried feed is blown up from the dryer by a pneumatic lift system (Walker et al., 2010). It is caught in acrylic bags and dropped into bins above the flash furnace reaction shaft. It is fed from these bins onto drag or screw (preferred) conveyors and sometimes into a loss-in-weight feeder (Reed & Coleman, 2019; Romppanen et al., 2019). From the loss-in-weight feeder, it flows on an air slide to the concentrate burner. The rate of feeding is adjusted by varying the speed of the conveyors below the bins. The use of the loss-in-weight feeder provides an accurate measurement of the mass flow of feed to the furnace. Older feed systems were based on volumetric flow. A recent innovation is FeedGuard, which continually measures the evenness of the concentrate flow to the burner. This allows quick detection of blockages and other problems and thus quicker response.

Bin design is critical for controlled feeding of the flash furnace. Fine dry flash furnace feed tends to hang up on the bin walls or flood into the concentrate burner. This is avoided by mass flow bins (Dudley, 2009) that are steep enough and smooth enough to give even flow throughout the bin.

6.2.4 Oxygen plant

The principal oxygen plant in an Outotec flash smelter is usually a cryogenic liquefaction/distillation unit producing 200–1100 tonnes oxygen per day. It delivers 92–99.95 vol.-% O_2 industrial oxygen gas (0.8–2 bar gauge) to the flash furnace (Deneys et al., 2007). Some smelters also use molecular sieve oxygen plants (vacuum pressure swing absorption or pressure swing absorption) to supplement their liquefaction/distillation oxygen. Molecular sieve plants are small (<200 tonnes oxygen/day) units that produce oxygen with a purity of around 90 vol.-% O_2; Ar is the main impurity. They are suitable for incremental additions to the output of the main oxygen plant.

Oxygen-enriched blast is prepared by mixing industrial oxygen and air as they flow to the concentrate burner. The oxygen is added through a diffuser (holed pipe) protruding into the air duct. The diffuser is located about six duct diameters ahead of the concentrate burner to ensure good mixing.

The rates at which oxygen and air flow into the concentrate burner are important flash furnace control parameters. They are measured by orifice or mass flow flowmeters and are adjusted by butterfly valves.

6.2.5 Blast heater (optional)

A few Metso Outotec flash furnaces use heated blast. The blast is heated typically to 100–450°C using hydrocarbon-fired shell-and-tube heat exchangers. Hot blast ensures rapid Cu–Fe–S concentrate ignition in the flash furnace. It also provides energy for smelting. Modern, highly oxygen-enriched flash furnaces use ambient (~30°C) blast. Concentrate ignition is rapid with this blast at all temperatures.

6.2.6 Waste heat boiler

Offgas leaves an Outotec flash furnace at about 1300 °C. Its sensible heat is recovered as steam in a horizontal waste heat boiler (Chapter 11).

6.2.7 Dust recovery and recycle system

Metso Outotec flash furnace offgases typically contain 0.1–0.2 kg of dust per Nm^3 of offgas. The dust consists of mainly oxides (e.g. Cu_2O, Fe_3O_4). It becomes sulfated by SO_2 and O_2 in the waste heat boiler and electrostatic precipitators (Chapter 11). This sulfated dust is collected and recycled to the flash furnace for Cu recovery. In many flash smelters, air or industrial oxygen are deliberately fed into the offgas to sulfate the dust (Davenport et al., 2001). Controlling the oxidation and sulfation behavior of the dust minimizes accretion formation in the uptake and waste heat boiler (Miettinen et al., 2010).

About 70% of the dust leaving the uptake drops out in the waste heat boiler. The remainder is caught in electrostatic precipitators (Chapter 11) where the particles are charged in a high voltage electrical field, caught on a charged wire or plate and periodically collected as dust "clumps" by rapping the wires and plates. Electrostatic precipitator exit gas contains ~0.1 g of dust per Nm^3 of gas (Parker, 1997).

The collected dust contains approximately 25% Cu. It is almost always recycled to the flash furnace for Cu recovery. It is removed from the boilers and precipitators by drag and screw conveyors, transported pneumatically to a dust bin above the flash furnace and combined with the dried feed just before it enters the concentrate burner.

6.3 Flash furnace operation

Tables 6.1 and 6.2 indicate that Metso Outotec flash furnaces

1. smelt up to 4200 tonnes per day of new concentrate
2. typically produce ~65% Cu matte (depending on the matte converting process)
3. use 45%–88% O_2 blast, sometimes slightly heated
4. burn hydrocarbon fuel to some extent, mostly in the reaction shaft.

This section describes how the furnaces operate.

6.3.1 Startup and shutdown

Operation of an Outotec flash furnace is begun by heating the furnace to its operating temperature with hydrocarbon burners or hot air blowers. The heating is carried out gently and evenly over a week or two to prevent uneven expansion and spalling of the refractories. Adjustable springs attached to fixed position I beams keep the walls and hearth under constant pressure during the heating. During initial heatup, paper is inserted between newly laid hearth bricks to burn out and compensate for brick expansion. Concentrate feeding is begun as soon as the furnace is at its target temperature. Full production is attained in a day or so. Shutdown consists of the following:

1. overheating the furnace for 7–10 days to melt solid buildups
2. starting hydrocarbon burners
3. stopping the concentrate burner

4. draining the furnace with hydrocarbon burners on
5. turning off the hydrocarbon burners
6. turning off the cooling water
7. allowing the furnace to cool at its natural rate.

6.3.2 Steady-state operation

Steady-state operation of a flash furnace entails the following:

1. feeding solids and blast at a constant rate
2. drawing SO_2-rich gas from the gas uptake at a constant rate
3. tapping matte from the furnace on a scheduled basis or as needed by the converter(s)
4. tapping slag from the furnace on a scheduled basis or when it reaches a prescribed level in the furnace.

The next section describes how steady-state operation is attained.

6.4 Control

The Outotec flash furnace operator must smelt concentrate at a steady, specified rate while

1. producing matte of specified Cu grade
2. producing slag of specified SiO_2 content
3. producing slag at specified temperature
4. maintaining a protective coating of magnetite-rich slag on the furnace interior (Fig. 6.3).

6.4.1 Concentrate throughput rate and matte grade controls

Basic Outotec flash furnace strategy is to charge dried concentrate mix to the furnace at a prescribed rate and to base all other controls on this rate. Having chosen concentrate feed rate, the flash furnace operator must next select the grade (% Cu) of his product matte, i.e. the extent of Fe and S oxidation. It is selected as a compromise between the following:

1. maximizing SO_2 evolution in the flash furnace (where it is captured efficiently) and
2. keeping enough Fe and S in the matte so that subsequent converting can operate autothermally while melting its required amount of Cu scrap and smelter recycle materials.

Physically, matte grade is set by adjusting the ratio of oxygen input rate in the blast to concentrate feed rate until the target matte composition is obtained. A large ratio results in extensive Fe and S oxidation and high-grade (% Cu) matte. A small ratio gives the opposite. Physically, the ratio is controlled by adjusting the rates at which air and oxygen enter the furnace, after setting a constant concentrate feed rate. A newly developed servo-actuator improves the accuracy of air flow control, which reduces maintenance costs (Romppanen et al., 2019).

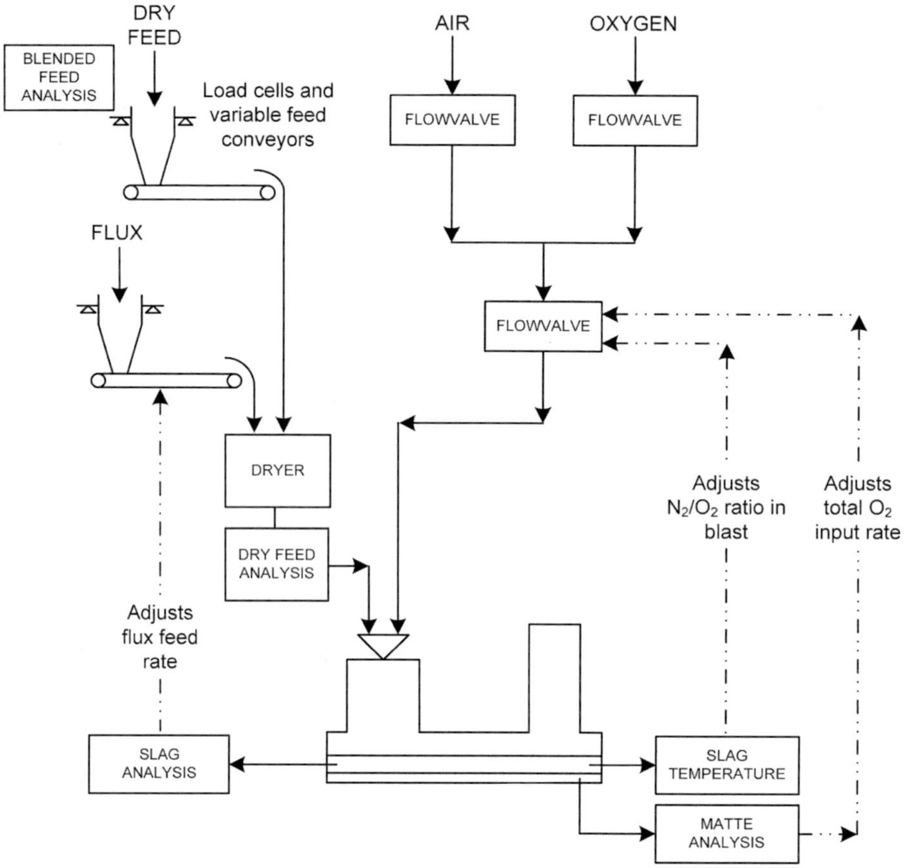

Figure 6.3 Example control system for Metso Outotec flash furnace. The three loops control slag temperature, slag composition, and matte composition. Slag temperature may also be controlled by adjusting reaction shaft hydrocarbon burner combustion rate. It is fine-tuned by adjusting settler burner combustion rates. Matte grades ±1.5% Cu and temperatures ±20°C are obtained.

6.4.2 Slag composition control

The iron oxide formed by concentrate oxidation is fluxed with SiO_2 to form liquid slag. The amount of SiO_2 is based upon the slag having a low solubility for Cu and sufficient fluidity for easy tapping and a clean matte/slag separation. An SiO_2/Fe mass ratio of 0.7−1.0 is used. It is controlled by adjusting the rate at which flux is fed to the solids feed dryer.

6.4.3 Temperature control

Matte and slag temperatures are measured as matte and slag flow from the furnace. Disposable thermocouple probes and optical pyrometers are used. Matte and slag temperatures are controlled by adjusting (a) the rate at which N_2 coolant enters the furnace (mainly in air) or (b) hydrocarbon burner combustion rates. Slag temperature is adjusted somewhat independently of matte temperature by adjusting settler hydrocarbon burner combustion rates.

Matte and slag temperatures are typically 1220 and 1250°C, respectively. They are chosen for rapid matte/slag separation and easy tapping. They are also high enough to keep matte and slag molten during transport to their destinations. Excessive temperatures are avoided to minimize refractory and cooling jacket wear.

6.4.4 Reaction shaft and hearth control (Davenport et al., 2001)

Long flash furnace campaign lives require that magnetite-rich slag be deposited in a controlled manner on the furnace walls and hearth. Magnetite slag deposition is encouraged by the following:

1. highly oxidizing conditions in the furnace
2. low operating temperature
3. low SiO_2 concentration in slag

It is discouraged by reversing these conditions and by adding coke or coal to the furnace. An ultrasonic—acoustic technique developed by Hatch has proven useful in monitoring hearth conditions (Sadri et al., 2016). Campaigns up to 11 years in duration have been achieved for flash smelting furnaces (Janney et al., 2007).

6.5 Impurity behavior

Flash furnace concentrates inevitably contain impurities from their original ore. They must be separated from Cu during smelting and refining. Table 6.3 shows that this is partially accomplished during flash smelting, as portions of the impurities report to the slag and offgas, while almost all the Cu reports to the matte. Important exceptions to this are gold, silver, and platinum group metals. They accompany Cu through to electrorefining where they are recovered as byproducts (Chapter 21). Most Ni also follows Cu.

Industrial impurity distribution is complicated by recycle of the following:

1. flash furnace and converter dusts
2. flash furnace slag concentrate
3. converter slag concentrate and (occasionally) molten converter slag
4. solid reverts from around the smelter
5. acid plant sludges

Nevertheless, Table 6.3 provides guidance as to how impurities distribute themselves during flash smelting.

Table 6.3 Distribution of elements in Outotec flash smelting (Davenport et al., 2001).

Element	% To matte	% To slag	% To offgas[a]
Cu	97	2	1
Ag	90–95	2–5	3–8
Au	95	2	3
As	15–40	05–25	35–80
Bi	30–75	05–30	15–65
Cd	20–40	05–35	25–60
Co	45–55	45–55	0–5
Ni	70–80	20–25	0–5
Pb	45–80	15–20	05–40
Sb	60–70	05–35	05–25
Se	85	05–15	0–5
Te	60–80	10–30	00–10
Zn	30–50	50–60	05–15

[a]collected as precipitated solids during gas cleaning.

6.5.1 Nonrecycle of impurities in dust

Impurities are also found in flash furnace dust (Table 6.3). This dust is usually recycled to the flash furnace for Cu recovery, so it is not usually an escape route for impurities. However, some flash smelters, including Chuquicamata, Kennecott (Kaur et al., 2010), Saganoseki, and Kosaka (Maeda et al., 1998), recover Cu from some of their dust hydrometallurgically rather than by recycle to the flash furnace. This allows the impurities to leave the smelter in leach plant residues. It is particularly effective for removing As, Bi, and Cd from the smelter. Lead is removed to a lesser extent.

6.5.2 Other industrial methods of controlling impurities

Flash smelters that treat several concentrates blend their high- and low-impurity concentrates to ensure that impurity levels in their product anodes are low enough for efficient electrorefining. Flash smelters that are dedicated to treating high-impurity concentrates control anode impurity levels by (a) minimizing dust recycle or (b) modifying converting and fire refining to increase impurity removal (Larouche et al., 2003).

6.6 Outotec flash smelting recent developments and future trends

Outotec flash furnace smelting capacity and campaign life has increased in recent years through enhancements in the design of the feed system, concentrate burner, and furnace cooling elements (Romppanen et al., 2019). Advances in process control and dust sulfation techniques have also contributed. The future development foreseen by Metso Outotec is use of its flash furnace for direct-to-copper smelting (Chapter 10) of lower grade concentrates and further use of continuous flash converting (Urbaniak et al., 2019; Zhou, 2013, Chapter 8). Both have the advantages of improved SO_2 capture and an SO_2-free workplace because they eliminate batch Peirce–Smith converting.

6.7 Inco flash smelting

Inco flash smelting blows industrial oxygen, dried Cu–Fe–S concentrate, SiO_2 flux, and recycle materials horizontally into a hot (~1250°C) furnace. Once in the furnace, the oxygen reacts with the concentrate to generate the following:

1. molten matte, 55%–60% Cu
2. molten slag, 1%–2% Cu
3. offgas, 60–75 vol.% SO_2.

The matte is tapped into ladles and sent to converting. The slag is tapped into ladles and sent to stockpile, with or without Cu-from-slag removal (Chapter 11). The offgas is water quenched, cleaned of dust, and sent to a sulfuric acid plant. The Inco flash furnace is also used to recover Cu from molten recycle converter slag. The slag is poured into the furnace via a steel chute and water-cooled door (Fig. 6.4).

At the start of 2010, there are four Inco flash furnaces in operation: at Almalyk, Uzbekistan (Ushakov et al., 1975); Hayden, Arizona (Marczeski & Aldrich, 1986); and Sudbury, Ontario (two furnaces; Carr et al., 1997; Humphris et al., 1997; Zanini et al., 2009). The Almalyk and Hayden furnaces smelt Cu–Fe–S concentrates. The Sudbury furnaces smelt Ni–Cu–Co–Fe–S concentrates to produce ~45% Ni + Cu + Co matte and ~1% Ni + Cu + Co slag.

6.7.1 Furnace details

The Inco flash furnace is made of high-quality MgO and $MgO-Cr_2O_3$ brick (Fig. 6.4). Its main components are as follows:

1. concentrate burners, two at each end of the furnace
2. end- and sidewall water-cooled copper cooling jackets
3. a central off-gas uptake
4. sidewall tapholes for removing matte
5. an endwall taphole for removing slag
6. an endwall chute for charging molten converter slag

6.7.2 Concentrate burner

Inco concentrate burners are typically ~0.25 m diameter, 1 m long, 1 cm thick stainless steel pipes, water cooled with an internal ceramic sleeve. They are fed with industrial oxygen, blown in horizontally, and dry feed, dropped in from above through an angled tube (Fig. 6.5).

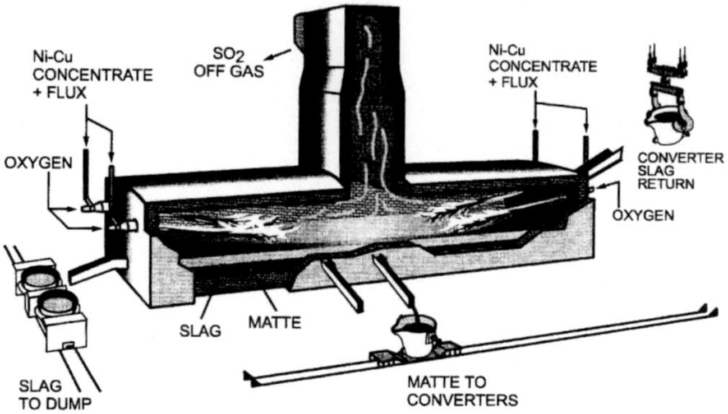

Figure 6.4 Inco flash furnace at Sudbury, Ontario (Queneau & Marcuson, 1996).

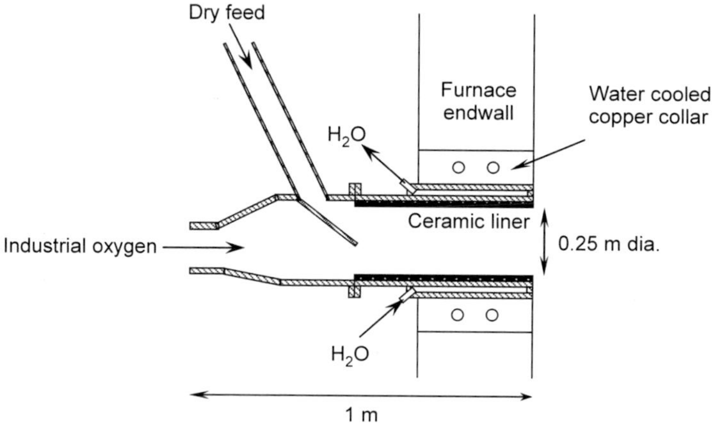

Figure 6.5 Details of Inco flash furnace concentrate burner.

The diameter of the burner barrel gives an oxygen/feed entry velocity of ~40 m per second. This velocity creates a concentrate/oxygen flame that reaches the central uptake. The burners are typically angled 7 degrees down and 7 degrees in so that the flame plays on the slag surface rather than on the roof and walls.

6.7.3 Water cooling

Inco furnace side and endwalls are fitted with water-cooled copper fingers, plates, and shelves to maintain the integrity of the furnace structure (McKenna, Newman, et al., 2010). As in the Outotec furnace, water cooling causes magnetite-rich slag to deposit on the furnace walls. This protects the bricks and water-cooled copper, extending furnace life. An Inco furnace can be operated for five or 6 years without reconstruction. Repairs are done annually and while the smelter is stopped for nonfurnace reasons, e.g.

acid plant maintenance. The end of a furnace campaign is usually caused by excessive refractory erosion, especially at the slag line.

6.7.4 Matte and slag tapholes

The Inco furnace is equipped with four sidewall matte tapholes and an endwall slag taphole. Each consists of a single-hole refractory block fitted in a water-cooled copper plate. Tapping and plugging are similar to Outotec flash furnace tapping and plugging (Section 6.1.5).

Reservoirs of slag (~0.5 m deep) and matte (0.2–1.0 m deep) are always kept in the furnace. The matte level is kept well below the slag taphole. This avoids accidental tapping of matte with slag. Matte tapping is rotated continuously between the four tapholes to keep matte flowing across the entire floor. This minimizes localized magnetite buildups on the furnace floor.

6.7.5 Gas uptake

The central off-gas uptake is brick or castable refractory backed by copper cooling jackets or stainless steel. A small amount of oxygen is injected into the uptake. It burns the elemental sulfur that is present in Inco furnace offgas. This prevents the sulfur from precipitating in downstream cooling and dust collection equipment.

6.7.6 Auxiliary equipment

The auxiliary equipment required for Inco flash smelting is the following:

1. oxygen plant (Section 6.2.4)
2. concentrate blending system (Section 6.2.1)
3. solids feed dryer
4. four concentrate burner feed systems
5. off-gas cooling system
6. dust recovery and recycle system
7. secondary gas collection and treatment system (Chapter 12)
8. sulfuric acid plant (Chapter 12).

6.7.7 Solids feed dryer (Carr et al., 1997)

Inco furnace solid feed (concentrate, flux, reverts, and recycle sludges) is dried to give an even flow of solids through the concentrate burners. This is essential to create a uniform oxygen concentrate suspension in the furnace. Natural gas–fired fluid bed dryers are used (Liu et al., 2007). The dryer feed is typically 10% H_2O; the dried product, ~0.2% H_2O. The fluidization gas is combustion gas plus air at ~330°C. A fixed bed of 1–10 mm gravel distributes the fluidization gas across the bed.

The fluidized dried feed particles and their surrounding gas are drawn at ~90°C through acrylic or aramid fabric tubes stretched on wire frameworks. About 99.99% of the concentrate is caught on the outside of the fabric. It is removed by reverse nitrogen or air pulsation and dropped into a large storage bin. From there, it is sent by conveyors or air slides to four dry charge bins above the furnace, one for each concentrate burner.

6.7.8 Concentrate burner feed system

The solid charge to the Inco flash furnace consists mainly of dried feed blend from the fluid bed dryer. It drops from its dry charge bins onto drag or screw conveyors which then drop it into the concentrate burners. The feed system may also include separate touch-up flux, touch-up revert, and converter dust bins. Material is drawn from these bins as needed to control flash furnace product temperature and slag composition. The rate of withdrawal from each bin is controlled by adjusting the speed of the drag or screw conveyor beneath the bin.

6.7.9 Off-gas cooling and dust recovery systems (Humphris et al., 1997)

Inco offgas is cooled and cleaned of dust in (a) a water spray evaporation cooler where the offgas is cooled from ~1230 to 80°C and where 90% of the entrained dust is removed as sludge or (b) scrubbers and wet electrostatic precipitators. The equipment is stainless steel to minimize corrosion. The offgas (60–75 vol.-% SO_2) is pulled through the equipment by fans, which push the gas onward to a sulfuric acid plant for SO_2 treatment.

Solids from the cooler and dust removal equipment contain ~35% Cu. The Cu is recovered by neutralizing and dewatering the sludge then recycling it through the concentrate dryer and flash furnace.

6.8 Inco flash furnace summary

The Inco flash furnace uses industrial oxygen (no air) to smelt Cu–Fe–S and Ni–Cu–Co–Fe–S concentrates. It produces high Cu and high Ni–Cu–Co mattes. It introduces dry feed and industrial oxygen through four horizontal burners and removes SO_2 offgas through a central gas uptake. The offgas is water quenched and sent to a sulfuric acid plant to capture its SO_2.

Very little nitrogen enters the Inco furnace, so its blast and offgas handling systems are small. Also, the strong offgas is ideal for SO_2 capture.

Inco flash furnace slag can contain less than 1% Cu, which can be discarded without Cu recovery treatment. This gives it a cost advantage over most other modern smelting techniques. If desired, converter slag can also be recycled through the furnace for Cu recovery. However, this procedure upsets an otherwise steady process.

6.9 Inco versus Outotec flash smelting

There are many more Outotec flash furnaces than Inco flash furnaces. Outotec advantages include the following:

1. higher concentrate throughput rates (double that of the Inco flash furnace)
2. single concentrate burner in place of four Inco burners
3. recovery of off-gas heat in a waste heat boiler as useful steam
4. dry dust recycle
5. engineering and operational support

6.10 Summary

Metso Outotec flash smelting accounts for around half of all Cu matte smelting. It is also used in three locations for direct-to-copper smelting and in two locations for continuous converting.

It blows oxygen, air, dried concentrate, flux, and particulate recycle materials as a well-dispersed mixture into a hot reaction shaft. Smelting reactions are extremely fast under these conditions. Outotec flash furnaces smelt ~4000 tonnes of new concentrate per day.

Modern Outotec flash furnaces operate with high oxygen blast and very little hydrocarbon fuel. Most of the energy for heating and melting comes from Fe and S oxidation. This operation also gives strong SO_2 offgas from which SO_2 can be captured efficiently as sulfuric acid.

Metso Outotec flash furnaces are operated under automatic control to give constant temperature, constant composition products at a rapid rate and with minimum energy consumption. Matte and slag compositions are controlled by adjusting the ratios of O_2 input rate to concentrate feed rate, and of flux input to concentrate input. Product temperatures are controlled by adjusting the N_2/O_2 ratio of the input blast and the hydrocarbon fuel combustion rate.

Wide adoption of Outotec flash smelting is due to its efficient capture of SO_2, its rapid production rate, and its small energy requirement. Its only limitation is its inability to smelt scrap.

References

Björklund, P., Ranki, T., & Miettinen, E. (2013). Recent experiences from implementing dynamic process control and monitoring in the flash smelting process. In *Proceedings of copper 2013* (pp. 139–157). Santiago: IIMCH.

Carr, H., Humphris, M. J., & Longo, A. (1997). The smelting of bulk Cu–Ni concentrates at the Inco Copper Cliff smelter. In C. Diaz, I. Holubec, & C. G. Tan (Eds.), *Nickel–cobalt 97, vol. III: Pyrometallurgical operations, environment, vessel integrity in high–intensity smelting and converting processes* (pp. 5–16). Montreal: CIM.

Chen, S. L., & Mansikkaviita, H. (2006). The beneficial effects of feeding dry copper concentrate to smelting furnaces and development of the dryers. In R. T. Jones (Ed.), *Southern African pyrometallurgy 2006* (pp. 265–271). Johannesburg: SAIMM.

Davenport, W. G., Jones, D. M., King, M. J., & Partelpoeg, E. H. (2001). *Flash smelting, analysis, control and optimization* (2nd ed.). Warrendale, PA: TMS.

Deneys, A., Dray, J., Mahoney, W., & Barry, M. (2007). Industrial gases for copper production. In A. E. M. Warner, C. J. Newman, A. Vahed, D. B. George, P. J. Mackey, & A. Warczok (Eds.), *Cu2007, vol. III (book 2): The Carlos Díaz symposium on pyrometallurgy* (pp. 271–288). Montreal: CIM.

Dudley, L. V. (2009). Secrets of the one–dimensional convergence hopper. In J. Liu, J. Peacey, M. Barati, S. Kashani–Nejad, & B. Davis (Eds.), *Pyrometallurgy of nickel and cobalt 2009* (pp. 577–589). Montréal: CIM.

Fagerlund, K., Lindgren, M., & Jåfs, M. (2010). Modern flash smelting cooling systems. In *Copper 2010, volume 2: Pyrometallurgy I* (pp. 699–711). Clausthal–Zellerfeld, Germany: GDMB.

Hoshi, M., Toda, K., Motomura, T., Takahashi, M., & Hirai, Y. (2010). Dryer fuel reduction and recent operation of the flash smelting furnace at Saganoseki smelter & Refinery after the SPI Project. In *Copper 2010, volume 2: Pyrometallurgy I* (pp. 779–791). Clausthal–Zellerfeld, Germany: GDMB.

Humphris, M. J., Liu, J., & Javor, F. (1997). Gas cleaning and acid plant operations at the Inco Copper Cliff smelter. In C. Diaz, I. Holubec, & C. G. Tan (Eds.), *Nickel–cobalt 97, vol. III: Pyrometallurgical operations, environment, vessel integrity in high–intensity smelting and converting processes* (pp. 321–335). Montreal: CIM.

Janney, D., George–Kennedy, D., & Burton, R. (2007). Kennecott Utah Copper smelter rebuild: Over 11 million tonnes of concentrate smelted. In A. E. M. Warner, C. J. Newman, A. Vahed, D. B. George, P. J. Mackey, & A. Warczok (Eds.), *Cu2007, vol. III (book 2): The Carlos Díaz symposium on pyrometallurgy* (pp. 431–443). Montreal: CIM.

Jastrzebski, M. U., Lamoreux, A., Gonzalez, T., & Veenstra, R. (2013). In pursuit of improved flash–smelting burner performance. In *Proceedings of copper 2013* (pp. 413–427). Santiago: IIMCH.

Jin, J.-Y., Lee, C.-H., & Shin, S.-J. (2016). 35 years of challenges for operational excellence in the flash smelting process in Onsan. In *Proceedings of copper 2016* (pp. 397–408). Tokyo: MMIJ.

Jokilaakso, A., & Taskinen, P. (2019). Towards a comprehensive model of the flash furnace. In *Proceedings of the 58th conference of metallurgists hosting the international copper conference 2019*. Montreal: CIM (Article 593961).

Kaur, R., Nexhip, C., Wilson, M., & George–Kennedy, D. (2010). Minor element deportment at the Kennecott Utah Copper smelter. In *Copper 2010, volume 6: Economics, process control, automatization and optimization* (pp. 2415–2432). Clausthal–Zellerfeld, Germany: GDMB.

Kennedy, M. W., MacRae, A., Jones, R. T., Kolbeinsen, L., Nos, P., & Filzweiser, A. (2015). Some considerations for safer furnace cooling. In *COM 2015, Proceedings of the conference of metallurgists* (p. 15). Montreal: CIM.

Larouche, P., Harris, R., & Wraith, A. (2003). Removal technologies for minor elements in copper smelting. In C. Diaz, J. Kapusta, & C. Newman (Eds.), *Copper 2003–cobre 2003, volume IV–pyrometallurgy of copper (book 1): The Herman Schwarze symposium on copper pyrometallurgy* (pp. 385–404). Montreal: CIM.

Liu, J., Vahed, A., Yotingco, R., Macnamara, B., & Warner, A. E. M. (2007). Study of elemental sulfur formation in the fluid bed dryer baghouse. In A. E. M. Warner, C. J. Newman, A. Vahed, D. B. George, P. J. Mackey, & A. Warczok (Eds.), *Cu2007, vol. III (book 2): The Carlos Díaz symposium on pyrometallurgy* (pp. 289–299). Montreal: CIM.

Maeda, Y., Inoue, H., Hoshikawa, Y., & Shirasawa, T. (1998). Current operation in Kosaka smelter. In J. A. Asteljoki, & R. L. Stephens (Eds.), *Sulfide smelting'98: Current and future practices* (pp. 305–314). Warrendale, PA: TMS.

Marczeski, W. D., & Aldrich, T. L. (1986). *Retrofitting Hayden plant to flash smelting*. Warrendale, PA: TMS. Paper A86–65.

McKenna, K., Newman, C., Voermann, N., Veenstra, R., King, M., & Bryant, J. (2010). Extending copper smelting and converting furnace campaign life through technology. In *Copper 2010, volume 3: Pyrometallurgy II* (pp. 971–986). Clausthal–Zellerfeld, Germany: GDMB.

McKenna, K., Voermann, N., Veenstra, R., & Newman, C. (2010). High intensity cast cooling element design and fabrication considerations. In *Copper 2010, volume 3: Pyrometallurgy II* (pp. 987—1002). Clausthal—Zellerfeld, Germany: GDMB.

Miettinen, E., Ahokainen, T., & Eklund, K. (2010). Management of copper flash smelting off—gas line gas flow and oxygen potential. In *Copper 2010, volume 3: Pyrometallurgy II* (pp. 1003—1012). Clausthal—Zellerfeld, Germany: GDMB.

Nagai, K., Kawanaka, K., Yamamoto, K., & Sasai, S. (2013). Development of side—blowing concentrate burner in flash smelting process at Sumitomo Toyo smelter. In *Proceedings of copper 2013* (pp. 297—310). Santiago: IIMCH.

Otaka, S., Kobayashi, J., Yamamoto, K., Kawanaka, K., Mori, K., Sasaki, Y., & Hattori, Y. (2016). Flash smelting operation at the Toyo smelter & Refinery. In *Proceedings of copper 2016* (pp. 421—432). Tokyo: MMIJ.

Parker, K. R. (1997). *Applied electrostatic precipitation*. London: Chapman & Hall.

Queneau, P. E., & Marcuson, S. W. (1996). Oxygen pyrometallurgy at Copper Cliff – a half centure of progress. *JOM, 48*(1), 14—21.

Reed, M. E., & Coleman, M. (2019). Flash furnace feed system developments. In *Proceedings of the 58th conference of metallurgists hosting the international copper conference 2019* (Article 587988).

Romppanen, J., Björklund, P., Suikkanen, P., Eklund, K., Ranki, T., Lahtinen, M., & Jyrkönen, S. (2019). Outotec flash smelting technology development. In *Proceedings of the 58th conference of metallurgists hosting the international copper conference 2019*. Montreal: CIM (Article 594836).

Sadri, A., Ying, W. L., Walters, G., & Szyplinski, P. (2016). Monitoring a flash smelting furnace hearth by a non-destructive (NDT) testing method. In *Proceedings of copper 2016* (pp. 559—569). Tokyo: MMIJ.

Talja, J., Chen, S., & Mansikkaviita, H. (2010). Kumera technology for copper smelters. In *Copper 2010, volume 3: Pyrometallurgy II* (pp. 1183—1197). Germany: Clausthal—Zellerfeld, GDMB.

Urbaniak, R., Gabka, T., Poks, P., & Garycki, L. (2019). Comparison of two sizes of FSFs in the Glogów Copper smelter. In *Proceedings of the 58th conference of metallurgists hosting the international copper conference 2019*. Montreal: CIM (Article 596538).

Ushakov, K. I., Bochkarev, L. M., Ivanov, A. V., Shurchov, V. P., Sedlov, M. V., & Zubarev, V. I. (1975). Assimilation of the oxygen flash smelting process at the Almalyk plant. *Tsvetnye Metally, 16*(2), 5—9.

Walker, P., Coleman, M., & Money, G. (2010). Energy efficient and reliable pneumatic conveying solutions at Aurubis. In *Copper 2010, volume 6: Economics, process control, automatization and optimization* (pp. 2517—2526). Clausthal—Zellerfeld, Germany: GDMB.

Zanini, M., Wong, D., & Cooke, D. (2009). Quality control improvements in bessemer matte production at the vale Inco Copper Cliff smelter. In J. Liu, J. Peacey, M. Barati, S. Kashani—Nejad, & B. Davis (Eds.), *Pyrometallurgy of nickel and cobalt 2009* (pp. 351—370). Montréal: CIM.

Zhou, S. (2013). Application of rotation—suspension smelting process and pulse cyclone burner in Xiangguang Copper. In *Proceedings of copper 2013* (pp. 995—1004). Santiago: IIMCH.

Further reading

Ahokainen, T., & Eklund, K. (2007). Flexibility of the Outokumpu flash smelting for low and high grade concentrates — evaluation by CFD modeling. In A. E. M. Warner, C. J. Newman, A. Vahed, D. B. George, P. J. Mackey, & A. Warczok (Eds.), *Cu 2007, vol. III (book 2): The Carlos Díaz symposium on pyrometallurgy* (pp. 343—356). Montreal: CIM.

CHAPTER 7

Bath matte smelting processes

Chapter 5 described flash smelting, the predominant worldwide technology for producing copper mattes. The advantages of flash smelting are well known and the technology is well established. However, flash smelting also has disadvantages. The biggest is its use of fine, dry concentrate particles as feed. Fine particles react faster, which is desirable. However, they also settle less quickly. As a result, flash furnaces generate considerable quantities of dust. To reduce this dust generation, a large settling area is built into flash furnaces. This increases the size of the vessel, and thus its cost.

Bath smelting, a technological alternative to flash smelting, has been subject of renovated interest in the last 15 years, initially with the growth of Top Submerged Lance (TSL) plants, followed by the development of new smelting technologies in China such as Bottom Blowing Smelting (BBS) and Side Blowing Furnace (SBF).

Bath smelting process involves the reaction of copper concentrate (dry or wet) with oxygen. The main smelting reactions will take place in the molten matte, molten slag, or partially on top of the molten bath, depending on the selected technology. The main technologies using bath smelting principles are as follows:

- Noranda and El Teniente submerged tuyeres,
- Vanyukov submerged tuyeres,
- TSL: ISASMELT™ and Metso Outotec Ausmelt®,
- Mitsubishi Process,
- Chinese smelting processes:
  - Bottom Blowing Smelting (SBS),
  - Side blowing smelting process (SBF).

Bath smelting technologies operates producing intermediate copper matte grades (55% ~ 65% Cu) or high copper matte grade (70% ~ 75%Cu). Intermediate copper matte grade will be normally processed in a two-stage batch converting process. High copper matte grades will be processed in a single batch converting stage. However, this approach also generates the disadvantage of a highly oxidized converter slag with low fluidity and excess metallic copper entrainment, as discussed in Section 8.3.

Extensive S oxidation from continuous smelting allows a more efficient SO_2 capture by virtually eliminating a slag blowing stage in the subsequent converting process. Teniente, Noranda, and the newly developed Chinese bath smelting technologies gain this SO_2 advantage from the violent stirring created by submerged injection of blast. The stirring dissolves and suspends magnetite in slag, preventing excessive deposition on the furnace

Extractive Metallurgy of Copper, Sixth Edition
ISBN 978-0-12-821875-4, https://doi.org/10.1016/B978-0-12-821875-4.00002-X

refractories even under the highly oxidizing conditions of super high grade matte production.

This chapter describes the above-mentioned option except for the Mitsubishi process that will be described in Chapter 9 as part of the continuous copper making processes.

7.1 Submerged tuyere: Noranda and Teniente processes

Cu–Fe–S concentrate is (a) dried and blown into the furnace through four to six dedicated submerged tuyeres or (b) thrown moist (~8% H_2O) with flux, recycle materials, and scrap onto the slag surface through an end wall.

The products of the processes are high-grade molten matte (72%–74% Cu, 3%–7% Fe, and ~20% S), slag (6%–10% Cu, 1200–1250°C), and offgas with 20–26 vol. % SO_2 (before dilution). The matte is sent to Peirce–Smith converters for a single blowing copper making stage (in same cases combination with Flash furnace matte from another reactor, allows to have a traditional two stage converting process), the slag is sent to a Cu recovery process, and the offgas is sent to cooling, dust recovery, and a sulfuric acid plant for sulfur recovery.

All or most of the energy for heating and melting the charge come from Fe and S oxidation, i.e.,

$$\begin{array}{ccccccc} CuFeS_2(s) & + & O_2(g) & \rightarrow & Cu-Fe-S(\ell) & + & SO_2(g) + \text{heat} \\ \text{injected concentrate} & & \text{in injected} & & \text{molten matte} & & \text{in offgas} \\ \text{in matte/slag bath} & & \text{oxygen}-\text{air blast} & & & & \end{array} \tag{7.1}$$

Natural gas, coal, or coke may be burnt to supplement this heat.

The Teniente and Noranda smelting processes use large, 4–5 m diameter × 14–26 m long, cylindrical furnaces (Figs. 7.1 and 7.2; Tables 7.1, 7.2 and 7.4). The furnaces always contain layers of molten matte (70%–74% Cu) and slag. Oxygen for concentrate oxidation is provided by blowing enriched air through submerged tuyeres into the molten matte layer.

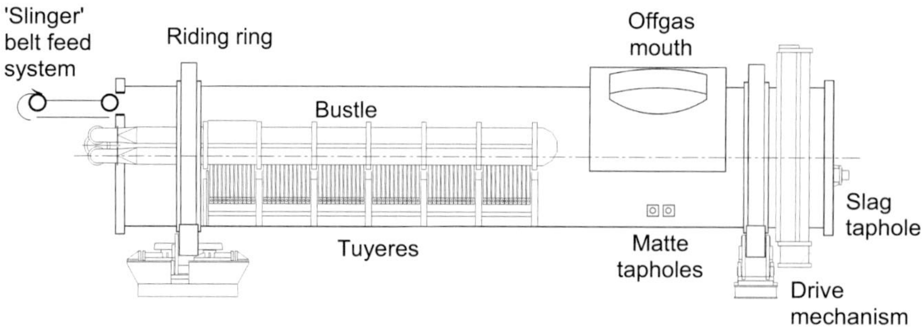

Figure 7.1 Schematic view of the Noranda smelting process.

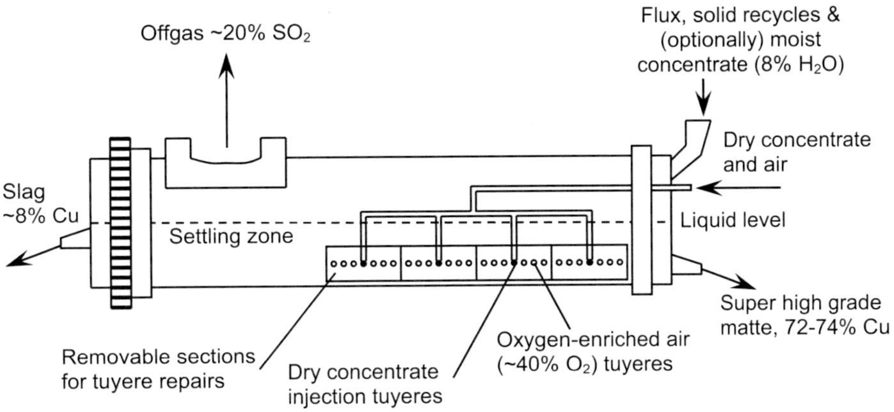

Figure 7.2 Schematic of Teniente smelting furnace, ~20 m long.

Table 7.1 Teniente and Noranda smelting furnaces around the world.

Smelting furnace type	Location	Length × diameter, m	Annual Cu production, kton per year
Teniente	Chuquicamata, Chile	23 × 5	~300 from Teniente furnace
	Potrerillos, Chile	22 × 5	190
	Paipote, Chile	14 × 4	90
	Las Ventanas, Chile	15 × 4	130
	Teniente, Chile	22 × 5	400 from 2 Teniente furnaces
	La Caridad, Mexico	21 × 4.5	~140 from Teniente furnace
Noranda	Noranda, Canada	21 × 5	220
	Altonorte, Chile	26 × 5	390

The preponderance in Chile is notable. Chuquicamata and La Caridad also have a flash furnace.
(Adapted from International Copper Study Group, 2019).

7.1.1 Noranda process (Prevost et al., 2007; Zapata, 2007)

The Noranda process was initially conceived as a direct-to-copper smelting process. The furnace at Noranda produced molten copper from 1973 to 1975. It was switched to high-grade matte production to (a) lower impurity levels in the smelter's anode copper and (b) increase smelting rate. Noranda furnaces now produce super high (70%–74% Cu), low-sulfur (~20% S) matte. This maximizes SO_2 production in the continuous smelting furnace and minimizes SO_2 production in discontinuous Peirce–Smith converters, maximizing SO_2 capture as sulfuric acid.

The Noranda furnace is a 0.1 m thick horizontal moly-steel barrel lined inside with about 0.5 m of magnesia–chrome refractory (Zapata, 2007). Industrial furnaces are ~5 m diameter and 21–26 m long. They have 54 to 66 submerged tuyeres (5 or 6 cm diameter) along the length of the furnace (Fig. 7.1).

Table 7.2 Operating details of two Noranda smelting furnaces (Carrasco, 2020; Prevost, 2020; Zapata, 2007).

Smelter	Noranda, Canada	Altonorte, Chile
Startup date	**1973**	**2002**
Furnace details		
Length × diameter, m	21.3 × 5.1	26.4 × 5.3
Slag layer thickness, m	0.3–0.6	0.3–0.5
Matte layer thickness, m	0.9–1.15	1.2
Active slag tapholes	1	1
Active matte tapholes	1	2
Auxiliary burners	0	2 (diesel) not for using as heat support during blowing
Campaign life, years	2.5	2.5
Tuyere details		
Tuyeres (total)	54	66
Diameter, cm	5.4	6.4
Active air blast tuyeres	44	49
Concentrate injection tuyeres	0	6
Type of charge	100% to top of bath	95% through tuyeres, 5% to top of bath
Feed, t/d (dry basis)		
New concentrate	1900–2200 (~27% Cu)	3000–3300 (26%–29% Cu)
Silica flux	300–400	350
Slag concentrate	80–260	440 (37% Cu)
Recycle dust	50–75 + converter dust	20 (solid residue from dust leaching)
Reverts	140–440	300–500
Other	None	Solid converter slag (250 included in reverts)
Tuyere blast details		
Volume% O_2	38–45	40–44
Flowrate per tuyere, Nm^3/min	18.4–27.3	20–24
Products, t/d		
Matte, t/d	650–900 (71%–73% Cu; 3%–5% Fe)	1350 (72%–73% Cu, 3.5%–4.5% Fe)
Slag, t/d	1300–1450 (3%–4% Cu)	2450 (7% Cu)
Mass% SiO_2/mass% Fe	0.6–0.8	0.7
Cu recovery, Noranda slag	Solidification/flotation	Solidification/flotation
Cu recovery, converter slag	Molten, to Noranda converter	Solid to Noranda reactor, 30%–35% Cu

Table 7.2 Operating details of two Noranda smelting furnaces (Carrasco, 2020; Prevost, 2020; Zapata, 2007).—cont'd

Smelter	Noranda, Canada	Altonorte, Chile
Startup date	**1973**	**2002**
Offgas, 1000 Nm^3/h	55–70 (before dilution)	62–72
Volume % SO_2, leaving furnace (wet)	15–20 (before dilution)	30–32
Dust production, t/day	70–100	30–40
Matte/slag/offgas T, °C	1210/1230/1300	1230/1220/1200
Hydrocarbon consumption, kg/t of charge	0–6, coke	3–5 metallurgical coke in solid charge

The Altonorte furnace injects almost all of its concentrates (dried) through five concentrate injection tuyeres.

Noranda smelting entails the following:

(a) continuously blowing oxygen-enriched air *blast* (40 volume% O_2, 1.6 bar gauge) through submerged tuyeres into the molten matte

(b) continuously blowing dried concentrate and blast through ~5 special submerged tuyeres into the matte layer (Antofagasta, Chile) and/or continuously feeding moist concentrate through an endwall onto the slag bath surface (Rouyn-Noranda, Canada)

(c) continuously drawing offgas through a large mouth and hood at the top of the furnace

(d) intermittently tapping matte and slag through submerged tapholes.

Flux, reverts, scrap, and coal/coke are also thrown onto the top of the molten slag bath. Molten converter slag is also occasionally recycled through the Noranda furnace mouth.

The blast tuyeres are periodically cleared by breaking blockages with a steel bar. Noranda and Teniente use Gaspe tuyere punchers (see Chapter 8 for Gaspe machine description).

Figs. 7.3 and 7.4 show the Noranda Reactors at Horne and Altonorte smelters, respectively.

The furnace is equipped with a rotation mechanism. It is used to correctly position the tuyere tips in the molten matte layer and to roll the tuyeres above the liquids during maintenance and repair. It also automatically rolls the tuyeres above the liquids in the event of a power failure or other emergency.

7.1.2 Reaction mechanisms

The reaction mechanisms in the Noranda furnace are as follows:

(a) sulfide concentrate is blown into the matte layer or thrown by "slinger" belt onto the slag surface

Figure 7.3 Noranda smelting furnace at Horne Smelter. *(Photograph courtesy of Mr. Yves Prevost).*

(b) the concentrate particles enter the matte layer where they melt and are oxidized by tuyere O_2

(c) Fe oxides from the oxidation react with SiO_2 flux to form slag, which rises to the top of the bath

(d) SO_2 from the oxidation rises through the bath and leaves the furnace along with N_2 from the tuyere blast and CO_2 and water vapor from hydrocarbon combustion.

Other parts of the charge such as scrap, sludges, and recycle materials melt and undergo oxidation and slagging. Oxides rise to the slag layer, while copper and precious metals (mostly from scrap) descend and join the matte layer.

7.1.3 Separation of matte and slag

Matte and slag are intimately mixed in the tuyere region. They are allowed to separate in a quiet tuyere-free zone at the taphole end of the furnace (Fig. 7.1). Matte falls, SO_2/N_2

Figure 7.4 View of the Noranda Altonorte reactor. *(Photograph courtesy of Glencore).*

gas rises, and slag forms a top layer dilute enough in Cu for tapping from the furnace. It contains ~6% Cu, 30% dissolved and 70% as entrained matte. It is tapped from the furnace and sent for Cu recovery by solidification/comminution/flotation (Chapter 10).

7.1.4 Impurity behavior

Table 7.3 describes impurity behavior during Noranda smelting. It shows that harmful impurities report mainly to slag and offgas. Most As, Bi, Pb, and Sb can be removed from the Cu circuit by not recycling offgas solids (recovered dust) to smelting or converting.

Table 7.3 Impurity distribution during Noranda smelting.

Element	% to matte	% to slag	% to offgas
As	8	12	80
Bi	9	12	79
Ni	77	22	1
Pb	13	13	74
Sb	15	31	54
Zn	6	84	10

(Adapted from Harris, 1999).

7.1.5 Scrap and residue smelting

The feed to the Noranda furnace at Noranda, Quebec, includes 10+% end-of-use scrap (Coursal et al., 2007). The scrap includes precious metal and copper-containing slags, ashes, residues (up to 14% moisture), precious metal catalysts, wire cables, jewelry, telephone scrap, automobile parts, and precious metal computer and electronic scrap. However, recycling of beryllium (a potential carcinogen) and radioactive materials are strictly forbidden.

Tuyere blast stirring in the Noranda furnace rapidly melts these materials and causes their precious metals and Cu to be rapidly absorbed in matte. Also, the high temperature and intensity of smelting cause potentially harmful organic compounds to be oxidized completely to CO_2 and H_2O.

7.1.6 Operation and control

Noranda smelting is started by heating the furnace with hydrocarbon burners. Molten matte is then poured in through the furnace mouth (tuyeres elevated). Once a meter or so of molten matte is in place, the tuyere "blast" is started and the tuyeres are rolled into the molten matte to begin oxidation and heat generation. Concentrate and flux feeding is then started and normal smelting is begun. About a week is taken to heat the furnace, provide the molten matte, and attain full production.

The initial molten matte is prepared by melting matte pieces or high-Cu concentrate in a converter or unused furnace in the smelter. Smelting is terminated by inserting hydrocarbon burners into the furnace, stopping smelting, and pouring slag then matte out the furnace mouth.

7.1.7 Control (Zapata, 2007)

Once steady operation has been reached, the furnace is controlled to

(a) smelt concentrates, scrap, and other metal-bearing solids at the company's prescribed rates

(b) produce matte and slag of prescribed composition and temperature

(c) maintain prescribed depths of matte and slag in the furnace.

Matte composition is controlled by adjusting the ratio of oxygen input in the blast to the rate of solid feed input. Increasing this ratio increases matte grade (by increasing Fe and S oxidation) and vice versa. It is often altered by adjusting solid feed input rate at a constant O_2-in-blast injection rate. This gives constant rate SO_2 delivery to the sulfuric acid plant.

Matte/slag temperature (1200−1250°C) is controlled by altering the ratio of hydrocarbon combustion rate to solid feed input rate. The ratio is increased to raise temperature and vice versa. It is altered by adjusting coal/coke feed rate and natural gas combustion rate. Matte/slag temperature may also be controlled by adjusting the N_2/O_2 ratio of the blast.

Slag composition is controlled by adjusting the ratio of flux input rate to solid feed input rate. The target SiO_2/Fe ratio is ~0.7.

In addition, the mix of metal-bearing solid feed to the furnace is controlled to keep impurity levels in the matte below preset values. This is done to avoid excessive impurity levels in the product anodes.

Feed rates and O_2 input rate are monitored continuously. Matte samples are taken every hour (analyses being returned 15 min later), and slag samples every 2 h. Bath temperature is monitored continuously with optical pyrometers (Fig. 7.5) sighted through two tuyeres (Zapata, 2007).

Matte and slag depths are monitored hourly with a 2.5 cm diameter vertical steel bar. This is done to (a) ensure that there is enough matte above the tuyeres for efficient O_2 utilization and (b) produce an even blast flow by maintaining a constant static liquid pressure at the tuyere tips. The depths are adjusted by altering matte and slag tapping frequency.

7.1.8 Production rate enhancement

The smelting rate of the furnace near Antofagasta, Chile, has more than doubled since 2003. Most of the increase has been due to increased furnace blowing availability, now over 90%. This in turn is the result of improved control over matte and slag compositions and temperatures, and better taphole dependability, which improves control over matte and slag depths (Zapata, 2007).

7.2 Teniente smelting

The Teniente technology was developed by Caletones Smelter in Chile with the first unit starting operation in 1977 (Campos et al., 1993). The process evolved from testwork done in Peirce—Smith converters to smelt copper concentrates. In its origin, the process involved a hybrid operation processing molten matte from reverberatory furnaces and wet feed added from the side wall of the furnace. However, with the decommissioning of reverberatory furnaces and the incorporation of solid injection technology, the process evolved to dry concentrate injection into the molten matte using special injection tuyeres using ultradense injection systems (Coleman et al., 2008).

The Teniente smelting shares many features with Noranda smelting (Carrasco et al., 2007; Moyano et al., 2010). It

(a) uses a cylindrical furnace with submerged tuyeres (Fig. 7.2)
(b) blows oxygen enriched air through the tuyeres into molten matte
(c) feeds dry concentrate through dedicated tuyeres
(d) (sometimes) charges moist concentrate onto its matte/slag surface
(e) produces high Cu, low Fe, and low S matte, which it sends to Peirce—Smith converting for copper making (i.e., Fe and S oxidation).

Table 7.4 gives operating details of two Teniente smelting furnaces.

Table 7.4 Mechanical and operating details of Teniente smelting furnaces at Caletones, Chile (Sánchez, 2020) and La Caridad, Mexico (Mariscal, 2020).

Smelter	Codelco, Caletones, Machali, Chile	Grupo Mexico, La Caridad, Nacozari, Mexico
Startup date	**1986**	**1997**
Furnace details, inside brick		
Length × diameter, m	21.1 × 4.4	19.5× 3.7
Slag layer thickness, m	0.5–0.7	0.3
Matte layer thickness. m	0.9–1.2	0.8
Active slag tapholes	1	1
Active matte tapholes	1	1
Temperature measurement	Optical pyrometer	0
Auxiliary burners	0	0
Time between tuyere line repairs	1 year	1.0–1.5 years
Time between complete relines	2 years	2.0–2.5 years
Refractory consumption	450 tonnes/full campaign	0.4 kg per tonne of Cu
Tuyere details		
Tuyeres (total)	45	44
Diameter, cm	7.33	6.35 (air blast/8.89 (concentrate)
Active air blast tuyeres	40	40
Concentrate injection tuyeres	5	4
Feed, t/d (dry basis)		
Dry concentrate through tuyeres	2000–2400	1020
% Cu in dry concentrate	27–29	21–23
Quartz flux	100–200	50–55
% SiO_2 in flux	96–98	85–98
Other, t/d	150–300 solid reverts 100–200 molten slags	100–130 solid reverts
Tuyere blast details		
Flowrate per tuyere, Nm^3/min	14–17	18–22
Volume, % O_2	35–38	31–32
Oxygen consumption per tonne of concentrate	300–330 kg industrial oxygen, 95.5% O_2	160–170 kg industrial oxygen, 98% O_2
Products, t/d		
Matte, t/d	800–950 (72%–75% Cu)	378 (68%–72% Cu)
Slag, t/d	1400–1700, 6%–8% Cu	730 (4%–8% Cu)
Mass% SiO_2/mass% Fe	0.45–0.59	0.47

Table 7.4 Mechanical and operating details of Teniente smelting furnaces at Caletones, Chile (Sánchez, 2020) and La Caridad, Mexico (Mariscal, 2020).—cont'd

Smelter	Codelco, Caletones, Machali, Chile	Grupo Mexico, La Caridad, Nacozari, Mexico
Startup date	**1986**	**1997**
Cu-from-slag recovery method	Flotation	Electric furnace
Cu recovery, converter slag	Recycle to Teniente converter	Recycle to electric furnace
Volume% SO_2, leaving furnace	23–26	21–22
Matte/slag/off-gas temps, °C	1220/1240/1250	1225/1260/1265
Hydrocarbon fuel	0	0

Data courtesy of Codelco and Grupo Mexico.

7.2.1 Process description

Teniente furnaces are 4–5 m diameter and 14–23 m long. The furnace barrels are steel, ~0.05 m thick, lined with about 0.5 m of magnesia–chrome refractory. The furnaces have 35–50 tuyeres (5 or 6 cm diameter) along 65% of their length. The remaining 35% of the furnace length is a quiet settling zone.

All Teniente furnaces blow dry concentrate into the furnace through four or six dedicated tuyeres (Moyano et al., 2010; Rosales et al., 2010). Flux, recycle materials, and (sometimes) moist concentrate are charged onto the matte/slag surface. Reactions are similar to those in the Noranda furnace. The principal products of the process are molten matte (72%–74% Cu, 3%–7% Fe, ~21% S), molten Fe–silicate slag (6%–10% Cu), and offgas (~25 vol.-% SO_2).

The operational procedure to start the furnace after a refractory maintenance shutdown involves the following sequence (Moyano et al., 2010):

(a) preheating the furnace with hydrocarbon burners according to a predefined heating curve,
(b) charging molten matte to the furnace, keeping tuyeres outside the molten bath,
(c) starting blowing oxygen-enriched air through the tuyeres, rotating them into the molten bath,
(d) starting normal feeding of concentrate, flux, and recycles, generating the matte and slag layers to the normal operating levels,
(e) Increasing feed rate until achieving steady production state,
(f) starting tapping matte and slag as required.

The initial charge of matte required for the start-up is normally from another furnace in operation (when more than one smelting line is in operation) or from a stand-by furnace (i.e., Peirce–Smith Converter) able to melt solid matte or smelting high-grade concentrate. The startup to full production process takes about 1 week.

7.2.2 Steady operation and process control

Steady operation of a Teniente furnace consists of the following (Morrow & Gajaredo, 2009; Moyano et al., 2010):

(a) continuous injection of dried concentrate and air through four to six dedicated submerged tuyeres
(b) continuous blowing of oxygen-enriched air (~40 vol.-% O_2) through submerged tuyeres
(c) continuous surface charging of flux and solid recycle materials onto the bath surface
(d) continuous withdrawal of offgas
(e) intermittent tapping of matte and slag through submerged tapholes
(f) occasional recycling of molten converter matte through the furnace mouth.

7.2.2.1 Temperature control

Operating parameters are controlled to produce matte and slag of specified compositions at a given operating temperature, normally defined to ensure proper slag fluidity and furnace refractory integrity.

Liquid temperature is measured by optical pyrometers through two of the tuyeres or onto the slag tapping stream. Slag temperature (~1240°C) is controlled by adjusting revert feed rates and blast oxygen enrichment level. Reducing the enrichment means that more N_2 is input, which acts as a coolant. Temperature is typically controlled within about ±20°C (Morrow & Gajaredo, 2009).

7.2.2.2 Slag and matte composition control

Matte and slag compositions are measured by on-site X-ray analysis. Results are available 15−120 min after a sample is taken.

Slag composition is controlled by adjusting flux feed rate. It is controlled to an SiO_2/Fe ratio of 0.65. This, plus good temperature control, gives a slag Fe_3O_4 content of 15%−20%, which maintains a protective (but not excessive) layer of solid magnetite on the furnace refractory.

Matte grade is controlled by adjusting the ratio of total O_2 input to concentrate feed rate. This ratio controls the degree of Fe and S oxidation, hence matte composition.

7.2.2.3 Matte and slag depth control

Matte and slag depths are measured frequently by inserting a steel bar vertically from above. Matte depth is controlled to give ~0.5 m of matte above the tuyeres. This ensures efficient use of tuyere O_2.

Heights of matte and slag above the tuyeres are also controlled to be as constant as possible. This gives a constant static pressure of liquid above the tuyeres, hence a constant flow of blast. The heights are kept constant by adjusting matte and slag tapping frequencies.

Table 7.5 Impurity distribution during Teniente smelting.

Element	% to matte	% to slag	% to offgas
As	6	7	87
Bi	23	40	37
Ni	80	19	1
Pb	22	25	53
Sb	19	30	51
Se	58	39	1
Zn	11	85	4

(Adapted from Harris, 1999).

7.2.3 Impurity distribution

Table 7.5 shows impurity behavior during Teniente smelting. As with Noranda smelting, As, Bi, Pb, Sb, and Zn are largely removed in slag and offgas. Se is removed less efficiently.

Teniente impurity removal appears to be slightly less effective than Noranda impurity removal (Table 7.3). This may, however, be due to differences in furnace feeds and measurement techniques.

7.2.4 Campaign life and hot tuyere repairing

The campaign lives of Teniente and Noranda furnaces are 1–2 years. Refractory wear in the tuyere region is often the limiting factor.

Most Teniente furnaces mount their tuyeres in four detachable panels (Carrasco et al., 2007; Guzman et al., 2007). These panels can be detached and replaced without cooling the furnace. This significantly improves furnace availability.

7.2.5 Furnace cooling

Chapters 6 and 9 show that flash and vertical lance furnaces need to be cooled by many copper water jackets and sprays. Teniente and Noranda furnaces use very little water cooling due to their simple barrel design and submerged oxidation reactions. This cooling simplicity is a significant advantage.

7.2.6 Off-gas heat recovery

Teniente and Noranda furnaces cool their offgases by water evaporation rather than by heat recovery boilers.

Improved smelting control and increased waste heat boiler (WHB) reliability may make heat recovery boilers economic for future Teniente and Noranda furnace retrofits. The steam will be especially valuable for steam drying of tuyere injection concentrates and heap leach lixiviants.

7.3 Vanyukov submerged tuyere smelting

Vanyukov smelting is performed in stationary, submerged tuyere furnaces (Fig. 7.6). It was developed in the former Soviet Union by A.V. Vanyukov at the National University of Science and Technology "MISIS." The technology is currently in use in Kazakhstan (Balkhash smelter) Russia (Norilsk Cu; Sredneuralsk), and Uzbekistan (Almalyk smelter) (Table 7.6).

The application of the Vanyukov process to copper was tested in 1974 in Balkhash smelter and finally implemented in 1985. The process since then has been through several developments and optimizations such as modifications to reduce dust carryover, modification of the off-gas stream to increase capacity, increase of the slag reduction area, and increase in number of electrodes (Ospanov, 2020).

The process entails the following (Tarasov & Paretsky, 2003):

(a) charging moist concentrate (up to 8% H_2O), reverts, flux, and occasionally lump coal through two roof ports (Fig. 7.6)

(b) blowing oxygen enriched air (50–95 volume% O_2, 9 bar gauge) into the molten slag layer through tuyeres on both sides of the furnace. The tuyeres are located ~0.5 m below the slag surface.

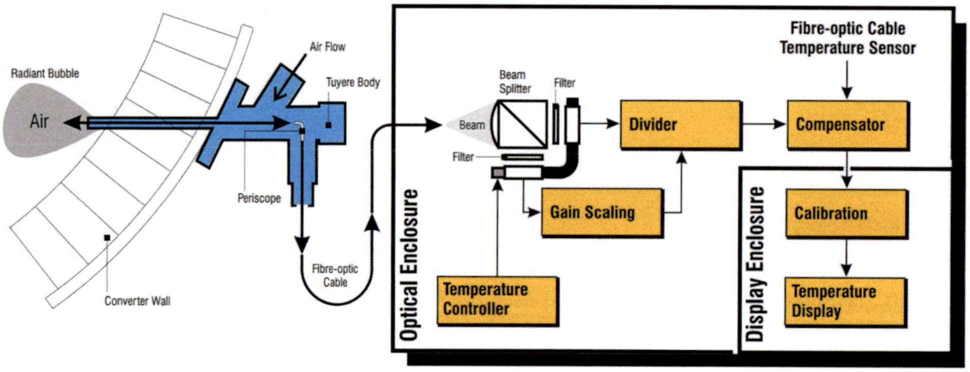

Figure 7.5 Schematic of in-tuyere optical pyrometer. Its main component is an optical fiber which transmits a color image of in-furnace matte to a distant computer for temperature interpretation. Its other key component is a periscope optical fiber holder which automatically lowers when the Gaspé puncher moves into position to punch this tuyere. Industrially, ball valves (Adapted from Heath and Sherwood, 2021) are used to prevent blowback rather than simple flap valves.

Table 7.6 Operating details of Vanyukov furnaces at Balkhash, Kazakhstan, Norilsk, Siberia, and Sredneuralsky, Urals, 2003 (Ramachandran et al., 2003; Tarasov & Paretsky, 2003; Balkash smelter data, personal communication with Dr. Yerzhan Ospanov, Kazakhmys Corporation LLC, Ospanov, 2020).

Smelter	Balkhash smelter	Norilsk Cu smelter	Sredneuralsky Cu
Number of Vanyukov furnaces	2	3	1
Copper production, tonnes per year	190,000		35,000
Furnace volume, inside brick, m^3	39; 39	36; 39; 6	35
Campaign life, years	1.5–2.5	5; 5; 4	5
Feed details			
Concentrate composition			
% Cu	12–22	19–23	13–15
% Fe	20–30	36–40	26–30
% S	28–35	28–34	35–39
Flux as % of concentrate	15–20	18/17/–	13
Fuel type	Fuel oil, coal	Natural gas	Natural gas
Blast details			
Number of operating tuyeres	46/22–34	~16	~16
Tuyere diameter, cm	3.6	~2	~2
Volume% O_2 in blast	75–92	75; 75; 94	65
Blast temperature	Ambient	Ambient	Ambient
Production details			
Operating temperature, °C	1250–1300	1320; 1320; 1250	1280
Matte composition			
% Cu	45–55	65; 55; 74	45
% Fe	21–36	8; 13; 3	30
% S	22–26	24; 24; 22	25
% Cu recovery to matte	96.2	97.3	96.2
slag composition			
% Cu	<0.9	1.2; 0.9; 2.0	0.7
% Fe	31–38	48; 45; 25	30
% SiO_2	29–31	29; 29; 35	32
% CaO	2–4.5	2; 2; 8	3
Offgas production, 1000 Nm^3/h	30–35	75; 50; 6	35

Continued

Table 7.6 Operating details of Vanyukov furnaces at Balkhash, Kazakhstan, Norilsk, Siberia, and Sredneuralsky, Urals, 2003 (Ramachandran et al., 2003; Tarasov & Paretsky, 2003; Balkash smelter data, personal communication with Dr. Yerzhan Ospanov, Kazakhmys Corporation LLC, Ospanov, 2020).—cont'd

Smelter	Balkhash smelter	Norilsk Cu smelter	Sredneuralsky Cu
Volume% SO_2 in offgas	N.A.	40; 35; -	25
Temperature °C	1100–1200	1300; 1300; 1300	1290
Dust production, % of solid charge	1.5–2.5	0.5; 0.7; 0.6	0.9
Fuel (gas) requirement	3%–11% of total heat balance	2%–3% of total heat requirement	1%–2% of total heat requirement

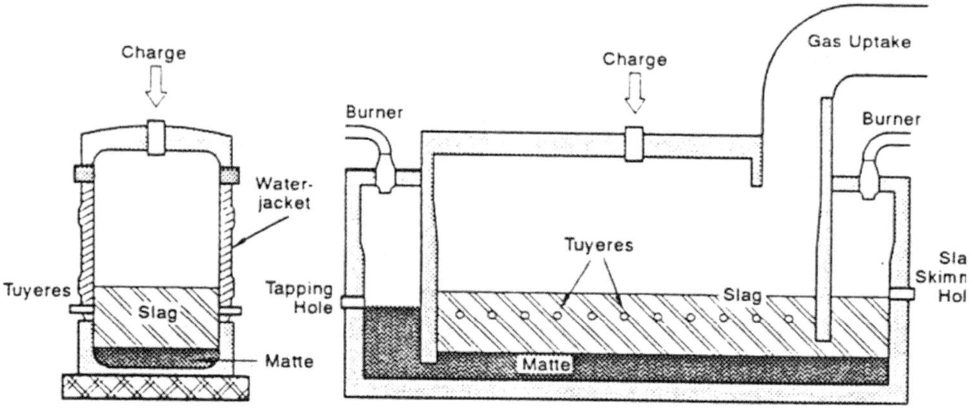

Figure 7.6 Sketch of Vanyukov matte smelting furnace.

Smelting is continuous. The furnace always contains layers of molten matte and slag. The smelting reactions are similar to those in Noranda and Teniente smelting furnaces. Matte and slag are tapped intermittently through tapholes at opposite ends of the furnace. Weirs are provided to give quiet matte/slag separation near the slag taphole (Fig. 7.7).

Matte grade is 45%–74% Cu, and slag Cu content is 0.7%–2%. The off-gas strength is 25–40 vol.-% SO_2, depending upon blast oxygen enrichment and hydrocarbon combustion rate.

7.3.1 Stationary furnace

Unlike the rotatable Noranda and Teniente furnaces, the Vanyukov furnace is stationary. The advantages of this are a directly connected gas collection system and no moving parts. The disadvantage is that the Vanyukov furnace cannot lift its tuyeres above the slag for maintenance and repair or in a blower emergency. This may be the reason that the process is not more widespread.

7.3.2 Operational challenges at Balkhash smelter (Ospanov, 2020)

Some of the operational challenges currently faced by the Vanyukov technology at Balkhash smelter are as follows:

- Processing of relatively low sulfur content concentrates requiring additional heat source
- Relatively high content of gangue in the concentrate (7% Fe_3O_4) requiring additional heat source
- Need to add natural gas and/or coal to compensate heat balance of the process (increasing operating cost)
- Need of additional flux to slag gangue material

Partially, the problem of supplementary heat source has been overcome by the implementation in 2011 of an oil heating devices.

7.4 Top Submerged Lance

In 1971, researchers at the Australian Commonwealth Scientific and Industrial Research Organisation began investigating the use of top-lancing technology for injecting coal into tin slags to improve reduction kinetics (Arthur & Partington, 2007; Collis, 2002; Floyd, 2002). This research led to the development and application of the air-cooled top entry submerged Sirosmelt lance to the processing of a wide range of raw materials. CSIRO licensed the Sirosmelt technology to Mount Isa Mines (now Glencore) and Ausmelt (Now Metso Outotec Ausmelt®), and they continued engineering and developing the technology to copper, lead, and other commodities. The TSL technology has found commercial application worldwide, becoming a significant factor copper smelting, particularly since 1995. Their share of the copper processing market is about 15% (International Copper Study Group, 2019) Chapter 5. Table 7.7 shows the list of current TSL operating plants (Metso Outotec Ausmelt, 2020; ISASMELT™, 2020). Processing capacities described in the table correspond to the original design. Some plants have upgraded their capacities since commissioning.

7.4.1 Basic operations

Ausmelt® and ISASMELT™ TSL copper smelting processes entail dropping moist solid feed into a tall cylindrical furnace while blowing oxygen-enriched air through a vertical lance into the mixture of molten matte and slag known as the *bath* (Matusewicz et al., 2007). The use of an advanced process control system results in the furnace operation being largely automated (Alvear Flores, Hunt, & Zhang, 2010). Bath temperature is controlled to $\pm 10°C$ from the set point by use of on-line temperature monitoring and advanced process control strategies (Arthur & Simon, 2005).

Table 7.7 Reference list of TSL Metso Outotec Ausmelt® and ISASMELT™ operating plants treating copper bearing materials (Metso Outotec Ausmelt, 2020; ISASMELT™, 2020).

Technology provider	Country	Customer	Startup year	Type	Capacity tpa feed
Metso: Outotec Ausmelt® TSL	Russia	OJSC Svyatogor	Under construction	Primary Cu	580,000
	China	Hudulao NonFerrous Metals Group Co Ltd	2014	Primary Cu	500,000
	China	Xinjiang Wuxin Copper Co Ltd.	2013	Primary Cu	575,000
	China	Yunnan Tin Corporation Ltd	2012	Primary Cu	450,000
	South Korea	Global Resources & Materials (GRM)	2011	Secondary Cu	110,000
	China	Daye NonFerrous Metals Co Ltd	2010	Primary Cu	1,500,000
	China	Jilin Nickel Group	2009	Primary Ni–Cu	275,000
	Namibia	Dundee Precious Metals Namibia[a]	2008	Primary Cu	240,000
	China	Chifeng Jinjian Copper Group	2008	Primacy Cu	480,000
	China	Jinchuan Non-Ferrous Metals Corporation	2008	Primary Ni–Cu	1,000,000
	Japan	Dowa Mining Company Ltd	2007	Secondary Cu	150,000
	Russia	Russian Copper Company	2006	Primary Cu	550,000
	South Korea	Korea Zin Company Ltd	2004	Cu residues	70,000
	South Africa	Amplats Management Services (Pty) Ltd	2002	Cu–Ni matte	213,000
	China	Zhong Tiao Shan Non Ferrous Metal Company	1999	Primary Cu	200,000

Table 7.7 Reference list of TSL Metso Outotec Ausmelt® and ISASMELT™ operating plants treating copper bearing materials (Metso Outotec Ausmelt, 2020; ISASMELT™, 2020).—cont'd

Technology provider	Country	Customer	Startup year	Type	Capacity tpa feed
Glencore Technology ISASMELT™	Zambia	Kansanshi Mining	2019	Cu converting	175,000
	Zambia	Kansanshi Mining	2015	Primary Cu	1,200,000
	China	Yunnan Copper Liangshan	2011	Primary Cu	500,000
	Kazakhstan	Kazzinc JSC	2011	Primary Cu	290,000
	Zambia	Yunnan Copper	2009	Primary Cu	350,000
	China	Yunnan Copper	2009	Primary Cu	500,000
	Peru	Southern Peru Corporation	2007	Primary Cu	1,200,000
	Zambia	Mopani Cu Mines	2006	Primary Cu	650,000
	India	Vedanta	2005	Primary Cu	1,200,000
	China	Yunnan Copper	2002	Primary Cu	800,000
	Germany	Aurubis AG	2002	Secondary Cu	150,000
	Belgium	Umicore Precious Metals	1995	Complex Cu–Pb	360,000
	Australia	Glencore Mount-Isa Mines	1992	Primary Cu	1,000,000
	USA	Freeport McMoRan	1992	Primary Cu	700,000

Capacities represent original design according to technology providers. Current processing capacities could be higher.
[a]originally started in 1997 as a lead smelter, transformed into a copper smelter in 2008.

The products of the process are a matte/slag mixture and strong SO_2 offgas (~25% SO_2). The matte/slag mixture can be tapped periodically or continuously into a fuel-fired or electric settling furnace for separation. The choice of separating furnace is decided largely by the local cost of electricity. Fig. 7.8 shows the Southern Peru Copper Corporation Ilo Smelter ISASMELT™ furnace with the rotary furnaces during the construction of the plant in 2005. The settled matte (~60% Cu) is sent to conventional converting. The slag (0.7% Cu) is discarded.

The offgas leaving the molten bath is postcombusted to ensure total oxidation of sulfur, volatiles, and any CO remaining in the top of the furnace. In Ausmelt® furnaces, this is done using the outermost pipe in the lance assembly, known as a *shroud*. ISASMELT™ furnaces accomplish this by injecting air in the freeboard area of the furnace or in the WHB (Herrera et al., 2013).The postcombusted gas enters the vertical upcomer of the WHB radiation section where gases are cooled to the appropriate temperature, minimizing the formation of sticky dust in the first section of the radiation

(a)

(b)

Figure 7.7 Tapping operations in a Vanyukov furnace. (A) Slag tapping from the electric furnace and (B) matte tapping from the furnace syphon into ladles *(Photos courtesy of Dr. Yerzhan Ospanov, Kazakhmys Corporation LLC, 2020).*

section. The downcomer section can be connected to s spray cooler or to a convection section. The spray cooler or the convection section will bring the off-gas temperature down to the inlet requirements of the electrostatic precipitator (350°C). After dedusting in the electrostatic precipitator, the gases are conveyed to the gas cleaning and the acid plant, respectively.

Most of the energy for smelting comes from oxidizing the concentrate charge. Additional energy is provided by combusting (a) oil, gas, or coal fines blown through the vertical lance and (b) coal fines in the solid charge. TSL furnaces are versatile, can treat a wide range of copper concentrates materials, and use a variety of fuels, depending on the most cost-effective option available at the site.

Figure 7.8 ISASMELT™ and Rotary Holding furnaces during erection and construction of the Southern Peru Copper Corporation Ilo Smelter in 2005. The furnace rests on top of a concrete base. *(Photo courtesy of Enrique Herrera, Ilo Smelter, Southern Peru Copper Corporation).*

7.4.2 Feed materials

Ausmelt®/ISASMELT™ feed is moist concentrate, flux, and recycle materials, sometimes pelletized (Tables 7.8 and 7.9). Drying of the feed is not necessary because the smelting reactions take place in the matte/slag bath rather than above it. Moist feed also decreases dust evolution. About 1% of the feed is captured as dust.

Oxygen enrichment air is blown into the molten bath with a standard selection of 50–65 vol.-% O_2. This initial range allows plants to be expanded by means of additional oxygen using the same WHB and similar total off-gas flows.

TSL furnaces are normally designed to operate just below autothermal conditions to allow space for proper temperature control. This is achieved by adding small amounts of additional fuel (solid, liquid, or gas) with the feed (coal) or through the lance (natural gas, or diesel).

7.4.3 The TSL furnace and lances

Fig. 7.9 shows a typical TSL furnace. It is a vertically aligned steel barrel, 3.5–5 m in diameter, and 12–16 m high. The largest furnace processes well over 5000 tonnes of concentrate per day. It is lined inside with chrome–magnesite refractory, and can be backed with copper water cooling blocks (Table 7.8 and 7.9). Its roof consists of water-cooled copper slabs or steel panels (Binegar, 1995; Voltura, 2004). Tapping is done through one or more water-cooled tapping blocks (ISASMELT™) or a weir system (Reuter & Matusewicz, 2010; Ausmelt, 2005). The weir system allows continuous operation of the furnace, providing more stability to the operation of the subsequent settling furnace.

Table 7.8 Operating data for ISASMELT™ primary copper smelting furnaces (Burrows, 2011; de Vries, Hunt, Dyussekenov, Milovanov, & Chikashi, 2016).

Plant	Mount Isa, Australia	Kunming, China	Miami, USA	Ilo, Peru	Sterlite, India	Mufulira, Zambia	Ust-Kamenogorsk, Kazakhstan #	Kansanshi
Capacity (t/y **concentrate)**	1,000,000	800,000	750,000	1,200,000	1200 000	650,000	290,000	1,200,000
Furnace details								
Diameter, inside brick (m)	3.75	4.4	3.75	4.4	4.4	4.4	3.6	5.5
Molten bath depth (m)	1–2	1–2	1–2	1–2	1–2	1–2	1–2	2
Number of tapholes	2	2	2	2	2	1	1	2
Furnace lining	Refractory only	Refractory only	Water-cooled copper staves behind slag line refractory	Refractory only	Water-cooled copper staves behind slag line refractory	Refractory only	Refractory only	Refractory only
Lance details								
Nominal bore (mm)	450	350	400	450	400	300	250	450
Volume% O_2 in lance air	60–65	45–50	45–50	60–67	80–90	60–70	50–55	Variable to adjust total lance flow
Hydrocarbon fuel	Natural gas	Diesel	Natural gas	#2 fuel oil	Furnace oil	Diesel	Diesel	Diesel

Feed details								
Solid fuel	Coal or coke breeze	Coal	Nil	Coal	Pet coke	Coal	Coal	Not required
Feed mixing	Pelletiser	Multiple pelletisers	Paddle mixer	Paddle mixer	Conveyed direct to furnace	Conveyed direct to furnace	Shaft mixer	Paddle mixer
% Cu in concentrate	22–26	18–22	25–29	25–29	26–31	28–32	22–26	22–23
% H_2O in concentrate	9	9	9	9	8	9	9	9
Matte/slag product								
Destination	RHF (2)	EF	EF	RHF (2)	RHF (2)	EF	EF	EF
Temperature (°C)	1190	1180	1185	1185	1185	1180	1180	1170
Matte grade (% Cu)*	60–63	52–55	55–60	60–65	60–65	60–65	55–60	57–59
Slag SiO_2:Fe	0.85	0.85	0.65	0.75	0.75	0.85	0.8	0.85

Table 7.9 Operating data for Mesto Outotec Ausmelt® smelting furnaces (Metso Outotec Ausmelt, 2020; Schlesinger, King, Sole, & Davenport, 2011).

Smelter	Zhong Tiao Shan, China	Tongling, China	RCC, Russia	JinJian, China	Huludao, China	NCS, Namibia	Daye, China	YTCL, China	Xinjiang China
Startup date	1999	2003	2006	2008	2010	2008	2010	2011	Under design
Furnace details									
Diameter, m	4.4	4.4	5	4	4.4	4.4	5	4.4	5
Height, m	~12	~14	~14	~15	~12	~12	~18	~16	~16
Bath depth, m	1.2	~1.5	~1.5	~1.5	~1.5	~1.5	~1.5	~1.5	~1.50
Number of tapholes	Weir	Weir	Weir	Weir	Weir	Weir	Weir	Weir	Weir
Auxiliary burners	Yes	Yes	Yes	Yes	Yes	Yes	Yes	Yes	Yes
Furnace cooling system	Partial	Partial	Partial	Partial	Partial	Partial	Partial	Partial	Partial
Lance details									
Lance pipe diameter, mm		~400	~500	~300	~400	~300	~500	~400	~400
Construction material	Mixed steel	Mixed steel	Mixed steel	Mixed steel	Mixed steel	Mixed steel	Mixed steel	Mixed steel	Mixed steel
Tip penetration into bath, m	~0.3	~0.3	~0.3	~0.3	~0.3	~0.3	~0.3	~0.3	~0.3
Overall length, m	16	21	22	19	20	15	25	23	23
Feed details, t/y	200,000	700,000	550,000	480,000	500,000	140,000	1,500,000	450,000	550,000
Type of charge		Moist, agglomerated	Moist, agglomerated	Moist, agglomerated	Moist, agglomerated	Moist, agglomerated	Moist, agglomerated	Moist, agglomerated	Moist, agglomerated
% H_2O in charge		~10	~10	~10	~10	~10	~10	~10	~10
% Cu (dry basis)	17–22	25	14–23	26	22	25	20	22	19

Lance inputs									
Vol.-%O_2 in blast	40	40	60	50	40	30	60	40	45
Hydrocarbon fuel	Coal	Coal	Natural gas	Coal	Oil (LFO)	Oil	Coal	Coal	Coal
Products									
Matte/slag mixture destination	Electric furnace	Electric furnace	Rotary furnace	Electric furnace	Electric furnace	Reverberatory furnace	Electric furnace	Electric furnace	Electric furnace
Off-gas destination	Acid plant	Acid plant	Acid plant	Acid plant	Acid plant	Spray cooler	Acid plant	Acid plant	Acid plant
Liquids/offgas temperatures, °C	(12,001,300)/1180	1180/1100	1180/1100	1180/1100	1180/1100	1180/1100	1180/1100	1180/1100	1180/1100
Liquid products after settling									
Matte, t/d	250	1120	710	800	700	240	1750	540	600
Grade (% Cu)	60	50	40	50	50	50	55	60	56
Slag, t/d									
%Cu, approx.	0.6	0.6	1	0.6	0.6	1	0.6	0.6	0.6
%SiO_2/%Fe, approx.	0.8	0.7	0.7	0.7	0.7	0.7	0.7	0.7	0.7
Destination	Granulation/discard	Granulation/concentrator	Mill/concentrator	Granulation/discard	Granulation/discard	Granulation/concentrator	Granulation/discard	Granulation/discard	Granulation/discard

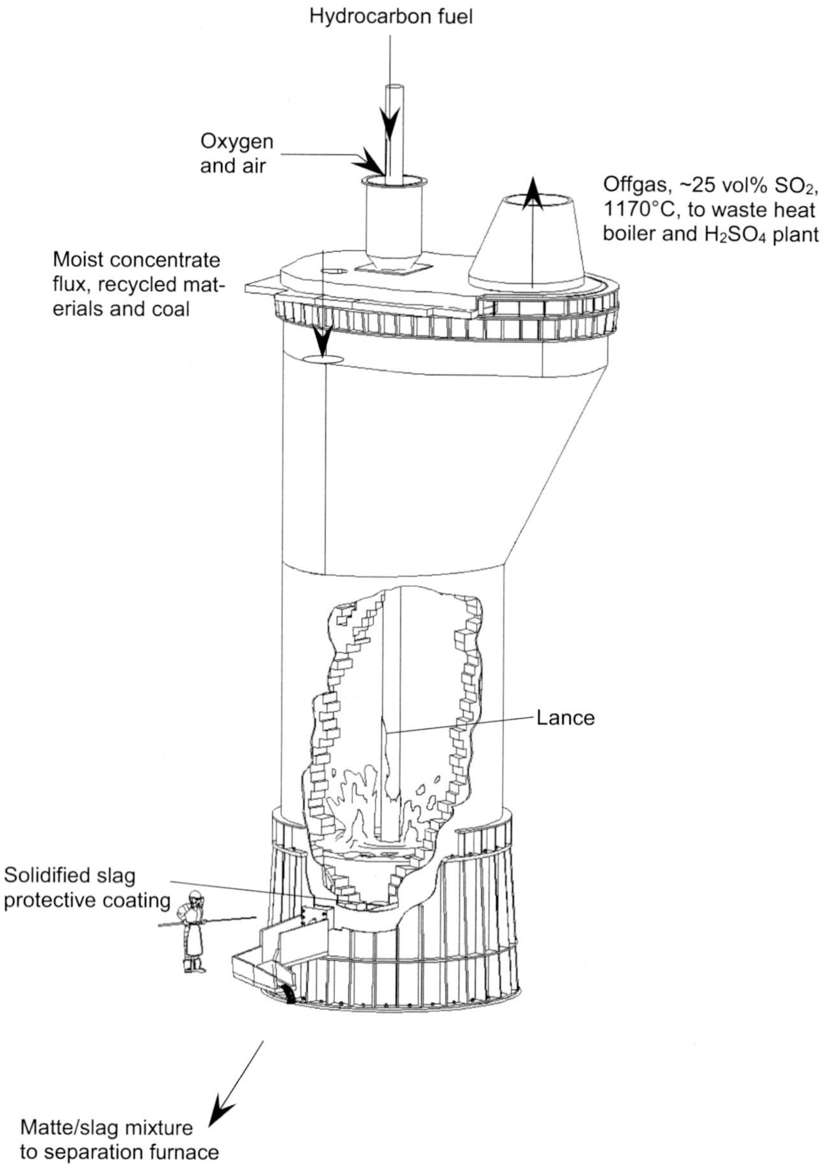

Figure 7.9 Cutaway view of ISASMELT™ furnace, 2001. A furnace is typically ~3.5 m diameter and 12 m high. It smelts up to 3000 tonnes of new concentrate per day. The outside of the furnace is often water cooled with copper cooling blocks. The main product of the furnace is a mixture of molten matte and slag, which is sent to an electric or gas-fired matte/slag separation furnace.

Fig. 7.10 shows a TSL lance. It consists of a stainless-steel outer pipe (up to 0.5 m diameter) for oxygen-enriched air and a carbon steel inner pipe for oil or natural gas. The outer pipe is normally immersed about 0.3 m into the furnace slag. The inner pipe ends about 1 m above the slag surface. The lance pressure is low enough to allow

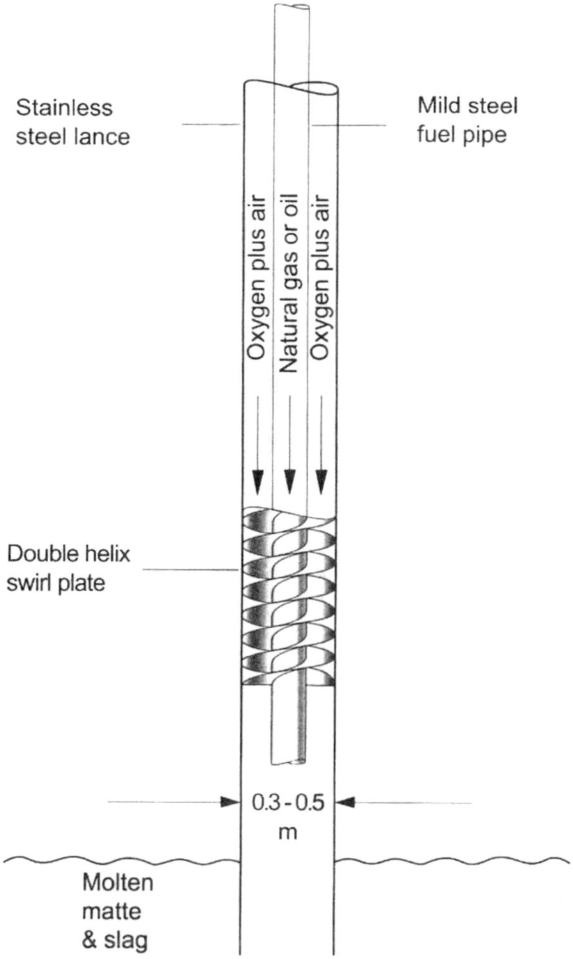

Figure 7.10 Sketch of Isasmelt lance tip. Vanes for swirling blast down the lance tip are shown. The swirling gives rapid heat transfer from lance to blast, causing a protective layer of slag to freeze on the outside of the lance tip (Solnordal & Gray, 1996).

use of a single-stage blower, rather than a compressor (Alvear Flores, Hunt, & Zhang, 2006; Bill et al., 2002). The Ausmelt® lance assembly includes an additional outer pipe for oxygen addition, which does not extend into the bath.

The lance tip is cooled by swirling the enriched air blast in the annulus between the pipes (Bill et al., 2002). The swirled gas extracts heat from the outside pipe and causes a protective slag layer to freeze on the pipe surfaces. Helical swirl vanes in the annulus are used to create this effect.

Despite this, the immersed lance tip slowly erodes away. The lance is lowered to compensate for this erosion. It is removed and replaced when ~1 m has eroded. Better operational practice has extended lance replacement time from 1 week to as many as 4 in some cases (Arthur & Partington, 2007; Bill et al., 2002; Reuter & Matusewicz, 2010; Walqui et al., 2006). The used lances are refurbished for reuse by welding a new 1 m section of outside pipe to the bottom tip, a process that takes less than an hour. Lance wear is minimized by avoiding excessive immersion (exposure to the matte layer) and limiting the oxygen content of the blast (nitrogen cooling effect).

7.4.4 Smelting mechanisms

TSL smelting is different from flash smelting and bath tuyere injection. The main difference is the oxygen transfer mechanism under which the matte formation occurs. It is commonly believed that dissolved magnetite in the slag serves as a catalyst of sorts for the overall process. Alvear Flores, Hunt, & Zhang (2010) propose this reaction sequence as follows:

$$6CuFeS_2 + 18Fe_3O_4 \rightarrow 54\ FeO + 7SO_2 + 3Cu_2S + 2FeS + 4FeO \quad (7.2)$$

$$54\ FeO + 9O_2 \rightarrow 18Fe_3O_4 \quad (7.3)$$

The proposed mechanism intends to demonstrate that the oxidation process is achieved through a controlled oxidation of the slag. As a result, a minimum amount of spinel (magnetite) is required bs the process (5% according to Binegar, 1995). TSL furnaces require therefore to operate just under slag liquidus temperature to ensure the proper amount of % solid in the slag. The operating temperature will vary depending on the slag composition and will normally between 1150 and 1200°C. TSL furnaces also operate with silica/iron ratio 0.65–0.85 for ISASMELT™, and 0.6–0.7 for Ausmelt®. Operating under the liquidus temperature also allows to refractory wear and achieve large smelting campaigns. Mount Isa Mines has reported achieving campaign in the proximity of 4 years (Edwards et al., 2013).

7.4.5 Impurity elimination

Bill et al. (2002) report impurity elimination into slag and offgas during Isasmelt smelting as 91% As, 85% Cd, 76% Bi, 68% Zn, 61% Sb, 60% Tl, 44% Pb, and 29% Te. Arsenic retention in the matte is less than half of that obtained using other smelting technology (Alvear Flores, Hunt, & Zhang, 2006), making TSL smelting technology especially suitable for high arsenic concentrates. Elimination of As, Bi, Pb, and Sb is encouraged by decreasing oxygen enrichment. Off-gas solids are recycled to the smelting furnace for Cu recovery, so they are not usually an escape route for impurities. TSL furnaces have been successfully commissioned to process complex materials in Belgium (Alvear Flores, Hourn, & Salas, 2013) and Kazakhstan (Burrows et al., 2013, 2018) and Namibia (Nolte & Kruger, 2019).

7.4.6 Startup and shutdown

Smelting is started by the following:

(a) preheating the furnace
(b) slowly charging about 2 m of solid slag pieces
(c) melting the slag, using the lance as an oxy-fuel burner
(d) immersing the smelting lance in the molten slag
(e) beginning normal concentrate smelting.

At least 1 day is required. Smelting is terminated by stopping the solid feed, draining the furnace, and turning off the lance. The furnace is then allowed to cool at its natural rate.

Steady operation consists of continuous feeding of solid charge through the roof feed port and continuous blowing of oxygen-enriched air into the molten bath. The furnace is computer controlled to give a specified concentrate smelting rate while producing matte and slag of desired composition and temperature. Matte/slag temperature is sensed by thermocouples embedded in the furnace walls (MacLeod et al., 1997). It is controlled by adjusting the rate at which fossil fuel is supplied through the lance.

Matte and slag compositions are determined by X-ray fluorescence analysis of samples from each matte/slag tap. The compositions are controlled by adjusting the O_2/concentrate and flux/concentrate ratios in the furnace input.

7.5 Chinese bath smelting technology developments: SKS-BBS process and side-blow smelting

During the last 10 years, Chinese technologists have reported several innovations in the nonferrous metals processing, with the following copper bath smelting processing technologies developed and implemented at industrial scale:

- SKS, BBS (Bottom Blowing Smelting),
- Submerged lance smelting,
- Side-blown smelting.

Fig. 7.11 shows a schematic view of the SKS-BBS furnace (Lile et al., 2016) . The SKS-BSB was firstly implemented at demonstration scale (10,000 tpy copper) at Sin Quyen Copper Smelter in Vietnam, followed by a fully commercial scale plant Shandong Dongying (Bin & Suping, 2019). Nine smelters are reported to have implemented this bath smelting technology in 2019 (Bin & Suping, 2019). The submerged lance smelting is a variation of the SKS-BBS technology developed as part of a two-stage continuous smelting process by Dongying Fangyuan. The process is described in Chapter 9.

The Side-blown smelting principle is a variation of the Vanyukov furnace principle, in which a stationary furnace has enriched air blown into the molten matte (Baiyin furnace, Jinfeng furnace).

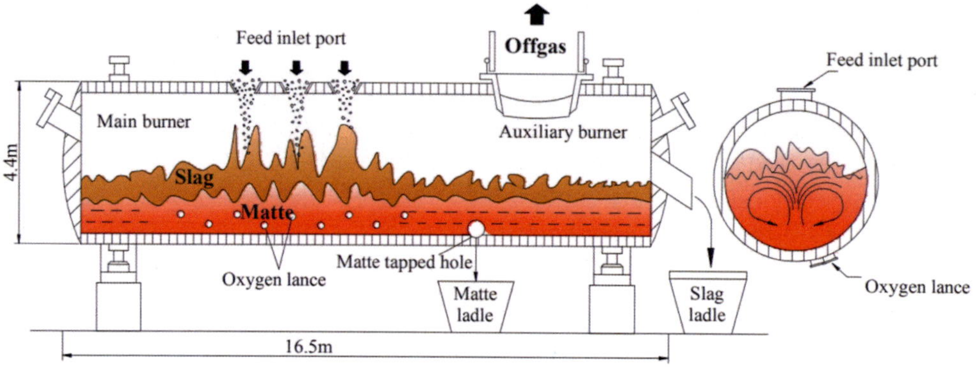

Figure 7.11 Schematic view of the Bottom Blowing Smelting (BBS) furnace (Lile et al., 2016). *(Reprinted from E proceedings Cu 2016, with permission from the Mining and Materials Processing Institute of Japan).*

In addition, SKS-BBS and Side-blown technologies have been also coupled with continuous copper converting technologies, such as Bottom Blowing Converting and aimed to replace batch converting and to develop continuous production of blister copper from copper concentrates. These efforts are discussed in Chapter 9.

7.5.1 SKS-BBS process

The oxygen bottom-blown smelting process was originally developed at Shuikoushan smelter and hence it was named SKS process. Fig. 7.11 shows a schematic view of the SKS-BBS furnace. Wet feed is continuously fed to the furnace from feed ports located on top of the furnace shell. The horizontally positioned furnace shell dimensions can vary depending on the processing capacity with lengths between 16.5 and 30 m and diameters between 4.4 and 5.8 m (Li, 2016; Wang et al., 2016). The furnace is lined with 380 mm thick chrome—magnesia bricks with water cooling only applied to the tapholes.

The SKS-BBS process entails the following:

(a) Continuously blowing oxygen-enriched air through the lances, located in the bottom of the furnace shell, into the molten matte.

(b) Continuously feed of wet concentrate through feedports located on top of the furnace shell.

(c) Continuously drawing offgas though a furnace uptake, located on one side of the furnace.

(d) Intermittently tapping matte and slag from water-cooled tapholes located on opposite sides of the furnace shell

Flux, coal, and reverts are also added, as required through the feed ports.

The SKS-BBS process has been applied for processing of relatively low copper content materials, containing significant amounts of minor oxides. This feature, combined with the use of high-pressure injection, has allowed the use of high oxygen enrichment.

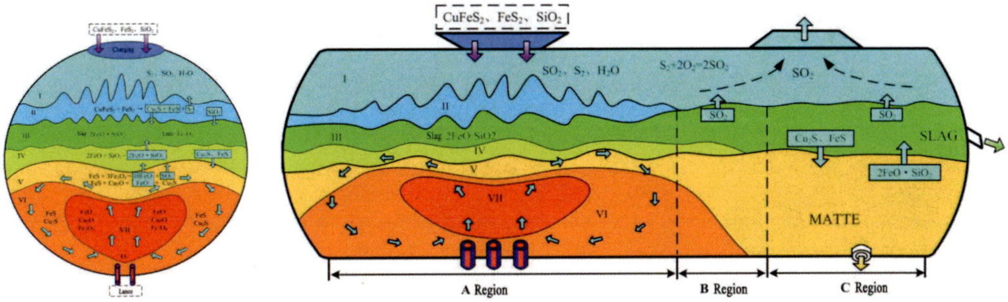

Figure 7.12 SKS-BBS reaction mechanism proposed by Wang et al. (2016). Model proposes seven different zones within the reactor, with different chemical reactions and physical separation processes taking place. *(Reprinted from E proceedings Cu 2016, with permission from the Mining and Materials Processing Institute of Japan).*

7.5.2 SKS-BBS reaction mechanisms

The main reactions involved in the SKS-BBS process are similar to those of the Noranda and El Teniente processes when feed was added from the top of the furnace. Wang et al. (2016) proposed a combination of a cross-section mass transfer and at a longitudinal section multiphase interface mass transfer behavior to explain the fundamental metallurgical aspects of the SKS-BBS process. The model, shown in Fig. 7.12, shows the proposed cross-section and longitudinal model with six different zones across the reactor. Table 7.10 presents the main characteristics attributed to each reactor zone. According to Wang, the model is based on thermochemical fundamental analysis and practical observations in the SKS plants. However, plant data with longitudinal and vertical sampling to validate the model are not reported in this publication.

7.5.3 SKS-BBS refractory campaign

The SKS-BBS technology is still novel and there are not enough plant data available at the time of this edition to understand how long the refractory campaign could last. However, use of high pressure—enriched air and low operating temperature (under liquidus temperature, with solids in suspension) ensures a relatively long refractory campaigns. Gonzalez reported SKS-BBS campaigns of 1.5 years and 36 months from one operation (Gonzalez, White, & Walters, 2019). New information should be available in the coming years as smelters have been recently commissioned, to confirm reported data.

Chen et al. (2016) reported analysis of accretions formed at the bottom of the furnace, mainly consisting of spinel (olivine phase with entrained matte) and glass phases. The sampled accretions were from the following:

- Vicinity of the oxygen lance: Formed at high temperature and high oxygen partial pressure in equilibrium with refractory material,
- Side wall settling area: Formed at lower temperature showing high-Cu spinel, magnetite spinel, and silica crystals in a crystalline structure (SiO_2)

Table 7.10 Proposed SKS-BBS reaction mechanism, Wang et al. (2016).

Model	Main objective of the area	Main reactions
Cross section Mass transfer model		
Gas layer	Combustion of sulfur and volatilization of water and other minor sulfides	$H_2O(l) = H_2O(g)$ $S_2(g) + 2\ O_2(g) = 2\ SO_2(g)$ Oxidation of volatilized minor sulfides
Mineral decomposition transition layer	Pyritic decomposition of sulfides	$4\ CuFeS_2(s) = 2\ Cu_2S(s) + 4\ FeS(s) + S_2(g)$ $2\ FeS_2(s) = 2\ FeS(s) + S_2(g)$
Slag layer	Settling of matte droplets through fayalitic slag phase	Fayalitic-rich area with matte droplets settling down toward the matte phase
Slag formation layer	Allows reaction of flux with iron oxide to allow slag formation	$2\ FeO(s) + SiO_2(s) = 2\ FeO \cdot SiO_2(l)$
Matte formation transition layer	Generation of copper matte	$FeS(l) + Cu_2O(l) = FeO(s) + Cu_2S(l)$ $3\ Fe_3O_4(s) + FeS(l) = 10\ FeO(s) + SO_2(g)$ $10\ Fe_2O_3(s) + FeS(l) = 7\ Fe_3O_4(s) + SO_2(g)$ $FeS(l) + 3\ Fe_3O_4(s) + SiO_2(s) = 5(2\ FeO \cdot SiO_2)\ (l) + SO_2(g)$
Weak oxide layer	Polycomponents move back to the reaction zone and enhance transport of sulfur compounds to the reaction zone	
Strong oxide layer	Oxygen transfer from gas to molten phase allowing oxidation of iron and other minor sulfides	$2\ FeS(l) + 3O_2(g) = 2\ FeO(l) + 2\ SO_2(g)$ $2\ Cu_2S(l) + SO_2(g) = 2\ Cu_2O(l) + 2\ SO_2(g)$ $6\ FeO(s) + O_2(g) = 2\ Fe_3O_4(s)$
Longitudinal section model and multiphase interface transfer model		
Reaction region	Stirring of compounds generated in the strong oxide layer Strong circulation of materials generated in the different zones of the cross-sectional area Gradual settling of matte	
Separation transition region	Reaction completion for those reactions not completed in adjacent regions	
Liquid phase clarification area	Area with less agitation allowing final settling and separation of phases	

7.5.4 SKS-BBS lances (Bin and Suping, 2019; Li, 2016; Xiaohong, Kefei, Shuangjie, & Xin, 2016)

Fig. 7.13 shows a typical design of an SKS-BBS lance (Xiaohong, Kefei, Shuangjie, & Xin, 2016). High pressure, enriched air required for the smelting of the materials is added via lances (similar to tuyeres) located on the bottom of the furnace. Oxygen is injected through the outer ring, while oxygen flows through the inner ring of the lance. The pressure is between 0.5 and 0.8 MPa. The number of lances varies depending on the processing capacity. Typical intermediate size furnaces will have between five and nine lances in service and large furnaces will have up to 28 lances. The lance is of two-sleeved pipes, with inner pipe flowing oxygen and outer air. The injection pressure is in the range of 0.5—0.8 Mpa and the oxygen enrichment between 70% and 75%. The relatively high injection pressure allows the formation of accretions that help protecting the lance tip.

The SKS-BBS lances have the following characteristics:

- Lance operates submerged on the matte phase
- It has double casing and the inner tube has a constant section area
- Separation between the lances is between 0.8 and 1.3 m
- There are two channels between the inner and outer tube with cooling air flowing between the channels
- Head of the lance is manufactured of high-quality materials to allow longer service life
- Lance must be monitored on regular basis to ensure proper performance

In larger SKS-BBS plants, number of lances increase to 28 units.

Figure 7.13 SKS-BBS lance typical design, according to Xiaohong, Kefei, Shuangjie, & Xin (2016). *(Reprinted from E proceedings Cu 2016, with permission from the Mining and Materials Processing Institute of Japan.).*

7.5.5 SKS-BBS operating parameters

Table 7.11 shows selected design and operating parameters of SKS-BBS furnaces in China (Bin & Suping, 2019; Wang, Liu, Yang, Tang, & Liao, 2019; Kapusta & Watt, 2019; Yan, 2016). Plant capacities fluctuate between 600,000 and 1,000,000 t/y of copper concentrates. The furnace operates on high Fe/SiO_2 ratio decreasing the amount of generated slag. There is a slight temperature profile between the matte, slag, and off-gas phases. Slag temperature seems to be selected just under the liquidus temperature, ensuring adequate proportion of solids in the slag to protect the refractories.

7.5.6 SKS-BBS minor elements distribution (Li, 2016; Lile et al., 2016)

Table 7.12 shows the distribution of minor elements for the SKS-BBS process. To compare the data with other technologies, process conditions associated with the partition are required. However, in general, the SKS-BBS process shows a large volatilization of arsenic and lead. The two references show discrepancies in the deportment of bismuth and lead.

Table 7.11 Selected typical operating parameters for the SKS-BBS and SBS processes (Extracted from Yan, 2016 and Bin, 2019).

Parameter					
Technology	SKS-BBS	SKS-BBS	SBF	SBF	BYSBF(*)
Typical smelter	Yuguang gold	SDIC Jincheng Metallurgical Co., Ltd.	Chifeng smelter	Heding smelter	Baiyin Nonferrous Group
Smelting capacity, t/y	550,000	660,000	680,000	1,300,000 (**)	850,000 (***)
Concentrate composition					
Cu, %	22–24	15–20	18–22	25	18
Fe, %	28–29	24–26	23–27	-	-
S, %	26–27	18–20	25–28	-	-
Feed rate, tph	86	110–130	83		
Oxygen enrichment, $\%O_2$	72–73	70–73	85	80	75
Fe/SiO_2	1.7–1.8	1.7–1.8	-	1.6–1.8	-
Matte Grade	72–73	70–72	75	57	59
Cu in slag	-	1.5–2.5	2	-	0.7

(*)to be commissioned by 2021.
(**)production from two furnaces.
(***)estimated from copper content and 150 kt/y of cathode production.

Table 7.12 Impurity distribution in the SKS-BBS smelting process (extracted from Lille, 2016: Li, 2016).

Element	% To matte	% To slag	% To offgas
As	3–6	4–12	82–93
Bi	15–19	11–45	40–70
Pb	15–55	20–24	19–65
Sb	12–15	65–71	17–20
Zn	18	65	17

7.5.7 Side-blown smelting process (Wang, Liu, Yang, Tang, & Liao, 2019)

The Side-blown smelting process (Side Blown Furnace, SBF) has been applied for the processing of copper concentrates and industrial solid wastes (Wang, Liu, Yang, Tang, & Liao, 2019). The SBF concept, originally developed to treat high lead slags, has been gradually applied to the processing of a wide range of raw materials, including zinc waste materials, industrial wastes, slags, among others. The SBF process has the following main characteristics:

(a) Mixed feed materials are continuously fed to the furnace top
(b) Enriched air is blown into the molten slag (Jifeng version) or matte (Baijin version) via tuyeres located on the side of the furnace
(c) The narrow cross-section allows to have a vigorous agitation to ensure fast and complete reactions

Some of the main features of the SBF process are as follows:

- Simple feed preparations as drying or pelletizing are not required
- Low dust ration (below 2%)
- Moisture of feed between 7% and 10% (as received)
- Low copper content in slag
- High oxygen enrichment (up to 85%)
- Coal is added to maintain heat balance and control magnetite levels

In terms of reaction mechanism, Ling (2019) listed a series of chemical reactions to describe the SBF process. The proposed reactions are, in general, similar to those reported for the SKS-BBS process. Typical selected operating parameters for SBF plants are shown in Table 7.11

7.5.8 Baijin and Jifeng SBF furnace design

Fig. 7.14 and 7.15 show typical structures of the SBF Baijin and Jifeng furnaces (Bin & Suping, 2019; Wang, Liu, Yang, Tang, & Liao, 2019). The SBF furnace is a stationary vessel that can be divided in three main areas (bottom to top view):

- Hearth: Made of fused Mg–Cr refractory bricks. This area also has water-cooled jackets at the bath level
- Middle section consisting of the bath and postcombustion areas

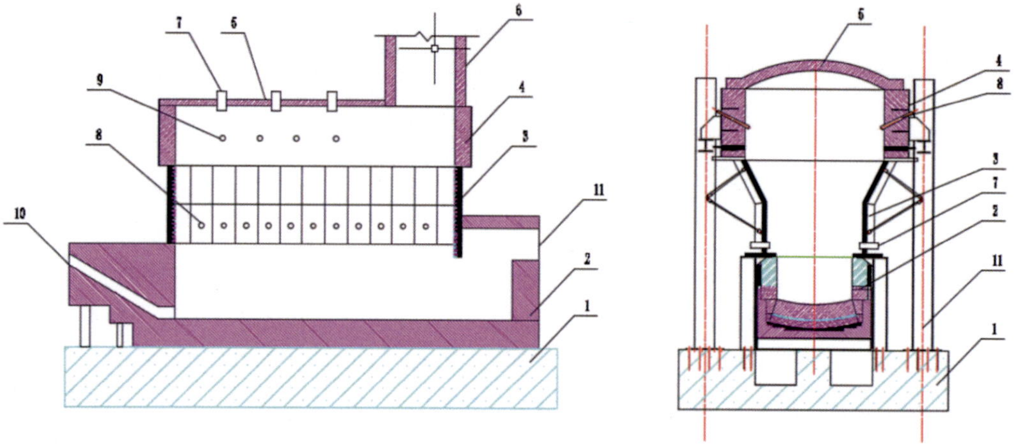

Figure 7.14 Baijin type side-blown smelting furnace schematics (Wang, Liu, Yang, Tang, & Liao, 2019). Figure originally published in Proceedings of the 58th Conference of Metallurgists Hosting the International Copper Conference 2019. *(Reprinted with permission of the Canadian Institute of Mining, Metallurgy and Petroleum).*

- Top section: Also includes a steel water-cooled jacket around the feedport area

The nozzles are located on both side of the furnace and are made of stainless steel and red metal copper composites. They are installed within the copper jacket structure.

A water-cooled partition wall is located at the uptake shaft near the charging ports. Matte and slag discharging areas are located at opposite ends of the furnace. Matte is discharged via a syphon and slag via overflow. Emergency discharge holes are also provided at both ends of the furnace in case of emergency-operational requirements.

The main differences between the Baijin and Jifeng designs are the following:

- The Jifeng design blows the enriched air into the molten slag Fig. 7.15
- The Baijin design blows the enriched air into the molten matte Fig. 7.14. This design also incorporates a settling area in addition to the smelting area, achieving lower copper contents in the slag (0.80% Cu)
- The Baiyin process does not have a postcombustion tuyere line in their design

The SBF process has been also coupled with the horizontal multilance top blowing furnace as part of a continuous copper making process (Ling, 2019).

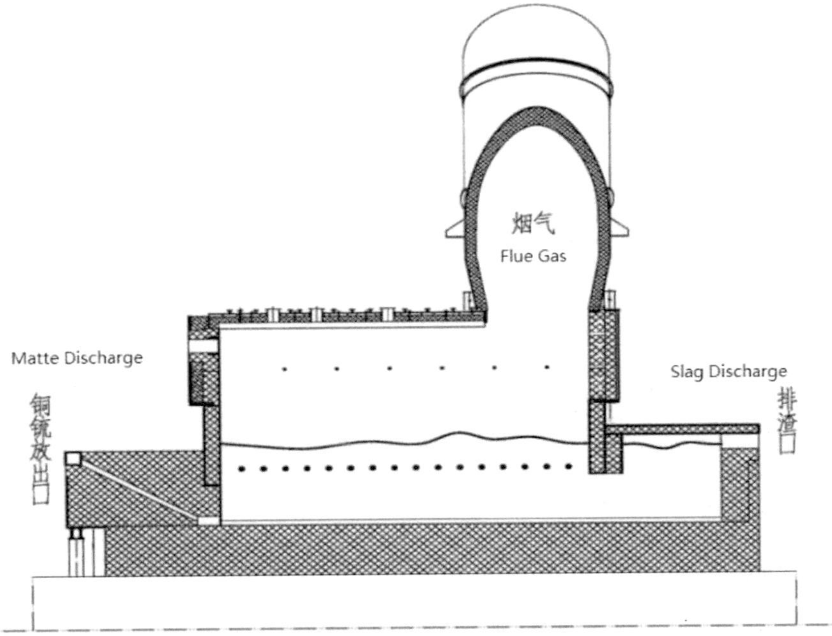

Figure 7.15 Jifeng type side-blown smelting furnace schematics (Bin et al., 2019). Figure originally published in Proceedings of the 58th Conference of Metallurgists Hosting the International Copper Conference 2019. *(Reprinted with permission of the Canadian Institute of Mining, Metallurgy and Petroleum).*

7.5.9 Baijin and Jifeng SBF typical operating parameters

Table 7.11 shows some selected operating parameters for the SBF furnaces at Chifeng and Heding smelters (Bin & Suping, 2019; Wang, Liu, Yang, Tang, & Liao, 2019; Kapusta & Watt, 2019; Yan, 2016).

7.6 Concluding remarks

The main characteristic of bath smelting processes is the transfer of oxygen into molten phase to produce copper matte. Oxygen can be transferred to the slag as in case of TSL processes, Vanyukov, and the SBF-Baijin process or to the molten bath as in case of the SKS-BBS, Noranda, El Teniente, and SBF Jifeng processes.

Bath smelting processes are flexible and capable of processing a variety of materials from low- to high-grade copper. They can operate from medium (35%) to very high (90%) oxygen enrichment conditions. Depending on the process, they can produce matte grades from 55% to 74% copper.

Submerged blowing of blast causes violent stirring of the matte/slag bath. This results in rapid melting and oxidation of the furnace charge. It also prevents excessive deposition of solid magnetite in the furnace even under highly oxidizing conditions. The violent stirring also permits extensive smelting of scrap and reverts.

7.6.1 Tuyere lance processes

SKS-BBS, SBF, Teniente, and Noranda are submerged tuyere (or lance) smelting processes. They oxidize Fe and S by blowing oxygen-enriched air through submerged tuyeres into a matte-slag bath. The principal product is super high grade matte, 72%–74% Cu.

The Teniente and Noranda use horizontal refractory-lined cylindrical furnaces with a horizontal line of submerged tuyeres. The furnaces are rotatable so that their tuyeres can be rolled out of the liquids when blowing must be interrupted.

The SKS-BBS process has a similar furnace design injecting enriched oxygen via lances located on the furnace shell bottom.

The SBF process is a stationary vessel with two different design, injected enriched air via tuyeres into the slag (Baijin process) or matte (Jifeng process).

The SKS-BBS process has experienced a significant growth in the last 10 years, with several plants installed in China.

Compared with the traditional Noranda and El Teniente bath smelting processes, the SKS-BBS process uses high oxygen enrichment at relatively high injection pressure. The process also operates at high Fe/SiO_2 ratios, achieving low copper content in slag. The process operates under relatively low operating temperatures achieving larger furnace refractory campaigns compared with the Noranda and El Teniente processes.

A third submerged tuyere copper smelting technology is Vanyukov smelting. It uses stationary furnaces and is used in Russia and Kazakhstan.

7.6.2 TSL processes

TSL Metso Outotec Ausmelt® and ISASMELT™ smelting processes are done in vertically aligned cylindrical furnaces 3.5–5 m diameter and 12–16 m high. The smelting entails the following:

(a) dropping moist concentrate, flux, and recycle materials into a molten matte/slag bath in a hot furnace

(b) blowing oxygen-enriched air through a vertical lance into the matte/slag bath.

Most of the energy for smelting is obtained from oxidizing the Fe and S in the concentrate.

The principal product of the furnace is a matte/slag mixture. It is tapped into a hydrocarbon fired or electric settling furnace. The products after settling are 60% Cu matte and 0.7% Cu slag. The main advantages of the process are as follows:

(a) its small footprint, which makes it easy to retrofit into existing smelters

(b) its ability to process a variety of feed materials, and

(c) its small evolution of dust.

References

Alvear Flores, G. R. F., Hourn, M., & Salas, M., J.C. (2013). Xstrata Technology's approach for the processing oc copper bearing materials. In *Copper 2013 pyrometallurgy, volume III, Nickolas Themelis symposium on pyrometallurgy and process engineering* (pp. 389–400). Santiago: IIMCH.

Alvear Flores, G. R. F., Hunt, S. P., & Zhang, B. (2006). Copper Isasmelt — dealing with impurities. In F. Kongoli, & R. G. Reddy (Eds.), *Sohn international symposium, vol. 8: International symposium on sulfide smelting 2006* (pp. 673–685). Warrendale, PA: TMS.

Alvear Flores, G. R. F., Hunt, S. P., & Zhang, B. (2010). Feasibility to profitability with copper ISASMELTTM. In , *Pyrometallurgy: Vol. 2. Copper 2010* (pp. 615–630). Germany: GDMB, Clausthal-Zellerfeld.

Arthur, P. S., & Partington, P. J. (2007). Latest developments with copper ISASMELT™. In A. E. M. Warner, et al. (Eds.), *(Book 2): The Carlos Díaz symposium on pyrometallurgy: vol.III. Cu2007* (pp. 3–15). Montreal: CIM.

Arthur, P. S., & Simon, S. P. (2005). Isasmelttm – 25 years of continuous evolution (2005). In M. Nilmani, & W. J. Rankin (Eds.), *Sustainable developments in metals processing: Proceedings of the John Floyd international symposium, Melbourne, Vic., July 3-6* (pp. 73–94).

Ausmelt. (2005). Ausmelt TSL Technology: Capabiltity & experience in copper smelting & converting. *Ausmelt Global Environmental Technologies*, 22 September 2005.

Bill, J. L., Briffa, T. E., Burrows, A. S., Fountain, C. R., Retallick, D., Tuppurainen, J. M. I., Edwards, J. S., & Partington, P. (2002). Isasmelt — mount Isa copper smelter performance update. In R. L. Stephens, & H. Y. Sohn (Eds.), *Sulfide smelting 2002* (pp. 181–193). Warrendale, PA: TMS.

Binegar, A. H. (1995). Cyprus Isasmelt start–up and operating experience. In W. J. Chen, C. Díaz, A. Luraschi, & P. J. Mackey (Eds.), *Pyrometallurgy of copper: vol. IV. Copper 95–Cobre 95* (pp. 117–132). Montreal: CIM.

Bin, T., & Suping, Y. (2019). Application and development of modern copper metallurgical technologies in China's copper industry. In *E-proceedings copper 2019 pyrometallurgy*. Vancouver: CIMM. Paper 588809. ISBN-978-1-926872-44-5.

Burrows, A. (2011). *Personal communication.*

Burrows, A. S., Alvear Flores, G. R. F., & Tynybayev, A. T. (2013). Smelting of Kazakhstan concentrates at Ust-Kamenogorsk using a copper ISASMELT™ furnace. In , *Nickolas Themelis symposium on pyrometallurgy and process engineering: vol. III. Copper 2013 pyrometallurgy* (pp. 39–48). Santiago: IIMCH.

Burrows, A., & Azekenov, T. (2018). Ust-kamenogorsk metallurgical complex: A silent achiever. In B. Davis, et al. (Eds.), *Extraction 2018. The minerals, metals & materials series*. Cham: Springer. https://doi.org/10.1007/978-3-319-95022-8_29.

Campos, R., & Torres, L. (1993). Caletones smelter: Two decades of technological improvements. In C. A. Landolt (Ed.), *The Paul Queneau international symposium extractive metallurgy of copper, niquel and cobalt, volume II: Copper and nickel operations* (pp. 1441–1460). Warrendale: TMS.

Carrasco, C., Bobadilla, J., Duarte, G., Araneda, J., & Rubilar, S. (2007). Evolution of the Teniente converter – Caletones smelter. In A. E. M. Warmer, C. J. Newman, A. Vahed, D. G. George, P. J. Mackey, & A. Warczok (Eds.), *(Book 2): The Carlos Diaz symposium of pyrometallurgy: Vol. III. Copper–Cobre 2007* (pp. 325–337). Montreal: CIM.

Carrasco, J. (2020). *Personal Communication*. Glencore, Altonorte Smelter.

Chen, M., Zhao, B., Cui, Z., & Wei, C. (2016). Investigation of the accretions in bottom blown copper smelting furnace. In *Copper 2016 pyrometallurgy*. Kobe: MMIJ. Paper PY19-4.

Coleman, M., & Money, G. (2008). Increasing capacity and productivity in the metals market through pneumatic conveying and process injection technologies. In *Lead and zinc 2008* (pp. 1–16). The South African Institute of Mining and Metallurgy.

Collis, B. (2002). *Fields of discovery: Australia's CSIRO*. Crows Nest, Australia: Allen & Unwin.

Coursol, P., Mackey, P. J., Prevost, Y., & Zamalloa, M. (2007). Noranda Process reactor at Xstrata Copper – impact of minor slag components (CaO, Al_2O_3, MgO, ZnO) on the optimum %Fe/SiO_2 in slag and operating temperature. In A. E. M. Warmer, C. J. Newman, A. Vahed, D. G. George, P. J. Mackey, & A. Warczok (Eds.), *(Book 2): The Carlos Diaz symposium of pyrometallurgy: vol. III. Copper–Cobre 2007* (pp. 79–93). Montreal: CIM.

de Vries, D., Hunt, S., Dyussekenov, N., Milovanov, M., & Chikashi, H. M. (2016). Design, commissioning and operation of an ISASMELT furnace at the kansanshi copper smelter. In *Copper 2016 pyrometallurgy. Kobe: MMIJ. Paper PY 6-2.*

Edwards, J. S., Telford, P., & Yunus, L. (2013). Coming of age — 21 Years commercial operation of copper Isasmelt™ at Mount Isa. In , *Nickolas Themelis aymposium on pyrometallurgy and process engineering: vol. III. Copper 2013 pyrometallurgy* (pp. 367—374). Santiago: IIMCH.

Floyd, J. M. (2002). Converting an idea into a worldwide business commercializing smelting technology. *Metallurgical and Materials Transactions B, 36B*, 557—575.

Glencore Technology. (2020). https://www.isasmelt.com/en/installations/Documents/isasmelt-Installations.pdf.

Gonzalez, T., White, M., & Walters, G. (2019). In *Proceedings Copper 2019 Pyrometallurgy*. CIMM. Paper 602421. ISBN-978-1-926872-44-5.

Guzman, G., Gomez, Z., & Silva, A. (2007). Automatic flux feeding into El Teniente converter. In A. E. M. Warmer, C. J. Newman, A. Vahed, D. G. George, P. J. Mackey, & A. Warczok (Eds.), *(Book 2): The Carlos Diaz symposium on pyrometallurgy: vol. III. Copper—Cobre 2007* (pp. 301—313). Montreal: CIM.

Harris, C. (1999). Bath smelting in the Noranda process reactor and the El Teniente process converter compared. In D. B. George, W. J. Chen, P. J. Mackey, & A. J. Weddick (Eds.), *Copper 99eCobre 99, Vol. V: Smelting operations and advances* (pp. 305—318). Warrendale, PA: TMS.

Heath and Sherwood. (2021). *Smelter tuyere line management products*. https://heathandsherwood64.com/pdf/Tuyere/HS_Tuyere_Pyrometer.pdf. (Accessed 1 October 2021).

Herrera, E., & Mariscal, L. (2013). Controlling SO_3 formation in the off-gases process from the ISASMELT™ furnace at Southern Peru Ilo smelter. In , *Nickolas Themelis symposium on pyrometallurgy and process engineering: vol. III. Copper 2013 Pyrometallurgy* (pp. 61—74). Santiago: IIMCH.

International Copper Study Group. (2019). *ICSG Directory of Copper Mines and Plants*. Lisbon, Portugal: Published by ICSG.

ISASMELT™. (2020). *Copper Nickel smelter Instalations*. https://www.isasmelt.com/en/download/Brochures/ISASMELT%20Brochure.pdf. (Accessed 1 September 2020).

Li, B. (2016). Development of oxygen bottom-blowing copper smelting & converting technology. In *Copper 2016 pyrometallurgy*. Kobe: MMIJ. Paper PY21-3.

Kapusta, J., & Watt, L. (2019). The 2019 copper smelting survey. In *E-proceedings copper 2019 pyrometallurgy*. Vancouver: CIMM, ISBN -978-1-926872-44-5. Paper 592732.

Lile, L., Wang, W., Tian, Q., & Guo, X. (2016). Multiphase equilibrium modelling study on the oxygen bottom blowing copper smelting (SKS) process. In M. M. I. J. Kobe (Ed.), *Copper 2016 pyrometallurgy*. Paper PY21-2.

Ling, W. (2019). The utilization of copper continuous smelting, converting and fire-refining processes in China. In *E-proceedings copper 2019 Pyrometallurgy*. Vancouver: CIMM. Paper 595879. ISBN-978-1-926872-44-5.

MacLeod, I. M., Harlen, W., Cockerell, R. A., & Crisafulli, S. (1997). Advanced temperature control for a Copper Isasmelt plant. In *ECC 97, paper TH-M-H-6*. http://www.cds.caltech.edu/conferences/related/ECC97/proceeds/251_500/ECC316.PDF.

Mariscal, L. (2020). *Mexicana de Cobre*. SA de CV; Personal Communication.

Matusewicz, R., Hughes, S., & Hoang, J. (2007). The Ausmelt continuous copper converting (C3) process. In A. E. M. Warner, C. J. Newman, A. Vahed, D. B. George, P. J. Mackey, & A. Warczok (Eds.), *(Book 2); the Carlos Díaz symposium on pyrometallurgy: vol. III. Cu2007* (pp. 29—46). Montreal: CIM.

Metso Outotec Ausmelt. (2020). Metso Outotec Ausmelt® TSL Reference List. *Outotec*.

Morrow, A. B., & Gajaredo, M. (2009). Control of copper Teniente smelting units using integrated advanced control technologies. In H. Chang, L. Ryan, & S. Kennedy (Eds.), *Process control applications in mining and metallurgical plants* (pp. 187—198). Montreal: CIM.

Moyano, A., Rojas, F., Caballero, C., Font, J., Rosales, M., & Jara, H. (2010). The Teniente converter: A high smelting rate and versatile reactor. In , *Pyrometallurgy II, GDMB, Clausthal—Zellerfeld, Germany: vol. 3. Copper 2010* (pp. 1013—1023).

Nolte, M., & Kruger, B. (2019). Advances in complex concentrate smelting at the DPM Tsumeb smelter. In *E-proceedings copper 2019 pyrometallurgy*. Vancouver: CIMM. Paper 594846. ISBN-978-1-926872-44-5.

Ospanov, Y. (2020). *Personal communication*.

Prevost, Y. (2020). *Personal communication*. Glencore Horne Smelter.

Prevost, Y., Letourneau, P., Perez, H., Lind, P., & Lavoie, A. (2007). Improving flexibility: Recent developments at the Horne smelter. In A. E. M. Warmer, C. J. Newman, A. Vahed, D. G. George, P. J. Mackey, & A. Warczok (Eds.), *(Book 1): The Carlos Diaz symposium on pyrometallurgy: vol. III. Copper—Cobre 2007* (pp. 167—179). Montreal: CIM.

Ramachandran, V., Lehner, T., Diaz, C., Mackey, P. J., Eltringham, T., Newman, C. J., et al. (2003). Primary copper production e a survey of operating world copper smelters. In C. Diaz, J. Kapusta, & C. Newman (Eds.), *Copper 2003eCobre 2003, Vol. IV, Book 1: Pyrometallurgy of copper, the Hermann Schwarze symposium on copper pyrometallurgy* (pp. 3—119). Montreal: CIM.

Reuter, M. A., & Matusewicz, R. (2010). *Personal communication*.

Rosales, M., Font, J., Fuentes, R., Moyano, A., Rojas, F., Caballero, C., & Mackay, R. (2010). A fluid-dynamic review of the Teniente Converter. In , *Pyrometallurgy II: vol. III. Copper 2010* (p. 1095e1113). ClausthalZellerfeld, Germany: GMDB.

Sánchez, G. (2020). *Codelco Chile, Caletones Smelter, Personal Communication*.

Schlesinger, M. E., King, M. J., Sole, K. C., & Davenport, W. G. (2011). Extractive. Metallurgy of Copper(5th (ed.). Elsevier.

Solnordal, C. B., & Gray, N. B. (1996). Heat transfer and pressure drop considerations in the design of Sirosmelt lances. *Metallurgical and Materials Transactions B, 27B*(4), 221—230.

Tarasov, A. V., & Paretsky, V. M. (2003). Development of autogenous copper smelting processes in Russia and CIS countries. In C. Diaz, J. Kapusta, & C. Newman (Eds.), *Pyrometallurgy of copper (Book 1) the Hermann Schwarze symposium on copper pyrometallrgy: vol. IV. Copper 2003—Cobre 2003* (pp. 173—187). Montreal: CIM.

Voltura, S. A. (2004). Continuous improvements at the phelps Dodge Miami mining smelter. *Minerals and Metallurgical Processing, 21*, 158—163.

Walqui, H., Noriega, C., Partington, P., & Alvear, G. (2006). SPCC's 1200 TPA copper Isasmelt. In F. Kongoli, & R. G. Reddy (Eds.), *International symposium on sulfide smelting 2006: vol. 8. Sohn international symposium* (pp. 199—209). Warrendale, PA: TMS.

Wang, Q., Guo, X., & Tian, Q. (2016). Reaction mechanism in an oxygen bottom blown copper smelting furnace. In *Copper 2016 pyrometallurgy*. Kobe: MMIJ. Paper PY21-1.

Wang, W., Liu, T., Yang, W., Tang, B., & Liao, W. (2019). Industrial application of side-blowing furnace smelting process in China Nerin. In *E-proceedings copper 2019 pyrometallurgy*. Vancouver: CIMM. Paper 547828. ISBN-978-1-926872-44-5.

Xiaohong, H., Kefei, C., Shuangjie, F., & Xin, X. (2016). Optimization design for the oxygen lance of bottom-blowing smelting furnace. In *Copper 2016 pyrometallurgy*. Kobe: MMIJ. Paper PY21-4.

Yan, J. (2016). Recent operation of the bottom-blowing copper smelting and continuous copper converting technologies. In *Copper 2016 pyrometallurgy*. Kobe: MMIJ. Paper PY16-2.

Zapata, R. (2007). Continuous reactor Altonorte smelter. In A. E. M. Warmer, C. J. Newman, A. Vahed, D. G. George, P. J. Mackey, & A. Warczok (Eds.), *(Book 1): The Carlos Diaz symposium on pyrometallurgy: vol. III. Copper—Cobre 2007* (pp. 141—153). Montreal: CIM.

Suggested reading

Arthur, P. S. (2006). ISASMELT - 6,000,000 t/y and rising. In F. Kongoli, & R. G. Reddy (Eds.), *Sohn International symposium on Advanced processing of Metals and materials, Volume 8 — International symposium on Sulfide smelting 2006* (pp. 275—290). Pennsylvania: The Minerals, Metals and Materials Society: Warrendale.

Player, R., Fountain, C. R., Nguyen, T. V., & Jorgensen, F. R. (1992). Top-entry submerged injection and the Isasmelt technology. In J. K. Brimacombe, P. J. Mackey, G. J. W. Kor, C. Bickert, & M. G. Ranade (Eds.), *Proceedings of the Savard/Lee International symposium on Bath smelting* (pp. 215—229). Pennsylvania: The Minerals, Metals and Materials Society, Warrendale.

Wood, J., Hoang, J., & Hughes, S. (2017). Energy Efficiency of the Outotec® Ausmelt process for Primary copper smelting. *JOM, 69*, 1013—1020. https://doi.org/10.1007/s11837-017-2322-x

Yan, J. (2019). On the use of copper concentrates blended with complex gold concentrate for gold recovery using the SKS smelting process. In *E-proceedings copper 2019 pyrometallurgy*. Vancouver: CIMM, ISBN -978-1-926872-44-5. Paper 586813.

CHAPTER 8

Converting of copper matte

8.1 Introduction

As explained in Section 4.4, copper converting is the oxidation of molten Cu—Fe—S matte to form molten blister copper (99% Cu). It entails oxidizing Fe and S from the matte with oxygen-enriched air or air blast.

The main raw materials for converting are

- Cu—Fe—S matte (molten and/or solid) from the smelting stage
- flux to allow slag formation with the oxidized iron
- oxygen for the oxidation (air, or oxygen-enriched air)
- additional cooling agents to manage heat generation and increase copper production, in the form of Cu-bearing materials such as solidified Cu-bearing reverts and copper scrap.

The main products of copper matte converting are

1. molten blister copper, which is sent to fire refining and electrorefining,
2. molten slag, which is sent to Cu recovery, and
3. SO_2-bearing offgas, which is sent to cooling, dust removal, and H_2SO_4 manufacture.

The heat for converting is supplied entirely by Fe and S oxidation, i.e., the process is autothermal. As heat is released through the converting process, several other inputs are fed to the process to allow a proper heat management.

The converting process can be carried out on a batch or continuous basis. Table 8.1 compares some relevant fundamental aspects of batch and continuous copper converting processes.

A batch process involves addition of molten matte to a reactor, addition of silica flux and gradual oxidation in one or two stages, depending on the iron content of the copper matte:

- Slag blowing, meant to oxidize iron sulfide from the matte, and
- Copper blowing, to remove sulfur from copper sulfide, generating blister copper.

Air (often enriched) is blown into molten matte through submerged tuyeres (Figs. 8.1 and 8.2).

For smelters processing intermediate grade matte (55%—65% Cu), slag blowing is done in sequential stages. The reactor is charged with an initial load, starting the oxidation process for removal of iron. It is then loaded once again, after a first blowing to repeat the process and increase the total amount of material in the reactor, until achieving the right amount of high-grade matte (bath level), to continue with the copper blowing stage.

Extractive Metallurgy of Copper, Sixth Edition
ISBN 978-0-12-821875-4, https://doi.org/10.1016/B978-0-12-821875-4.00008-0

Table 8.1 General characteristics of copper converting processes.

Parameter	Batch	Continuous
Oxygen and sulfur potential	Gradual change to achieve required condition represented by sulfur content in blister copper	Fixed to achieve required blister quality
Temperature	Variable, managed by the addition of cooling agents 1180−1250°C	Fixed between 1230 and 1280°C depending on the selected process
Stages	Two stages: • Slag blowing • Copper blowing One stage: • Combined single blowing for high matte	One stage
Molten phases in equilibrium	Slag blowing: Cu matte and slag End Copper blowing: Cu and slag	• Two phases: Slag and blister copper • Three phases when saturated with sulfur: Cu, Cu_2S, and slag,
Initial matte grade	• 60%−65% standard matte grade • 70%−75% matte grade when receiving matte from bath smelting reactors operating at high matte grade conditions.	• 68%−75%
Charging method	• Batch	• Continuous • Semi-Continuous for and those feeding molten and molten/solid matte combination
Oxygen transfer mechanism	• Slag blowing: Oxygen directly reacts with FeS in the matte phase • Copper blowing: Oxygen directly reacts with Cu_2S in the white metal phase	• Flash converting: Oxygen reacts with matte in the reaction shaft forming Cu, Cu_2O, and FeO_x • Bath converting via tuyeres: Oxygen is transferred to the metallic phase • TSL option: Oxygen is transferred to the slag phase • Mitsubishi process: Molten matte is converted by copper oxide from the slag. Oxygen addition allows oxidation of copper to keep level of oxidation in the slag.
Slag system	• $FeO_x{-}SiO_2{-}Cu_2O$	• $CaO{-}FeO_x{-}Cu_2O$ system • $FeO_x{-}SiO_2{-}Cu_2O$ (double saturation) • $CaO{-}FeO_x{-}SiO_2{-}Cu_2O$

Table 8.1 General characteristics of copper converting processes.—cont'd

Parameter	Batch	Continuous
Cu in slag	• 5% in slag from slag blowing • Above 30% for copper blowing • Above 30% when processing high-grade copper matte	• Between 10% and 20% depending on the degree of oxidation and process • Some new Chinese technology have higher copper contents
Technology options	• Peirce–Smith converter • Hoboken converter • ISASMELT™ batch converting • Metso-Outotec Ausmelt® batch converting • Submerged lance converting refining (SLCR)	• Flash converting • Mitsubishi converting • Noranda converting • Ausmelt C3 • ISACONVERT™ • Bottom blowing converting (BBRc) • El Teniente continuous converting (not implemented) • Top blow multilance converting process (China)
Additional materials	• Copper scrap • WEEE • Au-bearing materials as fluxing agent • Reverts	• Copper scrap • WEEE • Au-bearing materials as fluxing agent • Reverts
Slag processing	Slag recycled to smelting unit or separately processed to recover copper (normally flotation)	Slag recirculated to the smelting and converting unit
Blister quality	• 300–500 ppm sulfur	• Two condensed phases: 2000 to 6000 ppm S • Three condensed phases: 8000 to 10,000 ppm S

Smelters processing high copper content matte grade (above 75%) use a single copper blowing stage, with a highly oxidized rich copper containing slag poured at the end of the converting process.

The copper continuous converting process, on the other hand, has a different operating strategy. The reactor operates at a fixed oxygen–sulfur potential, corresponding to the required sulfur level in blister. This condition is defined by the copper content in the slag in equilibrium with the blister copper, aimed to instantaneously remove iron and sulfur and produce the required blister copper. These two operating philosophies are described in Chapter 4.

The main two fluxes used for converting processes are

- Silica: Used for batch and continuous converting
- Lime or lime–silica mixture: Used for continuous converting

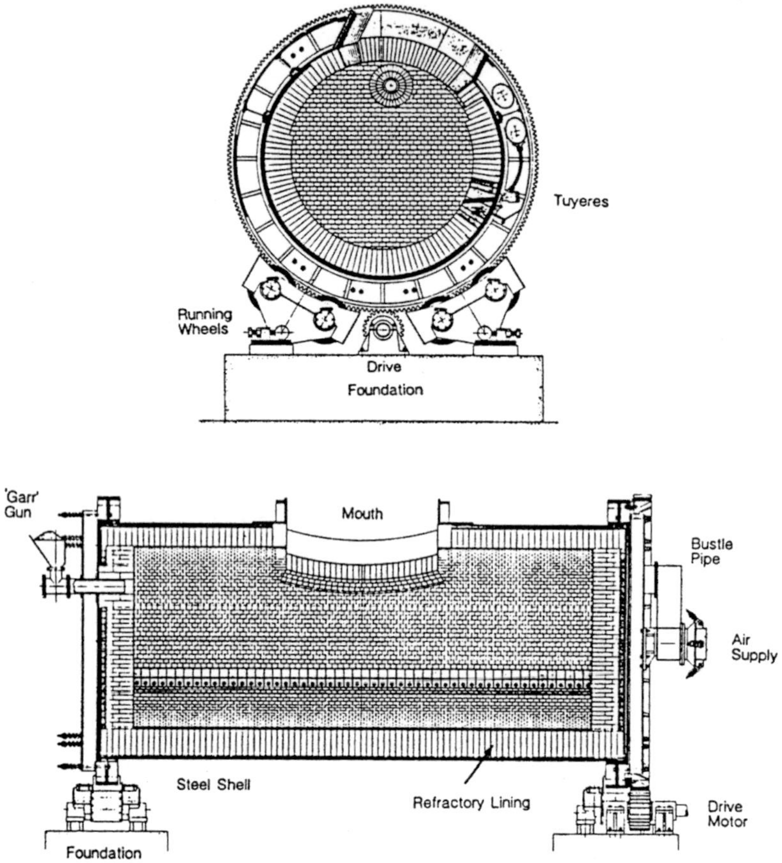

Figure 8.1 Peirce–Smith Converter. Note the blast supply, rotating seal and bustle pipe connected to the tuyeres. Flux and Cu-bearing reverts are added from bins above the offgas uptake or by air gun (*Garr Gun*) through an end of the converter. The converter rotates around its long axis in order to lift its tuyeres out of the liquids when it is not blowing and submerge them correctly in the matte during blowing, Fig. 8.2. *(Converter drawing courtesy Harbison–Walker Refractories, Pittsburgh, PA).*

8.2 Technology options for batch and continuous copper converting

Most of the smelters are currently using the Peirce–Smith batch converting process. Even though this technology is more than 100 years old, smelters are still investing in upgrading these installations and adding ancillary supporting systems to improve their environmental performance and meet tighter regulations (Gas Cleaning & Energy Solutions, 2020; Ramírez, 2019). Hoboken converters, an alternative batch converting technology, are currently in use at Caraiba Metais and Miami Smelter.

A new variation of batch copper converting is the Dongying–Fangyuan converting–refining process, to be discussed in Chapter 9 as part of continuous copper processes. In

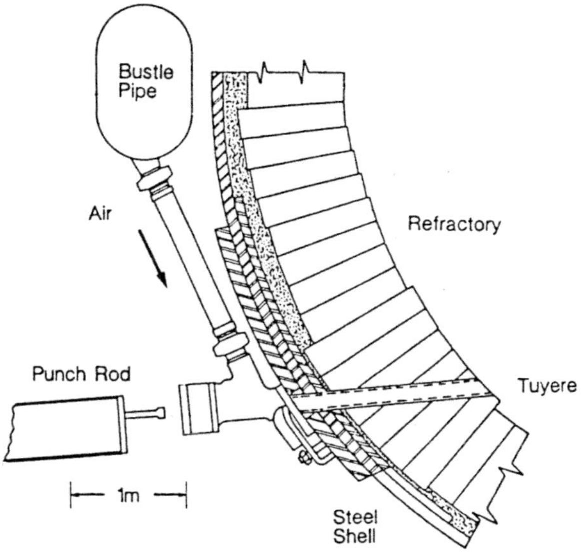

Figure 8.2 Details of Peirce–Smith converter tuyere. The tuyeres are nearly horizontal during blowing. Blast pressure is typically 1.2 bar (gauge) at the tuyere entrance. *(From Vogt, J. A., Mackey, P. J., & Balfour, G. C. (1979). Current converter practice at the Horne smelter. In: R. E. Johnson (Ed.),* Copper and nickel converters *(pp. 357–390). Warrendale, PA: TMS. Reprinted by permission of TMS).*

this process, the gradual oxidation is managed until the copper achieves sulfur and oxygen contents of anode copper quality, making it suitable for casting and electrorefining (Cui et al., 2016).

Continuous copper converting, aimed to improve productivity, reduce cost, and improve environmental performance, has been a permanent focus of attention in the copper industry.

In 1985, Kennecott and Outokumpu reported their initial testwork to develop flash converting. The testwork involved the exploration of several process conditions to convert ground copper matte (55%–70% Cu) to blister copper (Asteljoki et al., 1985). The concept was finally industrially implemented at Kennecott Utah Copper Smelter in 1995. Noranda (now Glencore) capitalized their experience in the development of the Noranda Process (also tested for a direct to blister process), with pilot test work conducted between 1991 and 1993, engineering design, construction, and commissioning of the Noranda Converter in 1997 (Boisvert et al., 1998). The Mitsubishi process, on the other hand, explored the option of combining their continuous converting process developed with the Noranda reactor at Port Kembla smelter in Australia in 2000 (West et al., 2002). Top Submerged Lance technology providers Ausmelt (now Meto Outotec Ausmelt) and Glencore also reported their efforts to implement their converting processes since early 2000, with new processes tested and industrially demonstrated in China (Wood, Matusewicz, & Reuter, 2009) and Zambia (Hunt & de Vries, 2019). Despite these efforts, no more than five plants were using continuous converting processes in 2010. However, after 2010 and following the increasing copper demand driven by the Chinese economy, several new plants have been commissioned in China. Table 8.2 shows the current list of the smelters using continuous converting technologies (Bin & Suping, 2019; ICSG, 2017; Kapusta, 2019b; Yongcheng et al., 2019). Most of these plants use flash converting and newly developed Chinese processes such as bottom blowing converting and top blowing converting (Feng & Huilan, 2019).

Table 8.2 Continuous copper converting processes.

Process	Operating plant	Plant capacity, kt Cu/y
Flash Converting	Rio Tinto Kennecott Copper Smelter	200
Cyclone Converting	Jinguan Nonferrous Metals Co, Fangchenggang	400
	copper smelter	400
	Chinalco Southeast Copper Indutries, Ningde smelter	300
	Sinogold copper smelter	400
	Yanggu Xiangguang, Liaocheng, Yanggu smelter	400
	Tongling Non-Ferrous Metals Group, Tongling smelter	300
	Henan Zhongyuan Gold Smelter Co, Sanmenxia smelter	
Noranda Converting	Horne smelter	220
Mitsubishi C furnace	Naoshima smelter	228
	Birla Copper Dahej smelter	212
	Gresik Copper Smelter	260
	LS-Nikko	300
Top Blowing Converting	Nanguo Copper smelter	300
	Yunnan Copper Chifeng smelter	400
	Yantai Pengui Copper	100
Metso Outotec Ausmelt® C3	Zhongtiaoshan	100
	Yunnan Tin Copper	100
ISACONVERT™[a]	Kansanshi	80
BBC[c]	Yuguang Gold and Lead Co., Henan smelter	130
	Jincheng Metallurgical Corp. Henan smelter	40
	Baotou Huading Copper Smelter	100
	Dongying–Fangyuan[b]	280

[a] Process demonstrated at the site.
[b] Converting process can be considered continuous from the point of view of receiving molten matte from the smelting unit.
[c] According to Yongcheng (2019), five copper smelters in China have applied the BBC process.

Kapusta, J. (2019b). The 2019 copper smelter survey. In: *Proceedings of the 58th conference of metallurgists hosting the international copper conference 2019* (*Article 592732*). Montreal: CIM; International Copper Study Group. (2017). Smelter data base. https://www.icsg.org/.php/component/jdownloads/finish/166-other/1575-icsg-directory-of-mines-and-plants-2011-03?Itemid=0. Yongcheng, Z., Bing, L., & Kai, L. (2019). Engineering and production practice of double bottom–blowing continuous copper smelting process with "Three connected furnaces" arrangement. *Proceedings of the 58th conference of metallurgists hosting the international copper conference 2019* (*Article 592917*). Montreal: CIM.

8.3 Batch converting

8.3.1 Batch converting chemistry

The overall batch converting process may be described by the schematic reaction:

$$\underset{\substack{\text{molten}\\\text{matte}}}{Cu-Fe-S} + \underset{\substack{\text{in air and}\\\text{oxygen}}}{O_2} + \underset{\text{in flux}}{SiO_2} \rightarrow Cu^{\circ}(\ell) + \underset{\substack{\text{molten slag with}\\\text{some solid } Fe_3O_4}}{\left\{\begin{matrix} 2FeO\cdot SiO_2 \\ Fe_3O_4 \end{matrix}\right\}} + SO_2 \quad (8.1)$$

For intermediate copper matte grades, converting takes place in two stages. The first stage is the *slag-forming* stage when Fe and S are oxidized to FeO, Fe_3O_4, and SO_2 by reactions like

$$FeS(l) + 1.5\ O_2(g) \rightarrow FeO(s) + SO_2(g) \quad (8.2)$$

$$3FeS(l) + 5O_2(g) \rightarrow Fe_3O_4(s) + 3SO_2(g) \quad (8.3)$$

The melting points of FeO and Fe_3O_4 are 1377 and 1597°C, so silica flux is added to form a liquid slag with FeO and Fe_3O_4.

$$2FeO(l) + SiO_2(s) \rightarrow 2FeO\ SiO_2(l) \quad (8.4)$$

The slag-forming stage is finished when the Fe in the matte has been lowered to about 1%. The principal product of the slag-forming stage is impure molten Cu_2S (*white metal*) at ~1200°C.

The second stage is *copper making* when the sulfur in Cu_2S is oxidized to SO_2. Copper is not appreciably oxidized until it is almost devoid of S. Thus, the blister copper product of converting is low in both S and O (0.001%–0.03% S, 0.1%–0.8% O). Nevertheless, if this copper were cast, the S and O would form SO_2 bubbles or blisters, which give blister copper its name.

For high-grade matte, the converting process is carried out in a single combined slag-copper blowing stage. The chemistry is similar, but activities of magnetite and copper oxide are higher, increasing the amount of solids in equilibrium with the liquid slag and the copper content in the slag. The implications of this operational strategy are discussed in Section 8.3.

Industrially, matte is charged to the converter in several steps, with each step followed by oxidation of FeS from the charge. This is represented by several steps of increasing (oxidation) and reduction of matte grade (fresh copper matte added) during the slag blowing. Slag is poured from the converter after each oxidation step and a new matte addition is made (Fig. 8.3). In this way, the amount of Cu in the converter gradually increases until it is sufficient (100–250 tonnes Cu as molten Cu_2S) for a final copper making blow. At this point, the Fe in the matte is reduced by oxidation to about 1%, a final slag is removed, and the resulting white metal is oxidized to molten blister copper. The converting process is terminated when copper oxide begins to appear in samples of the molten copper.

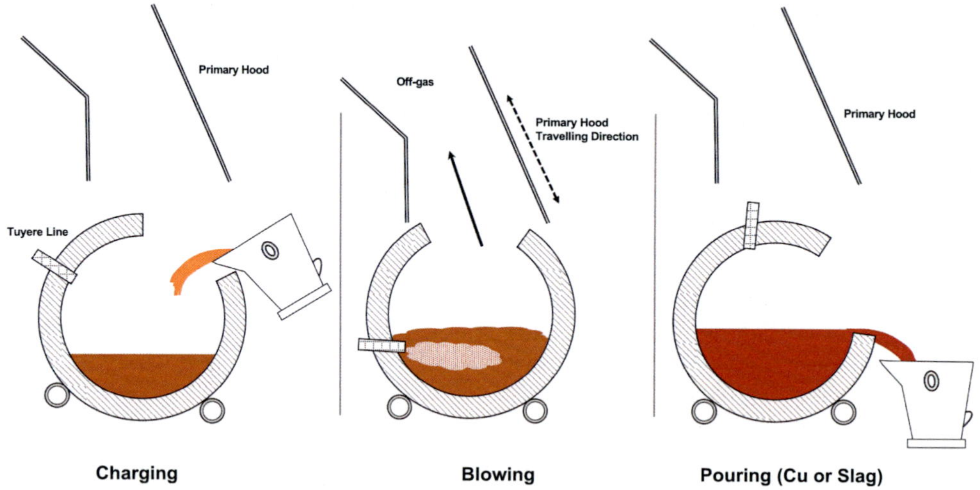

Figure 8.3 Positions of a Peirce–Smith converter for charging, blowing, and skimming. SO_2 gas escapes the system unless the hooding is tight. *(Redrawn from Boldt, J. R., & Queneau, P. (1967). The winning of nickel. Toronto: Longmans Canada Ltd; courtesy Inco Limited).*

The molten copper is poured from the converter into ladles and transported by crane to a fire-refining furnace (Chapter 12) for S and O removal and casting of anodes. A start-to-finish converting cycle typically lasts 6–12 h (Tables 8.3 and 8.4).

Table 8.3 Dimensions and production details of three Peirce–Smith converter operations.

Smelter	Hibi Kyodo Smelting Co. Tamano, Japan	Sumitomo Toyo, Japan	Pan Pacific Copper Saganoseki, Japan
Number of converters			
Total	3	4	4
Hot	2	3	3
Blowing at one time	1	2	2
Converter details			
Diameter × length, inside shell, m	3.96 x 12.7	4.2 × 11.9	4.2 × 11.5
Number of tuyeres			
Total	60	58 (No. 1–3) 60 (No. 4)	56
active	60		
Tuyere diameter, mm	50		49.5
Usual blast rate per converter			
Slag blow, Nm^3/min	730-780	660	650
Copper blow, Nm^3/min	730–770	700	650

Table 8.3 Dimensions and production details of three Peirce–Smith converter operations.—cont'd

Smelter	Hibi Kyodo Smelting Co. Tamano, Japan	Sumitomo Toyo, Japan	Pan Pacific Copper Saganoseki, Japan
Usual vol-% O_2 in blast			
Slag blow	26–28	23–24	22.3
Copper blow	22–23	21	22.3
SO_2 in offgas, vol-%	6–10	15–18	6–12
Production details (per converter)			
Inputs, tonnes/cycle			
Molten matte	205	215 (62% Cu)	(63% Cu)
Source	Outotec flash furnace + ESCF	Outotec flash furnace	Outotec flash furnace + ESCF
Other inputs, tonnes			
Slag blow	17 ladle skulls	22	
Copper blow	49 Cu scarp	75	
Outputs, tonnes/cycle			
Blister copper	214	208	
Slag	65	69	
Slag average wt-% Cu	5.2	6.5	8.4
Slag $\%SiO_2/\%Fe$	0.41	0.45	0.46
Cycle time			
Usual converter cycle time, hours	10	13.6	4.9
Slag blow, hours	1.3	1.9	
Copper blow, hours	3.3	32	
Campaign details			
Time between tuyere line repairs, days	158	120	210
Copper produced between tuyere line repairs, tonnes	78,631	64,500	
Time between complete converter relines, years	–	–	
Refractory consumption, kg/tonne Cu	0.9	1.3	1.8

Table updated with the following sources: Suzuki, Y., Hoshino, Y., Yamamoto, K., Mori, K., Kawanaka, Y., & Hattori, Y. (2016). Improvements in PS converter operation at Toyo smelter and refinery. In: *Proceedings of copper 2016* (pp. 678–690). Tokyo: MMIJ; Katamata, K., Sasaki, T., Motomura, T., & Chida H. (2016). Operational excellence in Peirce–Smith converting at Saganoseki smelter. In: *Proceedings of copper 2016* (pp. 666–677). Tokyo: MMIJ; Taniuchi (2016); private communications with JX Nippon Mining & Metals Corporation. (2020); Sumitomo Metal Mining Co. Ltd. (2020) and Mitsui Mining & Smelting Co. Ltd, & Metals Sector (2020).

Table 8.4 Dimensions and production details of three Peirce–Smith converter operations.

Smelter	Aurubis Hamburg, Germany	Boliden Rönnskär, Sweden	CODELCO Caletones, Chile
Number of converters			
Total	3	3	4
Hot	2	2	3
Blowing at one time	2	1	–
Converter details			
Diameter × length, inside shell, m	4.6 × 12.2	4.5 × 13.4	Three 4.5 × 10.6 One 4.0 × 10.6
Number of tuyeres			
Total	62	60	48
Active	60	54	46
Tuyere diameter, mm	50	50	63.5
Usual blast rate per converter			
Slag blow, Nm^3/min	700	740	Only copper blow
Copper blow, Nm^3/min	700–800	830	600
Usual vol-% O_2 in blast			
Slag blow	23	23.5	None
Copper blow	23	2422	21
SO_2 in offgas, vol-%	8–14	5–12	15
Production details (per converter)			
Inputs, tonnes/cycle			
Molten matte	300 (63% Cu)	350 (58% Cu)	200 (74.3% Cu)
Source	Outotec flash furnace + ESCF	Flash-electric-Kaldo furnace	Teniente and slag cleaning furnaces
Other inputs, tonnes			
Slag blow	15 ladle skulls + 10 secondaries	25 ladle skulls + 15 secondaries	None
Copper blow	75 Cu scrap	70-110 Cu scrap	35 reverts
Outputs, tonnes/cycle			
Blister copper	260–300	270–310	145
Slag	70	140–160	30
Slag average wt-% Cu	4	6	25
Slag SiO_2/Fe	0.55	0.75	–
Cycle time			
Usual converter cycle time, hours	9.5	7.5	7 to 7.5
Slag blow, hours	2	2	None
Copper blow, hours	5	4	5

Table 8.4 Dimensions and production details of three Peirce–Smith converter operations.—cont'd

Smelter	Aurubis Hamburg, Germany	Boliden Rönnskär, Sweden	CODELCO Caletones, Chile
Campaign details			
Time between tuyere line repairs, days	60	70	30 tuyere line (180 tuyere line + body)
Copper produced between tuyere line repairs, tonnes	55,000	32,000	11,200
Time between complete converter relines, years	1	0.85	2.0
Refractory consumption, kg/tonne Cu	1.9	4	4.5

Table updated with: Private communication with (Aurubis, 2020; Boliden, 2020).

8.3.2 Copper making reactions

Blowing air and oxygen into molten white metal creates a turbulent Cu_2S–copper mixture. The products of oxidation in this mixture are SO_2, molten copper and copper oxide. The molten copper is dense and fluid. It quickly sinks below the tuyeres. The most probable copper making reactions are

$$Cu_2S + 1.5O_2 \rightarrow Cu_2O + SO_2 \tag{8.5}$$

$$Cu_2S + 2Cu_2O \rightarrow 6Cu(l) + SO_2 \tag{8.6}$$

though some copper may be made directly by

$$Cu_2S + O_2 \rightarrow 2Cu(1) + SO_2 \tag{8.7}$$

In principle, there are three sequential steps in copper making, as indicated on the Cu–S phase diagram (Fig. 8.4).

1. Blowing of air and oxygen into the Cu_2S removes S as SO_2 to give S-deficient white metal, but no metallic copper. The reaction for this step is

$$`Cu_2S + xO_2 \rightarrow Cu_2S_{1-x} + xSO_2 \tag{8.8}$$

It takes place until the S is lowered to 19.6% (Fig. 8.4, point b, 1200°C).

2. Subsequent blowing of air and oxygen creates a second liquid phase, metallic copper (1% S, point c), to appear. It appears because the average composition of the liquids is now in the liquid–liquid immiscibility region. The molten copper phase is dense and sinks to the bottom of the converter (Fig. 8.5). Further blowing oxidizes additional S from the Cu_2S, and the amount of molten copper increases at the expense of the white metal, according to reaction (8.6). As long as the combined average

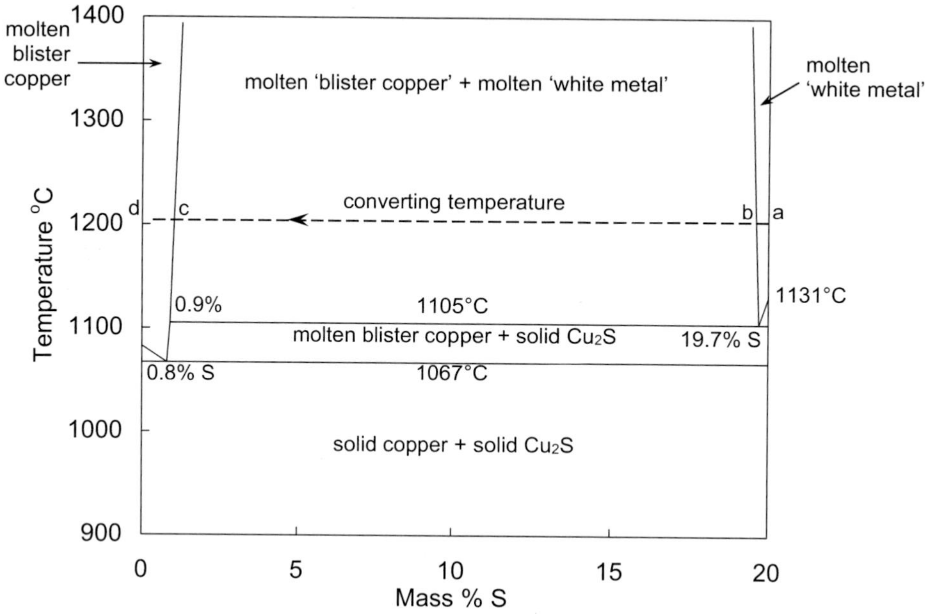

Figure 8.4 Cu—S equilibrium phase diagram showing copper making reaction path (a, b, c, d, 1200°C) (Sharma & Chang, 1980).

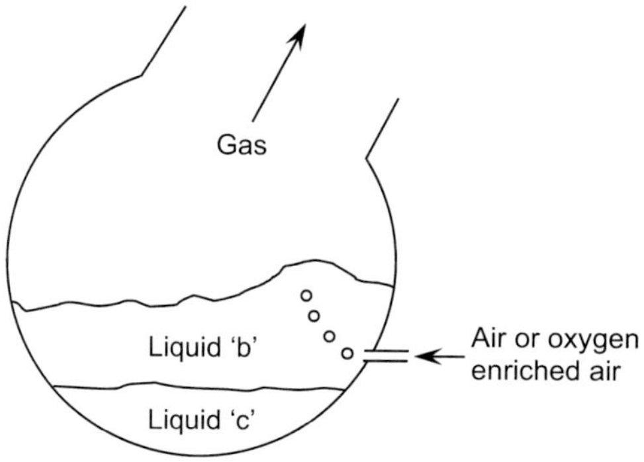

Figure 8.5 Sketch of Peirce—Smith converter and its two immiscible liquids during the copper making stage of converting. In practice, the liquid "b" region is a Cu_2S—Cu—Cu_2O—gas foam/emulsion from which metallic copper "c" descends and SO_2 and N_2 ascend. The immiscibility of copper and Cu_2S is due to their different structures; copper is metallic while Cu_2S is a semiconductor. *(After Peretti, E. A. (1948). An analysis of the converting of copper matte.* Discussions of the Faraday Society, *4, 179—184).*

composition of the system is in the immiscibility range, the converter contains both white metal (19.7% S) and molten copper (1% S). Only the proportions change.

3. Eventually the white metal becomes so S deficient that the sulfide phase disappears and only molten copper (1% S) remains. Further blowing removes most of the

Table 8.5 Distribution of impurity elements during Peirce–Smith converting of low and high-grade mattes (Mendoza & Luraschi, 1993; Vogt et al., 1979).

	54% Cu matte feed			70% Cu matte feed		
	Distribution %			Distribution %		
Element	To blister copper	To converter slag	To converter offgas	To blister copper	To converter slag	To converter offgas
As	28	13	58	50	32	18
Bi	13	17	67	55	23	22
Pb	4	48	46	5	49	46
Sb	29	7	64	59	26	15
Se	72	6	21	70	5	25
Zn	11	86	3	8	79	13

remaining S (point d). Great care is taken during this period to ensure that the copper is not overoxidized to Cu_2O. This care is necessary because Cu_2S is no longer available to reduce Cu_2O back to Cu by reaction (8.5).

Step (1) is very brief, because of the reduced requirements for S oxidation. Step (3) is also brief. Its beginning is marked by a change in the converter flame color from clear to green when metallic copper begins to be oxidized in front of the tuyeres. This tells the converter operator that the copper blow is nearly finished.

8.3.3 Elimination of impurities during converting

The principal elements removed from matte during converting are Fe and S. However, many other impurities are partially removed as vapor or in slag. Table 8.5 shows some minor element distributions. The outstanding feature of the data is that impurity retention in the product blister copper increases significantly with increasing matte grade (%Cu in matte). This is because high-Cu mattes have less blast blown through them, and they form less slag.

The table also shows that significant amounts of impurities report to the offgas. They are eventually collected during gas cleaning as part of the generated dust or in effluents from the acid plant. The collected dust contains sufficient Cu to be recycled to the smelting furnace. However, such recycle returns all impurities to the circuit as well. For this reason, some smelters treat the dusts for impurity removal before they are recycled (Mendoza and Luraschi, 1993). Bismuth, in particular, is removed because it causes brittleness in the final copper anodes and it can be a valuable byproduct.

8.4 Industrial Peirce–Smith converting operations

Modern industrial Peirce–Smith converters are typically 4.5 m diameter by 12 m long. They consist of a 5 cm steel shell lined with ~0.5 m of magnesite–chrome refractory

Table 8.6 Representative analyses of converter raw materials and products, mass%.

	Cu	Fe	S	O	As	Bi	Pb	Sb	Zn	Au	Ag
Matte	45–75	3–30	20–23	1–3	0–0.5	0–0.1	0–1	0–0.5	0–1	0–0.003	0–0.3
White metal (Cu_2S)	79	~1	~20	<1							
Blister copper	~99	0.001–0.3	0.001–0.3	0.1–0.8	0–0.2	0–0.03	0–0.5	0–0.1	0	0–0.004	0–0.5

The data are from industrial surveys and Ohnson, R. E., Themelis, N. J., & Eltringham, G. A. (1979). A survey of worldwide copper converter practices. In: R. E. Johnson (Ed.), *Copper and nickel converters* (pp. 1–32). Warrendale, PA: TMS; Pannell, D. G. (1987). A survey of world copper smelters. In: J. C. Taylor & H. R. Traulsen (Eds.), *World survey of nonferrous smelters* (pp. 3–118). Warrendale, PA: TMS; Lehner, T., Ishikawa, O., Smith, T., Floyd, J. Mackey, P., & Landolt, C. (1993). The 1993 survey of worldwide copper and nickel converter practices. In: J. D. McCain & J. M. Floyd (Eds.), *Converting, fire refining and casting* (pp. 1–8). Warrendale, PA: TMS.

brick. Converters of these dimensions treat 600–1000 tonnes of matte per day to produce 500–900 tonnes of copper. Typical compositions of converter raw materials and products are shown in Table 8.6.

A smelter will have two to five converters depending on its overall smelting capacity. Oxygen-enriched air or air is blown into a converter at 600–850 Nm^3/min and 1.2 bar gauge. It is blown through a single line of 50 mm diameter tuyeres, 40 to 60 per converter. It enters the matte 0.5–1 m below its surface, nearly horizontal (Lehner et al., 1993).

The flowrate per tuyere is about 14 Nm^3/min at a velocity of 100–130 m per second. Blowing rates above about 17 Nm^3/min/tuyere cause slopping of matte and slag from the converter (Johnson et al., 1979). High blowing rates without slopping are favored by deep tuyere submergence in the matte (Richards et al., 1986).

About half of the operating Peirce–Smith converters enrich their air blast with industrial oxygen, up to ~29 vol-% O_2.

8.4.1 Tuyeres

Peirce–Smith tuyeres are carbon steel or stainless steel pipes embedded in the converter refractory (Fig. 8.2). They are joined to a distribution *bustle pipe* which is affixed the length of the converter and connected through a rotatable seal to a blast supply flue. The blast air is pressurized by electric or steam driven blowers. Industrial oxygen is added to the supply flue just before it connects to the converter.

Steady flow of blast requires periodic clearing (*punching*) of the tuyeres to remove matte accretions which build up at their tips, especially during the slag blow (Fig. 8.6; Bustos et al., 1984, 1988). Punching is done by ramming a steel bar completely through the tuyere. It is usually done with a Gaspé mobile carriage puncher which runs on rails behind the converter. Fig. 8.7 shows (a) an automatic Gaspé tuyere puncher and (b) an operator control desk. Gaspé can be (Hinsberger & Marinigh, 2020; Marinigh, 2009)

- Manual: With an operator protective cab mounted on the back of the machine with an on-board manual control to traverse the punching machine and punch the tuyeres,

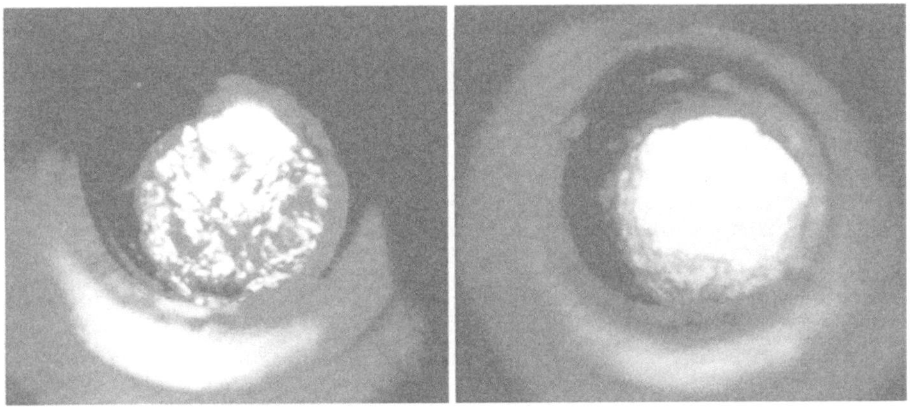

Figure 8.6 Photograph showing buildup of accretion at the interior end of a Peirce–Smith converter tuyere (Bustos et al., 1984). Left, tuyere is nearly blocked; right, the accretion has dislodged spontaneously. Bustos et al. (1988) report that accretion "tubes" are formed in front of the tuyeres. They also indicate that tuyere blockage is discouraged by high matte temperature and oxygen-enrichment of the blast. This is particularly important near the end of the slag blow and the start of the copper blow. Clear tuyere conditions at the beginning of the copper blow often give free blowing conditions (without punching) during most or all of the copper blow. *(Photograph courtesy of Dr. Alejandro Bustos, Air Liquide).*

- Semiautomatic: In addition to the features of the manual machine, offers the option to have a CCTV camera to provide remote control of the puncher from a remote location, as shown in Fig. 8.7B. Alignment between tuyere puncher and tuyere is fully automated. The traversing drive uses a DC servomotor capable of storing the location of tuyeres relative to the punching machine. Punching cylinder control is manually controlled by the operator,
- Automatic: Machine not equipped with operator cab. In addition to the characteristic of the semiautomatic machines, all punching cylinder control is fully automated. CCTV camera is also useful for verifying alignment of the recorded tuyere location.

Punching machine area in front of the converter must be safely isolated to prevent accidents as automatic machines may start operating and move at any time. Access gates to the punching platform area are equipped with safety switches. Opening a gate disables the puncher's MCR. With the MCR deenergized, the puncher is isolated from all electrical power. Also, deenergizing the MCR automatically bleeds all air pressure from the puncher. A reset and start-up procedure is required to reinitiate the punching process.

An alternative to the Gaspé puncher is the Kennecott 4B–5 system, which consists of a pneumatic cylinder actuated punch rod contained in a housing bolted to each tuyere body (Pasca et al., 2005).

(a)

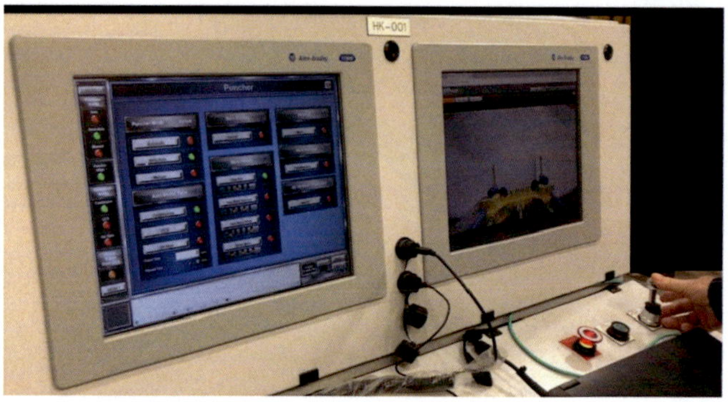

(b)

Figure 8.7 (A) Automatic Gaspé tuyere puncher allowing operator to work in a safe and comfortable environment. Machine offers automatic, fast, accurate positioning. (B) Operator Control Desk. HMI (left), Puncher Mounted CCTV (right), and Manual Controls. Operator Control Desk is typically located inside the Converter Control Room in a clean, comfortable environment for the Puncher Operator (Hinsberger & Marinigh, 2020).

8.4.2 Offgas collection

Peirce—Smith converters offgas is collected by a steel hood (usually water cooled) which fits as snugly as possible over the converter mouth. This is shown as the primary hood in Fig. 8.8 (Mori et al., 2009). The gas then passes through a waste heat boiler or water-spray cooler, electrostatic precipitators, before being directed to a sulfuric acid plant. Peirce—Smith converter offgases contain ~8 vol.-% SO_2 (slag blow) to ~10 vol.-% SO_2 (copper blow) after cooling and dust removal (Tables 8.1 and 8.2).

Also shown in Fig. 8.8 is the secondary hood and fugitive gas collection system. This system collects dilute SO_2 gas not captured by the primary hood. Primary and secondary hood system designs vary considerably. Fig. 8.9 shows an example of a primary hood system with water-cooled smooth panels (Brännström & Hedlund, 2013). All Peirce—Smith converters have a primary gas capture system, with most of them having these days also secondary gas capture systems.

The incorporation of effective secondary and sometimes even tertiary gas capture systems on Peirce—Smith converters has been driven by the need to meet more strengthen

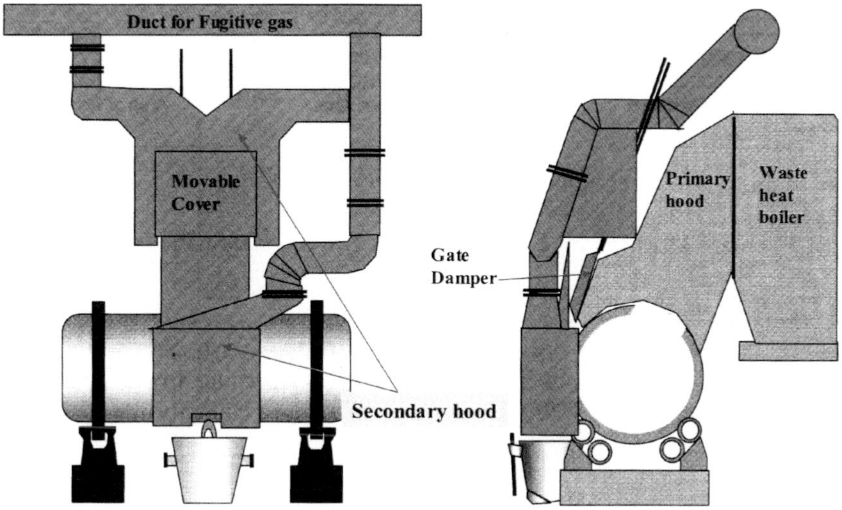

Figure 8.8 Offgas collection systems for a Peirce–Smith converter at the Sumitomo Toyo smelter (Mori et al., 2009).

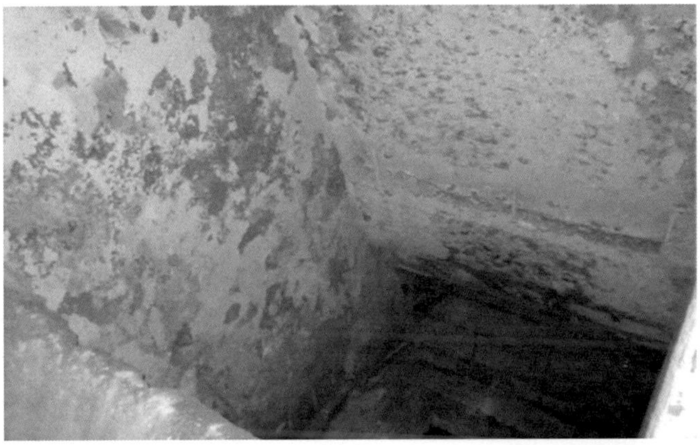

Figure 8.9 Outotec water panel cooled hood internal view during a shutdown after (Brännström & Hedlund, 2013). *(Reprinted with permission of the Southern African Institute of Mining and Metallurgy, SAIMM).*

environmental regulations aimed to improve sulfur capture, and improve the working conditions within the smelters. Examples of this trend have been the new regulations imposed for operating smelters in Chile, that are required to capture over 95% of sulfur (Ramirez, 2019), and for any new smelter to be erected in the country over 98% (Libertad y Desarrollo, 2012). The state of Arizona, in USA, also imposed a new standard in 2012, requiring copper smelters to reduce sulfur emissions during a 24-hour period. Based on this new regulation, FMI Miami Smelter and Asarco Hayden Copper developed new solutions to improve their sulfur capture. Fig. 8.10 shows the new installed Peirce–Smith converters and hoods at Asarco Hayden (Russell et al., 2016).

Figure 8.10 View of the Asarco Hayden smelter new converters with upgrade hood system (Russell et al., 2019). Published in Proceedings of the 58th Conference of Metallurgists Hosting the International Copper Conference 2019. *(Reprinted with permission of the Canadian Institute of Mining, Metallurgy and Petroleum).*

8.4.3 Temperature control

All heat required for maintaining the converter liquids at their specified temperatures results from Fe and S oxidation, i.e., from reactions like

$$FeS + 1.5O_2 \rightarrow FeO + SO_2 + heat \quad (8.9)$$

$$Cu_2S + O_2 \rightarrow 2Cu(l) + SO_2 + heat \quad (8.10)$$

Converter temperature can be raised by oxygen enrichment, which reduces the rate at which N_2 coolant enters the converter and can be lowered by increasing the amount of revert and scrap copper added.

8.4.4 Choice of temperature

Representative liquid temperatures during converting are 1200°C for the input matte and blister copper, and 1220°C for the skimmed slag. The high temperature during the middle of the cycle is designed to give rapid slag formation and fluid slag with a minimum of entrained matte. It also discourages tuyere blockage (Bustos et al., 1987). An upper limit of about 1250°C is imposed to prevent excessive refractory wear.

8.4.5 Temperature measurement

Converter liquid temperature is measured by means of an optical pyrometer sighted downwards through the converter mouth or a two-wavelength optical pyrometer periscope sighted through a tuyere (Marinigh, 2009). The tuyere pyrometer (Fig. 8.11) is preferred because it sights directly on the matte rather than through a dust-laden

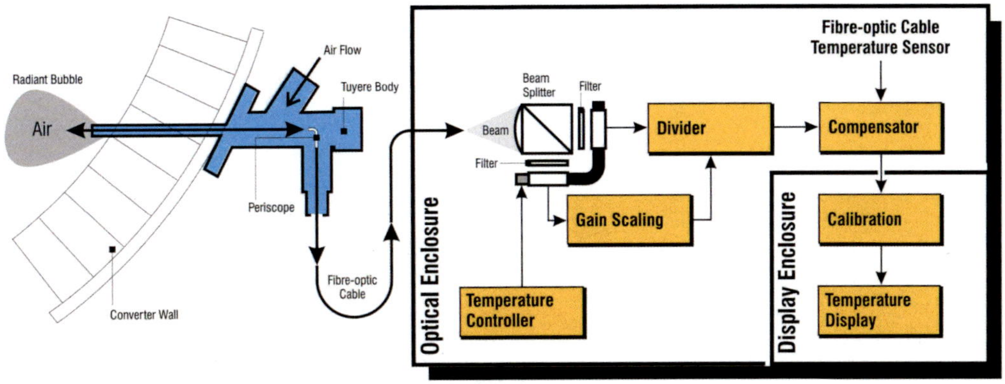

Figure 8.11 Tuyere pyrometer schematic (Heath and Sherwood, 2021).

atmosphere. RHI Magnesita has developed a continuous online temperature measurement system based on the use of a digital fiber optic pyrometer, which is embedded in a refractory housing (Vukovic et al., 2016).

8.4.6 Slag and flux control

The main objective of creating a slag in the converter is to liquefy newly formed solid FeO and Fe_3O_4 so they can be poured from the converter. SiO_2-bearing flux (e.g., quartz, quartzite, sand) is added for this purpose. Enough SiO_2 flux is added to produce an SiO_2/Fe ratio of ~0.5. The selection of the SiO_2/Fe ratio also defines the process temperature at which the converter operates to keep a liquid slag with an adequate amount of Fe_3O_4 to protect refractories (coating) without compromising copper recovery due to excessive matte entrainment. Acceptable Fe_3O_4 levels are typically 12%–18% (Eltringham, 1993). Thermodynamic and process modeling packages can support process engineers selecting suitable process conditions for the slag blowing stage. Fig. 8.12 shows the liquidus line for different Fe/SiO_2 ratio as a function of the remaining Fe content in matte and the temperature (Cardona et al., 2013).

Some smelters use Au- and Ag-bearing siliceous material as converter flux. The Au and Ag dissolve in the matte and proceed with copper to the electrorefinery where they are profitably recovered. These smelters tend to maximize flux input. Most smelters, however, use just enough flux to obtain an appropriately fluid slag. This minimizes flux cost, slag handling, and the expense of recovering Cu from the slag.

8.4.7 Slag formation rate

Flux is added through chutes above the converter mouth or via a high pressure air gun (*Garr Gun*) at one end of the converter. It is added at a rate that matches the rate of Fe oxidation (usually after an initial several-minute delay while the converter heats up). The

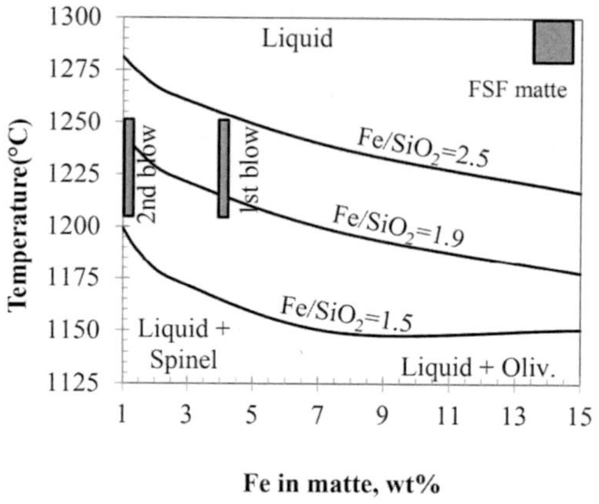

Figure 8.12 Example of process calculation showing the impact of Fe content in matte and operating temperature for different Fe/SiO_2 ratios (Cardona et al., 2013). *(Reprinted with permission of the Instituto de Ingenieros de Minas de Chile (Chilean Institute of Mining Engineers), IIMCH).*

flux is commonly crushed to 1–5 cm diameter. Sand (0.1 cm) is used in some smelters. Rapid reaction between O_2, matte, and flux to form liquid slag is encouraged by

1. high operating temperature,
2. steady input of small and evenly sized flux (Schonewille et al., 1993),
3. deep tuyere placement in the matte (to avoid overoxidation of the slag),
4. the vigorous mixing provided by the Peirce–Smith converter,
5. reactive flux.

Casley et al. (1976) and Schonewille et al. (1993) report that the most reactive fluxes are those with a high percentage of quartz (rather than tridymite or feldspar).

8.4.8 End point determinations

8.4.8.1 Slag blow

The slag-forming stage is terminated, and slag is poured from the converter when there is about 1% Fe left in the matte. Further blowing causes excessive Cu and solid magnetite in slag. The blowing is terminated when

1. metallic copper begins to appear in matte samples or when X-ray fluorescence shows 76%–79% Cu in matte (Mitarai et al., 1993)
2. the converter flame turns green from Cu vapor in the converter offgas
3. PbS vapor (from Pb in the matte feed) concentration decreases and PbO vapor concentration increases (Persson et al., 1999).

8.4.8.2 Copper blow

The copper making stage is terminated the instant that copper oxide begins to appear in copper samples. Copper oxide attacks converter refractory, so it is avoided as much as possible. The copper blow is ended, and metallic copper is poured from the converter when

1. copper oxide begins to appear in the samples
2. SO_2 concentration in the offgas falls because S is nearly gone from the matte (Shook et al., 1999)
3. PbO concentration in the offgas falls and CuOH concentration increases (H from moisture in the air blast; Persson et al., 1999).

8.5 Batch converting of high matte grades

Low iron content copper mattes are produced in bath smelting reactors such as El Teniente, Noranda, or the newly developed Chinese bottom blowing furnace. Due to the low iron content, and the relatively high oxygen potential required to continue the converting process, iron oxidation cannot be controlled in the same way as in a traditional slag blowing process, to allow a controlled slagging of iron and formation of fayalite by the addition of silica.

Cardona et al. (2013) reported the efforts carried out at Atlantic Copper to control iron removal during the converting of intermediate matte grades (Fig. 8.12). Using a thermochemical modeling approach, they concluded that the first slagging operation should be ideally conducted at iron contents over 4% to minimize excess soluble copper deportment to slags requiring further cleaning steps in electric furnaces. For the second slag blowing they recommended Fe/SiO_2 ratios under 1.9 to minimize spinel concentration and therefore increased entrained and soluble copper losses. However, in practice, this fine adjustment can be challenging due to operational constraints and unsuitable process control. In this case, the additional iron left in the matte will be responsible for additional generation of "mushy" slag during copper blowing.

Table 8.7 compares some relevant metallurgical parameters for the traditional two-stage batch converting process operated at Atlantic Copper and for single-stage batch converting process as mainly conducted in Chilean smelters. The main difference can be appreciated in the high-copper slag achieved in the conversion of high matte grades.

Warczok et al. (2001) reported that batch converting of high-grade copper mattes in the Pierce–Smith converter faced difficulties with the proper slagging of iron oxides. To overcome this operational challenge, smelters were overblown during a single copper making converting stage (combination of slag and copper blowing), allowing the formation of a liquid slag. However, this combined single stage led to high content of copper oxide in the converter slag that must be recycled to the smelting unit, creating new operational and environmental issues. They recommended a separate pyrometallurgical treatment of the converter slag to increase the overall smelter recovery and decrease impurity content in blister copper.

Table 8.7 Comparison between two-stage and one-stage batch converting.

Parameter	Two-stage batch converting		One-stage batch converting	
Initial matte grade	58%–65% Cu		68%–75% Cu	
Initial Fe content	10%–14%		5%–10%	
Typical Fe content before second matte charge	7%–9%		3%–5%	
Typical Fe content end of slag blowing	1.5%–4%		Not applicable	
Typical slag composition end of slag blowing (includes entrained copper matte)	Al_2O_3	0.8	Not applicable	
	ZnO	1.5		
	CaO	0.4		
	MgO	0.2		
	Cu (total)	1.0–4.0		
	Fe_3O_4	23–30		
	SiO_2	18–33		
Fe/SiO_2 slag blowing	1.5–3.0		3.0–4.0	
Typical slag bulk composition end of copper blowing (excludes entrained metallic copper)	Al_2O_3	0.5	Al_2O_3	8.1
	ZnO	0.9	ZnO	0.0
	CaO	0.2	CaO	1.0
	MgO	0.1	MgO	0.3
	Cu_2O	30–60	Cu_2O	55.0
			Fe_2O_3	18.0
	SiO_2	14–16	SiO_2	14.0
Blister composition	Cu	99.0–99.3	Similar to two-stage converting	
	O, ppm	4000–7000		
	S, ppm	100–300		
Cu in slag	Soluble: Iin oxide form Solid: Spinel		Soluble: in oxide form Solid: Spinel	
Operational strategy to reduce copper losses	Control Fe content in matte to 2% (Cardona et a., 2013)		Recycling of converting slag to smelting unit to reoxidize and recover copper. However, strategy might not be totally effective (Warczok et al., 2001)	
Operational strategy to increase slag fluidity	Overblow to solubilize magnetite with copper oxide		Overblow to solubilize magnetite with copper oxide	

Table 8.7 Comparison between two-stage and one-stage batch converting.—cont'd

Parameter	Two-stage batch converting	One-stage batch converting
Impact	Optimization of iron content in matte allows minimization of copper oxide generation to solubilize magnetite, reducing generation of converter reverts	Overblow increase copper oxide content and copper oxide attack to refractory materials
Other options to solubilize slag	Addition of silica to solubilize formed spinel phases at the end of converting. However, overall slag amount is increased	Addition of lime to solubilize copper oxide. Slag amount increases but entrained metallic copper is reduced (Henao et al., 2019)

Henao and Moyano (2016, (Henao et al., 2019) also reported the operational copper converting difficulties of Chilean copper smelters due to the processing of high-grade copper mattes (68%–75% Cu), with the generation of a semisolid converting slag. They recommended the addition of CaO to reduce the formation of Fe–Cu oxides.

In general, batch converting of high-grade copper mattes might be disadvantageous due to generation of highly oxidized slags, and increased reverts, containing high amount of copper oxide.

8.6 Oxygen enrichment of Peirce–Smith converter blast

Most smelters enrich their converter blast during part or all of the converting cycle. The advantages of O_2 enrichment are

1. oxidation rate is increased for a given blast input rate
2. SO_2 concentration in offgas is increased, making gas handling and acid making cheaper
3. the amount of N_2 coolant entering the converter per kg of O_2-in-blast is diminished.

The reduction of N_2 is important because it permits (a) generation of high temperatures even with high-grade mattes, which have less FeS "fuel," (b) rapid heating of the converter and its contents, and (c) increased melting of valuable coolants such as reverts and copper scrap.

The only disadvantages of high-O_2 blast are the cost of the oxygen, and the high reaction temperature that results at the tuyere tip. This leads to rapid refractory erosion in the tuyere area. This erosion is discouraged by blowing at a high velocity, which promotes tubular accretion formation and pushes the reaction zone away from the tuyere tip (Bustos et al., 1988). However, use of high pressure can overcome this disadvantage.

On balance, the advantages of O_2 enrichment outweigh the refractory erosion disadvantages, especially in smelters which wish to convert high mattes, maximize converting rate, especially if converting is a production bottleneck, and maximize melting of solids such as flux, reverts, and scrap.

The present upper practical limit of oxygen enrichment seems to be about 29 vol.-% O_2. Above this level, refractory erosion becomes excessive. This is because strong tubular accretions do not form in front of the tuyeres above 29 vol.-% O_2, causing the O_2-matte reactions to take place flush with the tuyere tip and refractory. Sonic high-pressure blowing is expected to permit higher oxygen levels (Hills et al., 2007).

8.7 Maximizing converter productivity

The production rate of a converter is maximized by

1. charging high Cu grade (low FeS) matte to the converter (Fig. 8.13)
2. blowing the converter blast at its maximum rate (including avoidance of tuyere blockages)
3. enriching the blast to its maximum feasible O_2 level
4. maximizing O_2 utilization efficiency (deep tuyere submergence)
5. minimizing idle time along with efficient crane scheduling for material transfers
6. maximizing campaign life (Section 8.4.3)

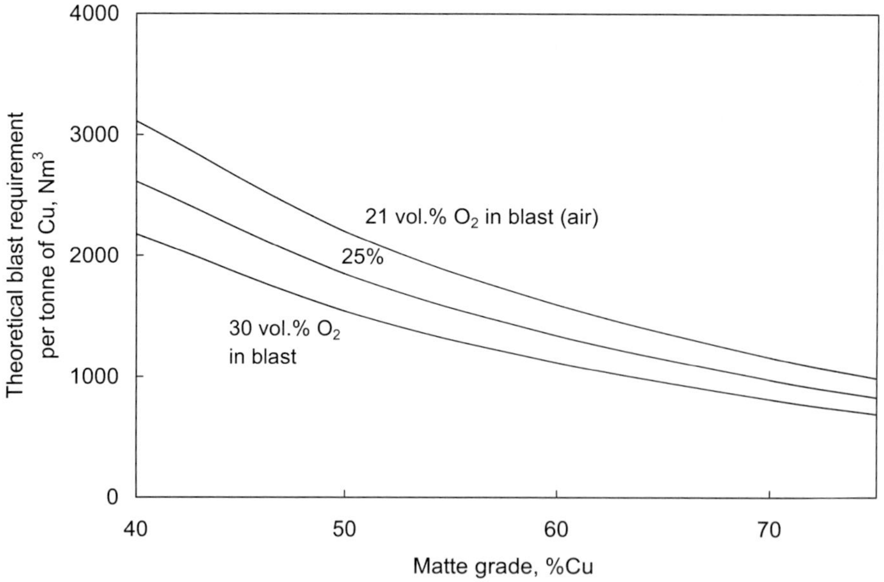

Figure 8.13 Theoretical air and oxygen-enriched air blast requirements for converting Cu_2S–FeS mattes to copper. Blast requirement decreases with increasing matte grade and %O_2 in blast. 100% O_2 efficiency is assumed.

High-grade matte contains little FeS, so it requires little O_2 (and time) to convert (Fig. 8.13). Rapid blowing of blast, a high %O_2 in blast, and high O_2 utilization efficiency all lead to rapid oxidation. However, the blast rate should be maintained at a level that prevents bath slopping, which leads to excessive wave motion on the surface and uncontrolled splashing (Kapusta, 2010). This splashing results in rapid accretion growth on the converter mouth, ultimately decreasing availability. Efficient scheduling, high crane availability, and shortening the time for transferring matte, slag and blister copper through optimizing ladle size also increases converter productivity (Mori et al., 2009).

8.7.1 Maximizing solids melting

An important service of the Peirce–Smith converter is melting of valuable solids using the heat from the converting reactions. Typical solids include Cu-bearing revert materials, scrap copper, and Au- and Ag-bearing flux. Cu concentrate is also melted in several smelters. Melting of solids is maximized by (a) maximizing blast O_2 enrichment and (b) blowing the converter at a rapid rate with the tuyeres deep in the matte. This maximizes reaction rate, and thus heat production rate (at an approximately constant heat loss rate from the converter).

The solids are added steadily to avoid excessive cooling of the converter liquids. This is easily done with flux and reverts, which can be crushed and added at controlled rates from storage bins above the converter. Scrap copper, on the other hand, is often large and uneven in shape. It is usually added in batches by crane with the converter in the charging position. This has the disadvantages that blowing must be stopped and the large batch of scrap may excessively cool the converter liquids.

Several converters have conveyor systems which feed large pieces of copper (scrap anodes and purchased blister copper) at a steady rate during blowing (Fukushima et al., 1988; Maruyama et al., 1998). This avoids excessive cooling and maximizes the converter's scrap melting capability. Up to 30% of a converter's blister copper product comes from copper scrap (Fukushima et al., 1988; Pannell, 1987).

8.7.2 Smelting concentrates in the converter

Most smelters melt scrap copper and solid reverts in the Peirce–Smith converter. Several smelters also smelt dried concentrates in their converters by injecting the concentrates through the tuyeres (Kawai et al., 2005; Mast et al., 1999). The technology is well-proven and has two advantages: (a) it can increase smelter capacity without major investment in a larger smelting furnace, and (b) it can lengthen the converting blow and improve removal of impurities, especially bismuth and antimony.

8.7.3 Maximizing campaign life

Converters produce 50,000–90,000 tonnes of blister copper before they must be taken out of service for tuyere refractory replacement. The replacement takes about 2 weeks and it is done many times before the converter must be completely relined (*shelled*). Copper production per tuyere-refractory replacement period (campaign life) has continued to increase due to

a. improved, higher quality refractories
b. higher-grade matte feeds (requiring less blowing per tonne of Cu)
c. better temperature measurement and control.

The most suitable refractories for the copper converting duties are fired magnesia—chromite bricks. Industrial evidence suggests that oxygen enrichment up to 25% O_2 enhances converter productivity without shortening campaign life. This is especially true if converter blowing rates are high (Verney, 1987). Enrichment above this level should be tracked to determine the optimum from the points of view of converter productivity and campaign life.

Suzuki et al. (2016) reported their efforts to extend the converter campaign life by reducing oxygen enrichment during the slag blowing stage, managing bath levels to reduce splashing to the waste heat boiler section and conducting a proactive management of spring hammers in the waste heat boiler. Combination of these actions allowed Toyo smelter to reach over 400 batches per campaign.

8.8 Recent improvements in Peirce—Smith converting

8.8.1 Shrouded sonic injection (Kapusta, 2019a)

Shrouded sonic injection has been of long-term interest to copper smelters; however, despite its potential reported benefits only a few smelters have adopted the technology (Kapusta, 2010).

The previous edition of this book reported that the ALSI (Air Liquide Shrouded Injector) technology was successfully demonstrated in the Thai Copper Smelter Peirce—Smith converters (Pagador et al., 2009). The process was able to

a. oxidize matte using 30%—60% O_2 blast, thereby increasing converter productivity and its ability to melt solids (Fig. 8.14)
b. eliminate the need to punch the converter (Section 8.4.1)
c. minimize refractory wear in the tuyere area.

The tuyere used to achieve these objectives consists of two concentric pipes: the inner pipe for oxygen-enriched air blast (30%—60% O_2) and the annulus for nitrogen. The purpose of the nitrogen is (a) to cool the circumference of the tuyere tip and (b) to protect the refractory around the tuyere by building up an accretion of solidified matte/slag.

The blast and nitrogen are blown in at high pressure, 4—8 bar gauge. This prevents the accretion from bridging across the tuyere and eliminates the need for punching.

Following the implementation in Thai Copper, the ALSO technology was implemented in the Copper Nickel Operations of Falconbridge, Canada, and Lonmin Platinum in South Africa. In addition to the traditional copper and copper—nickel matte converting applications, Glencore Nickel developed a new process using sonic injection by combining the slag-making and slag-cleaning processes, traditionally conducted in separated vessels into a single vessel operation (Hybrid Vessel) (Kapusta, 2019a).

8.8.2 Scrap injection

Pneumatic injection of electronic scrap (<12 mm) at 1 tonne per hour into a converter through a tuyere has been successfully implemented on a commercial scale (Coleman & Money, 2009). This avoids charging through the mouth, allowing particulate (soot) and dioxins to be captured by the primary gas handling system.

8.8.3 Converter shell design

The design of the steel converter shell has been improved through the use of curved disk heads integrated with the shell (Chen et al., 2009). This design eliminates the gap between the head and the shell, minimizing the risk of a molten metal leak.

8.8.4 Improvements to batch productivity

Katamata et al. (2016) reported the efforts at Saganoseki smelter to increase the batch productivity by modifying the smelter operational strategy from 4 hot—2 blow to 3 hot—2 blow converters. This was achieved by extending the campaign life from 220 to 350 batches, increasing the blast rate and matte grade, modifications to the waste heat boiler to increase energy recovery and improving the fugitive gas collection.

8.9 Alternatives to Peirce–Smith converting

Peirce–Smith converting has been used for over 100 years, and accounts for over 90% of Cu matte converting. Around 250 converters are in operation worldwide in the combined copper, nickel, and PGM industries (Price et al., 2009). This is due to its simplicity and high chemical efficiency. Peirce–Smith converting has problems, however:

1. it leaks SO_2-bearing gas into the workplace during charging and pouring
2. it leaks air into its offgas between its mouth and gas-collection hood, producing a relatively weak SO_2 gas
3. it operates batchwise, giving uneven flow of SO_2 offgas into the sulfuric acid plant.
   These deficiencies are addressed by several different alternative converters:

1. The Hoboken or siphon converter, a Peirce–Smith converter with an improved gas-collection system (four units operating, 2010)
2. Flash converting, which oxidizes solidified crushed matte in a small Outotec flash furnace (two units operating, 2010)
3. Noranda continuous converting, which uses submerged tuyeres to blow oxygen-enriched air into matte in a Noranda-type furnace (one unit operating, 2020)

4. The Mitsubishi top-blown converter, which blows oxygen-enriched blast onto the molten matte surface via vertical lances (four units operating, 2020)
5. Metso Outotec Ausmelt® TSL (top submerged lance) batch converting (two units operating for matte converting, 2020)
6. ISASMELT™ (top submerged lance) batch converting (one unit operating for matte converting, 2020)
7. ISACONVERT™ continuous converting (one unit operating for matte converting, 2020)
8. Bottom blowing converting (five units in operation in China)
9. Top Blown multilance converting (four units in operation in China)

In the 2011 edition, the Codelco—Chile Continuous Converting process was also mentioned (Moyano et al., 2009). However, since then no new developments have been reported.

8.9.1 Hoboken converter

The Hoboken converter collects its offgas through an axial flue at one end of the converter (Binegar & Tittes, 1993; Pasca et al., 2005). A gooseneck is provided to allow the offgas (but not the liquids) to enter the flue. The offgas is collected efficiently. Considerable care must be taken to prevent buildup of splash and dust in the gooseneck.

In 2020, four Hoboken converters were operating at the Freeport McMoRan smelter in Miami, Arizona, and three Hoboken converters were operating in Caraiba Metais, in Brazil. Three Hoboken converters were commissioned in 2004 at the Thai Copper Industries smelter in Thailand (Pagador et al., 2009). The converter achieved design capacities and were able to increase the consumption of reverts (Fig. 8.14) taking advantage of the use of sonic injection. Unfortunately, the smelter has since shut down for other reasons.

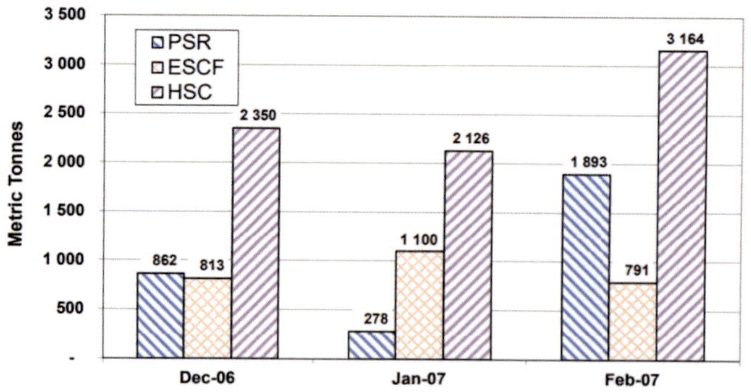

Figure 8.14 Consumption of reverts per vessel at TCI between December 2006 and February 2007 (Pagador et al., 2009), showing high reprocessing rates for the Hoboken converters within 2 months of commissioning. *ESCF*, Electric slag cleaning furnace; *HSC*, Hoboken syphon converter; *PSR*, Primary smelting reactor.

8.9.2 Flash converting (Fig. 8.15; Table 8.8)

Flash converting uses a small Outotec flash furnace to convert solidified/crushed matte (50 μm) to molten metallic copper (Davenport et al., 2001; Kojo et al., 2009). Flash converting entails

1. tapping molten 70% Cu matte from a smelting furnace
2. granulating the molten matte to ~0.5 mm granules in a water torrent
3. crushing the matte granules to 50 μm followed by drying
4. continuously feeding the dry crushed matte to the flash converter with 80 vol.-% O_2 blast and CaO flux
5. continuously collecting offgas
6. periodically tapping molten blister copper and molten calcium ferrite slag.

The first flash converting furnace was commissioned in 1995 at Rio Tinto Kennecott Utah Copper in Magna, UT, USA. Kennecott helped Outotec develop the process known as Kennecott–Outotec Flash Converting (Kojo et al., 2009). It is simply referred to as flash converting in this text. At 2020, five more flash converting furnaces have been commissioned in China, compared with 2010.

The uniqueness of the process is its use of particulate solid matte feed. Preparing this feed involves extra processing, but it is the only way that a flash furnace can be used for converting.

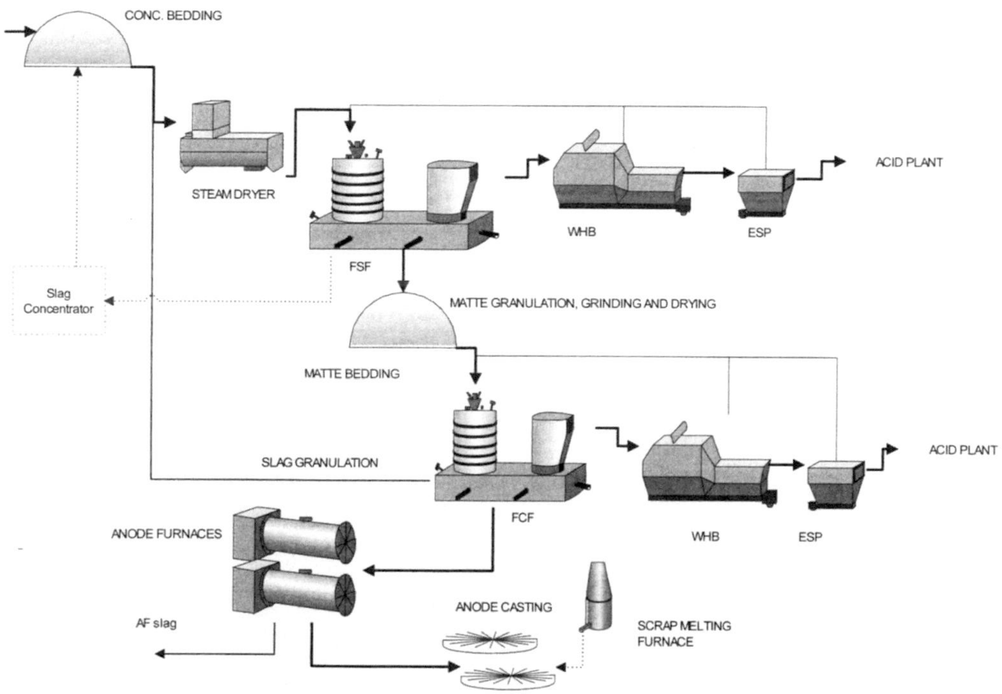

Figure 8.15 Sketch of Outotec flash smelting and flash converting process (Kojo et al., 2009).

Table 8.8 Physical and operating details of Kennecott's flash converter (2010).

Smelter	Rio Tinto Kennecott Utah Copper Magna, UT USA
Flash converter startup date	1995
Size, inside brick, m	
Hearth: w × l × h	5.1x17.16x1.92
Reaction Shaft	
Diameter	4.6
Height above settler roof	6.7
Gas uptake	
Diameter	3
Height above settler roof	8.7
Slag layer thickness, m	0.3
Copper layer thickness, m	0.41
Active Copper tapholes	8 tapholes +2 drain holes
Active Slag tapholes	3
Particulate matte burners	1
Feed, tonnes/day	
Granulated/crushed matte (dry)	1200–1800
Cu in matte, wt.-%	67–70
CaO flux	24–41
Recycle flash converter dust	82–200
Blast	
Blast temperature, °C	ambient
vol-% O_2	82–88
Input rate, 1000 Nm^3/h	8–12
Oxygen input rate, tonnes/day	260–425
Products	
Copper, tonnes/day	1100–1500
%S in copper	0.12–0.4
%O in copper	0.15–0.45
Slag, tonnes/day	200–400
%Cu in slag	17.0–26.0
%CaO/%Fe	0.24–0.36
Offgas, 1000 Nm^3/h	22–46
vol-% SO_2 in offgas	22–30
Copper/slag/offgas temperatures, °C	1215/1260/1320
Fuel inputs (natural gas)	
Hydrocarbon fuel burnt in reaction shaft, Nm^3/h	88–330
Hydrocarbon fuel into settler burners, Nm^3/h	85–300

A benefit of the solid matte feed is that it unlinks smelting and converting. A stockpile of crushed matte can be built while the converting furnace is being repaired, and then depleted while the smelting furnace is being repaired.

8.9.2.1 *Chemistry*

Flash converting is represented by the (unbalanced) reaction

$$\underset{\substack{\text{solidified}\\ \text{matte}}}{Cu-Fe-S} + \underset{\substack{\text{in oxygen}\\ \text{air blast}}}{O_2} \rightarrow Cu^{\circ}(\ell) + \underset{\substack{\text{in molten}\\ \text{calcium ferrite}\\ \text{slag}}}{Fe_3O_4} + SO_2 \tag{8.11}$$

Exactly enough O_2 is supplied to make metallic copper rather than Cu_2S or Cu_2O.

The products of the process are

1. molten copper (0.2% S, 0.3% O)
2. molten calcium ferrite slag (~16% CaO) containing ~20% Cu
3. sulfated dust, ~0.1 tonnes per tonne of matte feed
4. 30–40 vol-% SO_2 offgas.

The molten copper is periodically tapped and sent forward to pyro- and electrorefining. The slag is periodically tapped, water-granulated, and sent back to the smelting furnace. The offgas is collected continuously, cooled in a waste heat boiler, cleaned of its dust, and sent to a sulfuric acid plant. The dust is recycled to the flash converter and flash smelting furnace.

8.9.3 Choice of calcium ferrite slag

The flash converter uses calcium ferrite slag. This slag is fluid and shows little tendency to foam (Chapter 5). It also absorbs some impurities (As, Bi, Sb, but not Pb) better than SiO_2 slag. It is, however, somewhat corrosive and poorly amenable to controlled deposition of solid magnetite on the converter walls and floor.

8.9.4 No matte layer

There is no matte layer in the flash converter. This is shown by the 0.2% S content of its blister copper—far below the 1% S that would be in equilibrium with Cu_2S matte. The layer is avoided by adjusting the ratio of O_2 input to matte feed rate to slightly favor Cu_2O formation rather than Cu_2S formation. The matte layer is avoided to minimize the possibility of SO_2 formation (and slag foaming) by the reactions:

$$\underset{\text{in slag}}{2\,Cu_2O} + \underset{\text{in matte}}{Cu_2S} \rightarrow 6\,Cu + SO_2 \tag{8.12}$$

$$\underset{\text{in slag}}{2\,CuO} + \underset{\text{in matte}}{Cu_2S} \rightarrow 4\,Cu + SO_2 \tag{8.13}$$

$$\underset{\text{in slag}}{2\,Fe_3O_4} + \underset{\text{in matte}}{Cu_2S} \rightarrow 2\,Cu + 6\,FeO + SO_2 \tag{8.14}$$

beneath the slag (Davenport et al., 2001).

8.9.4.1 Productivity

Kennecott's flash converter treats ~1700 tonnes of 70% Cu matte and produces ~1300 tonnes of blister copper per day. It is roughly equivalent to three large Peirce–Smith converters.

8.9.4.2 Flash converting summary

Flash converting is an extension of the successful Outotec flash smelting process. It was first implemented at Kennecott Utah Copper in 1995. Campaign life has steadily increased to around 5 years through improvements in the cooling systems and improved process control (Janney et al., 2007; McKenna et al., 2010). Since the commissioning of the second flash converter in 2007 at the Yanggu Xiangguang copper smelter in China, five additional flash converters have been commissioned in China (Bin & Suping, 2019).

The process has the disadvantages that (a) it requires granulated and crushed matte, which requires extra energy, and (b) it is not yet well adapted to melting scrap copper. On the other hand, it has a simple, efficient matte oxidation system and efficiently collects offgas and dust.

8.9.5 Submerged tuyere Noranda continuous converting

Noranda continuous converting developed from Noranda submerged tuyere smelting. It uses a rotary furnace (Fig. 8.16) with

1. a large mouth for charging molten matte and large pieces of scrap
2. an endwall slinger and hole for feeding flux, reverts and coke
3. a second large mouth for drawing offgas into a hood and acid plant
4. tuyeres for injecting oxygen-enriched air into the molten matte
5. tapholes for separately tapping molten matte and slag
6. a rolling mechanism for correctly positioning the tuyere tips in the molten matte.

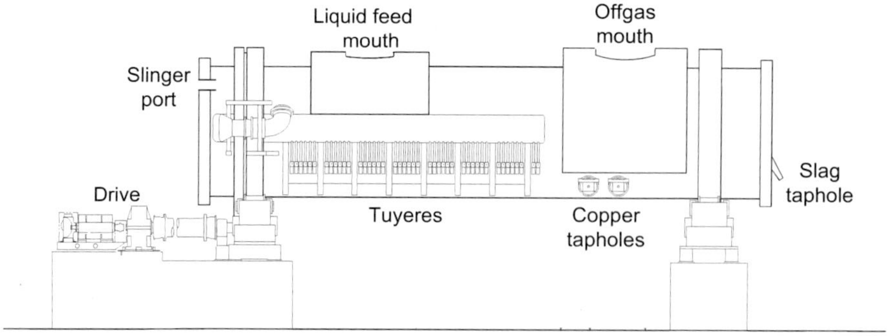

Figure 8.16 Sketch of Noranda continuous submerged tuyere converter. The furnace is 19.8 m long and 4.6 m diameter. It converts matte from a Noranda smelting furnace.

The converter operates continuously and always contains molten copper, molten matte (mainly Cu_2S), and molten slag. It blows oxygen-enriched air continuously through its tuyeres and continuously collects ~20% SO_2 offgas. It taps copper and slag intermittently. It presently operates at the Glencore Horne (formerly Noranda Horne) smelter in Quebec (Table 8.9).

Table 8.9 Physical and operating details of Noranda continuous submerged tuyere converting (Prevost, 2020; Ramachandran et al., 2003).

Smelter	**Xstrata Copper Horne Smelter**
Noranda converter startup date	1997
Noranda converter details	
Shape	Horizontal rotating cylinder
Diameter × length, inside, m	4.6 × 19.8
Tuyeres	42
Diameter, mm	63.5
Feeds, tonnes/day	
Molten matte from Noranda smelting furnace	679
Silica flux	130
"coolants" (solid matte, smelting furnace slag concentrate, internal and external reverts)	225
Blast	
vol-% O_2	38–43
Total input rate, 1000 Nm^3/hour	20.4
Products	
Copper, tonnes/day	640
%Cu/%S	98/1.3
Slag, tonnes/day	297
%Cu in slag	13.2
SiO_2/Fe	1.49
Cu-from-slag recovery method	solidification/flotation
Offgas leaving furnace, thousand Nm^3/hour	30
vol-% SO_2	20

8.9.5.1 Chemical reactions

The Noranda converting process is a continuous converting process with three molten phases in equilibrium: copper saturated with sulfur, copper sulfide, and slag. The process controls matte and O_2 input rates to always have matte (mainly Cu_2S) in the furnace. It is this matte phase that is continuously oxidized by tuyere-injected O_2.

The constant presence of this matte is confirmed by the high S content (~1.3%) in the molten copper product.

8.9.5.2 Reaction mechanisms

Reactions in the Noranda continuous converter are as follows:

1. a ladle of molten ~70% Cu matte (5%–10% Fe, ~22% S) is poured into the furnace, mixing with the molten matte layer between copper and slag.
2. this matte is oxidized by O_2 in the tuyere blast by the reactions

$$\underset{\text{in molten matte}}{3\,FeS} + \underset{\text{in tuyere blast}}{5O_2} \rightarrow Fe_3O_4 + 3\,SO_2 \tag{8.15}$$

$$3\,Fe_3O_4 + FeS \rightarrow 10\,FeO + SO_2 \tag{8.16}$$

$$2\,FeO + \underset{\text{flux}}{SiO_2} \rightarrow \underset{\text{molten slag}}{2\,FeO\cdot SiO_2} \tag{8.17}$$

then (Prevost et al., 1999)

$$\underset{\text{in molten matte}}{Cu_2S} + \underset{\text{in tuyere blast}}{O_2} \rightarrow 2\,Cu^\circ(\ell) + SO_2 \tag{8.18}$$

3. as the matte phase is continuously consumed, drops of molten slag rise and drops of molten copper fall below the tuyeres to the molten copper layer.
4. the matte layer is replenished with Cu, Fe, and S by the next ladle of matte feed.

Slag, matte, gas, and copper are intimately mixed in emulsion form in the converter's tuyere zone, so the above reaction scheme is an oversimplification. Nevertheless, the concept of slag formation, copper formation, matte consumption, and intermittent matte replenishment is probably correct.

8.9.5.3 Silicate slag

Noranda continuous converting uses SiO_2 slag rather than the CaO slag used for Mitsubishi and flash converting. This is because (a) the Cu_2S layer in the Noranda converter tends to reduce magnetite so that magnetite solubility (in CaO-base slag) is not critical and (b) SiO_2 slag is cheaper, less corrosive, and more easily controlled than CaO slag.

8.9.5.4 Control

The critical control parameters in Noranda continuous converting are (a) matte temperature and (b) matte layer position and thickness (to ensure that tuyere O_2 blows into the matte rather than the slag or copper).

Matte temperature is measured continuously with a Noranda tuyere two-wavelength optical pyrometer (Prevost et al., 1999). It is adjusted by increasing or decreasing the rate at which solid coolants (solid matte, slag concentrate, reverts, etc.) are charged to the converter. Natural gas combustion rate and coke addition rate are also used to control temperature. Matte layer thickness is controlled by adjusting the ratio of O_2 input to matte feed rate. A high ratio decreases matte mass (hence matte layer thickness), a low ratio the opposite. Matte layer position is controlled by adjusting the amount of copper below the matte. It is altered by adjusting the frequency at which copper is tapped from the furnace. Blowing of O_2 into the slag is avoided. It tends to overoxidize the slag, precipitate magnetite, and cause slag foaming. It is avoided by controlling copper and matte layer thicknesses as described above.

8.9.5.5 Noranda converting summary

The Noranda continuous converter is a compact, highly productive, submerged tuyere converting process. It charges its matte via ladle through a large mouth, which is also used for charging large pieces of scrap copper. It produces 1.3% S molten copper which is sent to a desulfurizing furnace prior to pyro- and electrorefining.

8.10 Top submerged lance converting

Top submerged lance converting technology options are

1. Metso Outotec Ausmelt® (batch) and C3 continuous converting,
2. Glencore Technology: ISASMELT™ (batch converting) and ISACONVERT™ continuous converting.

8.10.1 Metso Outotec Ausmelt converting (Wood & Hughes, 2016)

The Ausmelt process for matte conversion was initially developed for batch converting. The batch process has similarities to the Peirce—Smith batch converting, as converting is done in two stages. The process has the following characteristics:

- Liquid and/or solid granulated copper matte is partially oxidized to remove iron,
- After achieving a high matte grade, matte addition is stopped and slag is removed,
- Oxygen-enriched air is injected in the proximity of the white metal layer to remove sulfur and produce blister copper,
- Once all copper sulfide has been converted to copper, the lance is removed from the bath, to allow effective separation of copper and slag,
- Copper is tapped, and the remaining slag can be used as a seed to start the blowing for the next batch.

The Ausmelt batch converting technology was firstly implemented at Bindura smelter to process high-grade leaching residues. Following this implementation, the process was also implemented at Zhongtiaoshan in 1999, Birla Copper in 2003, and Yunnan Tin in 2012.

Following the development of the batch converting process, Metso Outotec Ausmelt® developed the Ausmelt C3 continuous copper converting process. This process uses a FeO_x-SiO_2-CaO slag to continuously convert copper mattes. It was industrially demonstrated at Zhongtiaoshan and Yunnan Tin copper smelters.

8.10.2 Glencore Technology ISASMELT™ batch converting and ISACONVERT™ continuous converting

The batch ISASMELT™ converting process has been applied in two different fields (Alvear et al., 2013). The first application is as part of the Precious Metals and Base Metals Operations flowsheet of Umicore Hoboken plant. The process involved a smelting stage in which the complex materials are smelted to form a copper matte. The silica-based Pb-rich slag produced during smelting is tapped and the remaining copper matte is converted to blister copper. All the precious minor elements are captured into the copper phase. Fig. 8.17 shows a view of the Hoboken plant, with the ISASMELT building standing in the center.

The second application of ISASMELT™ batch converting process is in the field of copper recycling as part of the Kayser Recycling System, developed by Hüttenwerke Kayser (Now Aurubis AG) at their Lünen plant, in Germany. In this process, black copper generated in a reductive smelting stage is converted to blister copper in the ISASMELT™ furnace. The process run successfully from 2002 until 2011, when Aurubis implemented a TBRC furnace to increase capacity.

The ISACONVERT™ process developed for the continuous converting of copper and copper–nickel mattes (Hunt & de Vries, 2019; Nikolic et al., 2009). The process involves continuous feeding of solid matte from the furnace roof into a molten bath.

Figure 8.17 View of the ISASMELT™ plant at the Umicore, Hoboken plant. The furnace runs a batch smelting and converting process, treating complex materials. *(Photo courtesy of Umicore Precious Metals).*

Figure 8.18 View of the ISACONVERT™ plant at First Quantum Minerals Kansanshi smelter. *(Photo courtesy of First Quantum Minerals).*

Enriched air is injected through the lance into the calcium ferrite slag phase. Copper and slag are tapped on intermittent basis from a taphole located on the base of the furnace. A view of the first ISACONVERT™ plant, commissioned in 2019 at Kansanshi smelter, Zambia is shown in Fig. 8.18.

8.11 Chinese continuous converting technologies

From 2013, Chinese technologists have been reporting the development of continuous converting technologies in China. Although there are several works describing fundamental aspects as well as advantages and disadvantages of the developed process, still not enough technical information is available to produce a sensible operational comparison with previous technologies. The main process developments described in this section are

- Bottom blowing continuous converting: technology developed using the bottom blowing smelting principles. The process has some similarities to the Noranda continuous copper converting and the El Teniente continuous copper converting (only demonstrated but not installed)
- Top-blown multilance continuous converting: Process has similarities to the Mitsubishi converting process, using calcium ferrite slags but at a higher degree of oxidation achieving lower sulfur contents in the blister.

In addition to the above two mentioned technologies, Dongying Fangyuan has developed the Submerged Lance Converting Refining process (SLCR). This process is discussed in Chapter 9.

8.11.1 Bottom blowing converting

Bottom blowing continuous converting (BBC) was developed by China ENFI Engineering Technology Co., Ltd. The BBC process main technical features are as follows:

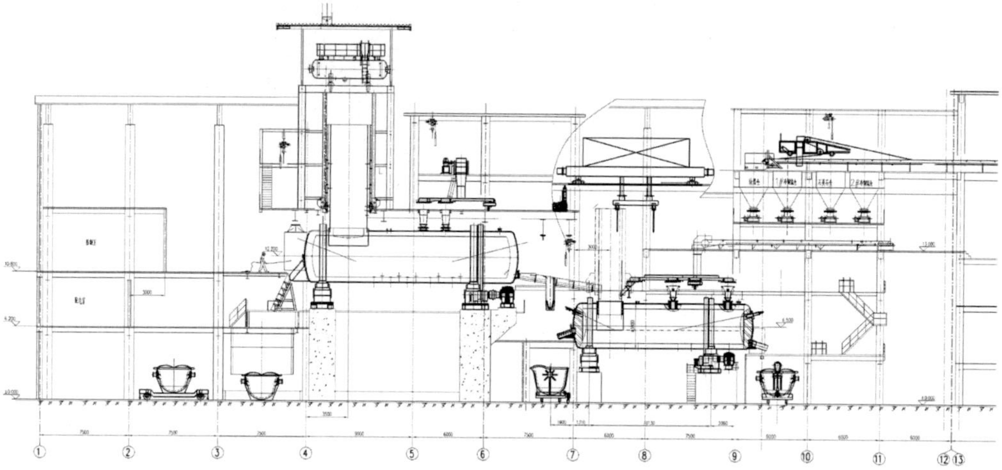

Figure 8.19 Schematic view of the Bottom Blowing Converter (left side furnace) connected to the bottom blowing smelting furnace (right side) (Hongfei, Yiongcheng, & Jianlong, 2019). *(Reprinted with permission of the Canadian Institute of Mining, Metallurgy and Petroleum).*

- Molten feed is fed via launder from the smelting reactor (additional cold matte might be required to balance heart requirements)
- Oxygen-enriched air is blown into the molten batch from furnace bottom via lances
- The process has three molten layers in equilibrium (copper, white metal, and slag) which defines the sulfur content in the blister (saturated with sulfur at the operating temperature)
- The use of enriched air and the continuous feeding reduces offgas volumes compared with batch process
- Blister and slag are intermittently tapped from the furnace

Fig. 8.19 shows a schematic layout of the Baotou Huading smelter. The furnace on the left side of the figure is the smelting furnace connected via launder to the BBC furnace (Hongfei et al., 2019). A bypass launder provides flexibility to continue smelting if the BBC is offline. The slag taphole and blister taphole are at opposite ends of the reactor. The figure also shows the solid feed bins delivering required materials (flux, reverts, matte) to the furnace top via a common conveyor and feed ports.

Typical operating parameters for the BBC converting are shown in Tables 8.10 and 8.11 (Yan, 2016). According to Ling (Ling, 2019), there were five plants in China using the BBC technology in 2019.

8.11.2 Top-blown multilance continuous converting technology

Top-blown multilance technology was developed by Jinfeng Company and Yunnan Copper Chifeng. The process layout involves the continuous feeding of copper matte

Table 8.10 Bottom blowing converting process operating parameters (Yan, 2016).

Smelter	Yuguang Gold and Lead
Noranda converter startup date	2014
BCC furnace details	
Shape	Horizontal rotating cylinder
Diameter × length, inside, m	4.1 × 18.0
Lances	11
Total	7
In operation	
Diameter, mm	N.A.
Feeds, tonnes/h	
Molten matte from Bottom Blowing Smelting furnace	21
Matte grade, %Cu	72–73
Fe/SiO_2 ratio in slag	1.1
Coal rate, % of feed	0.5–1.0
Dust rate, %	3.0–4.0
Operating temperature, °C	1220–1250
Blast	
vol-% O_2	45–55
Total input rate, 1000 Nm^3/h	4.0
Products	
%Cu/%S	98/0.8
%Cu in slag	10.0–12.0
SiO_2/Fe	1.38
SO_2 in offgas, %	~32

from the side-blown smelting furnace. According to Bin et al. (Bin and Suping, 2019), the process provides the following advantages:

- Continuous feeding from the smelting unit,
- Full utilization of the molten matte contained heat in the converting process,
- Good refractory campaign,
- High availability due to easing change of top lances,
- Process can treat copper scrap,
- High SO_2 concentration in offgases.

Four plants currently using top-blown continuous converting technology in China and Africa (Bin and Suping, 2019).

Table 8.11 Multilance top blowing converting parameters (Yan, 2016).

Smelter	Chifeng Yunnan copper smelter
Startup date	2015
Furnace details	
Shape	Horizontal rotating cylinder
Diameter × length, inside, m	4.1 × 18.0
Lances	11
Total	7
In operation	
Diameter, mm	N.A.
Feeds, tonnes/h	
molten matte from Bottom Blowing Smelting furnace	21
Matte grade, %Cu	72–73
Fe/SiO_2 ratio in slag	1.1
Coal rate, % of feed	0.5–1.0
Dust rate, %	3.0–4.0
Operating temperature, °C	1220–1250
Blast	
vol-% O_2	45–55
Total input rate, 1000 Nm^3/h	4.0
Products	
%Cu/%S	98/0.8
%Cu in slag	10.0–12.0
SiO_2/Fe	1.38
SO_2 in offgas, %	~32
vol-% SO_2	20

8.12 Summary

Converting is the second-half of the smelting/converting sequence by which most Cu–Fe–sulfide concentrates are made into metallic copper. The process oxidizes the Fe and S from molten smelting furnace matte with oxygen-enriched air or air to produce molten metallic copper. It is most often carried out in the cylindrical Peirce–Smith converter. The products of the process are

1. molten blister copper (99% Cu, 0.02% S and 0.6% O), which is sent forward to fire refining for final S and O removal, then anode casting
2. molten Fe-silicate slag (4%–8% Cu) which is sent to Cu recovery, then discard
3. SO_2-bearing offgas which is treated for heat, dust and SO_2 capture.

All of the heat for converting comes from Fe and S oxidation. Peirce–Smith converting is a batch process. It produces SO_2 intermittently and captures it somewhat inefficiently. Alternatives are

1. Hoboken batch converting
2. Kennecott–Outotec continuous flash converting
3. Noranda continuous submerged tuyere converting.
4. Mitsubishi continuous downward lance converting
5. Ausmelt top submerged lance (TSL) batch converting
6. ISASMELT top submerged lance (TSL) batch converting and ISACONVERT
7. Bottom blowing continuous converting
8. Top-blown multilance continuous converting technology

References

Alvear, F., Gerardo, R. F., & Nikolic, S. (2013). ISASMELT™ for recycling of valuable elements contributing to a more sustainable society. In A. Kvithyld, C. Meskers, R. Kirchain, G. Krumdick, B. Mishra, M. Reuter, C. Wang, M. Schlesinger, G. Gaustad, D. Lados, J. Spangenberger, et al. (Eds.), *Rewas 2013*. Springer. https://doi.org/10.1007/978-3-319-48763-2_11

Asteljoki, J. A., Bailey, L. K., George, D. B., & Rodolff, D. W. (1985). Flash converting — continuous converting of copper mattes. *Journal of Occupational Medicine, 37*(5), 20–23.

Aurubis, A. G. (2020). *Private Communication*.

Bin, T., & Suping, Y. (2019). Application and development of modern copper metallurgical technologies in China's copper industry (Article 598809). In *Proceedings of the 58th conference of metallurgists hosting the international copper conference 2019*. Montreal: CIM.

Binegar, A. H., & Tittes, A. F. (1993). Cyprus Miami Mining Corporation siphon converter operation, past and present. In J. D. McCain, & J. M. Floyd (Eds.), *Converting, fire refining and casting* (pp. 297–310). Warrendale, PA: TMS.

Boisvert, M., Janneteau, G., Landry, J. P., Levac, C. A., Perron, D., McGlynn, F., … Porretta, F. (1998). Design and Construction of the Noranda Converter at the Horne Smelter. In J. A. Asteljoki, & R. J. PA StephensTMSWarrendale (Eds.), Sulfide Smelting 1998 (pp. 569–583). TMS.

Boliden. (2020). *Private Communication*.

Brännström, P. G., & Hedlund, L. M. (2013). Environmental improvements for Peirce-Smith converters with Outotec's converter hood technology for primary and secondary gas capture. *Base metals conference* (pp. 139–150). Johannesburg: SAIMM.

Bustos, A. A., Brimacombe, J. K., & Richards, G. G. (1988). Accretion growth at the tuyeres of a Peirce–Smith copper converter. *Canadian Metallurgy Quarterly, 27*, 7–21.

Bustos, A. A., Brimacombe, J. K., Richards, G. G., Vahed, A., & Pelletier, A. (1987). Developments of punchless operation of Peirce–Smith converters. In C. Díaz, C. Landolt, & A. Luraschi (Eds.), *Copper 87, vol. 4: Pyrometallurgy of copper* (pp. 347–373). Alfabeta Impresores, Lira 140, Santiago, Chile.

Bustos, A. A., Richards, G. G., Gray, N. B., & Brimacombe, J. K. (1984). Injection phenomena in nonferrous processes. *Metallurgical and Materials Transactions B, 15B*, 77–79.

Cardona, N., Martin, A., Jiménez, F., Ríos, G., Mackey, P., & Coursol, P. (2013). Optimizing converter aisle operation at Atlantic copper smelter, Huelva, Spain. In R. Bassa, R. Parra, A. Luraschi, & S. Demetrio (Eds.), *IIMCH, Vol. III Nikolas Themelis symposium on pyrometallurgy and process engineeringProceedings of copper 2013* (pp. 163–183).

Casley, G. E., Middlin, J., & White, D. (1976). Recent developments in reverberatory furnace and converter practice at the Mount Isa Mines copper smelter. In J. C. Yannopoulos, & J. C. Agarwal (Eds.), *Extractive metallurgy of copper, volume 1: Pyrometallurgy and electrolytic refining* (pp. 117–138). Warrendale, PA: TMS.

Chen, S., Mansikkaviita, H., Rytkonen, M., & Kylmakorpi, I. (2009). Continuous improvement in Peirce Smith converter design — Kumera's approach. In J. Kapusta, & T. Warner (Eds.), *International Peirce—Smith converting centennial* (pp. 315—319). Warrendale, PA: TMS.

Coleman, M. E., & Money, G. (2009). Increasing capacity and productivity in the metals markets through pneumatic conveying and process injection technologies. In J. Kapusta, & T. Warner (Eds.), *International Peirce—Smith converting centennial* (pp. 217—230). Warrendale, PA: TMS.

Cui, Z., Wang, Z., Wang, H., & Wei, C. (2016). Two—step copper smelting process at Dongying Fangyuan. In *Proceedings of copper 2016* (pp. 1043—1052). Tokyo: MMIJ.

Davenport, W. G., Jones, D. M., King, M. J., & Partelpoeg, E. H. (2001). *Flash smelting: Analysis, control and optimization* (2nd ed.). Warrendale, PA: TMS.

Eltringham, G. A. (1993). Developments in converter fluxing. In J. D. McCain, & J. M. Floyd (Eds.), *Converting, fire refining and casting* (pp. 323—331). Warrendale, PA: TMS.

Feng, H., & Huilan, Y. (2019). Introduction of China's copper converting technologies (Article 591007). In *Proceedings of the 58th conference of metallurgists hosting the international copper conference 2019.* Montreal: CIM.

Fukushima, K., Baba, K., Kurokawa, H., & Yamagiwa, M. (1988). Development of automation systems for copper converters and anode casting wheel at Toyo smelter. In E. H. Partelpoeg, & D. C. Himmesoete (Eds.), *Process control and automation in extractive metallurgy* (pp. 113—130). Warrendale, PA: TMS.

Gas Cleaning & Energy Solutions. (2020). *Converting retrofit project* (Accessed 27 August 20) https://gcteng.com/gas-cleaning/.

Heath and Sherwood (2021). https://heathandsherwood64.com/pdf/Tuyere/HS_Tuyere_Pyrometer.pdf.

Henao, H., Kohnenkamp, E., Rojas, L., & Moyano, A. (2019). Experimental determination of the effect of CaO and Al_2O_3 in slags systems related to the conversion process of high copper matte grade. *Minerals, 9*(11), 716. https://doi.org/10.3390/min9110716

Henao, H. M., & Moyano, A. (2016). Conversion of mattes with high grade copper. In *Proceedings of copper 2016* (pp. 901—908). Tokyo: MMIJ.

Hills, I. E., Warner, A. E. M., & Harris, C. L. (2007). Review of high pressure tuyere injection. In A. E. M. Warner, C. J. Newman, A. Vahed, D. B. George, P. J. Mackey, & A. Warczok (Eds.), *Cu2007, volume III (book 1): The Carlos Díaz symposium on pyrometallurgy* (pp. 471—481). Montreal: CIM.

Hinsberger, S., & Marinigh, M. J. (2020). *Personal communication.*

Hongfei, L., Yiongcheng, Z., & Jianlong, W. (2019). Application Practice of Bottom Blowing Continuos Converting to Replace Copper P-S Converter. In *Copper 2019 E-Proceedings of Copper 2019 Pyrometallurgy*. CIM. Paper 594634. ISBN978-1-926872-44-5.

Hongfei, L., Yongcheng, Z., & Jianlong, W. (2019). Application practice of bottom blowing continuous converting to replace copper P—S converters (Article 594634). In *Proceedings of the 58th conference of metallurgists hosting the international copper conference 2019.* Montreal: CIM.

Hunt, S., & de Vries, D. (2019). Kansanshi copper smelter: The first four years of operation (Article 591109). In *Proceedings of the 58th conference of metallurgists hosting the international copper conference 2019.* Montreal: CIM.

International Copper Study Group. (2017). *Smelter data base.* https://www.icsg.org/.php/component/jdownloads/finish/166-other/1575-icsg-directory-of-mines-and-plants-2011-03?Itemid=0.

Janney, D., George—Kennedy, D., & Burton, R. (2007). Kennecott Utah Copper smelter rebuild: Over 11 million tonnes of concentrate smelted. In A. E. M. Warner, C. J. Newman, A. Vahed, D. B. George, P. J. Mackey, & A. Warczok (Eds.), *Cu2007, volume III (book 2): The Carlos Díaz symposium on pyrometallurgy* (pp. 431—443). Montreal: CIM.

Johnson, R. E., Themelis, N. J., & Eltringham, G. A. (1979). A survey of worldwide copper converter practices. In R. E. Johnson (Ed.), *Copper and nickel converters* (pp. 1—32). Warrendale, PA: TMS.

JX Nippon Mining & Metals Corporation. (2020). *Private Communications.*

Kapusta, J. (2010). Gas injection phenomena in converters — an update on buoyancy power and bath slopping. In *Copper 2010, volume 2: Pyrometallurgy I, Clausthal—Zellerfeld* (pp. 839—855). Germany: GDMB.

Kapusta, J. (2019a). Twenty—five years of sonic injection plant trials and implementation — let's take stock (Article 528496). In *Proceedings of the 58th conference of metallurgists hosting the international copper conference 2019.* Montreal: CIM.

Kapusta, J. (2019b). The 2019 copper smelter survey (Article 592732). In *Proceedings of the 58th conference of metallurgists hosting the international copper conference 2019.* Montreal: CIM.

Katamata, K., Sasaki, T., Motomura, T., & Chida, H. (2016). Operational excellence in Peirce—Smith converting at Saganoseki smelter. In *Proceedings of copper 2016* (pp. 666—677). Tokyo: MMIJ.

Kawai, T., Nishiwaki, M., & Hayashi, S. (2005). Copper concentrate smelting in Peirce—Smith converters at Onahama smelter. In A. Ross, T. Warner, & K. Scholey (Eds.), *Converter and fire refining practices* (pp. 119—123). Warrendale, PA: TMS.

Kojo, I., Lahtinen, M., & Miettinen, E. (2009). Flash converting — sustainable technology now and in the future. In J. Kapusta, & T. Warner (Eds.), *International Peirce—Smith converting centennial* (pp. 383—395). Warrendale, PA: TMS.

Lehner, T., Ishikawa, O., Smith, T., Floyd, J., Mackey, P., & Landolt, C. (1993). The 1993 survey of worldwide copper and nickel converter practices. In J. D. McCain, & J. M. Floyd (Eds.), *Converting, fire refining and casting* (pp. 1—8). Warrendale, PA: TMS.

Libertad y Desarrollo. (2012). Norma de emisiones para fundiciones de cobre: ¿Demasiadas exigencias?. In *Temas Públicos, N° 1.079.* https://lyd.org/wp-content/uploads/2012/09/TP-1079-NORMA-DE-FUNDICIONES.pdf (Accessed 01 September 20).

Ling, W. (2019). The Utilization of Copper Continuous Smelting, Converting and Fire-Refining Process in China. *E-Proceedings Copper 2019Pyrometallurgy*. Vancouver: CIMM. Paper 595879. ISBN-978-1-926872-44-5.

Marinigh, M. J. (2009). Technology and operational improvements in tuyere punching, silencing, pyrometry and refractory drilling equipment. In J. Kapusta, & T. Warner (Eds.), *International Peirce—Smith converting centennial* (pp. 199—215). Warrendale, PA: TMS.

Maruyama, T., Saito, T., & Kato, M. (1998). Improvements of the converter's operation at Tamano smelter. In J. A. Asteljoki, & R. L. Stephens (Eds.), *Sulfide smelting '98* (pp. 219—227). Warrendale, PA: TMS.

Mast, E. D., Arrian, V. J., & Benavides, V. J. (1999). Concentrate injection and oxygen enrichment in Peirce—Smith converters at Noranda's Altonorte smelter. In D. B. George, W. J. Chen, P. J. Mackey, & A. J. Weddick (Eds.), *Copper 99—Cobre 99, vol. V: Smelting operations and advances* (pp. 433—445). Warrendale, PA: TMS.

McKenna, K., Newman, C., Voermann, N., Veenstra, R., King, M., & Bryant, J. (2010). Extending copper smelting and converting furnace campaign life through technology. In *Copper 2010, volume 3: Pyrometallurgy II, Clausthal—Zellerfeld* (pp. 971—986). Germany: GDMB.

Mendoza, H., & Luraschi, A. (1993). Impurity elimination in copper converting. In J. D. McCain, & J. M. Floyd (Eds.), *Converting, fire refining and casting* (pp. 191—202). Warrendale, PA: TMS.

Mitarai, T., Akagi, S., & Masatoshi, M. (1993). Development of the technique to determine the end point of the slag—making stage in copper converter. In J. D. McCain, & J. M. Floyd (Eds.), *Converting, fire refining and casting* (pp. 169—180). Warrendale, PA: TMS.

Mitsui Mining & Smelting Co. Ltd, & Metals Sector. (2020). *Private Communication.*

Mori, K., Nagai, K., Morita, K., & Nakano, O. (2009). Recent operation and improvement at the Sumitomo Toyo Peirce—Smith converters. In J. Kapusta, & T. Warner (Eds.), *International Peirce—Smith converting centennial* (pp. 151—160). Warrendale, PA: TMS.

Moyano, A., Caballero, C., Toro, C., Morales, P., & Font, J. (2009). The validation of the Codelco—Chile continuous converting process. In J. Kapusta, & T. Warner (Eds.), *International Peirce—Smith converting centennial* (pp. 349—360). Warrendale, PA: TMS.

Nikolic, S., Edwards, J. S., Burrows, A. S., & Alvear, G. R. F. (2009). ISACONVERT™ — TSL continuous copper converting update. In J. Kapusta, & T. Warner (Eds.), *International Peirce—Smith converting centennial* (pp. 407—414). Warrendale, PA: TMS.

Pagador, R. U., Wachgama, N., Khuankla, C., & Kapusta, J. (2009). Operation of the air liquide shrouded injection (ALSI™) technology in a Hoboken siphon converter. In J. Kapusta, & T. Warner (Eds.), *International Peirce—Smith converting centennial* (pp. 367—381). Warrendale, PA: TMS.

Pannell, D. G. (1987). A survey of world copper smelters. In J. C. Taylor, & H. R. Traulsen (Eds.), *World survey of nonferrous smelters* (pp. 3—118). Warrendale, PA: TMS.

Pasca, O., Ushakov, V., & Welker, E. (2005). Hoboken converter performance improvements at the Phelps Dodge Miami smelter. In A. Ross, T. Warner, & K. Scholey (Eds.), *Converter and fire refining practices* (pp. 167—176). Warrendale, PA: TMS.

Persson, W., Wendt, W., & Demetrio, J. S. (1999). Use of optical on-line production control in copper smelters. In D. B. George, W. J. Chen, P. J. Mackey, & A. J. Weddick (Eds.), *Copper 99—Cobre 99, vol. V: Smelting operations and advances* (pp. 491—503). Warrendale, PA: TMS.

Prevost, I. (2020). *Glencore.* Personal Communication.

Prevost, Y., Lapointe, R., Levac, C. A., & Beaudoin, D. (1999). First year of operation of the Noranda continuous converter. In D. B. George, W. J. Chen, P. J. Mackey, & A. J. Weddick (Eds.), *Copper 99—Cobre 99, vol. V: Smelting operations and advances* (pp. 269—282). Warrendale, PA: TMS.

Price, T., Harris, C., Hills, I. E., Boyd, W., & Wraith, A. (2009). Peirce—smith converting — another 100 years? In J. Kapusta, & T. Warner (Eds.), *International Peirce—Smith converting centennial* (pp. 181—197). Warrendale, PA: TMS.

Ramachandran, V., Díaz, C., Eltringham, T., Jiang, C. Y., Lehner, T., Mackay, P. J., Newman, C. J., & Tarasov, A. (2003). Primary copper production — a survey of operating world copper smelters. In C. Díaz, J. Kapusta, & C. Newman (Eds.), *Copper 2003—cobre 2003, volume IV: Pyrometallurgy of copper (book 1) the Herman Schwarze symposium on copper pyrometallurgy* (pp. 3—119). Montreal: CIM.

Ramirez, J. P. (2019). *Cual es la situación de las fundiciones Chilenas?*. https://www.mch.cl/reportajes/la-situacion-las-fundiciones-chile/# (Accessed 27 August 20).

Ramírez, J. P. (2019). Cuál es la situación de las fundiciones en Chile?. In *Minería Chilena*. https://www.mch.cl/reportajes/la-situacion-las-fundiciones-chile/#. (Accessed 11 August 2020).

Richards, G. G., Legeard, K. J., Bustos, A. A., Brimacombe, J. K., & Jorgensen, D. (1986). Bath slopping and splashing in the copper converter. In D. R. Gaskell, J. P. Hager, J. E. Hoffmann, & P. J. Mackey (Eds.), *The Reinhardt Schuhmann international symposium on innovative technology and reactor design in extraction metallurgy* (pp. 385—402). Warrendale, PA: TMS.

Russell, M., Safe, P., Veek, A., Fernandez, A., & Parameswaran, K. (2019). Asarco LLC Hayden converter retrofit project, an update (Article 598518). In *Proceedings of the 58th conference of metallurgists hosting the international copper conference 2019*. Montreal: CIM.

Schonewille, R. H., O'Connell, G. J., & Toguri, J. M. (1993). A quantitative method for silica flux evaluation. *Metallurgical and Materials Transactions B, 24B*, 63—73.

Sharma, R. C., & Chang, Y. A. (1980). A thermodynamic analysis of the copper sulfur system. *Metallurgical and Materials Transactions B, 11B*, 575—583.

Shook, A. A., Pasca, O., & Eltringham, G. A. (1999). Online SO_2 analysis of copper converter offgas. In D. B. George, W. J. Chen, P. J. Mackey, & A. J. Weddick (Eds.), *Copper 99—cobre 99, vol. V: Smelting operations and advances* (pp. 465—475). Warrendale, PA: TMS.

Sumitomo Metal Mining Co. Ltd. (2020). *Private Communication*.

Suzuki, Y., Hoshino, Y., Yamamoto, K., Mori, K., Kawanaka, Y., & Hattori, Y. (2016). Improvements in PS converter operation at Toyo smelter and refinery. In *Proceedings of copper 2016* (pp. 678—690). Tokyo: MMIJ.

Taniuchi, Y., Nishima, K., & Hashiuchi, F. (2016) Improvements to the Peirce-Smith Converter Operation at Tamano Smelter. In *E Proceedings Copper 2016Pyrometallurgy*. MMIJ, Kobe, Paper PY7.3.

Verney, L. R. (1987). Peirce—Smith copper converter operations and economics. In C. Díaz, C. Landolt, & A. Luraschi (Eds.), *Copper 87, vol. 4: Pyrometallurgy of copper* (pp. 375—391). Santiago, Chile: Alfabeta Impresores.

Vogt, J. A., Mackey, P. J., & Balfour, G. C. (1979). Current converter practice at the Horne smelter. In R. E. Johnson (Ed.), *Copper and nickel converters* (pp. 357—390). Warrendale, PA: TMS.

Vukovic, G., Zivanovic, B., Drew, B., & Gamweger, K. (2016). Continuous online temperature measurement system. In *Proceedings of copper 2016* (pp. 1418—1426). Tokyo: MMIJ.

Warczok, A., Riveros, G., Mackay, R., Cordero, D., & Alvear, G. (2001). Effect of converting slag recycling into Teniente Converter on copper losses. In P. R. Taylor (Ed.), *EPD congress 2001* (pp. 431—444). Warrendale, PA: TMS.

West, R. C., Fahey, N. P., Maxwell, T. R., & Hay, S. M. (2002). Commissioning the port Kembla copper smelter. In R. J. Stephens, & H. Y. Sohn (Eds.), *Sulfide smelting 2002* (pp. 167—179). Warrendale, PA: TMS.

Wood, J., & Hughes, S. (2016). Future development opportunities for the Outotec® Ausmelt process. In *Proceedings of copper 2016* (pp. 361—372). Tokyo: MMIJ.

Wood, J., Matusewicz, R., & Reuter, M. A. (2009). Ausmelt C3 converting. In J. Kapusta, & T. Warner (Eds.), *International Peirce—Smith Converting Centennial* (pp. 397—406). Warrendale, PA: TMS. Published by TMS.

Yan, J. (2016). Recent operation of the oxygen bottom—blowing copper smelting and continuous copper converting technologies. In *Proceedings of copper 2016* (pp. 1015—1030). Tokyo: MMIJ.

Yongcheng, Z., Bing, L., & Kai, L. (2019). Engineering and production practice of double bottom—blowing continuous copper smelting process with "Three connected furnaces" arrangement (Article 592917). In *Proceedings of the 58th conference of metallurgists hosting the international copper conference 2019*. Montreal: CIM.

Suggested reading

Boisvert, M., Janneteau, G., Landry, J. P., Levac, C., Perron, D., McGlynn, F., Zamalloa, M., & Porreta, F. (1998). Design and construction of the Noranda converter at the Horne smelter. In J. A. Asteljoki, & R. L. Stephens (Eds.), *Sulfide smelting'98* (pp. 569–583). Warrendale, PA: TMS.

Homepage of copper area.(2020). https://www.copperarea.com/pages/asarco-plans-100-million-upgrade-of-hayden-smelter-to-meet-new-epa-sulfur-dioxide-emission-limits-project-not-expected-to-disrupt-operation-of-plant-or-employees-hours/. (Accessed 1 September 2020).

Kapusta, J., & Warner, T. (Eds.). (2009). *International Peirce–Smith converting centennial.* Warrendale, PA: TMS.

Kapusta, J. (2013). Sonic injection in bath smelting and converting; Myths, facts and dreams. In *COM 2013* (pp. 267–315). Montreal, Canada: CIM.

Larouche, P. (2001). *Minor elements in copper smelting and electrorefining.* M.Eng Thesis (pp. 118–119). Canada: McGill University http://digitool.library.mcgill.ca:8881/R/?func=dbin-jump-full&object_id=33978.

Ross, A., Warner, T., & Scholey, K. (Eds.). (2005). *Converter and fire refining practices.* Warrendale, PA: TMS.

Wood, J., Matusewicz, R., & Reuter, M. A. (2009). Ausmelt C3 converting. In J. Kapusta, & T. Warner (Eds.), *International Peirce–Smith converting centennial* (pp. 397–406). Warrendale, PA: TMS.

CHAPTER 9

Continuous copper making processes

As indicated in Chapter 5, continuous and/or direct production of copper has been also focusing of development in the copper making industry. In general, continuous copper concentrate smelting and converting has several advantages:

(a) ability to smelt all concentrates, including $CuFeS_2$ concentrates
(b) elimination of a batch converting process, improving process efficiencies, and reducing SO_2 collection and air infiltration difficulties
(c) continuous production of high SO_2-strength offgas, albeit from two sources
(d) relatively simple Cu-from-slag recovery
(e) minimal materials handling, as liquid transfer in the converter aisle can be eliminated.

Continuous copper production can be achieved in a single, two, or three processing stages. The main aspect to consider when selecting the appropriate number of stages is to fully understand the properties of the material to be processed. This means

- Knowing the full chemical and mineralogical composition of the materials
- Understanding the heat that will be released at different oxygen and sulfur potential conditions (matte or blister making conditions)
- Understanding slag making components of the feed and fluxing requirements
- Considering appropriate requirements for copper recovery

The above-mentioned aspects are part of the fundamental metallurgical and process design evaluation required to appropriate selection the most suitable configuration.

Continuous copper making can be achieved by

- Single-stage process
- Two-stage process
- Three-stage process

Single-stage direct to blister production has been successfully applied using flash smelting and converting principles with plant operating in Poland, Australia, and Zambia, with the latest addition been the second production line at Glogow smelter in Poland (Urbaniak et al., 2019).

The two-stage process is based on the continuous smelting of copper concentrate into an ultra-high matte grade (above 72% Cu) in a smelting reactor, followed by the converting—refining stage in a single reactor. This process was developed by Dongying-Fangyuan in China (Wang et al., 2017). Although the two-stage process has the novelty of combining converting and refining into a single process, from the fundamental metallurgical thermodynamic point of view, the converting—refining stage can be considered as a batch process in which the oxygen potential is gradually increased.

Extractive Metallurgy of Copper, Sixth Edition
ISBN 978-0-12-821875-4, https://doi.org/10.1016/B978-0-12-821875-4.00010-9

A three-stage process consists in the combination of smelting, converting, and refining in a cascade type arrangement, where materials flow from one furnace into the next one, and each one of these furnaces operates at a defined oxygen and sulfur potential condition. The most know process at this stage is the Mitsubishi process. The Mitsubishi process combines a three-stage continuous blister production process, in a cascade arrangement, for smelting, slag cleaning, and converting, followed by the traditional anode refining (Shibasaki & Hayashi, 1991). Plants are currently in operation in Japan, South Korea, India, and Indonesia.

In the same category of three-stage continuous copper production, Chinese technologists have echoed the Mitsubishi process principles, developing two new applications for continuous copper production:

- Double Bottom−Blowing Continuous Copper Smelting Process with "Three Connected Furnaces" arrangement (Yongcheng et al., 2019)
- Combination of oxygen-enriched side blow smelting technology and horizontal multilances top blow converting (Ling, 2019)

This chapter describes the above-mentioned continuous copper production processes.

9.1 Single-stage process: direct to blister flash process

The direct to blister route is generally restricted to concentrates which have chalcocite (Cu_2S) and bornite (Cu_5FeS_4) as primary minerals and relatively small amounts of pyrite (FeS_2), chalcopyrite ($CuFeS_2$), and silica (SiO_2). The combination of these materials allows process design engineers to develop a single-stage process that will manage the heat release in the furnace within the requirement of the overall process heat balance (including heat losses) and the overall slag amount generation. Giving the operating conditions required to produce copper in a single stage (high oxygen potential), a slag cleaning process is also required to recover the copper that is oxidized in the single-stage copper making process.

There are four copper flash smelters operations that operate direct to blister processes:

- Olympic Dam, Australia (Ranasinghe et al., 2010);
- Glogów II, Poland (Byszynski et al., 2010);
- Chingola, Zambia (Taskinen & Kojo, 2009);
- Glogow I, Poland (Urbaniak et al., 2019).

Fig. 9.1 shows the interior of the new Zambian direct-to-copper flash furnace. Figs. 9.2 and 9.3 show the flowsheet for the Olympic Dam and Glogów direct-to-copper flash furnace smelters, respectively. The main difference between the two processes is the additional process unit to process the Cu−Pb−Fe alloy produced in the electric furnace. Multimetal recovery is also ensured as lead is recovered from the converter slag in the lead plant.

Figure 9.1 New Chingola direct-to-copper flash furnace. The photo shows the top of the reaction shaft and the bottom tip of the concentrate burner. The diameter of the reaction shaft is ~6 m. *(Photograph by W.G. Davenport.)*

9.1.1 Advantages and disadvantages

The principal advantages of combining smelting and converting are

(a) Isolation of SO_2 emission to a single, continuous, SO_2-rich gas stream, ideal for capturing sulfur in sulfuric acid;

(b) Minimization of energy consumption;

(c) Minimization of capital and operating costs.

The main disadvantage of the process is the requirement for an additional copper recovery process. The increase in oxygen potential required to produce copper in a single stage also implies that about 25% of the copper contained in the copper concentrate will be oxidized and dissolved in the slag phase. This also means that the single-stage production process is restricted to low-Fe chalcocite and bornite concentrates, as these types of concentrates will generate less slag and therefore less copper loss. An expansion to high-Fe containing concentrates will dramatically increase the amount of slag and therefore the overall copper recovery cost.

Tables 9.1a—c describe the concentrates currently being smelted in direct-to-copper furnaces.

9.1.2 The ideal direct-to-copper process

Fig. 9.4 is a sketch of the ideal direct-to-copper process. The principal inputs to the process are concentrate, oxygen, air, flux, and recycles. The principal outputs are molten copper, low-Cu slag, and high-SO_2 offgas.

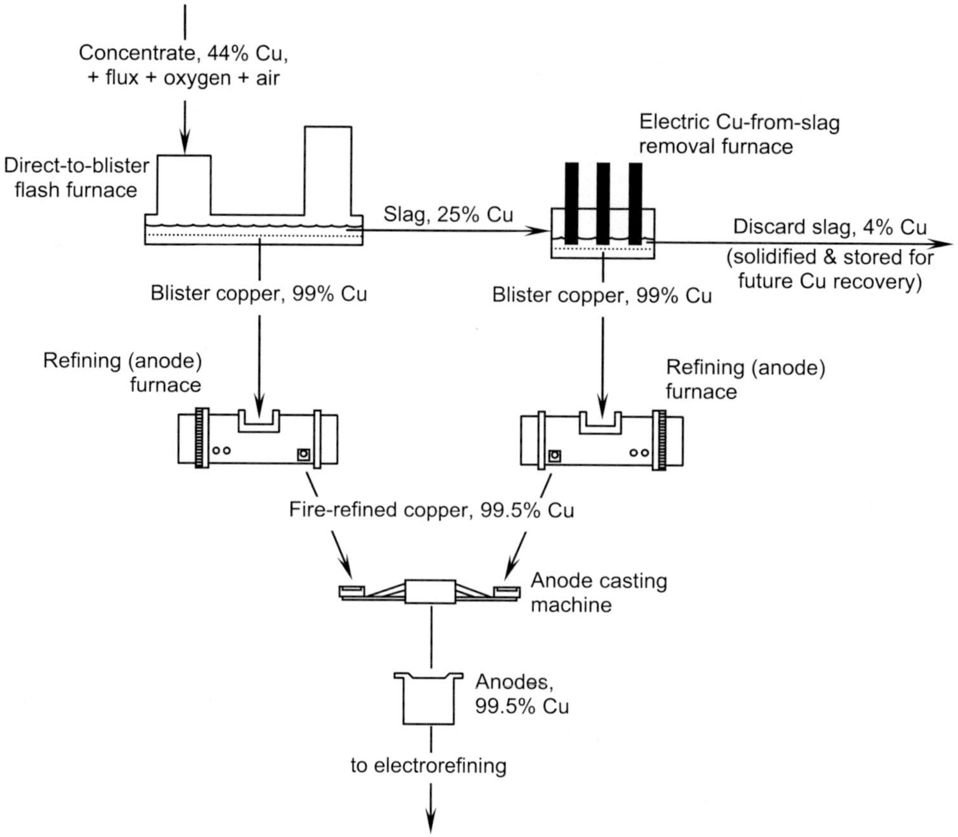

Figure 9.2 Simplified Olympic Dam direct-to-copper smelter flowsheet. The Outokumpu flash smelting furnace, electric Cu-from-molten slag recovery furnace, fire refining, and anode casting systems are notable. The slag is molten until it is solidified for storage (right). The copper is molten until it is cast as anodes (bottom). *(Modified from Ranasinghe et al., 2010.)*

The process is continuous and autothermal. With highly oxygen-enriched (low-N_2) blast, there is enough excess reaction heat to melt all the Cu-bearing recycle materials from the smelter and adjacent refinery, including scrap anodes. The remainder of this section indicates how close the process has come to this ideality.

9.1.3 Industrial single furnace direct-to-copper smelting

In 2020, single furnace direct-to-copper smelting is done by only one process: Outotec (formerly Outokumpu) flash smelting (Fig. 9.1). For several years the Noranda submerged—tuyere process (Chapter 7) also produced copper directly. This plant now produces high-grade (72%—74% Cu) matte. The change was made to avoid slag foaming (Section 9.1.7) and to increase impurity elimination.

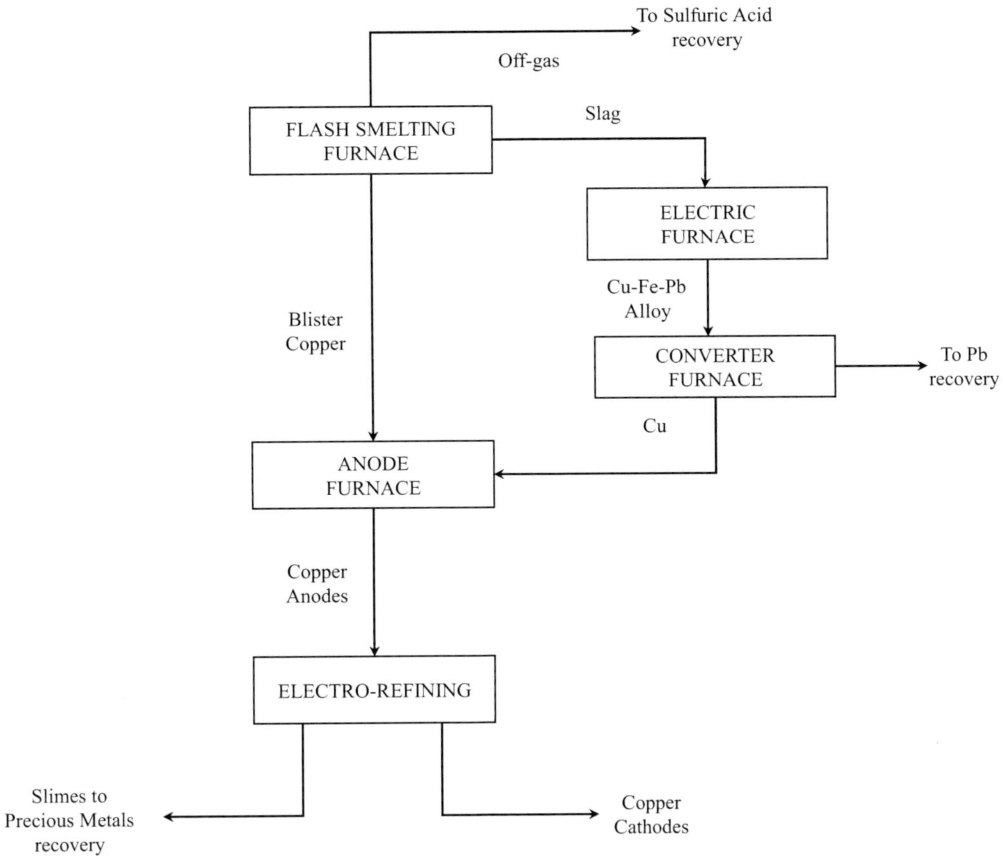

Figure 9.3 Schematic flowsheet of copper production and slag treatment at Glogow smelter modified from Urbaniak et al. (2019). The flowsheet includes a converting process to treat the Cu–Pb–Fe allow generated in the slag cleaning furnace. Multimetal recovery is ensured by synergies with the lead plant, where the slag from the converter is processed for lead recovery.

Table 9.1a Elements and equivalent minerals in Olympic Dam direct-to-copper flash furnace concentrate feed (Ranasinghe et al., 2010; Solnordal et al., 2006).

Element	Cu	Fe	S	SiO_2		
Mass%						
Minerals	44 Bornite Cu_5FeS_4	20 Digenite $Cu_{1.8}S$	25 Chalcopyrite $CuFeS_2$	5 Quartz SiO_2	Pyrite FeS_2	Hematite Fe_2O_3
Mass%						
	28–45	25–35	11–27	2–6	0.2–2	8–12

Table 9.1b Elements and equivalent minerals in Glogów direct-to-copper flash furnace concentrate feed (Czernecki et al., 2010; Dobrzanski & Kozminski, 2003).

Element	Cu	Fe	S	"Free" carbon	Pb	As	Other
Mass %							
Minerals	17–31 In chalcocite Cu_2S and bornite Cu_5FeS_4	2–7 In bornite Cu_5FeS_4 and hematite Fe_2O_3	8–11 Mainly in chalcocite and bornite	5–8 Carbon	1.5–3 In galena, PbS	0.05–4 In arsenopyrite, $FeAsS_2$, enargite, Cu_3AsS_4, tennantite	42–50 20% SiO_2 6% Al_2O_3 6% CaO[a] 4% MgO K_2O, Na_2O

[a]CaO, MgO, K_2O, and Na_2O are present as carbonates.

Table 9.1c Elements and equivalent minerals in Chingola direct-to-copper flash furnace concentrate feed (Syamujulu, 2007, which includes further details on mineralogy).

Element	Cu	Fe	S	SiO_2	other
Mass%					
	39 mainly in chalcocite, bornite, and chalcopyrite	8 mainly in bornite and chalcopyrite (minor in pyrite)	19 mainly in chalcocite bornite and chalcopyrite	18 mainly in feldspar and clay	10% Al_2O_3 (mainly in clay) 2% CaO 2% MgO

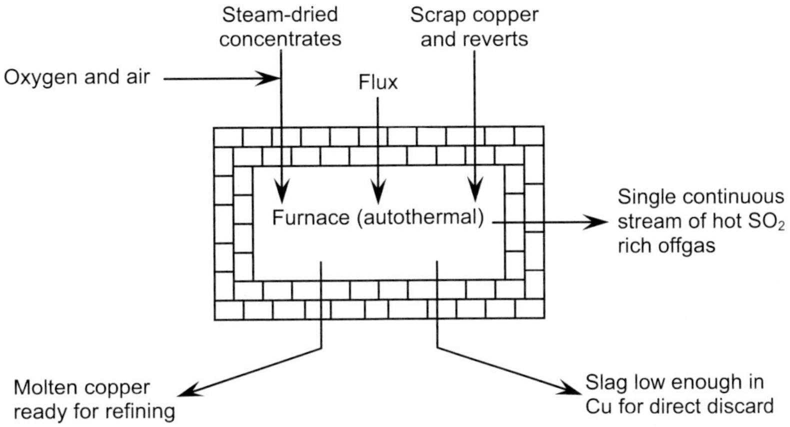

Figure 9.4 Ideal single-furnace copper making process. Ideally the copper is low in impurities, the slag is discardable without Cu-recovery treatment, and the offgas is strong enough in SO_2 for sulfuric acid manufacture. Little or no fossil fuel is required.

The products of direct-to-copper flash smelting (Tables 9.2 and 9.3) are impure molten copper (99% Cu, 1300°C), molten slag (12%–28% Cu, ~1310°C), and offgas (16–41 vol-% SO_2, 1250–1400°C). As with conventional matte flash smelting, the

Table 9.2 Compositions of slags from direct-to-copper flash smelting. The large amounts of CaO, MgO, and Al_2O_3 in Glogów slags are notable. They originate in the feed concentrates (Table 9.1b) and give a low Cu oxide solubility to the slag (Somerville, Chen, Alvear, & Nikolic, 2016; Taskinen & Kojo, 2009).

Smelter								
Element/compound	**Cu**	**Fe**	**SiO_2**	**CaO**	**MgO**	**Al_2O_3**	**K_2O**	**Na_2O**
Mass %								
Olympic Dam, Australia	23–28	33	18	0	0	3.5		
Glogów, Poland	12–15	6	31	14	6	9	Yes	Yes
Chingola, Zambia	17–20	17–29	28–32	5	3–7	5–7		

Table 9.3 Details of Chingola and Glogow direct-to-copper Metso Outotec flash furnaces (2010).

Smelter	Konkola Cu Mines, Chingola, Zambia, 2020	KGHM Polska Miedz Glogów II Poland, 2020
Startup date	2008	1978 modernized 2010
Size, inside brick, m		
Hearth: w × l × h	7.4 × 24 × 2.4	8.9 × 25.6 × 3.8
Reaction shaft		
Diameter	~6	7.55
Height, above settler roof	6	9.1
Gas uptake		
Diameter	4	6.7
Height above settler roof	8	12.3
Slag layer thickness, m	0.7–1.35	0.5
Copper layer thickness, m	0–0.7	0.7
Active slag tapholes	3	5
Active copper tapholes	8	13
Concentrate burners	1	1
Settler burners	10	Normally none
Feed details, tonnes/day		
New concentrate (dry)	1440–2600, 31%–38% Cu	2200–2400 (27% Cu)
Oxygen	400–700	
Silica flux		Self-fluxing
Recycle flash furnace dust	0–300	415
Other	80–130 CaO in flux	115 desulfurizing dust
Blast details		
Blast temperature, °C	Ambient	Ambient
Volume% O_2	95	68
Flowrate, 1000 Nm^3/h	22	32
Production details		
Copper production, tonnes/day	~1000	420
Composition	99.2% Cu, 0.5% O; 0.3% S	98.7% Cu; 0.01% Fe, 0.48% O; 0.31% Pb: 0.05% S
Temperature, °C	1280–1300	1300 maximum
Slag production, tonnes/day	900–1400	1060
Mass% Cu	17–20	14
Mass% SiO_2/mass%Fe	0.9	3.95
Temperature, °C	1300–1320	1310 maximum
Cu-from-slag recovery method	Electric furnace	Electric furnace
Offgas, 1000 Nm^3/h	26–46	47

Table 9.3 Details of Chingola and Glogow direct-to-copper Metso Outotec flash furnaces (2010).—cont'd

Smelter	Konkola Cu Mines, Chingola, Zambia, 2020	KGHM Polska Miedz Glogów II Poland, 2020
Volume% SO_2, leaving furnace	32−41	16
Temperature, °C	1400	1250
Dust production, tonnes/day	Boiler 150, ESP 80	250
Hydrocarbon fuel inputs, kg/h		
Burnt in reaction shaft	Oil, 150−650	Oil, 80
Burnt in settler burners	Oil, 3000	0

temperature in the smelting furnace is controlled by adjusting (a) the degree of oxygen enrichment of the blast (which changes the amount of N_2 coolant entering the furnace) and (b) the rate at which fossil fuel is burnt in the furnace.

The O_2 content of industrial direct-to-copper flash furnace blast is 65−95 vol.-% O_2, depending on the concentrate composition. Considerable fossil fuel is burnt in the reaction shaft and in settler burners (Table 9.3).

9.1.4 Chemistry

Direct-to-copper smelting takes place by the schematic (unbalanced) reaction:

$$\underset{\text{in concentrate}}{Cu_2S, Cu_5FeS_4(s)} + \underset{\text{in oxygen–enriched air}}{O_2(g)} + \underset{\text{in flux}}{SiO_2(s)} \xrightarrow{\sim 1300^\circ} \underset{\text{impure molten copper}}{Cu^\circ(l)} + \underset{\text{slag}}{FeO, Fe_3O_4, SiO_2(l)} + \underset{\text{in off-gas}}{SO_2(g)} \quad (9.1)$$

Just enough O_2 is supplied to produce metallic copper rather than Cu_2S or Cu_2O.

In practice, the flash furnace reaction shaft product is a mixture of overoxidized (oxide) and underoxidized (sulfide) materials. Individual particles may be overoxidized on the outside and underoxidized on the inside. The over- and underoxidized components react to give metallic copper, as shown in the reactions below:

$$\underset{\text{Over-oxidized Cu}}{2\,Cu_2O(s)} + \underset{\text{Under-oxidized Cu sulfide}}{Cu_2S(s)} \rightarrow \underset{\text{impure molten copper}}{6\,Cu^\circ(l)} + \underset{\text{in off-gas}}{SO_2(g)} \quad (9.2)$$

and

$$\underset{\text{Over-oxidized FeO}}{2\,Fe_3O_4(s)} + \underset{\text{Under-oxidized Cu sulfide}}{Cu_2S(s)} \rightarrow \underset{\text{impure molten copper}}{2\,Cu^\circ(l)} + \underset{\text{in silicate slag}}{6\,FeO(l)} + \underset{\text{in off-gas}}{SO_2(g)} \tag{9.3}$$

Industrially, the overall extent of Reaction (9.1) is controlled by (a) monitoring the Cu content of the product slag and the S content of the product copper and (b) adjusting the ratio of oxygen in the blast to concentrate fed to the furnace, based on these measured Cu-in-slag and S-in-copper values. An increasing Cu content in the slag is reversed by decreasing the O_2/concentrate ratio and vice versa. The copper slag content is kept between 12% and 28%.

9.1.5 Effect of slag composition on %Cu in slag

Table 9.2 gives the compositions of direct-to-copper flash furnace slags. The most notable feature is that the concentration of Cu in Glogów slag is half that of Olympic Dam slag. This is fortunate because Glogów produces about 2.5 tonnes of slag per tonne of copper (Table 9.3), while Olympic Dam produces about 1.4 t slag per tonne of copper.

The net result is that both smelters send about 25% of the copper in the feed to slag.

The process at Glogów produces a lower copper slag because its composition is essentially a Ca—Al—Mg—K—Na silicate slag (Table 9.3), which has a relatively low solubility for Cu oxide (Taskinen & Kojo, 2009).

Conventional smelting/converting smelters produce ~2.5 t slag per tonne of copper production, about the same as Glogów. However, the slags contain much less Cu (1%—2% for smelting; 4%—8% for converting, some it entrained rather than dissolved). Cu removal from these slags is quicker and cheaper than from the direct-to-copper smelting slags.

9.1.6 Industrial details

Operating details of two direct-to-copper flash furnaces are given in Table 9.3. The furnaces are similar to conventional flash furnaces. Differences are

(a) The hearths are deeply bowl-shaped to prevent molten copper from contacting the furnace sidewalls

(b) The hearths are more radically arched and compressed to prevent their refractory from being floated by the dense (7.8 t/m^3) molten copper layer (Brillo & Egry, 2003; Hunt et al., 1999)

(c) The furnace walls are extensively water-cooled and the hearth extensively air-cooled to prevent metallic copper from seeping too far into the refractories

(d) The bottom refractories are monolithic to prevent molten copper from seeping under the bricks, solidifying, and lifting them

(e) The copper tapholes are designed to prevent the out-flowing molten copper from enlarging the taphole to the point where molten copper contacts cooling water

At Olympic Dam, molten copper passes through magnesite—chrome brick (inside) and a silicon carbide insert and graphite insert (outside) (Hunt et al., 1999). The graphite insert is replaced after ~1200 t tapped copper and the silicon carbide insert is replaced after ~2400 t. Excessive copper flow (the result of an excessive taphole diameter) initiates earlier replacement.

9.1.7 Control

A direct to blister process is a full oxidation process in which compositions of the generated molten phases are controlled by adjusting the ratios of oxygen in the blast to concentrate feed rate and flux/concentrate feed rate.

In order to control an adequate fluidity of the molten products, process temperatures (molten and gas phases) are controlled by adjusting the oxygen-enrichment level of the blast (as represented by the N_2/O_2 ratio) and the rate at which fossil fuel is burnt in the furnace.

A key aspect to ensure a safe and reliable process is to avoid the formation of a matte layer in the reactor. This is achieved by ensuring having proper O_2/concentrate ratios, in other words, the right oxygen and sulfur potentials that will avoid the formation of a Cu_2S layer. This is also done to prevent the possibility of foaming slag out the top of the furnace (Hunt et al., 1999). A molten Cu_2S layer, once built up between the molten copper and molten slag layers, has the potential to react with slag by reactions such as (9.2) and (9.3), which can produce SO_2 beneath the slag layer.

The foaming problem is avoided by ensuring that the O_2/concentrate ratio is always at or above its set point, never below. This may lead to high Cu-in-slag levels but it avoids the potentially serious operational problems caused by foaming (Hunt et al., 1999). S-in-copper below ~1% S ensures that a Cu_2S layer does not form.

An unfortunate side effect of the above no-matte-layer strategy is a high Cu content in the slag, mainly as dissolved Cu_2O. It arises because there is no permanent layer of Cu_2S in the furnace to reduce Cu_2O to metallic copper, Reaction (9.2).

Simply stated, direct-to-copper smelting is operated in a slightly overoxidizing mode to prevent the foaming. The downside of operating this way is that 12%—28% Cu reports to the slag (Table 9.2).

The key process control aspect is therefore to define the right oxygen—sulfur potential setup that will guarantee no Cu_2S layer, acceptable copper content in slag and subsequently minimum possible sulfur in the generated copper phase.

9.1.8 Electric furnace Cu-from-slag recovery

All direct-to-copper smelters recover Cu from their slag in an electric slag—cleaning furnace. The slag flows from the direct-to-blister furnace directly into an electric furnace where it is settled for about 10 h under a 0.25 m blanket of metallurgical coke (Czernecki et al., 1999). The coke reduces the oxidized Cu from the slag by reactions such as

$$\underset{\text{dissolved in slag}}{Cu_2O(\ell)} + \underset{\text{in metallurgical coke}}{C} \xrightarrow{1300^\circ C} \underset{\text{impure molten copper}}{2Cu^\circ(\ell)} + \underset{\text{offgas}}{CO(g)} \tag{9.4}$$

Magnetite (molten and solid) is also reduced

$$\underset{\text{dissolved in slag and solid}}{Fe_3O_4(\ell,s)} + \underset{\text{in metallurgical coke}}{C} \xrightarrow{1300^\circ C} \underset{\text{dissolved in slag}}{3\,FeO(\ell)} + \underset{\text{offgas}}{CO(g)} \tag{9.5}$$

and some FeO is inadvertently reduced to Fe by the reaction:

$$\underset{\text{dissolved in slag}}{FeO(\ell)} + \underset{\text{in metallurgical coke}}{C} \xrightarrow{1300^\circ C} \underset{\text{dissolved in copper}}{Fe(\ell)} + \underset{\text{offgas}}{CO(g)} \tag{9.6}$$

The resulting Fe joins the newly reduced copper.

9.1.8.1 Glogów

The Cu content of the Glogów direct-to-copper slag is lowered from ~14% Cu to ~0.6% Cu in an 25000 kVA electric furnace. The molten product analyzes 70%–80% Cu, 5% Fe, and 15%–25% Pb, from Pb in the concentrate (Czernecki et al., 2010).

The Pb and Fe are oxidized in a Hoboken converter (Chapter 8), upgrading the metal to >98% Cu, 0.3% Pb. This product is sent to anode-making and electrorefining. The PbO-rich slag is sent to a lead smelter (along with lead- and zinc-rich dusts).

9.1.8.2 Olympic Dam

Olympic Dam lowers its direct-to-copper slag from 24% to ~4% Cu in its 15,000 kV A electric furnace (Hunt et al., 1999). This Cu content could be lowered further by using more coke and a longer residence time, but the copper product would then contain excessive radioactive ^{216}Pb and ^{210}Po, from the original concentrate (Ranasinghe et al., 2010). Instead, the slag is solidified and stored for future Cu recovery using solidification, comminution, and flotation (Fig. 9.2).

9.1.9 Cu-in-slag limitation of direct-to-copper smelting

The principal advantage of direct-to-copper smelting is isolation of SO_2 evolution to one furnace. The principal disadvantage of the process is its large amount of Cu-in-slag. Balancing these factors, it appears that direct-to-copper smelting is best suited to concentrates containing covellite (Cu_2S) and bornite (Cu_5FeS_4). Processing these concentrates produces little slag, so Cu recovery from slag is not too costly.

Direct-to-copper smelting will probably not be suitable for most chalcopyrite concentrates. These concentrates produce about 2.5 t slag per tonne of Cu, and the energy and cost of recovering Cu from their slag is considerable. Only about 60% of new Cu in the concentrate would report directly to copper; the other 40% would wind up in the slag.

Davenport et al. (2010) affirm this view. However, Outotec remains dedicated to direct-to-blister flash smelting for normal chalcopyrite concentrates, as well as for high-grade chalcocite and bornite concentrates (Tuominen & Kojo, 2005).

9.1.10 Direct-to-copper impurities

The compositions of the anode copper produced by the direct-to-copper smelters are given in Table 9.4. The impurity levels of the copper are within the normal range of electrorefining anodes (Chapter 14). The impurity levels could be reduced further by avoiding recycle of the flash furnace dust.

Table 9.4 Anode compositions from direct to copper smelters.

	Olympic Dam (2010)	Glogów II (2010)
Impurity	**ppm in copper**	**ppm in copper**
Ag	188	1800–3300
As	597	500–1900
Au	24	~2
Bi	182	2–94
Fe	11	23–263
Ni	81	680–1620
Pb	241	500–3800
S	12	5–19
Sb	36	20–290
Se	263	118
Te	75	NA

Impurities do not seem, therefore, to be a problem in the three existing direct-to-copper smelters. However, metallic copper is always present in the direct-to-copper furnace, ready to absorb impurities. For this reason, concentrates destined for direct-to-copper smelting should always be carefully tested in a pilot furnace before being accepted by the smelter.

9.2 Two-stage process: Dongying-Fangyuan process

9.2.1 General process description (Wang, Cui, Wei, & Wang, 2017; Wang, Zhixiang, Wang, Cui, & Huang, 2019)

The two-stage copper making process combines continuous smelting of copper concentrate into an ultra-high matte grade in a smelting reactor and the combined converting—refining process in a second reactor. The process is continuous from the chemical reaction point of view but batch from the tapping point of view, as the converting—refining process is gradually done by in a single reactor by managing the oxygen—sulfur potential with addition of air—oxygen—natural gas.

The main innovation of this process is the combination of the converting and refining duties in a single furnace. Two SLCRs work alternatively to guarantee the continuous discharge of matte from the SLS. The combined converting—refining process is done by combining and adjusting flows of nitrogen, natural gas, oxygen, and air through the injection lances. Slag generated in the process is tapped before starting the end part of the cycle. Gas flows and volumes are controlled to meet the different requirements of the converting and refining processes.

Offgas generated in both processes are captured and processed in a dedicated acid plant.

9.2.2 Dongying-Fangyuan process and plant technical description

A schematic view of the process, developed by Dongying-Fangyuan in China, and commissioned in 2015, is shown in Fig. 9.5. The process uses similar principles as the BBS process for smelting. The smelting furnace (SLS: **S**ubmerged **L**ance **S**melting) is continuously fed producing the ultra-high copper matte(>73%Cu) which is continuously discharged from the smelting furnace via a syphon and directed via launders to the SLCR converting—refining vessel (**S**ubmerged **L**ance **C**onverting **R**efining). The smelting slag is batch tapped and slow cooled for further copper recovery. Technical specifications for the Dongying-Fangyuan SLS and SLCR reactors are shown in Table 9.5.

9.2.2.1 SLS smelting furnace

Fig. 9.6 shows a schematic view of SLS furnace. The smelting furnace has a diameter of 5.5 m and length of 28.8 m. Blended materials are continuously fed into a single smelting

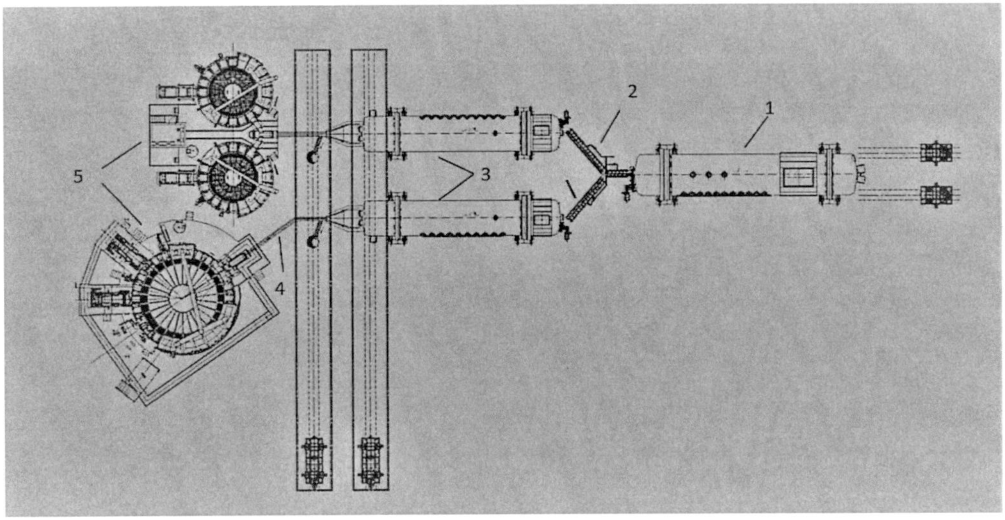

Figure 9.5 Top view of the plant layout of the two-stage copper smelting process at Dongying-Fangyuan. The smelter reactor (1) molten matte is continuously tapped via launders into the converting-refining reactor (2) operating in sequence. Anode quality copper is directly tapped via launders (4) into the casting wheels (5) (Zhixiang et al., 2016). Figure originally published in the Copper 2016 Proceedings, Kobe, Japan. *(Reprinted from E proceedings Cu 2016, with permission from the Mining and Materials Processing Institute of Japan.)*

reactor from three feed ports located on top of the furnace shell. Level and bath temperature can be measured from the top by inserting a dip bar and a temperature measuring device.

Twenty-three oxygen lances, arranged in two rows, supply the required amount of enriched air to the process. The lances are shrouded tubes. Oxygen is injected via the inner layer and air via the outer layer. This configuration of gases ensuring protecting the lance as air will shield the oxygen main jet.

Additional lances are also located in the vicinity of the slag end wall. According to Zjxiang (2018), the purpose of this extra lance row is to enhance fuming of volatile elements from the slag layer.

Offgas is collected via an uptake located on top of the slag end of the reactor. The copper matte is continuously discharged, via siphon, and directed into the SLCR furnace via a connecting cover launder to minimize emissions. Slag is tapped from two tapholes into ladle for slow cooling in pots.

9.2.2.2 SLCR converting—refining furnace

Two converting—refining vessels receive the ultra-high copper matte generated in the SLS process. The furnaces are 4.8 m in diameter and 23 m length. Enriched air is injected though 17 lances arranged in two rows.

Table 9.5 Selected Technical Information Dongying-Fangyuan two-stage process (Zhixiang et al., 2016; Wang et al., 2019).

Indicator	Value
SLS furnace	
Feed rate dry concentrate, t/h	175
Total wet feed, t/h	230
Fe/SiO_2	1.8–2.0
Cu in slag, %Cu	1.9–3.1
Dust generation, % of total feed	<2.0
Matte grade, %Cu	73–78
Operating temperature, °C	1180
Oxygen enrichment, $\%O_2$	73.0
SLCR furnace	
Matte flow, t/h	55–60
Batch capacity, t	400
Oxygen enrichment, $\%O_2$	22–26
Offgas volume, Nm^3/h	29,000–51,000
Temperature, °C	1200
Copper in slag, %Cu	<10.0
Anode composition,	
%Cu	99.1–99.3
%S, ppm	60–210
%As, ppm	20–500
%Zn, ppm	20–130
%Sb, ppm	70–150
%Pb, ppm	600–1100

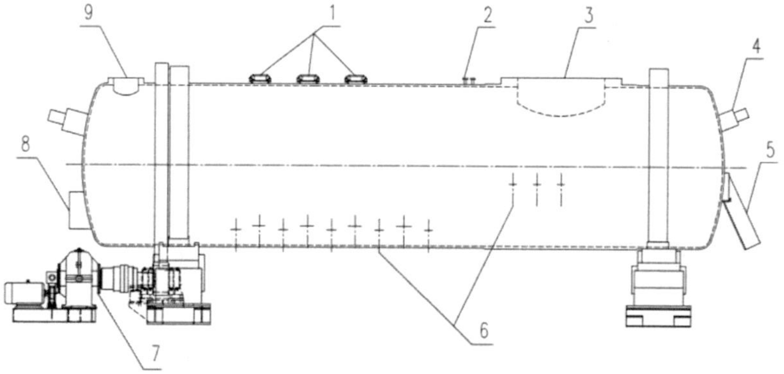

Figure 9.6 Schematic view of the SLS furnace. Main components of the furnace are (1) feed ports, (2) Dip bar and temperature measurement port, (3) offgas mouth, (4) burner, (5) slag taphole, (6) injection lances, (7) drive, (8) copper discharge port, (9) second offgas port (Zhixiang et al., 2016). Figure originally published in the Copper 2016 Proceedings, Kobe, Japan. *(Reprinted from E proceedings Cu 2016, with permission from the Mining and Materials Processing Institute of Japan.)*

The molten ultra-high matte grade is fed through a feeding port located at the center of the furnace end wall facing the SLS furnace. An additional feed port located on the top of the shell allows the addition of cold materials into the furnace. Offgases are evacuated from the furnace via an uptake located on the upper part of the shell close to matte feed port. The slag taphole is located at the same end of the copper pouring end. The furnace must rotate to tap the slag. Slag from the SLCR is recycled to the SLS furnace. Two SLCR furnaces alternatively operating ensure continuous reception of molten matte from the SLS furnace. The two following operating principles control the addition of molten matte and slag tapping from the SLCR furnace:

- Furnace will rotate to blowing position at a predefined matte level,
- When the slag reaches the required tapping level, the second SLCR will start its matte reception cycle. First SLCR furnace will be ready for casting

9.2.3 SLS-SLCR process control

Targeted compositions of ultra-high matte grade and anode copper are controlled by adjusting gas flows through the lances. Oxygen, air, natural gas, and nitrogen flows are adjusting as required to achieve the defined targets.

The SCLR cycle has the following steps:

- Feeding stage: Ultra-high matte reception from the SLS furnace: during this stage, converting starts
- Copper making/slag tapping stage: This process overlaps with the feeding stage for most of the cycle and only during the last part of the process, matte feeding is stopped, and the final blowing of copper is done. Slag is poured during this stage
- Reduction stage: Copper composition is adjusted to anode quality requirements
- Casting: Copper is finally tapped and casted into anodes. At the end of this stage, the furnace is prepared for the next batch.

9.3 The Mitsubishi process: introduction (Mitsubishi Materials, 2020)

The Mitsubishi process operates continuously, using three connected furnaces to smelt, settle, and convert copper concentrates. The development of the process started in 1959 in Naoshima Smelter, Japan, when trials were done to evaluate injection of materials into molten bath Technology development continued, and a 6000 t/y Cu pilot plant was commissioned in 1968 to further demonstrate the concept. The scale-up process continued with a semicommercial plant in 1970 in Onahama (18,000 t/y Cu), and a 48,000 t/y Cu first commercial installation in 1974 (Naoshima). Since then, the process has been implemented in Canada, India, Korea, and Indonesia. Elements of the Mitsubishi process have been also commercialized in combination with other processes. Examples of that are the Mitsubishi converting process in combination with the Noranda Process in Port Kembla, Australia, and the Mitsubishi smelting process in combination with a reverberatory furnace and Peirce—Smith converters in Onahama, Japan.

9.3.1 The Mitsubishi process

Fig. 9.7 shows the three furnaces of the Mitsubishi process (Ajima et al., 1999; Mitsubishi Materials Corporation, 2004). The S (smelting) furnace blows oxygen-enriched air, dried concentrates, SiO_2 flux and recycles into the furnace liquids via vertical lances. It oxidizes the Fe and S of the concentrate to produce ~68% Cu matte and Fe–silicate slag. Its matte and slag flow together into the electric slag cleaning furnace.

The CL (slag-cleaning) furnace separates the smelting furnace's matte and slag. Its matte flows continuously to the converting furnace. Its slag (0.7%–0.9% Cu) flows continuously to water granulation and sale or stockpile.

The C (converting) furnace blows oxygen-enriched air, $CaCO_3$ flux, and granulated converter slag coolant into the matte via vertical lances. It oxidizes the Fe and S in the matte to produce molten copper. The copper continuously exits the furnace into one of two holding furnaces for subsequent fire- and electrorefining. The slag (14% Cu) flows continuously into a water granulation system. The resulting slag granules are recycled to the smelting furnace for Cu recovery and the converting furnace for temperature control.

A major advantage of the process is its effectiveness in capturing SO_2. It produces two continuous strong SO_2 streams, which are combined to make excellent feed gas for sulfuric acid or liquid SO_2 manufacture. Also, the absence of crane-and-ladle transport of molten material minimizes workplace emissions.

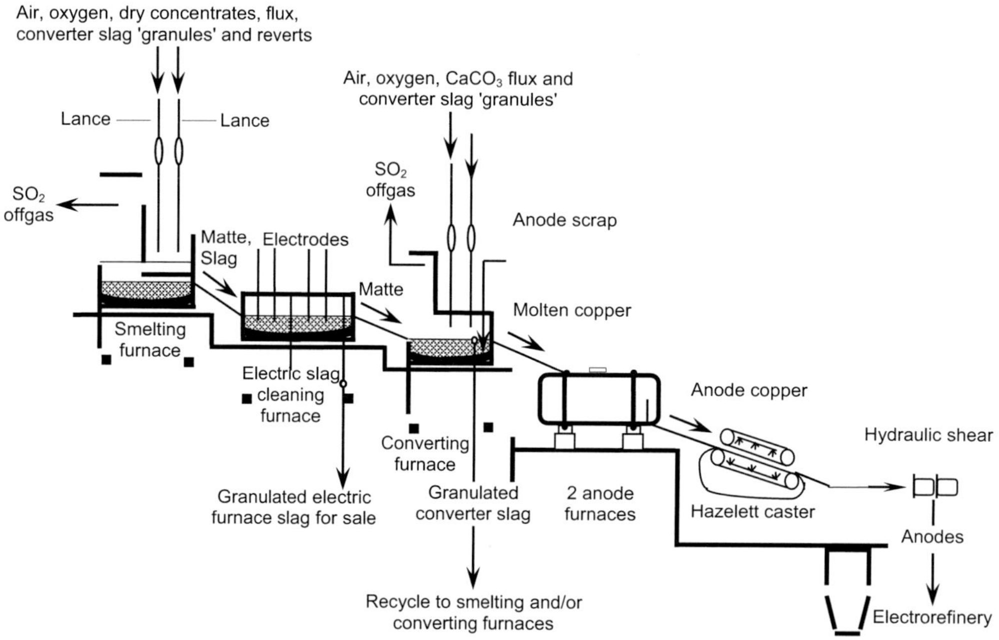

Figure 9.7 Mitsubishi process flowsheet and vertical layout at Gresik, Indonesia (Ajima et al., 1999). Note the continuous gravity flow of liquids between furnaces. The smelting furnace is about 15 m higher than the Hazelett caster. The smelting and converting furnaces each have 9 or 10 rotating lances. *(Fikry, A. M., Sundana, D., Kusunone, K. & Unno, Y. (2019) Modification of Feeding System for the Smelting Furnace and other Recent Improvements at Gresik Copper Smelter. In Proceedings of the 58th conference of metallurgists hosting the international copper conference 2019 (Article 594775). Montreal: CIM.)*

9.3.2 Smelting furnace details

Fig. 9.8 shows details of the Mitsubishi smelting furnace. Solid particulate feed and oxidizing gas are introduced through nine vertical lances in two rows across the top of the furnace. Each lance consists of two concentric pipes inserted through the furnace roof. The diameter of the inside pipe is 5 cm; the diameter of the outside pipe, 10 cm. Dried feed is air-blown from bins through the central pipe; oxygen-enriched air (55 vol.-% O_2) is blown through the annulus between the pipes. The outside pipes are continuously rotated (7–8 rpm) to prevent them from sticking to their water-cooled roof collars.

The outside pipes extend downward to ~0.7 m above the molten bath, the inside pipes to the furnace roof or just above. The outside pipes are high chromium steel (27% Cr); they burn back ~0.4 m per day and are periodically slipped downward to maintain their specified tip positions. New 3 m sections are welded to the top of the shortened pipes to maintain continuous operation. The inside pipes are 304 stainless steel. They do not wear back because their tips are high above the reaction zone.

Concentrate/flux/recycle feed meets oxidizing gas at the exit of the inside pipe. The mixture jets onto the molten bath to form a matte–slag–gas foam/emulsion in which liquids, solids, and gas react with each other to form matte and slag. These continuously overflow together through a taphole and down a sloped launder into the electric slag–cleaning furnace.

The offgas (25–30 vol.-% SO_2) from the oxidation reactions is drawn up a large uptake. It passes through waste heat boilers, electrostatic precipitators, and a wet gas cleaning system before being pushed into a sulfuric acid or liquid SO_2 plant.

The velocities of solids and gas leaving Mitsubishi smelting furnace lances are 130–150 m/s (Goto & Hayashi, 1998). Times of flight of the particles across the

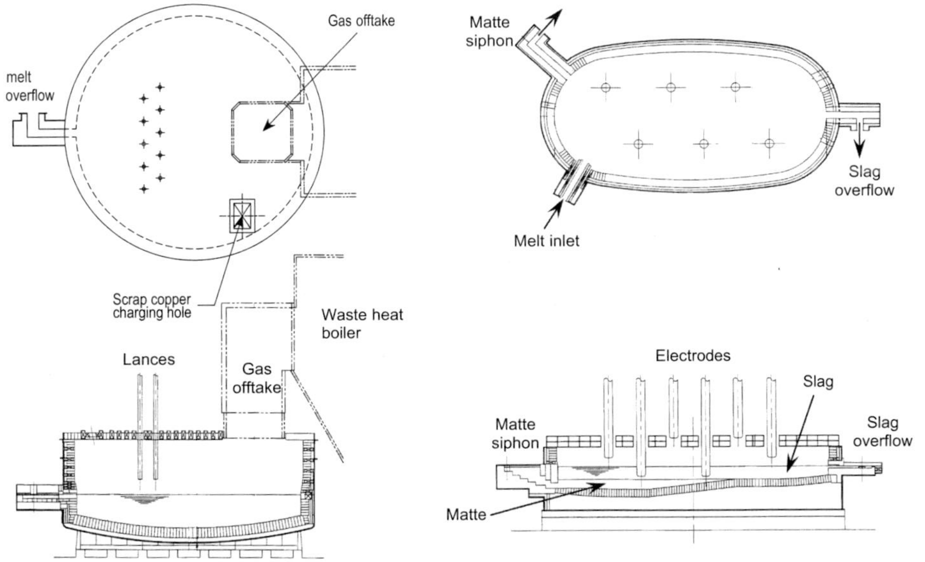

Figure 9.8 Details of Mitsubishi smelting and electric slag cleaning furnaces (Goto & Hayashi, 1998). Note particularly the siphon by which matte underflows the slag cleaning furnace. Most of the refractories are fused cast and direct-bonded magnesite chrome ($MgO–Cr_2O_3$).

0.7 m distance between lance tip and melt surface are, therefore, on the order of 10^{-3} to 10^{-2} s. The temperature rise of the gas/solid jet during this time is calculated to be ~50°C. This is not enough to cause ignition of the concentrate. Consequently, melting and oxidation of concentrate particles occurs entirely after entry into the gas—slag—matte foam/emulsion beneath the lances (Asaki et al., 2001).

Industrial evidence indicates that the smelting furnace contains mainly matte (1.2—1.5 m deep) with a gas/slag/matte foam/emulsion beneath the lances (Shibasaki & Hayashi, 1991). Away from the lances, SO_2/N_2 gas disengages from the foam and the matte and slag begin to separate. Newly formed slag (~0.05 m thick and containing some entrained matte) flows toward the taphole where it overflows. Matte also continuously overflows as new matte is made under the lances.

9.3.3 Electric slag cleaning furnace details

The electric slag—cleaning furnace (3600 kW) is elliptical with three or six graphite electrodes in one or two rows along the long axis. It accepts matte and slag from the smelting furnace and separates them into two layers—a bottom layer of matte 0.5—0.8 m thick, and a top layer of slag ~0.5 m thick. The residence times of the liquids are 1—2 h. These times plus electromagnetic stirring in the furnace allow the slag and matte to approach equilibrium. Passage of electricity through the slag ensures that the slag is hot and fluid. This, in turn, creates conditions for efficient settling of matte droplets from the slag.

The slag is the main route of Cu loss from the smelter. It is important, therefore, that the total amount of Cu-in-slag be minimized. This is done by

(a) maximizing slag residence time in the electric furnace (to maximize matte settling)
(b) keeping the slag hot, fluid, and quiescent
(c) minimizing slag mass (per tonne of Cu) by smelting high Cu grade concentrates and minimizing fluxing
(d) optimizing slag composition to minimize slag viscosity and density.

Matte continuously underflows from the bottom layer of the electric furnace to the converting furnace. A siphon and launder system is used. Slag continuously overflows through a taphole. It is granulated in flowing water and sold or stockpiled.

The purpose of the electrodes and electrical power is to keep the slag hot and fluid. Heat is obtained by resistance to electric current flow between the graphite electrodes in the slag, selectively heating the slag to 1250°C.

Only a tiny amount of offgas is generated in the electric furnace. It is collected from the slag taphole hood, drawn through an electrostatic precipitator, and vented to atmosphere.

9.3.4 Converting furnace details

The converting furnace continuously receives matte from the electric slag—cleaning furnace. It blows oxygen-enriched air blast (30—35 vol-% O_2), $CaCO_3$ flux, and converter slag granules onto the surface of the matte. It also receives considerable copper scrap, including scrap anodes.

It produces (a) molten copper (~0.7% S), (b) molten slag (14% Cu), and (c) SO_2 bearing offgas, 25—30 vol-% SO_2.

The molten copper continuously departs the furnace through a siphon and launder into one of two anode furnaces. The slag continuously overflows into a water-granulation system. The offgas is drawn through a large uptake, waste heat boiler, electrostatic precipitators, and a wet gas cleaning system before being pushed into a sulfuric acid or liquid SO_2 plant. The smelting and converting offgases are combined prior to entering the electrostatic precipitators.

The oxygen-enriched air and solids are introduced into the furnace through 10 lances like the smelting furnace lances, at a speed of ~120 m/s. The tips of the outside pipes are 0.3—0.8 m above the bath; the inside pipe tips above or just through the roof. The outer pipes are continuously rotated to prevent them from sticking to their roof collars. They are also slowly slipped downwards as they burn back.

9.3.4.1 Converting furnace slag

A unique feature of Mitsubishi converting is its CaO-based calcium ferrite slag (see Chapter 5). Early in the development of the process, it was found that blowing strong O_2 blast onto the surface of SiO_2-based slag made a crust of solid magnetite. This made further converting impossible. CaO, on the other hand, reacts with solid magnetite, molten Cu and O_2 to form a liquid $CaO-Cu_2O-FeO_n$ slag. The slag typically contains 12%—16% Cu (~60% as Cu_2O, balance Cu), 40%—55% Fe (~70% as Fe^{3+}, balance Fe^{2+}), and 15%—20% CaO. Its viscosity is ~10^{-1} kg/m·s (Wright et al., 2000).

9.3.4.2 Converting furnace copper

Mitsubishi process blister copper contains more sulfur than Peirce—Smith converter copper, ~0.7% versus ~0.02%. This is the result of operating with excess O_2 addition to avoid a permanent matte layer in the furnace. This prevents slag foaming. The only disadvantage of the higher S level is a longer oxidation period in the anode furnace. The S content can be lowered in the Mitsubishi converter by supplying more O_2, but this increases the amount of Cu in slag.

9.3.5 Optimum matte grade

The Cu grade of the matte produced by the smelting furnace (and flowing into the converting furnace) is a balance between:

(a) the amount of Cu lost in the discard slag from the electric furnace;

(b) the amount of granulated high-Cu slag which must be recycled back from the converting furnace to the smelting furnace; and

(c) the amount of coal which must be added to the smelting furnace and the amount of coolant which must be added to the converting furnace.

The optimum matte grade for the Mitsubishi process is 68% Cu. With this matte grade, the discard slag from the electric slag cleaning furnace contains 0.7%—0.9% Cu and converter slag recycle is 0.1—0.3 tonnes per tonne of concentrate feed. Selected design and operating parameters for two Mitsubishi plants are shown in Table 9.6 and 9.7 (Kaneda, 2020).

Table 9.6 Physical details of two Mitsubishi copper making systems. Personal Communication with Mitsubishi Materials (Kaneda, 2020).

Smelter	Mitsubishi Materials Corp. Naoshima, Japan	PT Smelting Co. Gresik, Indonesia
Furnace commissioning date	1991	1998
Copper production rate, tonnes/day	800	~870
Smelting furnace		
shape	Circular	Circular
Diameter × height (inside brick), m	10 × 4	10.1 × 4
Number of lances	10	12
Outside pipe diameter, cm	ID: 10	10
Inside pipe diameter, cm	ID: 5	6.25
Lance rotations per minute	7.7	6.5
Slag layer thickness, m	0.1	0.1
Matte layer thickness, m	1.4	1.4
Liquids, offgas temperature, °C	1240, 1240	1250, 1250
Electric slag cleaning furnace		
Power rating, kW	3600	2300 (No.1) 2100 (No.2)
Shape	Elliptical	Elliptical
Width × length × height, m	6.0 × 12.5 × 2.0	5.9 × 12.5 × 2.0
Electrodes		
Material	Graphite	Graphite
Number	6	6
Diameter, m	0.4	0.4
Immersion in slag, m	0.5	0.4
Voltage between electrodes, V	70 to 170	100 to 140
Current between electrodes, kA	6.0 to 7.5	9 (No.1); 7 (No.2)
Applied power, kW	2400	3100
Slag layer thickness, m	0.6	0.7
Matte layer thickness, m	0.45	0.5
Matte, slag, offgas temperatures, °C	1210, 1250, -	1230, 1270, -
Estimated slag residence time, hours	−1 to 2	−1
Converting furnace		
shape	Circular	Circular
Diameter × height, m	8.0 × 3.6	9.0 × 3.7
Number of lances	10	10
Outside pipe diameter, cm	ID: 10	10
Inside pipe diameter, cm	ID: 5	6.25
Lance rotations per minute	7.7	6.5
Slag layer thickness, m	0.16	0.15
Copper layer thickness, m	0.88	1.1
Copper, slag, offgas temperature, °C	1225, 1245, 1245	1220, 1235, 1230

Table 9.7 Operating data for two Mitsubishi copper making systems, Personal Communication with Mitsubishi Materials (Kaneda, 2020).

Smelter	Naoshima, Japan	Gresik, Indonesia
Smelting furnace		
Inputs, tonnes/day		
Concentrate	2080 (26% Cu)	3200 to 3500 (21%-23% Cu)
Silica flux	350	
Limestone flux	0	0
Granulated converting furnace slag	290	240
Smelting furnace dust	67	60
Converting furnace dust	61	60
Other:	210 (scrap and reverts)	No
Blast		
volume% O_2	66	72
Input rate, Nm^3/min	550	650−700
Oxygen, tonnes/day	820 (82% O_2)	850 to 950 (99.9% O_2)
Hydrocarbon fuel (coal, tonnes/day)	10	20
Output offgas		
volume% SO_2 (entering boiler)	39	45
Flowrate, Nm^3/min	550	650−700
Dust tonnes/day	67	60
Electric slag cleaning furnace		
Inputs	Matte and slag from smelting furnace	
Outputs		
Matte, tonnes/day	1080	1080−1290
Matte %Cu	65	67
Slag, tonnes/day	1660	2250
Slag %Cu	0.7	0.7
Slag %SiO_2/%Fe	1.02	1.0
Offgas, Nm^3/min (ppm SO_2)		
kWh consumption/tonne of slag	41	57
Converting furnace (autothermal)		
Inputs, tonnes/day		
Matte from electric furnace	1080	1080−1290
Limestone flux	75	65
Granulated converting furnace slag	260	336
Anode scrap copper	120	95 to 100
Purchased scrap copper	14	82
Blast		
Volume% O_2	42	36
Input rate, Nm^3/min	470	500 to 570

Continued

Table 9.7 Operating data for two Mitsubishi copper making systems, Personal Communication with Mitsubishi Materials (Kaneda, 2020).—cont'd

Smelter	Naoshima, Japan	Gresik, Indonesia
Oxygen, tonnes/day	330 (82%O_2)	170–240 (99.9%O_2)
Outputs		
Copper, tonnes/day	790	850 to 1000
Copper, %Cu	98.4	98.5
Copper, %O	0.3	0.3
Copper, %S	0.7	0.7
Slag, tonnes/day	290	576
Slag, %Cu	14	13
Slag, mass% CaO/mass% Fe	0.4	0.4
Offgas, Nm^3/min	460	490
Volume% SO_2	26	26
Dust, tonnes/day	61	60

9.3.6 Process control in Mitsubishi smelting/converting (Goto & Hayashi, 1998)

Bath temperature is the most important control parameter in the Mitsubishi process. Maintaining an optimum temperature ensures good slag and matte fluidity while at the same time minimizing refractory erosion. Typical operating temperatures at the Naoshima smelter are shown in Table 9.8.

Table 9.8 shows that the electric slag cleaning furnace slag is operated hotter than the smelting furnace matte/slag. This is done to avoid precipitation of solid Fe_3O_4 and/or SiO_2. Precipitation of Fe_3O_4 caused formation of a "muddy layer" between molten matte and molten slag. This layer prevents good matte/slag separation, and results in high matte entrainment in the discard slag, hence high Cu-in-slag losses. Precipitation of SiO_2 produces SiO_2 rich "floating solids." These solids (~10% Cu) do not settle into the matte layer and also result in high Cu-in-slag losses.

Process control at Naoshima combines quantitative and qualitative process information (Tables 9.9 and 9.10). The quantitative information is continuously inputted into an

Table 9.8 Maximum, minimum, and optimum operating temperatures in the Mitsubishi smelter at Naoshima.

	Minimum (°C)	Maximum (°C)	Optimum (°C)
Smelting furnace—matte/slag outlet	1200	1240	1220
Electric slag cleaning furnace—slag outlet	1220	1260	1240
Converting furnace—slag outlet	1210	1250	1230

Table 9.9 Quantitative process parameters in Mitsubishi smelting/converting. Process control has been improved using immersion thermocouples to continuously measure smelting furnace matte/slag temperature and converting furnace slag temperature. Molten copper temperature is measured with an optical fiber pyrometer.

Information	Adjustable or Result	Comments
Smelting furnace		
Melt temperature	Result	Continuous measurement by submerged K-type thermocouple at smelting furnace melt outlet.
Feedrate	Adjustable	Continuous monitoring of blended concentrate, silica, $CaCO_3$, coal, converter slag granules, Cu scrap, and dust input rates.
Blowing rate	Adjustable	Continuous monitoring of oxygen and air input rates.
Matte composition	Result	Hourly analysis by X-ray fluorescence analyzer (Cu and Pb). Samples are taken at electric slag cleaning furnace outlet.
Slag composition	Result	Hourly analysis by X-ray fluorescence analyzer (Cu, Fe, SiO_2, CaO, and Al_2O_3). Samples are taken at electric slag cleaning furnace outlet.
Converting furnace		
Slag temperature	Result	Hourly manual measurement by disposable K-type thermocouples.
Copper temperature	Result	Automatic measurement by submerged glass fiber optical pyrometer—every 15 min.
Feedrate	Adjustable	Continuous monitoring of $CaCO_3$, recycled converter slag granules, and Cu scrap input rates.
Matte flowrate	Result	Continuous measurement by infrared scanner at electric slag cleaning furnace matte siphon.
Blowing rate	Adjustable	Continuous measurement of oxygen and air input rates.
Slag composition	Result	Hourly sample and analysis (Cu, SiO_2, CaO, S, and Pb) by X-ray fluorescence analyzer.

Table 9.10 Qualitative process parameters in Mitsubishi smelting/converting.

Information	Adjustable or Result	Comments
Smelting furnace		
Melt flowrate and melt surface condition	Result	Visual inspection by remote camera and at the furnace.
Converting furnace		
Slag flowrate and slag surface condition	Result	Visual inspection by remote camera and at the furnace.
Matte inflow rate	Result	Visual inspection by remote camera and at the furnace.

expert type computer control system. The overall state of the process is best represented by the following quantitative variables:

(a) smelting furnace matte/slag outlet temperature
(b) converting furnace slag temperature
(c) matte grade
(d) %SiO_2 in electric slag cleaning furnace slag
(e) %CaO in electric slag cleaning furnace slag
(f) %Cu in converting furnace slag
(g) %CaO in converting furnace slag.

Table 9.9 shows how each parameter is controlled.

Control of the Mitsubishi process is a challenge for its operators. The difficulty lies in the fact that (a) there are many process variables to control and (b) they all interact with each other. When one variable goes out of control, the rest of the process is affected. For example, the solution to the problem of high viscosity slags, floating solids, and muddy layers has typically been to increase furnace temperature. Unfortunately, this increases refractory erosion, ultimately decreasing smelter availability.

However, recent development of sophisticated continuous melt temperature devices coupled with the use of an expert type computer control system has helped minimize process excursions. This has greatly increased Mitsubishi process availability.

9.3.7 The Mitsubishi process since 2000

Improvements in the Mitsubishi Process since 2000 have focused on two concerns: refractory life and process control. Refractory life is a challenge in the S furnace (due to the turbulence created by the injection lances), and in the C furnace (due to turbulence and the more corrosive nature of the calcium ferrite slag). Increasing the S-furnace lance diameter from 0.060 to 0.069 m reduced penetration depth and improved refractory life (Cho et al., 2006; Taniguchi et al., 2006), as did improved distribution control for the solid feed to create even penetration depth. C-furnace refractory life has been extended by adjusting the refractory in front of the water-cooled copper jacket at the bath line (Goto et al., 2002).

One of the challenges to successful operation of C furnace is magnetite control. Some magnetite is necessary to provide a protective coating for the refractories, but too much causes accretions to build up in the furnace, eventually blocking the siphon through molten blister copper is recovered (Hasegawa & Sato, 2003; Taniguchi et al., 2006). As a result, careful control of the slag chemistry is needed (Tanaka et al., 2003). Increasing the CaO content of the slag reduces accretions, but in turn requires better heat removal through the water jacket at the slag line. Plant operations at Naoshima now control the Fe/CaO ratio separately from the %Cu in the C-furnace slag; maintaining 2.2 < Fe/CaO < 2.45 seems to work best.

In terms on installations commissioned since 2000 the year, the Mitsubishi process was installed a Hindalco Birla Copper in 2005 with a capacity of 250,000 t/y of copper and also the S-furnace as standalone at Onahama, Japan, in 2007 with an initial capacity of 200,00 t/y (2012).

The Mitsubishi process has been also promoting the recycling business by combining a pretreatment incineration—melting plant in which copper containing recyclable materials are pretreated before adding them to the Mitsubishi process. The plant commissioned in 2004 was able to double its recyclable processing capacity by 2012. The combination of a pretreatment plant with the Mitsubishi process allowed a more efficient use of smelting capacity by reducing the e-scrap volume.

In 2016, Mitsubishi Materials also reported the completion of a second E-scrap center at Naoshima smelter allowing Naoshima smelter to increase the recycling capacity to 100,000 t/y.

In a similar fashion, Onahama smelter has increased its recyclable materials processing capacity (Nagata et al., 2019).

9.4 Other developments for continuous processing of copper

In addition to the Dongying-Fangyuan process described in Section 9.2, Chinese technologists have been working in the last 10 years in developing options to produce copper in a continuous flow of materials, arranging in cascade, smelting, and converting reactors. The following advances have been reported in the last 5 years:

- Double bottom bottom—blowing continuous smelting process (Yongcheng et al., 2019)
- Side—blowing smelting technology combined with horizontal bottom blowing converting (Ling, 2019)

9.4.1 Double bottom bottom—blowing continuous smelting process (SKS-BCC) (Yongcheng et al., 2019)

The double bottom—blowing continuous smelting process applies bath smelting principles using SKS type lances to inject oxygen into the molten bath. A simplified process flowsheet is shown in Fig. 9.9. A single smelting SKS furnace continuously smelts copper concentrate reacting with enriched air injected via 11 oxygen lances located on the bottom of the furnace shell. Molten matte and slag are tapped from tapholes located at opposite sides of the furnace shell end walls. Heat from the offgases is recovered in a waste heat boiler before dedusting (ESP) and cleaned prior to acid production in a Sulfuric Acid Plant.

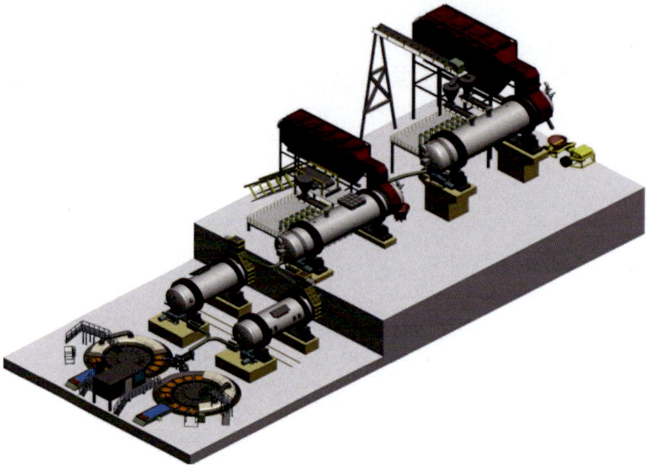

Figure 9.9 Simplified schematic arrangement of the SKS-BCC three-connected furnaces process (Yongcheng et al., 2019). Figure originally published in Proceedings of the 58th Conference of Metallurgists Hosting the International Copper Conference 2019. *(Reprinted with permission of the Canadian Institute of Mining, Metallurgy and Petroleum.)*

Molten matte continuously flows into the converting reactor (BCC). A bypass launder is also part of the arrangement to cool part of the generated matte and used as cold charge in the BCC for heat balance purposes.

The BCC furnace has seven oxygen lances and can be operated on a continuous mode with the SKS furnace or stand-alone only with cold matte is the smelting reactor is out of service. Generated blister and converter slag are tapped through dedicated tapholes. Offgases are processed in a similar fashion to the SKS smelting furnace. Converter slag is tapped into ladles for slow cooling and recycling into the SKS-BBS furnace. Copper is laundered into the anode furnaces for pyrorefining and anode production.

Table 9.11 shows the main design parameters of the SKS-BCC process, including chemical compositions of raw materials, and generated blister copper. The first SKS-BCC process was commissioned in 2018 at SDIC Jincheng Metallurgical Co. According to Yongcheng et al. (2019), the main features of the SKS-BCC process are

- Material intake flexibility: the process can treat a wide range of copper concentrate from 14% to 25%
- Faster ramp—up to design capacity: SDIC Jifeng plant reached design capacity in 4 months
- High energy efficiency: no cooling of copper matte
- Low emissions

Table 9.11 Selected Technical Information SKS-BCC process (Yongcheng et al., 2019).

Parameter	Value
SKS smelting furnace	
Smelting capacity, t/y	600,000
Feed rate, t/h	110–130
Furnace diameter, m	4.8
Furnace length, m	28.0
Oxygen lances	14
Installed	11
Operating	
Feed ports	2
Tapholes slag	1
Tapholes matte	1
Cu slag recovery method	Slow cooling, flotation
Bypass matte launder	1
Heat recovery	Waste heat boiler
Air flow, Nm^3/h	15,000–16,500
Oxygen flow, Nm^3/h	7000–8000
Oxygen enrichment, $\%O_2$	73
Copper in feed, %Cu	15–20
Cu/S ratio	0.75–1.1
Fe/S ratio	1.2–1.4
Fe/SiO_2 ratio in slag	1.7–1.8
As in feed, %As	0.2–0.4
Matte grade, %Cu	70–72
Product temperature (matte/slag/offgas)	1230/1200/1180
Copper in slag, %Cu	2.5
Dust generation, % of concentrate	2.5
BCC converting furnace	
Blister production, t/y	139,100
Furnace diameter, m	4.4
Furnace length, m	20.0
Oxygen lances	
Installed	10
Operating	7
Feed ports	
Liquid matte	1
Solid matte, flux and reverts	1
Tapholes slag	1
Tapholes matte	1
Cu slag recovery method	Slow cooling, recycling to SKS

Continued

Table 9.11 Selected Technical Information SKS-BCC process (Yongcheng et al., 2019).—cont'd

Parameter	Value
Heat recovery	Waste heat boiler
Air flow, Nm3/h	9000–12,000
Oxygen flow, Nm3/h	2800–3500
Oxygen enrichment, %O_2	22–45
Coal addition, t/h	0.5–1.0
Silica addition, t/h	0.8–1.5
Copper in slag, %Cu	12–14
Fe/SiO_2	1.1–1.3
Blister composition	
Cu	97.0–98.5
S	0.3–0.55
Product temperature (copper/slag/offgas)	1250/1220/1220
Casting, h	4
Preparation for next batch	2.5

Information on the anode refining process can be found in the cited reference for this section. Yongcheng also reported that further efforts were required to improve production stability, oxygen lance, and blister taphole performances.

9.4.2 Side-blowing smelting technology combined with horizontal bottom blowing converting

This technology configuration was implemented by Chifeng Yunnan Copper in 2015. According to Tang (Tang & Yao, 2019), the technology has been applied to four plants in China and Africa with a maximum smelting capacity of 200 t/h.

9.5 Summary

Continuous copper making processes can be achieved in a single, double, or three stages. The selection of the process depends on the type of material to be processed and the aimed final product.

9.5.1 Single-stage copper production

Direct-to-copper smelting is the smelting of concentrate directly to molten copper in one furnace. In 2011, it is practiced in three smelters: Olympic Dam (Australia), Glogów I and II (Poland), and Chingola (Zambia). All of these plants use an Outotec flash furnace.

The main advantage of the process is its restriction of SO_2 evolution to a single, continuous source of high SO_2-strength gas. In principal, the energy, operating, and capital costs of producing metallic copper are also minimized by the single-furnace process.

Metallic copper is obtained in a flash furnace by setting the ratio of oxygen input in the blast to concentrate feed rate to the point where all the Fe and S in the input concentrate are oxidized. The ratio must be precisely controlled to avoid the production of Cu_2S or Cu_2O. Avoiding formation of a molten Cu_2S layer in the furnace is critical. Reactions between Cu_2S layers and oxidizing slag may cause rapid SO_2 evolution and slag foaming.

Direct-to-copper flash smelting has proven effective for SO_2 capture. However, about 25% of the Cu in the input concentrate is oxidized, ending up as copper oxide dissolved in slag. This copper oxide must be reduced back to metallic copper, usually with coke. The expense of recovering Cu from slag will probably restrict future direct-to-copper smelting to concentrates which produce little slag. Processing of chalcopyrite concentrates are more suitable to be processed by multifurnace processes, either by conventional smelting and converting or by continuous multifurnace processing.

9.5.2 Two-stage copper production

The two-stage copper production process was developed by Dongying-Fangyuan. The process has been designed by continuous smelting (SLS) and the combination of the converting and refining duties in a single reactor (SLCR). The SLCR process can be defined a batch process in which gradual management of the oxygen potential, achieved by regulation of air, oxygen, nitrogen, and natural gas, allows the production of anode quality copper. The continuity of the process is achieved by arranging two SLCR furnaces operating in alternate cycles.

9.5.3 Three-stage copper production

The most known three-stage copper production process is the Mitsubishi process. The Mitsubishi process also performs smelting and converting in a molten bath. It uses three interconnected furnaces to continuously smelt copper concentrates, settle and separate matte from discard slag, and convert the matte to blister copper. When performing properly, continuous processing solves many of the problems created by more traditional copper smelting technology, most notably the elimination of ladle transfer of molten matte, the fugitive emissions associated with Peirce—Smith converting, and the uneven SO_2 offgas content that results from batch processing of copper concentrates.

Chinese technologists have developed two alternatives to the Mitsubishi process approach:

- Double bottom bottom–blowing continuous smelting process (Yongcheng et al., 2019)
- Side-blowing smelting technology combined with horizontal bottom blowing converting (Ling, 2019)

These processes have been recently commissioned. More process information will become available to evaluate their operational performance.

References

Ajima, S., Koichi, K., Kanamori, K., Igarashi, T., Muto, T., & Hayashi, S. (1999). Copper smelting and refining in Indonesia. In D. B. George, W. J. Chen, P. J. Mackey, & A. J. Weddick (Eds.), *Copper 99—cobre 99, vol. V: Smelting operations and advances* (pp. 57–69). Warrendale, PA: TMS.

Asaki, Z., Taniguchi, T., & Hayashi, M. (2001). Kinetics of the reactions in the smelting furnace of the Mitsubishi process. *Journal of Occupational Medicine, 53*(5), 25–27.

Brillo, J., & Egry, I. (2003). Density determination of liquid copper, nickel and their alloys. *International Journal of Thermophysics, 24,* 1155–1170.

Byszynski, L., Garycki, L., Gostynski, Z., Stodulski, T., & Urbanowski, J. (2010). Present and future modernization of metallurgical production lines of the Glogów copper smelter. In *Copper 2010, volume 2: pyrometallurgy I, GDMB* (pp. 631–647). Germany: Clausthal–Zellerfeld.

Cho, H. Y., Lee, K. H., Chang, J. S., Seo, W. C., & Kim, J. Y. (2006). Expansion and modernization of Onsan smelter. In F. Kongoli, & R. G. Reddy (Eds.), *Sohn international symposium, vol. 8: International symposium on sulfide smelting* (pp. 97–110). Warrendale, PA: TMS.

Czernecki, J., Smieszek, Z., Miczkowski, Z., Dobrzanski, J., Bas, W., Szwancyber, G., Warmuz, M., & Barbacki, J. (1999). The process flash slag cleaning in electric furnace at the Glogów II copper smelter. In F. Niemenen (Ed.), *9th international flash smelting Congress* (pp. 375–389). Espoo, Finland: Outokumpu Technology.

Czernecki, J., Smieszek, Z., Miczkowski, Z., Krawiec, G., & Gizicki, S. (2010). Problems of lead and arsenic removal from copper production in a one-stage flash-smelting process. In *Copper 2010, volume 2: pyrometallurgy I, GDMB* (pp. 669–683). Germany: Clausthal–Zellerfeld.

Davenport, W. G., Jones, D. M., King, M. J., & Partelpoeg, E. H. (2010). *Flash smelting, analysis, control and optimization* (2nd ed.). New York, USA: Wiley.

Dobrzanski, J., & Kozminski, W. (2003). Copper smelting in KGHM Polska Miedz S.A. In C. Díaz, J. Kapusta, & C. Newman (Eds.), *Copper/cobre 2003, volume IV, book 1: Pyrometallurgy of copper* (pp. 239–252). Montreal, Canada: CIM.

Goto, M., & Hayashi, M. (1998). *The Mitsubishi continuos process - a description and detailed comparison between commercial practice and metallurgical theory*. Tokyo: Mitsubishi Materials Corporation.

Goto, M., Konda, K., Muto, T., Yamashiro, A., & Matsutano, A. (2002). Gresik copper smelter and refinery — current operation and expansion plan. In R. L. Stephens, & H. Y. Sohn (Eds.), *Sulfide smelting 2002* (pp. 61–72). Warrandale, PA: TMS.

Hasegawa, N., & Sato, H. (2003). Control of magnetite behavior in the Mitsubishi process at Naoshima. In F. Kongoli, K. Itagaki, C. Yamauchi, & H. Y. Sohn (Eds.), *Yazawa international symposium, vol. II: High-temperature metals production* (pp. 509–519). Warrendale, PA: TMS.

Hunt, A. G., Day, S. K., Shaw, R. G., Montgomerie, D., & West, R. C. (1999). Start up and operation of the #2 direct-to-copper flash furnace at Olympic Dam. In F. Niemenen (Ed.), *9th international flash smelting Congress* (pp. 375–389). Espoo, Finland: Outokumpu Technology.

Kaneda, A. (2020). *Personal communications with Mitsubishi Materials Corporation*. Metallurgy Department, Metallurgy Division, Metals Company.

Ling, W. (2019). The utilization of copper continuous smelting, converting and fire-refining processes in China. Paper 595879. In *E-proceedings copper 2019 pyrometallurgy*. Vancouver: CIMM, ISBN 978-1-926872-44-5.

Mitsubishi Materials Corporation. (2004). *The Mitsubishi process: Copper Smelting for the 21st century*. http://www.mmc.co.jp/sren/MI_Brochure.pdf.

Mitsubishi Materials Corporation. (2020). *The Mitsubishi process: Copper smelting for the 21st century* (Accessed September 2020) https://www.mmc.co.jp/sren/History.htm.

Nagata, T., Ueda, I., & Ishikawa, S. (2019). Recent progress of the O-SR process technology. Paper 596538. In *E-proceedings copper 2019 pyrometallurgy*. Vancouver: CIMM, ISBN 978-1-926872-44-5.

Ranasinghe, D. J., Russell, R., Muthuraman, R., & Dryga, Z. (2010). Process optimization by means of heat and mass balance based modelling at Olympic Dam. In *Copper 2010, volume 3: Pyrometallurgy II, GDMB* (pp. 1063—1078). Germany: Clausthal—Zellerfeld.

Shibasaki, T., & Hayashi, M. (1991). Top blowing injection smelting and converting — the Mitsubishi process. In T. Lehner, P. J. Koros, & V. Ramachandran (Eds.), *Injection in process metallurgy* (pp. 199—213). Warrendale, PA: TMS.

Solnordal, C. B., Jorgensen, R. R. A., Koh, P. T. L., & Hunt, A. (2006). CFD modeling of the flow and reactions in the Olympic Dam flash furnace smelter reaction shaft. *Applied Mathematical Modelling, 30*, 1310—1325.

Somerville, M., Chen, C., Alvear, F, & Nikolic, S. (Eds.). (2016). Fluxing Strategies for the Direct to Blister Smelting of High Silica and Low Iron Copper Concentrates, *Advances in Molten Slags, Fluxes, and Salts: Proceedings of The 10th International Conference on Molten Slags, Fluxes and Salts (MOLTEN16) Edited by: Ramana G. Reddy, Pinakin Chaubal, P. Chris Pistorius, and Uday Pal TMS (The Minerals, Metals & Materials Society)*. Springer.

Syamujulu, M. (2007). Opportunities, problems and survival strategies from recent developments in the copper concentrate treatment and smelting practices at Vedanta's Konkola copper mines in the Zambian copperbelt. In A. E. M. Warmer, C. J. Newman, A. Vahed, D. G. George, P. J. Mackey, & A. Warczok (Eds.), *Copper/cobre 2007, volume III (book 1): The Carlos Diaz symposium of pyrometallurgy* (pp. 155—166). Montreal, Canada: CIM.

Tanaka, F., Iida, O., & Takeda, Y. (2003). Thermodynamic fundamentals of calcium ferrite slag and their application to Mitsubishi continuous copper converter. In F. Kongoli, K. Itagaki, C. Yamauchi, & H. Y. Sohn (Eds.), *Yazawa international symposium, vol. II: High-temperature metals production* (pp. 495—508). Warrendale, PA: TMS.

Tang, B., Yao, S. (2019) Application and development of modern copper metallurgical technologies in China's Copper Industry. In e-proceedings of copper 2019 pyrometallurgy. Paper 588809. ISBN978-1-926872-44-5.

Taniguchi, T., Matsutani, T., & Sato, H. (2006). Technological innovations in the Mitsubishi process to achieve four years campaign. In F. Kongoli, & R. G. Reddy (Eds.), *Sohn international symposium, vol. 8: International symposium on sulfide smelting* (pp. 261—273). Warrendale, PA: TMS.

Taskinen, P., & Kojo, I. (2009). Fluxing options in the direct-to-blister copper smelting. In M. Sánchez, R. Parra, G. Riveros, & C. Díaz (Eds.), *MOLTEN2009* (pp. 1139—1151). Santiago, Chile: Gecamin.

Tuominen, J., & Kojo, I. (2005). Blister flash smelting — efficient and flexible low-cost continuous copper process. In A. Ross, T. Warner, & K. Scholey (Eds.), *Converter and fire refining practices* (pp. 271—282). Warrendale, PA, USA: TMS.

Urbaniak, R., Gabka, T., Poks, P., & Garycki, L. (2019). Comparison of two sizes of FSFs in the Glogów smelter. In *E-proceedings copper 2019 pyrometallurgy*. Vancouver: CIMM, ISBN 978-1-926872-44-5. Paper 596538.

Wang, Z., Cui, Z., Wei, C., & Wang, H. (2017). Two-step copper smelting process at Dongying Fangyuan. In J. Y. Hwang, et al. (Eds.), *8th international symposium on high-temperature metallurgical processing. The minerals, metals & materials series*. Cham: Springer. https://doi.org/10.1007/978-3-319-51340-9_62

Wang, Z., Zhixiang, C., Wang, H., Cui, W., & Huang, Z. (2019). Development and improvement of submerged lance converting and refinign furnace of Fangyuan two-steps process. In *E-proceedings copper 2019 pyrometallurgy*. Vancouver: CIMM, ISBN 978-1-926872-44-5. Paper 600345.

Wright, S., Zhang, L., Sun, S., & Jahanshahi, S. (2000). Viscosity of calcium ferrite slags and calcium alumino-silicate slags containing spinel particles. In S. Seetharaman, & D. Sichen (Eds.), *Sixth international conference on molten slags, fluxes and salts*. Stockholm, Sweden: Division of Metallurgy, KTH. Paper number 059.

Yongcheng, Z., Bing, L., & Kao, L. (2019). Engineering and production practice of double bottom-blowing continuous copper smelting process with "three connected furnaces arrangement". In *E-proceedings copper 2019 pyrometallurgy*. Vancouver: CIMM, ISBN 978-1-926872-44-5. Paper 592917.

Zhixiang, C., Zhi, W., Haibin, W., & Chuanbing, W. (2016). Two-step copper smelting process at Dongying-Fangyuan. In *E-proceedings copper 2016 pyrometallurgy*. Kobe: MMIJ. Paper PY16-4.

Suggested reading

Gostynski, Z., & Haze, D. (2007). Flash smelting furnace of the KGHM Glogów copper plant — technological and process challenges as a driving force of its continuous modernization. In A. E. M. Warmer, C. J. Newman, A. Vahed, D. G. George, P. J. Mackey, & A. Warczok (Eds.), *Copper/cobre 2007, volume III (book 2): The Carlos Diaz symposium of pyrometallurgy* (pp. 233—243). Montreal, Canada: CIM.

Goto, M., & Hayashi, M. (1998). *The Mitsubishi continuous process — a description and detailed comparison between commercial practice and metallurgical theory*. Tokyo, Japan: Mitsubishi Materials Corporation.

Kojo, I., Lahtinen, M., & Miettinen, E. (2009). Flash converting — sustainable technology now and in the future. In J. Kapusta, & T. Warner (Eds.), *International Peirce-Smith converting centennial* (pp. 383—395). Warrendale, PA, USA: TMS.

Kucharski, M. (2007). Modeling of the direct-to-copper flash smelting process. In A. E. M. Warmer, C. J. Newman, A. Vahed, D. G. George, P. J. Mackey, & A. Warczok (Eds.), *Copper/cobre 2007, volume III (book 2): The Carlos Diaz symposium of pyrometallurgy* (pp. 159—171). Montreal, Canada: CIM.

Syamujulu, M., & Beene, G. (2003). The smelting characteristics of KCM concentrates. In C. Diaz, J. Kapusta, & C. Newman (Eds.), *Copper/cobre 2003, volume IV, book 1: Pyrometallurgy of copper* (pp. 325—337). Montreal, Canada: CIM.

Further reading

Mizuta, Y., Oguma, N., Iwahori, S., & Sato, H. (2012). Promotion of recycling busines by combination of a pre-treatment plant and the Mitsubihsi process at Naoshima smelter & refinery. In J. P. Downey, T. Battle, & J. S. White (Eds.), *International smelting technology symposium (incorporating the 6th Advances in sulfide smelting symposium)* (pp. 125—132). Warrendale, PA: TMS.

CHAPTER 10

Copper loss in slag

Pyrometallurgical production of molten copper generates two slags, smelting and converting. Smelting furnace slag typically contains 1%–2% Cu (Table 4.2). The percentage increases as matte grade increases (Font et al., 2013). Converter slag contains 4%–8% Cu (Table 7.2). Its percentage increases as converting proceeds, and the grade of the remaining matte increases.

Multiplying these percentages by the mass of each slag shows that a significant fraction of the Cu in the original concentrate is present in these slags. This fraction is increased by the production of higher-grade mattes in the smelting furnace. Because of this, the value of the copper in these slags is usually too high to justify the old practice of simply discarding them.

This chapter discusses the nature of copper in smelting and converting slags. It also describes strategies for minimizing the amount of copper lost from their disposal. The main strategies include the following:

(a) minimizing the mass of slag generated
(b) minimizing the percentage of Cu in the slags
(c) processing the slags to recover as much Cu as possible.

Slag processing can be divided into two types. The first is pyrometallurgical reduction and settling, performed in an electric or fuel fired slag cleaning furnace. The second is minerals processing of solidified slag, including crushing, grinding, and froth flotation, to recover Cu from the slag.

10.1 Copper in slags

The Cu in smelting and converting slags is present either as (a) dissolved Cu, present mostly as Cu^+ ions, or as (b) entrained droplets of matte. The dissolved Cu is associated either with O^{2-} ions (i.e., Cu_2O), or with S^{2-} ions (Cu_2S). Cu_2O becomes the dominant form of dissolved Cu at matte grades above 70% Cu (Acuña & Sherrington, 2005; Barnett, 1979; Font et al., 2013), due to the increased activity of Cu_2S in the matte. Higher Cu_2S activity pushes the reaction

$$Cu_2S\ \textit{(matte)} + FeO\ \textit{(slag)} \rightarrow Cu_2O\ \textit{(slag)} + FeS\ \textit{(matte)} \tag{10.1}$$

to the right. The solubility of sulfur in slags is also lower in contact with higher-grade mattes (Roghani et al., 2000). As a result, dissolved Cu in converter slags is present

Extractive Metallurgy of Copper, Sixth Edition
ISBN 978-0-12-821875-4, https://doi.org/10.1016/B978-0-12-821875-4.00013-4

mostly as Cu_2O (Font et al., 2013). Conversely, the dissolved Cu in smelting slags is present mostly as Cu_2S. This is due to the lower matte grades and oxygen potentials in the smelting furnace.

There are several sources of entrained matte in slags. The most obvious are droplets of matte that did not settle completely through the slag layer during smelting. The Hadamard–Rybczynski formula predicts the rate at which matte droplets will settle through molten slag (Warczok et al., 2007a):

$$V = (\rho_M - \rho_S)r^2/3\eta_S \tag{10.2}$$

In this expression, V is the settling rate of the matte droplets (m/s), g is the gravitational constant (9.8 m/s^2), ρ_M is the matte density (3900–5200 kg/m^3), ρ_S is the slag density (3300–3700 kg/m^3), η_S is the slag kinematic viscosity, and r is the radius (m) of the settling matte droplet. The expression is most accurate for systems with Reynolds numbers below 10 (i.e., droplet sizes below ~1 mm). Larger matte droplets settle at slower rates than predicted by Eq. (10.2). However, it is the settling rates of the smallest droplets that are of greatest concern.

Table 10.1 (Ip & Toguri, 2000) shows just how long matte droplets can take to settle. Besides droplet size, the biggest influences on settling rate are temperature and slag silica content. Higher temperatures and lower silica levels decrease slag viscosities, increasing settling rate (Zivkovic et al., 2020). A more reducing environment also encourages settling, by decreasing the Fe_3O_4(s) content of the slag.

In addition, matte grade has an impact on settling rates. Low-grade mattes have lower densities than high-grade mattes and therefore settle at slower rates (Fagerlund & Jalkanen, 1999).

Matte droplets can become suspended in smelter slags by several other mechanisms. Some are carried upward from the molten matte layer by gas bubbles generated by the following reaction (Cheng et al., 2019; Zivkovic et al., 2020):

$$3\ Fe_3O_4\ \textit{(slag)} + FeS\ \textit{(matte)} \rightarrow 10\ FeO\ \textit{(slag)} + SO_2 \tag{10.3}$$

Still others appear by precipitation from the slag in colder areas of the smelting furnace (Barnett, 1979). Converter slag returned to a smelting furnace also contains suspended

Table 10.1 Matte settling rates (mm/h) for $\rho_{matte} - \rho_{slag} = 1$ g/cm^3 (Ip & Toguri, 2000).

	Matte particle diameter		
Slag viscosity (Pa-s)	**2 μm**	**20 μm**	**40 μm**
0.05	0.157	15.690	62.578
0.08	0.098	9.806	39.224
0.1	0.078	7.845	31.379
0.12	0.065	6.537	26.149

matte droplets, which may not have time to completely settle (Cardona et al., 2011). As a result, entrained matte can represent from 50% to 90% of the copper contained in smelter slag (Acuña & Sherrington, 2005; Imriš et al., 2000). Fig. 10.1 shows the solubility of oxidic and sulfidic copper in smelter slag, calculated using thermodynamic modeling (Furuta et al., 2006) and compared with actual plant data. The difference between the two is entrained matte.

10.2 Decreasing copper in slag I: minimizing slag generation

It seems logical to suggest that decreasing the amount of Cu lost in smelting and converting slags could be accomplished by decreasing slag production. However, methods to decrease slag mass may do more harm than good. Possibilities include the following:

(a) *maximizing concentrate grades.* The less gangue in the concentrate, the less silica required to flux it, and the less overall slag generated. However, increasing concentrate grades may come at the expense of decreasing Cu recoveries in the concentrator.

(b) *adding less flux.* Adding less flux would decrease slag mass (desirable) and decrease its viscosity, making settling easier (also desirable). However, it would also increase the activity of FeO in the slag, leading to more dissolved Cu_2O by Reaction (10.1) (undesirable) and more magnetite (also undesirable; Mariscal & Herrera, 2009). The precipitation of solid magnetite raises slag viscosities and leads to higher slag copper contents as a result (Font et al., 2013; Kunze et al., 2003).

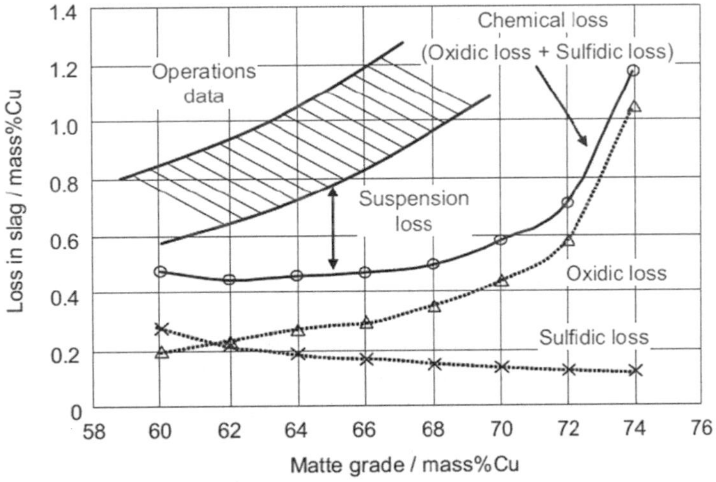

Figure 10.1 Thermodynamically predicted and actual loss of copper in smelter slag. The difference is mechanically entrained matte (suspension loss). *((Furuta et al., 2006). Used with permission of The Minerals, Metals & Materials Society.)*

10.3 Decreasing copper in slag II: minimizing Cu concentration in slag

The concentration of copper in smelting slag is minimized by the following:

(a) maximizing slag fluidity, principally by avoiding excessive Fe_3O_4 (s) in the slag and by keeping the slag hot (Zivkovic et al., 2020). Slag additions such as B_2O_3 and CaF_2 also help (Cui, Wang, Bian, Wei, & Zhao, 2018).

(b) changing slag chemistry. Pizarro et al. (2013) showed that adding CaO to generate olivine slags lowered dissolved copper levels.

(c) providing enough SiO_2 to form distinct matte and slag phases. This is especially a concern in high-alumina slags (Mariscal & Herrera, 2009; Sano et al., 2016).

(d) providing a large quiet zone in the smelting furnace.

(e) avoiding an excessively thick layer of slag (Zivkovic et al., 2020)

(f) avoiding tapping of matte with slag (Xia et al., 2005).

Metallurgical coke or coal may also be added to the smelting furnace to reduce Fe_3O_4 (*s*) to FeO (l) (Furuta et al., 2006).

10.4 Decreasing copper in slag III: pyrometallurgical slag settling/reduction

Conditions that encourage suspended matte droplets to settle to a matte layer are low viscosity slag, low turbulence, a long residence time, and a thin slag layer. These conditions are often difficult to obtain in a smelting vessel, particularly the necessary residence time. As desired matte grades increase, the problem gets worse, because the increase in concentrate feed rate decreases settling time further (Furuta et al., 2006). As a result, Cu producers have since the 1960s constructed separate furnaces specifically for *cleaning* smelting and converting slags. These furnaces have two purposes: (a) allowing suspended matte droplets to finish settling to the molten matte layer, and (b) facilitating the reduction of dissolved Cu oxide to suspended Cu sulfide drops.

Inputs to these furnaces vary considerably. Slag cleaning furnaces associated with bath-smelting units like the Ausmelt/Isasmelt or Mitsubishi smelting furnace accept an unseparated mixture of slag and matte and are required to do all the settling. Others accept converter slag and plant reverts in addition to smelter slag, requiring more emphasis on reduction (Moreno et al., 2003). Most commonly, these furnaces are fed only smelting furnace slag and are used primarily as a final settling furnace.

Fig. 10.2 illustrates a typical electric slag cleaning furnace (Barnett, 1979; Kennedy, 2012; Outotec, 2017). Heat is generated by passing electric current through the slag layer. AC power is used, supplied through three carbon electrodes. This method of supplying heat generates the least amount of turbulence, which improves settling rates. Most furnaces use self-baking Søderberg electrodes (Kunze et al., 2003) but a few use prebaked electrodes to avoid the need for baking control (Furuta et al., 2006).

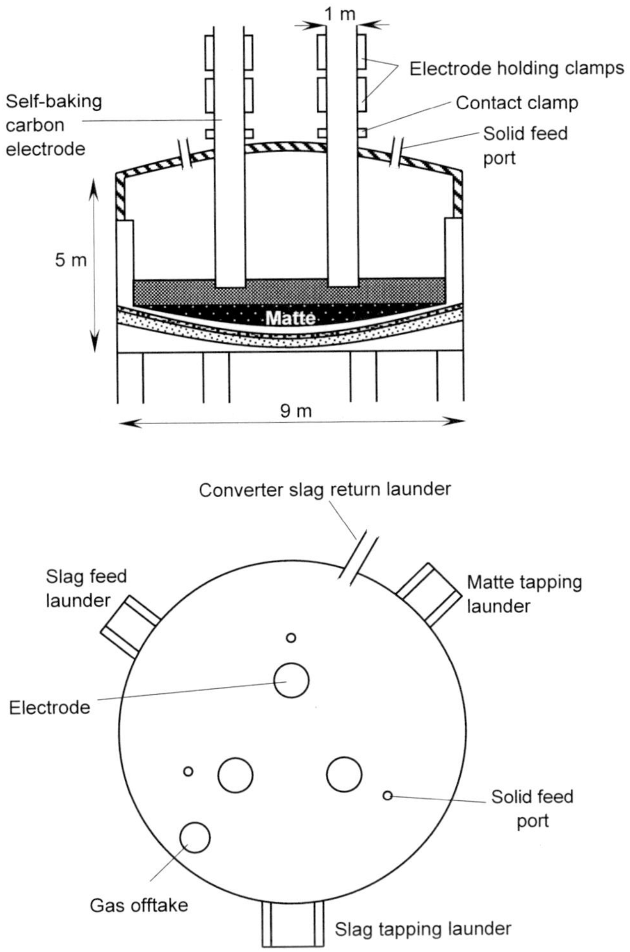

Figure 10.2 Electric slag cleaning furnace. A furnace of this size "cleans" 1000–1500 tonnes of slag per day.

The furnace sidewalls are typically cooled by external water jackets to minimize refractory erosion (SMS Group, 2013). However, embedded cooling elements are more effective, especially for slag cleaning furnaces processing high-copper slags (Kunze et al., 2003; Tuominen et al., 2007).

Table 10.2 compares the operating characteristics of seven electric furnaces. The choice of operating temperature represents a balance between the enhanced reaction kinetics of higher temperatures and the reduced refractory wear and lower copper losses of lower temperatures. Required capacities are set by the size of the smelting operation and

Table 10.2 Details of electric slag cleaning furnaces.

Smelter	Caraiba Metais, Dias d'Avila, Brazil	Norddeutsche, Affinerie, Hamburg	KGHM Polska Miedz, Poland	Konkola copper, Nchanga, Zambia	Codelco Las Ventanas, Chile	Mexicana de Cobre, Mexico, Furnace 1	Mexicana de Cobre, Mexico, Furnace 2
Slag details, tonnes/day							
Smelting furnace slag	880 OK flash furnace	1600 OK flash furnace	1100, flash furnace	1600, flash furnace	1000 Teniente furnace	900 OK flash furnace	740 Teniente furnace
% Cu	1.7	1–1.5	13.86	17	8–12	1.5–2.5	5
Converter slag		0	52		120	113	184
% Cu			18.76		30	8	8
Products							
Slag, % Cu	0.7	0.6–0.8	0.64	3	1.3	1.26	1.3
Matte, % Cu	65–70	65–70	Cu–Pb–Fe alloy, 77% Cu	99.28	70.5	70.3	70.5
Furnace details							
Shape	Circular	Circular	Circular	Circular	Circular	Circular	Circular
diameter, m	11	10.2	14.5	13.5	10	10	10
Power rating, MW	2–4	2–3	25	20	4.5	1.5–4.5	1.5–4.5
Electrodes	3	3	3	5	3	3	3
Material	Self-baking	Self-baking	Self-baking	Self-baking	Self-baking	Self-baking	Self-baking
Diameter, m	1	1	1.32	1.1	0.86	0.9	0.9

Operating details							
Slag residence time, hours	2–3	5	8	4	1.5–2.5	0.25–1	0.25–1
Energy use, kWh/tonne of slag	70	40–50	384	150	140	57	69
Reductant, kg/tonne of slag	Coke, 8.3	Coke, 4–5	Coke, 46	Coke, 21	Coke, 13	7.17 coke	7.32 coke
Slag layer thickness, m	0.97–1.4	1.5–1.8	<1	1–2	0.8–1.4	0.8–1.5	0.8–1.5
Matte layer thickness, m	0–0.45	0–0.4	0.4–1.0 (alloy)	1.2	0.1–0.5	0–0.2	0–0.2

the choice of input slags. Settling times are usually on the order of 1−5 hours. Typical energy use is 15−70 kWh per tonne of slag, depending upon furnace inputs, target % Cu, temperature, and residence time. Energy use is higher when the charge contains significant levels of reverts (Kunze et al., 2003). In addition to providing a reducing atmosphere, adding coke helps reduce electrode loss.

While some electric slag cleaning furnaces process only smelting furnace slag, others are fed a variety of materials. Several furnace operators input converter slag or solid reverts in addition to smelting slag. When this is done, a reducing agent is often required to reduce Cu oxide in the slag to Cu metal or Cu sulfide. Coal or coke is often added for this reduction (Coursol et al., 2012). Pyrite may also be added if additional sulfur is needed to form matte (Demetrio et al., 2000):

$$C + Cu_2O\ (\mathit{slag}) \rightarrow CO + 2\ Cu\ (\mathit{l}) \tag{10.4}$$

$$C + Cu_2O + FeS_2 \rightarrow Cu_2S + FeS + CO \tag{10.5}$$

Carbon additions also reduce solid magnetite in the slag to liquid FeO:

$$C + Fe_3O_4\ (\mathit{s}) \rightarrow CO + 3\ FeO\ (\mathit{slag}) \tag{10.6}$$

This decreases slag viscosity and improves settling rates (Furuta et al., 2006).

Ferrosilicon is occasionally used as a reducing agent (Shimpo & Toguri, 2000), especially in the Mitsubishi slag cleaning furnace. Recent initiatives in slag cleaning furnace practice have involved lance injection of solid reductants or gaseous reducing agents such as methane, to improve reduction kinetics (An et al., 1998; Sallee & Ushakov, 1999). Qu et al. (2020) recently demonstrated that hydrogen is also an effective alternative reducing agent.

Table 10.2 includes data for the electric slag cleaning furnace used by the direct-to-blister smelter at Glogów. This furnace is unusual in several regards, including the production of Cu−Pb−Fe alloy rather than matte, due to the lack of sulfide in the charged slag (Pietrzyk & Palimaka, 2016). This operation is discussed in detail in Chapter 9.

Recent research at the University of Chile has highlighted a number of potential improvements to the design and operation of electric furnaces for slag cleaning (SMS Group, 2013; Warczok et al., 2007a, 2007b). These include the following:

- the use of rectangular rather than circular furnace shape, particularly advantageous for continuous operation (SMS Group, 2013);
- replacing AC furnace power supply with DC, coupled with a magnetic field to enhance stirring and reduction kinetics (König et al., 2013; SMS Group, 2013);
- replacing coke with calcium carbide as a reductant, which improves kinetics and requires less furnace energy (Bydalek et al., 2015)

Using all of these innovations, it might be possible to reduce the copper content of the cleaned slag below 0.4 wt.-% (SMS Group, 2013). However, in practice, the additional cost of power and equipment requires most furnace operators to settle for a higher level.

Fuel-fired slag cleaning furnaces are used at a few smelters (Table 10.3). The foremost is the Teniente slag cleaning furnace, which is similar in design to a rotary fire refining furnace (Chapter 12; Demetrio et al., 2000; Mackay et al., 2000). It features injection of powdered coke and air into molten slag. In 2000, it began operating on a semicontinuous basis, generating slag with 0.75%–0.85% Cu. Semicontinuous operation improved

Table 10.3 Details of Teniente rotary hydrocarbon-fired slag settling furnace at Caletones, Chile.

Smelter	**Caletones, Chile**
Slag details	
Smelting furnace slag, tonnes/day	3000
% Cu	6–8
Converting furnace slag, tonnes/day	0
% Cu	
Products	
Slag, % Cu	1
Matte, % Cu	72
Matte destination	Peirce–Smith converters Teniente smelting furnace
% Cu recovery	85%
Furnace details	
Number of slag cleaning furnaces	4
Shape	Horizontal cylinder
Diameter inside refractory, m	4.6
Length inside refractory, m	3 × 10.7; 1 × 12.7
Number of reducing tuyeres	4
Tuyere diameter, cm	6.35
Operating details	
Slag residence time, hours	2
Reductant	Coal, oil, or natural gas
kg per tonne of slag	6
Slag layer thickness, m	1.4
Matte layer thickness, m	0.4
Fuel	Bunker C fuel oil
kg per tonne of slag	8.8

productivity by over 20%, while reducing coke consumption by half. The slag cleaning furnaces used by bath smelters are usually electric, but fuel-fired furnaces can be used for this purpose as well (Mariscal & Herrera, 2009).

10.5 Decreasing copper in slag IV: slag minerals processing

Several options are available for recovering Cu from converter slags. Pyrometallurgical cleaning in electric furnaces is quite common. Molten converter slag is also recycled to reverberatory smelting furnaces and Inco flash furnaces. Outotec and Teniente smelting furnaces occasionally accept some molten converter slag (Queirolo et al., 2016).

Cu is also removed from converter slags by slow solidification, crushing/grinding, and froth flotation (Queirolo et al., 2016). It relies on the fact that as converter slags cool, much of their dissolved Cu exsolves from solution by the following reaction (Jalkanen et al., 2002):

$$Cu_2O + 3\ FeO \rightarrow 2\ Cu\ (l) + Fe_3O_4 \qquad (10.7)$$

Reaction (10.7) is increasingly favored at low temperatures and can decrease the dissolved Cu content of converter slag to well below 0.5% (Imriš et al., 2000). After the slag has solidified, the exsolved copper and suspended matte particles respond well to froth flotation. As a result, converter slags have long been crushed, ground, and concentrated in the same manner as sulfide ores (Demetrio et al., 2000). The key to successful minerals processing of converter slags is ensuring that the precipitated grains of matte and metallic Cu are large enough to be liberated by crushing and grinding. This is accomplished by cooling the slag slowly to less than 1000 °C, then naturally to ambient temperature. Once this is done, the same minerals processing equipment and reagents that are used to recover Cu from ore can be used to recover Cu from slag (Table 10.4). Some smelting slags are also treated this way (Karimi et al., 2013).

Slag flotation is becoming less popular for processing smelter slags, for two reasons (Acuña & Sherrington, 2005). The first is the trend toward production of higher grade mattes. A higher fraction of the copper in these mattes is present as dissolved copper, rather than entrained matte. This copper is not recovered by flotation, but must be recovered in a settling furnace. The second is the increased throughput at many mills. This means that comminution and flotation equipment is less likely to be available for processing slags. A separate facility is required for slag mineral processing (Queirolo et al., 2016).

Table 10.4 Details of four slag flotation plants.

Smelter	Noranda, Québec	Saganoseki, Japan	Toyo, Japan	PASAR, Philippines
Slag inputs, tonnes/day				
Smelting furnace slag	1700	0	0	0
% Cu	6 (average)			
Converter slag	300	450	450	370
%Cu		8.33	6.5	10–15[a]
Products				
Slag concentrate, % Cu	42	21.8	28	29–33
Slag tailings, % Cu		0.65	0.4	0.5–0.6
Cu recovery, %	90–95	95	95	97–98
Operating details				
Solidification method	Ladle cooling with or without water sprays	~150 kg ingots on moving slag conveyor	~150 kg ingots on moving slag conveyor	
Cooling description		Cooled on slag conveyor	1 h in air then immersion in H_2O	
Crushing/grinding				
Equipment	80% semiautogenous grinding, 20% crushing, and ball milling	jaw crusher; cone crusher (twice); ball mill (twice)	Gyratory crusher; cone crusher (twice); ball mill	Jaw crusher; cone crusher; ball mills (primary and regrind)
Particle size after grinding	78% -44 μm	40%–50% -44 μm	90% -44 μm	65%–75% -45 μm

Continued

Table 10.4 Details of four slag flotation plants.—cont'd

Smelter	Noranda, Québec	Saganoseki, Japan	Toyo, Japan	PASAR, Philippines
Flotation				
Machinery	Mechanically agitated cells	Mechanically agitated cells	Mechanically agitated cells	Mechanical agitator Agitair 48, Jameson cell (Fig. 3.12)[b]
Flotation residence time	60 min		30 min (rougher + scavenger)	
Flotation reagents				
Promoter				NH_4 and Na dibutyl dithiophosphate
Collector	Thionocarbamate, SIPX	Na isopropyl xanthate, UZ200	Thionocarbamate, PAX	(a) Danafloat 245, Penfloat TM3 (b) K amyl xanthate
Frother	Propylene glycol	Pine oil, MF550	Pine oil	Pine oil NF 183
CaO?	No	No		Yes
pH	8–9	7–8	7–8	8.5–9.5
Energy use kWh/tonne slag	32.5	58.4	64	40

[a]"Nonmagnetic "white metal" (Cu_2S) pieces are isolated magnetically after crushing, leaving 5%–6.5% Cu in the ball mill feed slag.
[b]Switching to all Jameson cell.

10.6 Summary

Cu smelters produce two slags: smelting furnace slag with one to 2% Cu and converter slag with four to 8% Cu. Discard of these slags would waste considerable Cu, so they are almost always treated for Cu recovery.

Cu is present in molten slags as (*a*) entrained droplets of matte or metal and (*b*) dissolved Cu^+. The entrained droplets are recovered by settling in a slag cleaning furnace, usually electric. The dissolved Cu^+ is recovered by hydrocarbon reduction and settling of matte.

A second method of recovering this Cu from slag is slow—cooling/solidification, crushing/grinding, and froth flotation. Slowly cooled/solidified slag contains the originally entrained matte and Cu droplets plus matte and Cu which precipitate during cooling/solidification. These Cu-bearing materials are efficiently recovered from the solidified slag by fine grinding and froth flotation.

Electric furnace settling has the advantage that it can be used for recovering Cu from reverts and miscellaneous materials around the smelter. Slag flotation has the advantages of more efficient Cu recovery and the possibility of using existing crushing/grinding/flotation equipment.

References

Acuña, C. M., & Sherrington, M. (2005). Slag cleaning processes: A growing concern. *Materials Science Forum, 475—479*, 2745—2752.

An, X., Li, N., & Grimsey, E. J. (1998). Recovery of copper and cobalt from industrial slag by top-submerged injection of gaseous reductants. In B. Mishra (Ed.), *EPD Congress 1998* (pp. 717—732). Warrendale, PA: TMS.

Barnett, S. C. C. (1979). The methods and economics of slag cleaning. *Mining Magazine, 140*, 408—417.

Bydalek, A. W., Bydalek, A., Wolczynski, W., & Biernat, S. (2015). The concept of slag decopperization in the flash furnace process by use of complex reagents. *Archives of Metallurgy Materials, 60*, 319—322.

Cardona, N., Coursol, P., Mackey, P. J., & Parra, N. (2011). Physical chemistry of copper smelting slags and copper losses at the Paipote Smelter, Part I — thermodynamic modelling. *Canadian Metallurgy Quarterly, 50*, 318—329.

Cheng, X., Cui, Z., Contreras, M., Chen, M., Nguyen, A., & Zhao, B. (2019). Matte entrainment by SO_2 bubbles in copper smelting slag. *JOM, 71*, 1897—1903.

Coursol, P., Cardona Valencia, N., Mackey, P., Bell, S., & Davis, B. (2012). Minimisation of copperses in copper smelting slag during electric furnace treatment. *JOM, 64*, 1305—1313.

Cui, Z., Wang, Z., Bian, R., Wei, C., & Zhao, B. (2018). Application study on technology of reducing copper content in discarded slag. In B. Davis, M. Moats, S. Wang, D. Gregurek, J. Kapusta, T. P. Battle, … V. Ciminelli, et al. (Eds.), *Extraction 2018* (pp. 481—490). Warrendale, PA: TMS.

Demetrio, S., Ahumada, J., Ángel, D. M., Mast, E., Rosas, U., Sanhueza, J., Reyes, P., & Morales, E. (2000). Slag cleaning: The Chilean copper smelter experience. *JOM, 52*(8), 20—25.

Fagerlund, K. O., & Jalkanen, H. (1999). Some aspects on matte settling in copper smelting. In C. Díaz, C. Landolt, & T. Utigard (Eds.), *Smelting, technology development, process modeling and fundamentals: Vol. VI. Copper 99—Cobre 99* (pp. 539—551). Warrendale, PA: TMS.

Font, J., Moyano, A., & Itagaki, K. (2013). Copper dissolution in the FeO_x—SiO_2 base slag of the flash smelting furnace and the Teniente converter at the Chuquicamata smelter. In *Proceedings of copper 2013* (pp. 727—736). Santiago: IIMCH.

Furuta, S., Tanaka, S., Hamamoto, M., & Inada, H. (2006). Analysis of copper loss in slag in Tamano type flash smelting furnace. In F. Kongoli, & R. G. Reddy (Eds.), *Sohn international symposium: Vol. 8. International symposium of sulfide smelting 2006* (pp. 123–133). Warrendale, PA: TMS.

Imriš, I., Rebolledo, S., Sanchez, M., Castro, G., Achurra, G., & Hernandez, F. (2000). The copper losses in the slags from the El Teniente process. *Canadian Metallurgy Quarterly, 39*, 281–290.

Ip, S. W., & Toguri, J. M. (2000). Entrainment of matte in smelting and converting operations. In G. Kaiura, C. Pickles, T. Utigard, & A. Vahed (Eds.), *J.M. Toguri symposium: Fundamentals of metallurgical processing* (pp. 291–302). Montreal: CIM.

Jalkanen, H., Poijärvi, J., & Pajari, H. (2002). Slags of suspension smelting of chalcopyrite ores and copper matte converting. In R. L. Stephens, & H. Y. Sohn (Eds.), *Sulfide smelting 2002* (pp. 363–376). Warrendale, PA: TMS.

Karimi, N., Vaghar, R., Mohammedi, M., & Hashemi, S. A. (2013). Recovery of copper from the slag of Khatoonabad flash smelting furnace by flotation method. *Jounal of the Institution of Engineers (India): Series A, 94*, 43–50.

Kennedy, M. (2012). Electric slag furnace dimensioning. In J. P. Downey, T. P. Battle, & J. F. White (Eds.), *International smelting technology symposium* (pp. 279–290). Warrendale, PA: TMS.

König, R., Degel, R., Oterdoom, H., Schmidl, J., Kadereit, H., Specht, A., & Warczok, A. (2013). Highly efficient slag cleaning — Latest results from pilot–scale operation. In *Proceedings of copper 2013* (pp. 185–198). Santiago: IIMCH.

Kunze, J., Degel, R., & Warczok, A. (2003). Current status and new trends in copper slag cleaning. In C. Díaz, J. Kapusta, & C. Newman (Eds.), *(Book 1): Pyrometallurgy of copper: Vol. 4. Copper 2003–Cobre 2003* (pp. 459–474). Montreal: CIM.

Mackay, S. R., Cordero, C. D., Esteban, S. J., & Alvear, F. G. R. (2000). Continuous improvement of the Teniente slag cleaning process. In S. Seetharaman, & D. Sichen (Eds.), *Proc. 6th int. conf. molten slags, fluxes and salts (Paper 208)*. Stockholm: KTH.

Mariscal, L., & Herrera, E. (2009). Isasmelt™ slag chemistry and copper losses in the rotary holding furnaces slag at Ilo Smelter. In M. Sánchez, R. Parra, G. Riveros, & C. Díaz (Eds.), *Molten 2009* (pp. 1241–1250). Santiago: Gecamin.

Moreno, A., Sánchez, G., Worczok, A., & Riveros, G. (2003). Development of slag cleaning process and operation of an electric furnace at Las Ventanas smelter. In C. Díaz, J. Kapusta, & C. Newman (Eds.), *(Book 1): Pyrometallurgy of copper: Vol. 4. Copper 2003–Cobre 2003* (pp. 475–492). Montreal: CIM.

Outotec. (2017). *Outotec electric slag cleaning furnace*. https://www.outotec.com/products-and-services/technologies/smelting-and-converting/electric-slag-cleaning-furnace/. (Accessed 28 July 2020).

Pietrzyk, S., & Palimaka, P. (2016). Inclusions of metallic copper in slag from direct to blister process, before and after cleaning. In *Proceedings of copper 2016* (pp. 1390–1397). Tokyo: MMIJ.

Pizarro, C., Wastavino, G., Moyano, A., & Font, J. (2013). Copper losses control in flash smelting slag at Chuquicamata smelter. In *Proceedings of copper 2013* (pp. 1199–1211). Santiago: IIMCH.

Queirolo, C. H., Castillo, W. A., & Castillo, A. A. (2016). New slag flotation plant at Potrerillos Smelter, Codelco Chile. In *Proceedings of copper 2016* (pp. 1381–1389). Tokyo: MMIJ.

Qu, G., Wei, Y., Li, B., Wang, H., Yang, Y., & MacLean, A. (2020). Distribution of copper and iron components with hydrogen reduction of copper slag. *Journal of Alloys and Compounds, 824*. Paper 153910.

Roghani, G., Takeda, Y., & Itagaki, K. (2000). Phase equilibrium and minor element distribution between FeO_X–SiO_2–MgO–based slag and Cu_2S–FeS matte at 1573 K under high partial pressures of SO_2. *Metallurgical and Materials Transactions B, 31B*, 705–712.

Sallee, J. E., & Ushakov, V. (1999). Electric settling furnace operations at the Cyprus Miami Mining Corporation copper smelter. In D. B. George, W. J. Chen, P. J. Mackey, & A. J. Weddick (Eds.), *Smelting operations and advances: vol. V. Copper 99–cobre 99* (pp. 629–643). Warrendale, PA: TMS.

Sano, H., Nishikawa, K., Hong, J. H., & Sasaki, Y. (2016). Considerations for the influence of slag composition on the copper loss in slag. In *Proceedings of copper 2016* (pp. 570–579). Tokyo: MMIJ.

Shimpo, R., & Toguri, J. M. (2000). Recovery of suspended matte particles from copper smelting slags. In G. Kaiura, C. Pickles, T. Utigard, & A. Vahed (Eds.), *J.M. Toguri symposium: Fundamentals of metallurgical processing* (pp. 481–496). Montreal: CIM.

SMS Group. (2013). *SAF bulletin 2013*. www.sms-siemag.com/download/H4_303E_SAF_Bulletin_2013.pdf. (Accessed 4 August 2020).

Tuominen, J., Anjala, Y., & Björklund, P. (2007). Slag cleaning of Outokumpu direct–to–blister flash smelting slags. In A. E. M. Warner, C. J. Newman, A. Vahed, D. B. George, P. J. Mackey, & A. Warczok (Eds.), *(Book 2): The Carlos Díaz symposium on pyrometallurgy: Vol. III. Cu 2007* (pp. 339–350). Montreal: CIM.

Warczok, A., Riveros, G., Degel, R., Kunze, J., & Oterdoom, H. (2007a). Computer simulation of slag cleaning in an electric furnace. In A. E. M. Warner, et al. (Eds.), *(Book 2): The Carlos Díaz symposium on pyrometallurgy: Vol. III. Cu 2007* (pp. 367–378). Montreal: CIM.

Warczok, A., Riveros, G., Degel, R., Kunze, J., & Oterdoom, H. (2007b). Slag cleaning in circular and rectangular electric furnaces. In A. E. M. Warner, et al. (Eds.), *(Book 2): The Carlos Díaz symposium on pyrometallurgy: Vol. III. Cu 2007* (pp. 403–415). Montreal: CIM.

Xia, J. L., Ahokainen, T., Kankaanpää, T., Järvi, J., & Taskinen, P. (2005). Numerical modeling of copper droplet settling behavior in the settler of a flash smelting furnace. In *European metallurgical conference (EMC 2005)* (Vol. 3, pp. 1359–1372). Germany: GDMB. Clausthal–Zellerfeld.

Zivkovic, Z., Djordjevic, P., & Mitevska, N. (2020). Contribution to the examination of the mechanisms of copper loss with the slag in the process of sulfide concentrates smelting. *Mining, Metallurgy & Exploration, 37*, 267–275.

CHAPTER 11

Capture and fixation of sulfur (King et al., 2013)

About 85% of the world's primary copper originates in sulfide minerals. Sulfur is, therefore, evolved by most copper extraction processes. The most common form of evolved sulfur is sulfur dioxide (SO_2) gas from smelting and converting.

SO_2 is harmful to fauna and flora and is detrimental to the human respiratory system. It must be prevented from reaching the environment. Regulations for ground level SO_2 concentrations around copper smelters are presented in Table 11.1. Other regulations such as maximum total SO_2 emission (tonnes per year), percent SO_2 capture, and SO_2-in-gas concentration at point of emission also apply in certain locations.

In the past, SO_2 from smelting and converting was vented directly to the atmosphere. This practice is unacceptable, so most smelters capture a large fraction of their SO_2. It is almost always made into sulfuric acid, occasionally liquid SO_2 or gypsum. Copper smelters typically produce 2.5—4.0 tonnes of sulfuric acid per tonne of product copper, depending on the S/Cu ratio of their feed materials.

This chapter describes the following:

1. off-gases from smelting and converting
2. manufacture of sulfuric acid from smelter gases
3. recent and future developments in sulfur capture.

11.1 Off-gases from smelting and converting processes

Table 11.2 characterizes the off-gases from smelting and converting processes. SO_2 strengths in smelting furnace gases vary from about 70 vol.-% in Inco flash furnace gases to 1 vol.-% in reverberatory furnace gases. The SO_2 strengths in converter gases vary from about 40% in flash converter gases to 8%—12% in Peirce—Smith converter gases.

Off-gases from most smelting and converting furnaces are treated for SO_2 removal in sulfuric acid plants. The exception is off-gas from reverberatory furnaces. It is too dilute in SO_2 for economic sulfuric acid manufacture. This is the main reason reverberatory furnaces continue to be shut down.

The off-gases from electric slag cleaning furnaces, anode furnaces, and secondary gas collection hoods around the smelter are diluted in SO_2, <0.1%. These gases are usually vented to atmosphere. In densely populated areas, they may be scrubbed with basic solutions before being vented (Jones et al., 2019; Roy & Sardar, 2015; Shibata & Oda, 1990).

Extractive Metallurgy of Copper, Sixth Edition
ISBN 978-0-12-821875-4, https://doi.org/10.1016/B978-0-12-821875-4.00016-X

Table 11.1 Standards for maximum SO_2 concentration at ground level outside the perimeters of copper smelters.

Country	Time period	Maximum SO_2 + SO_3 (SO_x) concentration (parts per million)
Australia (Australian government, 2010)+	Yearly average	0.02
	Daily average	0.08
	1-hour average	0.2
Canada (CCME, 2019) (maximum acceptable level)	Annual average	0.005
	Daily average	0.070
	1-hour average	0.334
Chile (D.S.N° 113/02, 2010)	Yearly average	0.031
China (Class II)	Daily average	0.096
(MEP GB 3095–2012, 2012)	Yearly average	0.023
China (Class I)	Daily average	0.058
	1-hour average yearly	0.192
	average	0.008
	Daily average	0.015
	1-hour average	0.058
European Union (Council directive 1999/30/EC, 2010)	Yearly average	0.008
	Daily average	0.048
	1-hour average	0.13
Japan (MOE Japan, 2010)	Daily average	0.04
	Hourly average	0.1
U.S.A. (USEPA, 2020)	1-hour standard	0.075

Table 11.2 Characteristics of off-gases from smelting and converting processes. The data are for off-gases as they enter the gas-handling system.

Furnace	SO_2 concentration (volume%)	Temperature (°C)	Dust loading (kg/ Nm^3)	Destination
Inco flash furnace	50–75	1270–1300	0.2–0.25	H_2SO_4 occasionally liquid SO_2 plant
Outotec flash furnace	25–60	1270–1350	0.1–0.25	H_2SO_4 plant, occasionally liquid SO_2 plant
Outotec flash converter	35–40	1290	0.2	H_2SO_4 plant
Outotec direct-to-copper	43	1320–1400	0.2	H_2SO_4 plant

Table 11.2 Characteristics of off-gases from smelting and converting processes. The data are for off-gases as they enter the gas-handling system.—cont'd

Furnace	SO_2 concentration (volume%)	Temperature (°C)	Dust loading (kg/Nm^3)	Destination
Mitsubishi smelting furnace	30–35	1240–1250	0.07	H_2SO_4, occasionally liquid SO_2 plant
Mitsubishi converting furnace	25–30	1230–1250	0.1	H_2SO_4, occasionally liquid SO_2 plant
Noranda process	15–25	1200–1240	0.015–0.02	H_2SO_4 plant
Teniente furnace	12–25	1220–1250	0.015–0.02	H_2SO_4 plant
ISASMELT or Ausmelt furnace	20–25	1150–1220	~0.01	H_2SO_4 plant
Electric furnace	2–5	400–800		H_2SO_4 or liquid SO_2 plant or vented to atmosphere
Reverberatory furnace	1	1250	~0.03	Vented to atmosphere (made into gypsum in one plant)
Peirce–Smith converter	8–15	1200	0.05–0.1	H_2SO_4 plant or vented to atmosphere
Hoboken converter	12	1200		H_2SO_4 plant
Electric slag cleaning furnaces	0.1	800		Vented to atmosphere (occasionally scrubbed with basic solution)
Anode furnaces	<0.1	1000		Vented to atmosphere (occasionally scrubbed with basic solution or used as dryer dilution gas)
Gas collection hoods around the smelter	<0.1	50		Vented to atmosphere (occasionally scrubbed with basic solution)

Table 11.3 Typical copper smelter sulfur distribution.

Percent of incoming S in:	
Sulfuric acid	96–97
Gypsum (scrubbing)	1–3
Slag	1–2
Dust (to treatment)	0.2–2
Neutralized liquid effluent	0.3–2
Gaseous emissions (dryer, anode furnace and ventilation stacks, and/or acid plant tail gas)	0.2–1

11.1.1 Sulfur capture efficiencies

Table 11.3 shows the typical sulfur capture efficiency of a modern smelter with an Outotec flash furnace and Peirce–Smith converters. Gaseous emissions of sulfur compounds are ≤1% of the sulfur entering the smelter.

11.2 Sulfuric acid manufacture

Nearly all sulfuric acid manufactured from smelter off-gas is by the conventional contact type acid plant. Two copper smelters use Sulfacid to manufacture dilute sulfuric acid by treating the tail gas from their acid plants. The traditional contact type acid plant is described below. Sulfacid is described in Section 11.7.

Fig. 11.1 outlines the steps for producing sulfuric acid from SO_2-bearing smelter off-gas. The steps are as follows:

1. cooling and cleaning the gas
2. drying the gas with 93% sulfuric acid
3. catalytically oxidizing SO_2 to SO_3
4. absorbing this SO_3 into 98.5% H_2SO_4 sulfuric acid.

The strengthened acid from step (4) is then blended with diluted acid from step (2) and sent to market or used for leach operations.

Tail gases typically contain less than 0.5% of the sulfur entering the gas treatment system. They are cleaned of their acid mist, usually by candle filters (Begg Cousland, 2018), and discharged to the atmosphere. Several smelters scrub the remaining SO_2, SO_3, and H_2SO_4 mist with Ca/Na carbonate hydroxide solutions before releasing the gas to atmosphere (Jones et al., 2019; Kubo et al., 2007; Roy & Sardar, 2015; Shibata & Oda, 1990). Basic aluminum sulfate solution is also used (Oshima et al., 1997).

The following sections describe the principal sulfuric acid production steps and their purposes.

11.3 Smelter off-gas treatment

The key to high acid plant availability in any smelter is to efficiently operate and maintain properly designed gas cooling and cleaning systems as described in the next few sections.

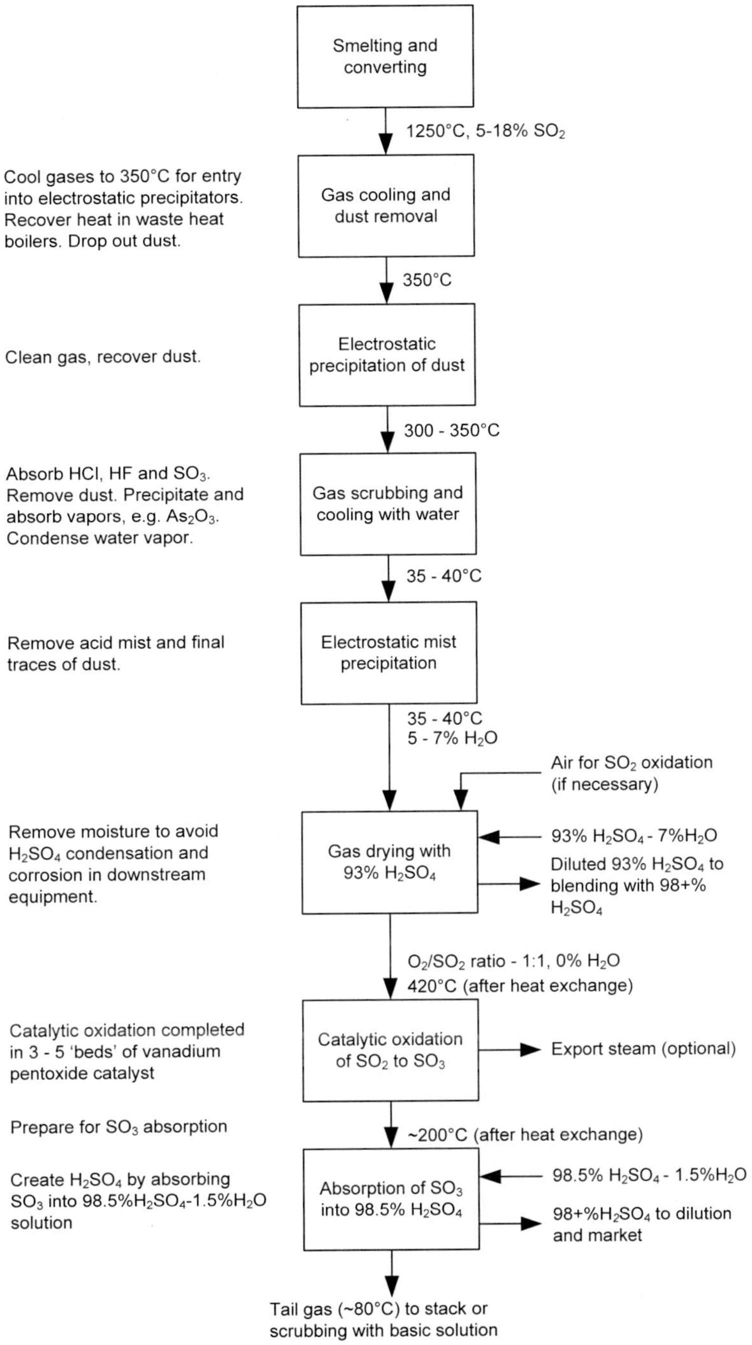

Figure 11.1 Flowsheet for producing sulfuric acid from smelting and converting gases.

11.3.1 Gas cooling and heat recovery

The first step in smelter off-gas treatment is cooling the gas in preparation for electrostatic precipitation of its dust. Electrostatic precipitators operate at about 350°C (Parker, 1997). Above this temperature, their steel structures begin to weaken. Below this temperature, there is a danger of corrosion by condensation of sulfuric acid from SO_3 and water vapor in the off-gas.

Gas cooling is usually done in waste heat boilers (Fig. 11.2), which not only cool the gas but also recover the heat in a useful form as steam (Köster, 2010; Kuang et al., 2016; Peippo et al., 1999). The boilers consist of the following:

1. a radiation section in which the heat in the gas is transferred to pressurized water flowing through 4 cm diameter tubes in the roof and walls of a large (e.g., 25 m long × 15 m high × 5 m wide) rectangular chamber
2. a convection section (typically 20 m long × 10 m high × 3 m wide) in which heat is transferred to pressurized water flowing through 4 cm diameter steel tubes suspended in the path of the gas.

The product of the boiler is a water/steam mixture (Clark et al., 2017). The water is separated by gravity and recirculated to the boiler. The steam is often superheated above its dew point and used for generating electricity. It is also used without superheating for concentrate drying and for various heating duties around the smelter and refinery.

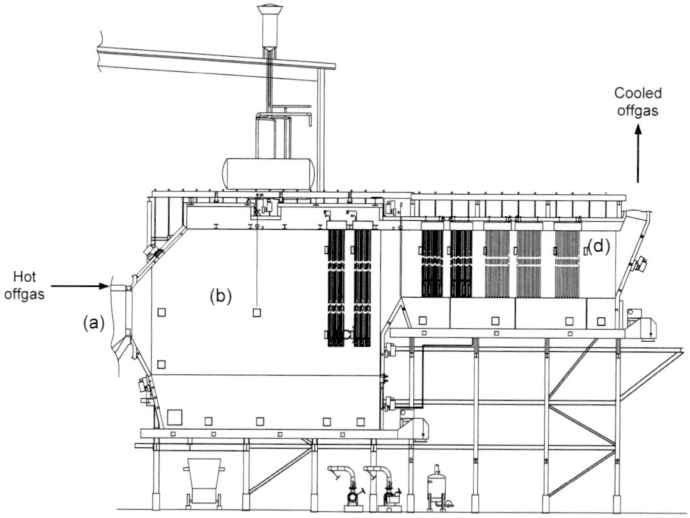

Figure 11.2 Waste heat boiler for the Rönnskär flash furnace (Peippo et al., 1999). Note, left to right, (a) flash furnace gas offtake, (b) radiation section with tubes in the walls, (c) suspended tube baffles in the radiation section to evenly distribute gas flow, (d) convection section with hanging tubes. Note also water tank above radiation section and dust collection conveyors below the radiation and convection sections. *(Peippo, R., Holopainen, H., & Nokelainen, J. (1999). Copper smelter waste heat boiler technology for the next millennium. In D.B. George, W.J. Chen, P.J. Mackey, & A.J. Weddick (Ed.), Copper 99—Cobre 99, Vol. V: smelting operations and advances (pp. 71—82), Warrendale, PA: TMS. Used with permission of The Minerals, Metals & Materials Society.)*

Dust falls out of waste heat boiler gases due to its low velocity in the large boiler chambers. It is collected and usually recycled to the smelting furnace for Cu recovery. However, this pattern has begun to change as the level of impurities in copper concentrates continues to grow. These impurities include volatile elements such as arsenic, fluorine, and mercury (Weidenbach et al., 2016), which wind up in the off-gas and ultimately in the dusts collected during off-gas processing. To separate the copper in the dust from these impurities, a leaching step is performed (Chen et al., 2012; Rios et al., 2019). This avoids impurity recycle to the smelting furnace, but produces a waste solid that smelters have to properly manage.

An alternative method of cooling smelter gas is to pass it through sprays of water. Spray cooling (evaporative cooling) lowers the gas temperature to 300–350°C and avoids the investment in waste heat recovery equipment (Hasselwander, 2019; Ng'ambi et al., 2016), but it wastes the heat in the gases. and increases the cooling duty in the downstream gas cleaning plant (Section 11.3.3). Dry dust is recovered, since the gases are not completely saturated. It is used primarily for Ausmelt, ISASMELT, Teniente, Noranda, and sometimes Peirce–Smith gases.

Water quenching (Section 11.3.3) can also be used to cool the gases leaving the furnace (Amiri & Alihosseinpour, 2015; Ng'ambi et al., 2016). All of the Cu-bearing dust ends up in a weak acid solution and must be recovered hydrometallurgically. Water quenching is used primarily for Inco and Peirce–Smith converter off-gases (Daum, 2000; Liu, 2006).

11.3.2 Electrostatic precipitation of dust

After cooling, the furnace gases are passed through electrostatic precipitators (Conde et al., 1999; Hasselwander, 2019; Parker, 1997) for more dust removal. The dust particles are caught by

1. charging them in the corona of a high voltage electric field
2. catching them on a charged plate or wire
3. collecting them by neutralizing the charge and shaking the wires or plates.

The precipitators remove 99+% of the dust from the incoming gas (Conde et al., 1999; Hasselwander, 2019). The dust consists primarily of oxide compounds (Okanigbe et al., 2017), and is usually resmelted to recover its Cu. About 70% of the dust is recovered in the cooling system and 30% in the electrostatic precipitators.

11.3.3 Water quenching, scrubbing, and cooling

After electrostatic precipitation, the gas is quenched with water and scrubbed to further remove dust and impurities (Hasselwander, 2019). This quenching (a) removes much of the remaining dust from the gas and (b) absorbs HCl, HF, SO_3, and vapor impurities such as As_2O_3. The gas is then further cooled to 35–40°C by direct contact with cool water in a packed tower or by indirect contact with cool water in a heat exchanger.

The gas leaves the cooling section through electrostatic mist precipitators to eliminate fine droplets of liquid remaining in the gas and remove the rest of the dust. This avoids fouling of downstream acid plant catalyst (Hanekom, 2017; Hasselwander, 2019; Kamino et al., 2016). Mist precipitators operate similarly to the electrostatic precipitators described in Section 11.3.2. They must, however, be (a) constructed of acid-resistant materials such as fiber-reinforced plastic, alloy steels, or lead and (b) periodically turned off and flushed with fresh water to remove collected solids.

11.3.4 Mercury removal

Smelters treating high-mercury concentrates often use the Boliden Norzink process to remove mercury from the gas leaving the electrostatic mist precipitators (Hasselwander, 2019; Hultbom, 2003). In the Boliden Norzink process, the gas passes through a water-based solution of dissolved mercuric chloride ($HgCl_2$) circulating over a packed tower to absorb gaseous elemental mercury and form solid Hg_2Cl_2 (calomel) by the following reaction:

$$HgCl_2(aq) + Hg^{\circ}(g) = Hg_2Cl_2(s) \tag{11.1}$$

About half of the scrubbing solution is withdrawn and regenerated with chlorine gas by the following reaction:

$$Hg_2Cl_2(s) + Cl_2(g) = 2\ HgCl_2(aq) \tag{11.2}$$

The overall reaction is the following:

$$2\ Hg^{\circ}(g) + Cl_2(g) = Hg_2Cl_2(s) \tag{11.3}$$

Mercury can be bled from the system as Hg_2Cl_2 (calomel) by solid/liquid separation or as elemental mercury by electrowinning:

$$HgCl_2(aq) + 2\ e^{-} = Hg^{\circ}(l) + Cl_2(g) \tag{11.4}$$

The Boliden Norzink process is efficient, typically removing more than 99.95% of the inlet mercury. This results in a product acid mercury concentration of 0.3–0.5 ppm. This acid will meet most quality specifications. However, if further mercury removal is required, then the gas can be passed through an additional filter that contains either PbS-coated pumice (Dowa process) or Se (Hasselwander, 2009; Lossin & Windhager, 1999). Mercury reacts with the Se to form HgSe and with PbS to form HgS. Product acid mercury concentrations below 0.1 ppm can be achieved with this additional step.

11.3.5 The quenching liquid, acid plant blowdown

The water from quenching and direct-contact cooling is passed through water-cooled heat exchangers and used again for quenching/cooling. It becomes acidic (from SO_3

absorption) and impure (from dust and vapor absorption). A bleed stream of this impure solution (acid plant *blowdown*) is continuously withdrawn and replaced with fresh water. The amount of bleed and water replacement is controlled to keep the corrosive H_2SO_4 content of the cooling water below about 10%. The quantity of bleed depends on the amount of SO_3 in the off-gas as it enters the water-quench system.

The acid plant blowdown stream is neutralized and either stored or treated for metal recovery (Ante, 2010; Evans et al., 1998).

11.4 Gas drying

11.4.1 Drying tower (Hanekom, 2017)

The next step in off-gas treatment is gas dehydration (drying). It is done to prevent unintentional H_2SO_4 formation and corrosion in downstream ducts, heat exchangers, and catalyst beds. The $H_2O(g)$ is removed by contacting it with 93% (occasionally 96%) acid. H_2O reacts strongly with H_2SO_4 to form hydrated acid molecules. The contacting is done in a counter-current packed tower filled with 5−10 cm ceramic saddles (Fig. 11.3). The sulfuric acid flows down over the saddles. The gas is drawn up by the main acid plant blower(s).

The liquid product of gas drying is slightly diluted 93% H_2SO_4 acid. It is strengthened with the 98+% acid produced by subsequent SO_3 absorption (Section 11.5.2). Most of the strengthened acid is recycled to the absorption tower. A portion is sent to storage and then to market.

11.4.1.1 Optimum absorbing and composition

The gas product of the drying tower contains typically 50−100 mg H_2O/Nm^3 of off-gas. It also contains small droplets of acid which it picked up during its passage up the drying tower. These droplets are removed by passing the dry gas through stainless steel or fiber mist eliminator pads or candles.

11.4.2 Main acid plant blowers

The dried gas is drawn into the main acid plant blowers, which push it on to $SO_2 \rightarrow SO_3$ conversion and acidmaking (DKL, 2019; Louie, 2010; Ng'ambi et al., 2016). One centrifugal blower is used (occasionally two in larger plants). The blower typically draws 1000−4000 kW each, and moves 150,000−380,00 Nm^3 of gas per hour (Lüdtke & Müller, 2003). The gas handling system is under a slight vacuum before the blowers (typically <0.07 bar gauge at the smelting furnace) and under pressure (0.3−0.5 bar gauge) after (Kamino et al., 2016).

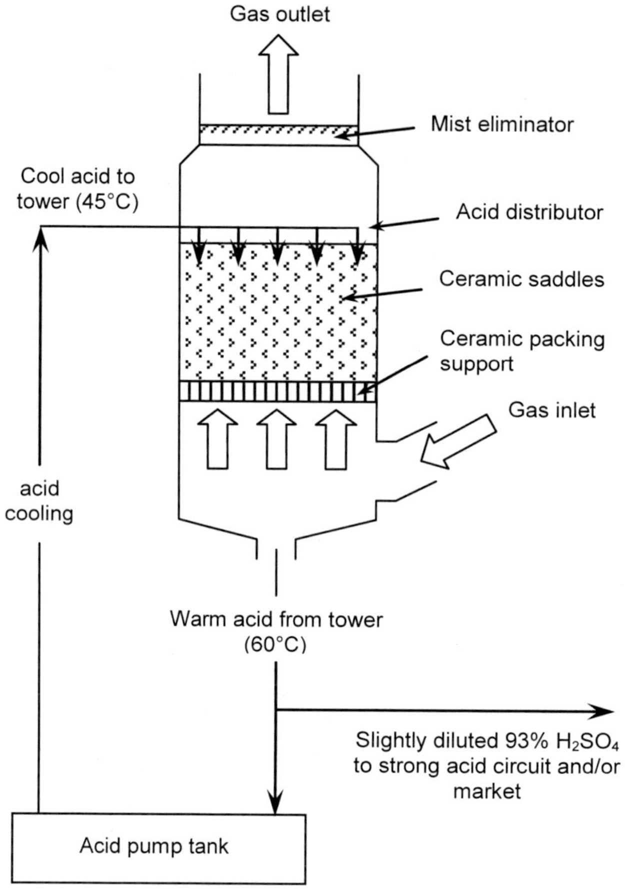

Figure 11.3 Drying tower and associated acid circulation and cooling equipment. Acid is pumped around the tubes of the acid–water heat exchanger to the top of the tower where it is distributed over the packing. It then flows by gravity downward through the packing and returns to the pump tank. The mist eliminator in the top of the tower is a mesh pad. In most SO_3 absorption towers this "pad" is usually replaced with multiple candle type mist eliminators.

11.5 Acid plant chemical reactions

11.5.1 Oxidation of SO_2 to SO_3

The SO_2 in the off-gas is oxidized to SO_3 in preparation for absorption in the water component of 98.5% H_2SO_4 acid. The oxidation reaction is as follows:

$$SO_2(g) + 0.5\ O_2(g) \rightarrow SO_3(g) \qquad (11.5)$$

This reaction is often referred to as SO_2 *conversion*. It is very slow without a catalyst, so the gas is always passed through V_2O_5–K_2SO_4 catalyst beds. The volumetric O_2/SO_2 ratio entering the catalyst beds is set at 1.0 or greater (by adding air, if necessary) to ensure near complete conversion of SO_2 to SO_3.

11.5.1.1 Catalyst reactions

Typical V_2O_5—K_2SO_4 based catalyst contains 5%—10% V_2O_5, 10%—20% K_2SO_4, 1%—5% Na_2SO_4, and 55%—70% SiO_2. It may also contain 5%—15% cesium sulfate (Cs_2SO_4) substituted for K_2SO_4. The active components of the catalyst are V_2O_5, K_2SO_4, Na_2SO_4, and Cs_2SO_4 (if present). The inactive material is SiO_2, which acts as a support for the active components.

V_2O_5—K_2SO_4 catalyst is a supported liquid-phase catalyst (Revelli, 2019). At the catalyst operation temperature, ~400°C, the active catalyst components (V_2O_5, K_2SO_4, Na_2SO_4, and Cs_2SO_4) exist as a film of molten salt solution on the solid inactive SiO_2 support. Oxidation of SO_2 to SO_3 in the presence of oxygen takes place by homogeneous reactions in this liquid film. Pores on the surface of the silica substrate provide the large surface area necessary for rapid SO_2 oxidation. The rate of catalytic SO_2 oxidation depends largely upon the rate of mass and heat transfer at the catalyst gas—liquid interface.

One of the most widely cited SO_2 conversion reaction mechanisms is that proposed by Mars and Maessen (1964, 1968). It is based on the experimental observation that during SO_2 conversion, the oxidation state of the vanadium ions in the catalyst changes between the V^{4+} and V^{5+}. This observation suggests that the reaction involves the following:

1. absorption of SO_2, reduction of vanadium ions from V^{5+} to V^{4+}, and formation of SO_3 from SO_2 and O^{2-} ions, i.e.,:

$$SO_2 + 2\ V^{5+} + O^{2-} \rightarrow SO_3 + 2\ V^{4+} \qquad (11.6)$$

and

2. absorption of oxygen, reoxidation of the vanadium ions and formation of O^{2-} ions:

$$0.5\ O_2 + 2\ V^{4+} \rightarrow 2\ V^{5+} + O^{2-} \qquad (11.7)$$

11.5.1.2 Industrial V_2O_5—K_2SO_4 catalysts (Fig. 11.4)

Catalyst is manufactured by mixing together the active components and substrate to form a paste which is extruded and baked at ~530°C into solid shapes. Ring-shaped (or

Figure 11.4 Photograph of catalyst pieces, courtesy Haldor Topsøe A/S. The outside diameter of the largest piece (far left) is 20 mm. Pellets are shown far right, rings in the middle and "daisy" shapes (modern) left and right of the rings.

ribbed ring) catalyst is the most commonly used shape because (a) it gives a small pressure drop in a catalyst bed and (b) its catalytic activity is only slowly affected by dust in the acid plant feed gas. A typical catalyst ring is 10 mm in diameter by 10 mm in length (Fig. 11.4).

11.5.1.3 Catalyst ignition and degradation temperatures

The ignition temperature at which the SO_2 conversion reaction begins with $V_2O_5-K_2SO_4$ catalyst is ~360°C. The reaction rate is relatively slow at this ignition temperature. Therefore, the gases entering the catalyst beds are heated to temperatures in the range of 400–440°C to ensure rapid SO_2 conversion.

Above 650°C, thermal deactivation of the catalyst begins. Two mechanisms for high temperature thermal deactivation have been proposed:

1. Silica in the substrate partly dissolves in the catalytic melt. This causes the thickness of the melt film to increase, which in turn blocks the pore structure, preventing gas access to the liquid phase inside the pores.
2. Sintering of the silica substrate closes pores, restricting gas access to liquid phase inside the pores.

Thermal deactivation proceeds slowly. $V_2O_5-K_2SO_4$ catalyst can usually be subjected to temperatures of 700–800°C for short periods without causing significant deactivation. However, long times at these temperatures reduce catalyst activity and decrease SO_2 conversion rate.

11.5.1.4 Cs-promoted catalyst

Substituting Cs_2SO_4 for K_2SO_4 in the active liquid component of the catalyst lowers the melting point of the liquid, providing higher reaction rates at lower temperatures. The lower melting point of Cs-promoted catalyst allows V^{4+} to remain in solution at a lower temperature, increasing its low temperature catalytic activity. Cs-promoted catalyst has an ignition temperature of ~320°C. Its typical operating temperature range is 370–500°C. Cs-promoted catalyst costs 2–2.5 times that of non-Cs-promoted catalyst. Therefore, its use is typically optimized by installing it only in the top half of the first and/ or last catalyst beds.

11.5.1.5 Dust accumulation in catalyst beds

Over time, dust, which inadvertently passes through the gas cleaning section, begins to build up in the catalyst beds. It blocks gas flow through the catalyst and increases the pressure that must be applied to achieve the acid plant's required gas flowrate. When the pressure drop in the catalyst beds becomes excessive, the acid plant must be shut down and the catalyst screened to remove the accumulated dust. Keeping off-gas cleaning apparatus in optimum operating condition is critical to maintaining acid plant availability.

11.5.1.6 $SO_2 \rightarrow SO_3$ conversion equilibrium curve

Oxidation of SO_2 to SO_3 proceeds further toward completion at lower temperatures. Fig. 11.5 shows the equilibrium curve for a gas containing 12% SO_2, 12% O_2, and balance N_2 at a total pressure of 1.2 bar. The equilibrium curve on the graph represents the maximum attainable conversion of SO_2 to SO_3 at a given temperature. This curve is also shown in Fig. 11.8. with reaction heat-up paths for each catalyst bed.

11.5.1.7 Absorption of SO_3 into H_2SO_4—H_2O solution

The SO_3 formed by the above-described catalytic oxidation of SO_2 is absorbed into 98.5% H_2SO_4 acid. The process occurs in a packed tower of similar design to a drying tower (Fig. 11.3). In absorption, SO_3 laden gas and sulfuric acid flow counter currently. The overall absorption reaction is the following:

$$SO_3(g) + H_2O(l) \rightarrow H_2SO_4(l) \quad (11.8)$$

It is not possible to manufacture sulfuric acid by absorbing sulfur trioxide directly into water. Sulfur trioxide reacts with water vapor to form H_2SO_4 vapor. This sulfuric acid vapor condenses as a mist of fine, submicron, droplets, which are practically impossible to coalesce. However, the theoretical vapor pressure of water over 98.5% H_2SO_4 is low ($<2\times10^{-5}$ bar at 80°C), avoiding this water vapor problem. The most likely absorption reactions are as follows:

$$SO_3(g) + H_2SO_4(l) \rightarrow H_2S_2O_7(l) \quad (11.9)$$

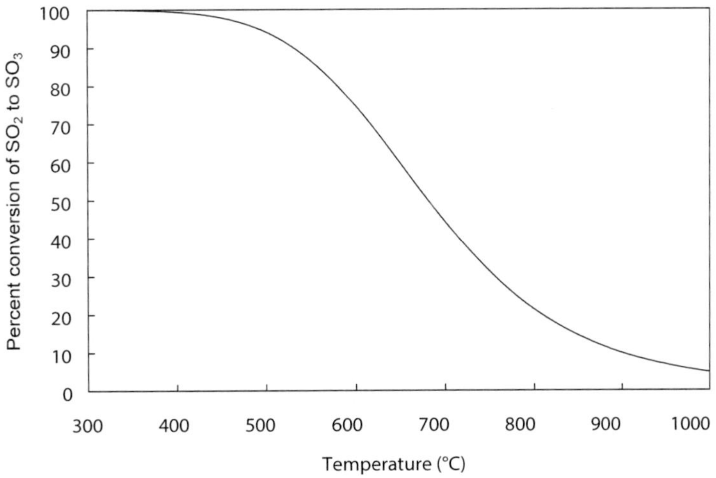

Figure 11.5 Equilibrium curve for SO_2 conversion for an initial gas composition of 12 vol.-% SO_2, 12% O_2, and 76% N_2 at a total pressure of 1.2 bar. The curve shows that higher SO_2 conversions are possible at lower temperatures.

followed by the following:

$$H_2O(l) + H_2S_2O_7(l) \rightarrow 2\ H_2SO_4(l) \quad (11.10)$$

Some SO_3 is undoubtedly absorbed directly by water according to Eq. (11.8). However, the preponderance of H_2SO_4 molecules in the absorbent means that absorption by Eqs. (11.9) and (11.10) probably predominates. SO_3 absorption is exothermic, so that the strengthened acid must be cooled before it is (a) recycled for further absorption or (b) sent to storage.

11.5.1.8 Optimum absorbing acid composition

The optimum absorbing acid composition is 98%–99% H_2SO_4. This is the composition at which the sum of the equilibrium partial pressures of H_2O, SO_3, and H_2SO_4 over sulfuric acid is at its minimum.

Below this optimum, H_2O vapor pressure increases and sulfuric acid mist may form by the reaction of $H_2O(g)$ and SO_3. This mist is difficult to coalesce, so it tends to escape the acid plant into the environment. Above this optimum, SO_3 and H_2SO_4 partial pressures increase. This also increases the release of sulfur compounds to the environment.

Acid plant flow rates and compositions are controlled to keep the absorbing acid in the 98%–99% range before and after SO_3 absorption.

11.6 Industrial sulfuric acid manufacture (Tables 11.4–11.6)

Fig. 11.6 shows a typical flowsheet for SO_2 conversion and SO_3 absorption. The acid plant is a 3:1 double absorption plant; this means that the gases pass through three catalyst beds before intermediate absorption, and then one catalyst bed before final absorption. Figs. 11.7 and 11.8 describe the process thermodynamically. The steps are as follows:

1. heating the incoming gas to the minimum continuous catalyst operating temperature (~410°C) by heat exchange with the hot gases from SO_2 oxidation
2. passing the hot gas through a first bed of catalyst where partial SO_2 conversion takes place and where the gases are heated by the heat of the SO_2 reaction
3. cooling the gas down by heat exchange with cool incoming gas
4. passing the cooled gas through a second bed of catalyst where more SO_2 conversion takes place and where the gases again become hot
5. repeating steps (3) and (4) with a third catalyst bed.

The gas from the third catalyst bed is cooled and its SO_3 absorbed into 98.5% H_2SO_4 acid. The exit gas from this absorption is then passed through a second set of heat exchangers, a fourth catalyst bed, and a second absorption tower. In some plants, initial absorption takes place after the gas passes through two catalyst beds and final absorption after the remaining two catalyst beds (2:2 configuration). Plants requiring very low SO_2 emissions sometimes have a fifth catalyst bed (3:2 configuration).

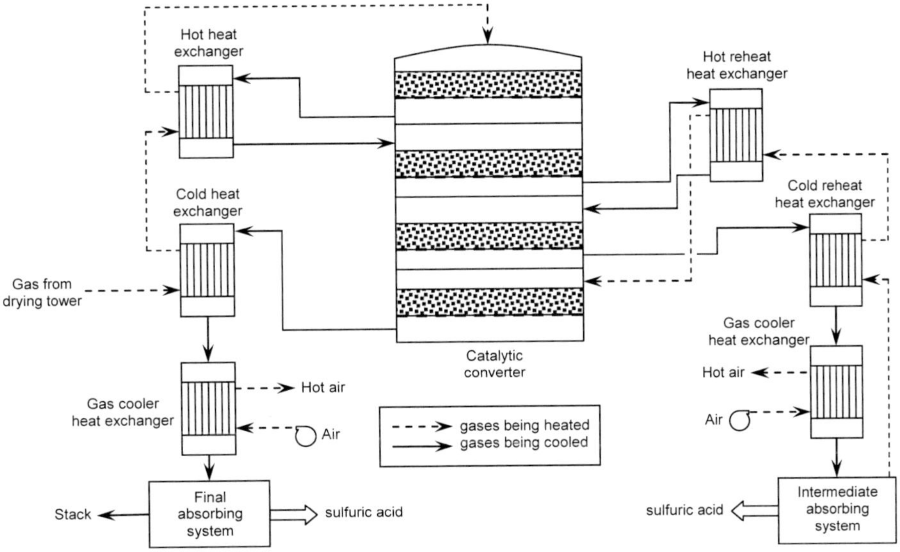

Figure 11.6 Flowsheet for a typical 3:1 metallurgical sulfuric acid plant. Catalyst beds with gas cooling between (to avoid overheating the catalyst) are notable. A hydrocarbon fuel-fired heat exchanger (not shown) is used to heat the feed gas during acid plant startup.

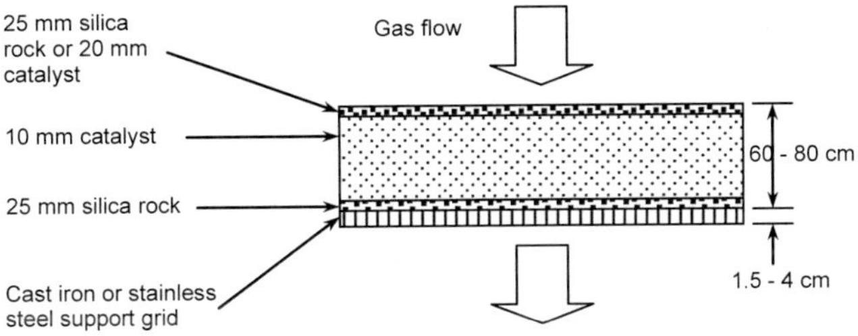

Figure 11.7 Catalyst bed showing steel support, catalyst, and silica rock. The bed is typically 8–14 m in diameter. The silica rock on the top of the bed distributes the gas into the catalyst, preventing localized channeling and short-circuiting through the bed.

The above description is for a *double absorption* plant which converts and absorbs >99.7% of the SO_2 entering the acid plant. Single absorption acid plants convert SO_2 to SO_3 in three or four catalyst beds followed by single absorption of SO_3 (Table 11.6). Their conversion of SO_2 to SO_3 is less complete with consequentially lower sulfur capture efficiencies (96%–98%; Amiri & Alihosseinpour, 2015). About 10%–15% of current sulfide smelter acid plants use single absorption design; the percentage is expected to decline in the future as SO_2 emission limits become tighter.

Table 11.4 Operating details of three double absorption sulfuric acid manufacturing plants.

	Aurubis Hamburg, Germany Line 1	Aurubis Hamburg, Germany Line 2	Aurubis Hamburg, Germany Line 3
Start-up date	1972	1972	1991
Manufacturer	Lurgi	Lurgi	Lurgi
Gas source	Outotec flash + Peirce–Smith converters	Outotec flash + Peirce–Smith converters	Outotec flash + Peirce–Smith converters
Single or double absorption	double	double	double
number of catalyst beds	4	4	5
intermediate absorption after bed?	2	2	3
Converter design			
materials of construction	Stainless steel	Stainless steel	Brick lined
first pass diameter, m	8	8	8.5
others, diameter, m	8	8	8.5
Catalyst bed loading			
bed 1, L	75,000	70,000	42,300
bed 2, L	58,000	58,000	46,200
bed 3, L	64,000	64,000	48,200
bed 4, L	64,000	64,000	46,200
bed 5, L	–	–	54,200
Catalyst type			
bed 1	BASF ring + 0.19m Cs cat	Haldor Topsøe ring +0.22m Cs cat	BASF ring + 0.19m Cs cat
bed 2	Haldor Topsøe ring catalyst	Haldor Topsøe ring catalyst	BASF ring catalyst
bed 3	BASF ring catalyst	Haldor Topsøe ring catalyst	BASF ring catalyst
bed 4	BASF Cs ring catalyst	Haldor Topsøe ring catalyst	BASF ring catalyst
bed 5	–	–	BASF ring catalyst

Table 11.4 Operating details of three double absorption sulfuric acid manufacturing plants.—cont'd

	Aurubis	Aurubis	Aurubis
	Hamburg, Germany	Hamburg, Germany	Hamburg, Germany
	Line 1	Line 2	Line 3
Gas into converter			
Flowrate, Nm^3/min	1830	1830	1830
SO_2, volume%	11	11	11
O_2, volume%	>12.1	>12.1	>12.1
CO_2, volume%	1830	1830	1830
Design SO_2 conversion	99.7–99.8	99.7–99.8	99.8
H_2SO_4 production rate			
Design, tonnes/day	1265	1265	1265
Actual, tonnes/day	Approximately 3000 total		
Products, %H_2SO_4	94, 96, 98	94, 96, 98	94, 96, 98
Other (oleum, liquid SO_2, etc.)		20% oleum	

Table 11.5 Operating details of three double absorption sulfuric acid manufacturing plants with heat recovery equipment. All three export steam to the smelter.

	Sumitomo	Sumitomo	Rio tinto
	Toyo, Japan	Toyo, Japan	Kennecott Utah Copper
	Line 1	Line 2	Magna, UT, USA
Start-up date	1971	2003	1995
Manufacturer	Sumitomo Chemical Engineering	Sumitomo Chemical Engineering	MECS
Gas source	Outotec flash + Peirce–Smith converters	Outotec flash + Peirce–Smith converters	Outotec flash smelting + Kennecott-Outotec flash converting
Single or double absorption	double	double	double
number of catalyst beds	5	5	4
intermediate absorption after bed?	4	4	3

Continued

Table 11.5 Operating details of three double absorption sulfuric acid manufacturing plants with heat recovery equipment. All three export steam to the smelter.—cont'd

	Sumitomo	Sumitomo	Rio tinto
	Toyo, Japan	Toyo, Japan	Kennecott Utah Copper
	Line 1	Line 2	Magna, UT, USA
Converter design			
Materials of construction	stainless steel + carbon steel	stainless steel	304H stainless steel
first pass diameter, m	12.5	12.3	14.1
others, diameter, m	12.5	12.3	14.1
Catalyst bed loading			
bed 1, L	40,000	45,000	38,500/77,000
bed 2, L	50,000	60,000	154,000
bed 3, L	65,000	70,000	40,000/152,000
bed 4, liters	104,000	75,000	288,800
bed 5, liters	105,000	110,000	
Catalyst type			
bed 1	Topsøe VK38	Topsøe VK38,Sud Chemie SulfoMax-RR	Haldor Topsøe VK59 Haldor Topsøe VK38
bed 2	Nihonshokubai 7s Monsanto T-516 Topsøe VK38	Monsanto LP110	Haldor Topsøe VK38
bed 3	Nihonshokubai R10 Monsanto T-516 Topsøe VK38	Monsanto LP110	Haldor Topsøe VK59 Haldor Topsøe VK48
bed 4	Topsøe VK48, Nihonshokubai R10	Monsanto LP1150	Haldor Topsøe VK69
bed 5	Nihonshokubai 7s, R10 BASF O4-111	Monsanto LP1150, BASFO4-111	–
Gas into converter			
Flowrate, Nm^3/min	2917 (maximum)	2833 (maximum)	3965
SO_2, volume%	12–13	12–13	14
O_2, volume%	12–14	12–14	13.8 (0.8 CO_2)

Table 11.5 Operating details of three double absorption sulfuric acid manufacturing plants with heat recovery equipment. All three export steam to the smelter.—cont'd

	Sumitomo Toyo, Japan Line 1	Sumitomo Toyo, Japan Line 2	Rio tinto Kennecott Utah Copper Magna, UT, USA
Heat recovery			
Waste heat boiler	After Bed 1, 21.2 t/h steam	After Bed 1, 21.2 t/h steam	After Bed 3
Superheater	None	None	After Bed 1
Economizer	After Bed 4	After Bed 4	None
Acid heat recovery	None	None	HRS
Steam produced (t/t H_2SO_4)	0.23	0.23	0.69
Design SO_2 conversion (%)	99.6	99.6	>99.5
H_2SO_4 production rate			
Design, tonnes/day	2254	2254	3500
Actual, tones/day	1775	1675	—
Products, %H_2SO_4	98, 95	98	94, 98
Other (oleum, liquid SO_2, etc.)	Sodium bisulfite	oleum	—

Table 11.6 Physical and operating of two single absorption sulfuric acid manufacturing plants, 2010. Design of the Mt. Isa plant is discussed by Daum (2000).

Smelter	Incitec Pivot Mt. Isa, Queensland Australia	Glencore Horne Smelter Rouyn-Noranda, Quebec
Start-up date	1999	1989
Manufacturer	Lurgi	Chemetics
Gas source	ISASMELT, 4 Peirce–Smith converters and sulfur burner	Noranda continuous process reactor and Noranda converter or Peirce–Smith converters
Single or double absorption	Single	Single
number of catalyst beds	3	3
Converter design		
Materials of construction	304 H SS (1 internal Hot heat exchanger)	304H SS (1 internal Hot heat exchanger)
first pass diameter, m	15	14.2
others, diameter, m	same	14.2

Continued

Table 11.6 Physical and operating of two single absorption sulfuric acid manufacturing plants, 2010. Design of the Mt. Isa plant is discussed by Daum (2000).—cont'd

Smelter	Incitec Pivot Mt. Isa, Queensland Australia	Glencore Horne Smelter Rouyn-Noranda, Quebec
Catalyst bed loading		
bed 1,L	162,300	83,000
bed 2, L	183,500	178,000
bed 3, L	211,366	216,000
Catalyst type		
bed 1	$K-V_2O_5$	Topsøe
bed 2	$K-V_2O_5$	Topsøe
bed 3	$Cs-K-V_2O_5 + K-V_2O_5$	Topsøe
Gas into converter		
Flowrate, Nm^3/min	6333	3583
SO_2, volume%	11.2 maximum 10.6 normal operating	6–10.5
O_2, volume%	not measured	11.0 (3.0 CO_2)
Heat recovery		
Waste heat boiler	1 after sulfur burner 1 between beds 2 and 3	None
Superheater	2 bundles after bed 3	None
Economizer	2 bundles after bed 3	None
Acid heat recovery	None	None
Steam produced (t/t H_2SO_4)	0.8	N/A
Design SO_2 conversion (%)	>98	96 (actual 97.5)
H_2SO_4 production rate		
Design, tonnes/day	3645	2400
Actual, tones/day	3636	2000
Products, %H_2SO_4	98.5	93.5
Other (oleum, liquid SO_2, etc.)	None	None

11.6.1 Catalytic converter

A catalytic converter typically houses three to five catalyst beds. It is usually made of 304H stainless steel. Fig. 11.7 shows the cross-section of a typical catalyst bed.

11.6.2 SO_2 → SO_3 conversion reaction paths

Figs. 11.8 and 11.9 show the schematic steady state %SO_2 conversion/temperature reaction path for a 12% SO_2, 12% O_2 gas flowing through a double absorption 3:1 sulfuric acid plant. The gas enters the first catalyst bed of the converter at about 410°C. SO_2 is oxidized to SO_3 in the bed, heating the gas to about 630°C. About 64% of the input SO_2 is converted to SO_3.

The gas from bed 1 is then cooled to 430°C in a heat exchanger (Fig. 11.8) and is passed through the second catalyst bed. There, a further 26% of the SO_2 is converted to SO_3 (to a total of 90%) and the gas is heated to about 520°C by the oxidation reaction. This gas is then cooled to 435°C in a heat exchanger and is passed through the third catalyst bed. A further 6% of the initial SO_2 is oxidized to SO_3 (to 96% conversion), while the temperature increases to about 456°C.

At this point, the gas is cooled to ~180°C and sent to the intermediate absorption tower, where virtually all (99.99%) of its SO_3 is absorbed into 98.5% H_2SO_4 sulfuric acid.

After this absorption, the gas contains about 0.5% SO_2. It is heated to 415°C and passed through the last catalyst bed in the converter (Fig. 11.9). Here, about 90% of its remaining SO_2 is converted to SO_3, leaving only about 0.025% SO_2 in the gas. This gas is again cooled to ~180°C and sent to the final SO_3 absorption tower. Overall conversion of SO_2 is approximately 99.8%, and sometimes up to 99.95%.

11.6.3 Reaction path characteristics

Figs. 11.8 and 11.9 show some important aspects of SO_2 conversion.

1. Conversion to SO_3 is maximized by a low conversion temperature, consistent with meeting the minimum continuous operating temperature requirement of the catalyst.

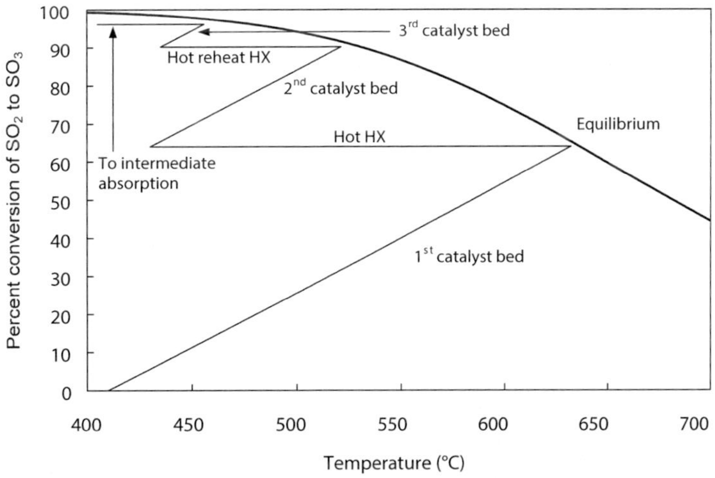

Figure 11.8 Equilibrium curve and first through third catalyst bed reaction heat-up paths. The horizontal lines represent cooling between the catalyst beds in the heat exchangers. The feed gas contains 12 vol.-% SO_2, 12% O_2, balance N_2 (1.2 bar, gauge, overall pressure).

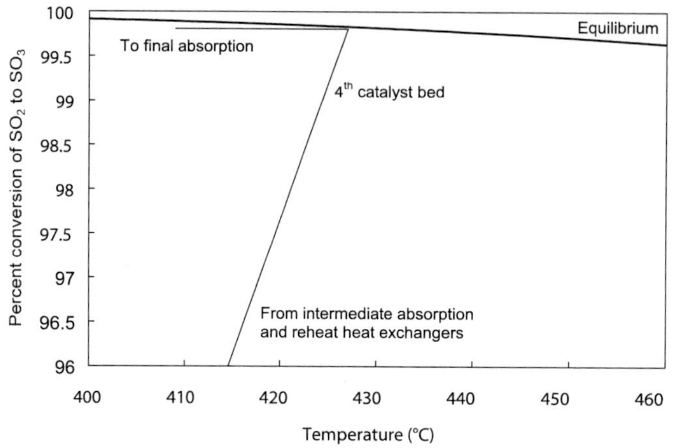

Figure 11.9 Equilibrium curve and fourth catalyst bed reaction heat-up path. Almost all of the SO_3 in the gas leaving the third catalyst bed has been absorbed into sulfuric acid in the intermediate absorption tower.

2. The maximum catalyst temperature is reached in the first catalyst bed where most of the SO_2 conversion takes place. This is where a low ignition temperature Cs catalyst can be useful. Catalyst bed temperature increases with increasing SO_2 concentration in the gas because SO_2 conversion energy release has to heat less N_2. Cs catalyst is expensive, so it is only used when low temperature catalysis is clearly advantageous.
3. Conversion of SO_2 to SO_3 after intermediate absorption is very efficient (Fig. 11.9). This is because (a) the gas entering the catalyst contains no SO_3 (driving Reaction (11.4) to the right) and because (b) the temperature of the gas rises only slightly due to the small amount of SO_2 being oxidized to SO_3.
4. Maximum cooling of the gases is required for the gases being sent to SO_3 absorption towers (~440–180°C), hence the inclusion of air coolers in Fig. 11.6.
5. Maximum heating of the gases is required for initial heating and for heating after intermediate absorption, hence the passage through several heat exchangers in Fig. 11.6.

11.6.4 Absorption towers

Double absorption sulfuric acid plants absorb SO_3 twice: after partial SO_2 oxidation and after final oxidation. The absorption is done counter-currently in towers packed with 5–10 cm ceramic saddles which present a continuous descending film of 98.5% H_2SO_4 acid into which rising SO_3 absorbs. Typical sulfuric acid irrigation rates, densities, and operating temperatures for absorption towers are shown in Table 11.7.

The strengthened acid is cooled in water-cooled shell and tube type heat exchangers. A portion of it is sent for blending with 93% H_2SO_4 from the gas drying tower to produce the grades of acid being sent to market. The remainder is diluted with blended acid and recycled to the absorption towers. These cross-flows of 98+ and 93% H_2SO_4 allow a wide range of acid products to be marketed.

Table 11.7 Typical sulfuric acid design irrigation rates and irrigation densities for drying and absorption towers (Guenkel & Cameron, 2000).

Tower	Sulfuric acid irrigation rate ((m^3/min)/(tonne of 100% H_2SO_4 produced/day)	Sulfuric acid irrigation density (m^3/min per m^2 of tower cross-section)	Sulfuric acid temperature (°C) Inlet/outlet
Drying tower	0.005	0.2–0.4	45/60
Intermediate absorption tower	0.01	0.6–0.8	80/110
Final absorption tower	0.005	0.4	80/95

11.6.5 Gas to gas heat exchangers and acid coolers

Large gas-to-gas heat exchangers are used to transfer heat to and from gases entering and exiting a catalytic converter. The latest heat exchanger designs are radial shell and tube. Older designs were single or double segmental type. Acid plant gas-to-gas heat exchangers typically transfer heat at 10,000–80,000 MJ/h. They must be sized to ensure that a range of gas flowrates and SO_2 concentrations can be processed. This is especially significant for smelters treating off-gases generated by batch type Peirce–Smith converters.

The hot acid from SO_3 absorption and gas drying is typically cooled in indirect shell-and-tube heat exchangers. The water flows through the tubes of the heat exchanger and the acid through the shell. The warm water leaving the heat exchanger is usually cooled in an atmospheric cooling tower before being recycled for further acid cooling.

Anodic protection of the coolers is required to minimize corrosion by the hot sulfuric acid (Halimi, 2020). Anodically protected coolers last 20–30 years. Recently, nonanodically protected acid coolers fabricated of high silicon stainless steel have been shown to last 10–15 years (DKL Engineering, 2016).

11.6.6 Grades of product acid

Sulfuric acid is sold in grades of 93%–98% H_2SO_4 according to market demand. The principal product in cold climates is 93% H_2SO_4 because of its low freezing point, −35°C (Veolia, 2020).

Oleum, H_2SO_4 into which SO_3 is absorbed, is also sold by several smelters. It is produced by diverting a stream of SO_3-bearing gas and contacting it with 98+ H_2SO_4 in a small absorption tower.

11.7 Alternative sulfuric acid manufacturing methods

11.7.1 Haldor Topsøe WSA

The WSA (Wet gas Sulfuric Acid) process, developed and commercialized in the early 1980s by Haldor Topsøe (Christensen, 2016; Louie, 2010), is used to treat wet, low concentration gases (0.2%–6.5% SO_2), such as those produced by sulfide ore and concentrate roasters. WSA treats the off-gas directly from upstream gas cleaning plants. No further drying is required, since the humidity present in the off-gas is used to hydrate the SO_3 generated in the converter and produce concentrated (98%) sulfuric acid. The heat generated during oxidation, hydration, and condensation is principally reused in the process to preheat the off-gas to the required temperature level, and also to generate high pressure superheated steam for process use (Fig. 11.10).

WSA plants can achieve greater than 99% SO_2 conversion (Louie, 2010; Rosenberg, 2006). A newly developed WSA–DC (Double Conversion) process improves conversion even more (Christensen, 2017) and allows autothermal operation in off-gases with SO_2 levels as low as 3%.

11.7.2 Sulfacid

Sulfacid is presently marketed by Carbon Process and Plant Engineering GmbH. About 20 installations worldwide produce weak sulfuric acid (10%–20% H_2SO_4) from very low concentration gases (<1.0% SO_2). The saturated gas is cleaned (<30 mg/dry Nm^3)

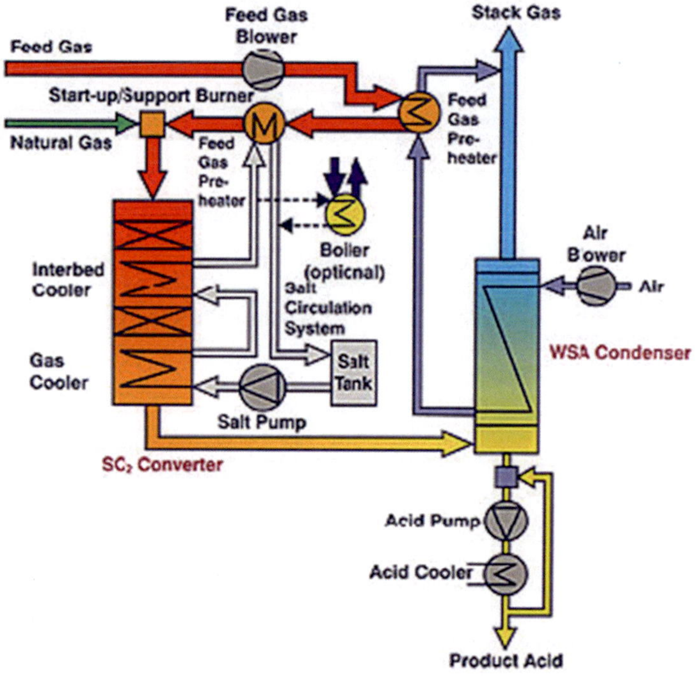

Figure 11.10 Haldor Topsøe WSA Process flow schematic. A molten salt mixture is used to transfer heat from the catalyst beds to the cold feed gas entering the plant. Excess heat removed from the catalyst beds is turned into useful steam in a small waste heat boiler. A direct fired hydrocarbon fuel burner can be used to supply additional heat to the catalyst beds.

before entering an activated carbon catalytic reactor where SO_2 reacts with O_2 and H_2O at 30−80°C (Kruger, 2004) to produce H_2SO_4 by the following reaction:

$$SO_2(g) + 0.5\ O_2(g) + n\ H_2O(l) \rightarrow H_2SO_4 \cdot (n-1)H_2O(l) \qquad (11.11)$$

The acid is intermittently washed from the catalyst, producing weak sulfuric acid. The cleaned gas is discharged to the atmosphere. Sulfacid plants are most economical when their dilute sulfuric acid product can be consumed in nearby operations, such as titanium dioxide manufacture or fertilizer production.

11.8 Recent and future developments in sulfuric acid manufacture

11.8.1 Maximizing feed gas SO_2 concentrations

Since the 1980s, smelting technology has shifted from reverberatory and electric furnace smelting to flash furnace and other intensive smelting processes. Oxygen enrichment of furnace blasts also increased significantly. An important (and desired) effect of these changes has been increased SO_2 strength in the gases that enter smelter sulfuric acid plants (Dijkstra et al., 2017).

SO_2 off-gases entering the drying tower now average 6%−18% SO_2. The low concentrations come from smelters using Peirce−Smith converters. The high concentrations come from direct-to-copper smelting and continuous smelting/converting smelters.

High SO_2 concentration gases contain little N_2. They heat up more than conventional smelter gas during passage through catalyst beds. This can lead to overheating and degradation of the $V_2O_5-K_2SO_4$ catalyst (650°C) and to weakening of the steel catalyst bed support structure (630°C). These two items limit the maximum strength of sulfuric acid plant feed gas in conventional acid plants to ~13% SO_2 (Daum, 2009; Dijkstra et al., 2017). This means that the feed gas entering the acid plant must be diluted with air, resulting in a larger gas volume to be treated and higher operating costs associated with the power required for the blower(s). Four approaches have been used to raise permissible SO_2 strength entering a sulfuric acid plant.

1. Installation of Cs-promoted catalyst in the first pass catalyst bed (Daum, 2009). This allows the bed inlet temperature to be operated at ~380°C, i.e., about 30°C cooler than conventional catalysts. This allows a larger temperature rise (and more SO_2 conversion) in the first bed without exceeding the bed outlet temperature limit.
2. Installation of a preconverter to lower the SO_2 concentration entering the first catalyst bed of the main converter (Daum, 2009; Dijkstra et al., 2017; Hattori & Sunamoto, 2016).
3. Recycle of SO_3-laden gas from the outlet of the third catalyst bed to the first catalyst bed. This is the LUREC process (Daum, 2009). It was developed by Outotec and installed in 2007 at the Yanggu Xiangguang Copper smelter acid plant in China (Outotec, 2014). The use of LUREC allows treatment of 18% SO_2 feed gas at the inlet to the first catalyst bed.

4. Originally developed by Bayer Technology Services (Hands & Connock, 2010), what is now known as the CORE process is now offered by Chemetics, a division of Worley (Dijkstra, 2017; Dijkstra et al., 2017). A CORE system is in operation at a copper smelter in Europe and has been treating 15%–17% SO_2 gas since 2017. CORE uses "active cooling" of a quasiisothermal reactor to remove the heat of the conversion reaction as fast as it can be generated. The coolant is either air for the original process or a recirculating molten salt for the larger-scale CORE-S process. CORE can either be used as an add-on to an existing acid plant or as a replacement facility. Fig. 11.11 shows the schematic for a CORE converter processing high-strength off-gas from a high-intensity smelter, diluted with weaker off-gas (e.g., from Peirce–Smith converters). CORE can process off-gas with up to 25% SO_2, and CORE-S up to 50% strength off-gas. Fig. 11.12 compares the methods of handling high-strength off-gas; the movement to high-intensity smelting processes will make this comparison increasingly important.

11.8.2 Maximizing heat recovery

Heat is generated during SO_2 conversion. In almost all sulfur burning acid plants and in some metallurgical acid plants, this heat is usually recovered into a useful form—steam. The hot gases exiting the catalyst beds are passed through boiler feed water economizers, steam superheaters, and waste heat boilers (Hattori & Sunamoto, 2016; Kubo et al., 2007). Many metallurgical acid plants capture SO_2 conversion and SO_3 absorption heat, but others remove their excess heat in air coolers if it cannot be recovered in a useable product stream (Louie, 2010). Some of the warm air produced by the air coolers has been used as concentrate dryer (rotary kiln type) dilution air (Hoshi, 2010; Kamino et al., 2016). This lowers kiln hydrocarbon fuel consumption.

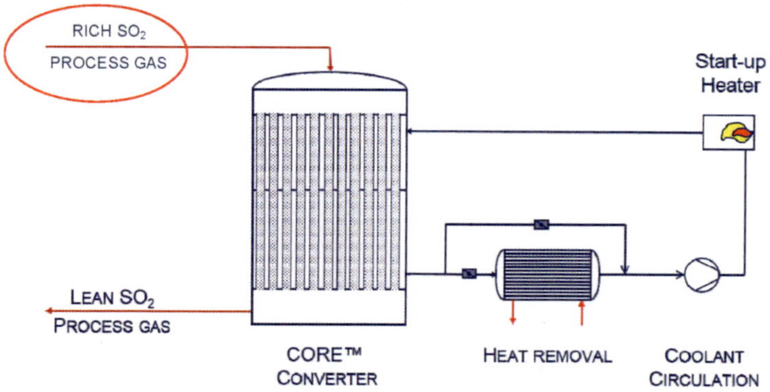

Figure 11.11 Process schematic of a stand-alone CORE unit for processing high-strength (<25% SO_2) cleaned off-gas (Dijkstra et al., 2017). Air or O_2 is used to adjust the O_2/SO_2 ratio. Illustration supplied courtesy of Chemetics Inc.

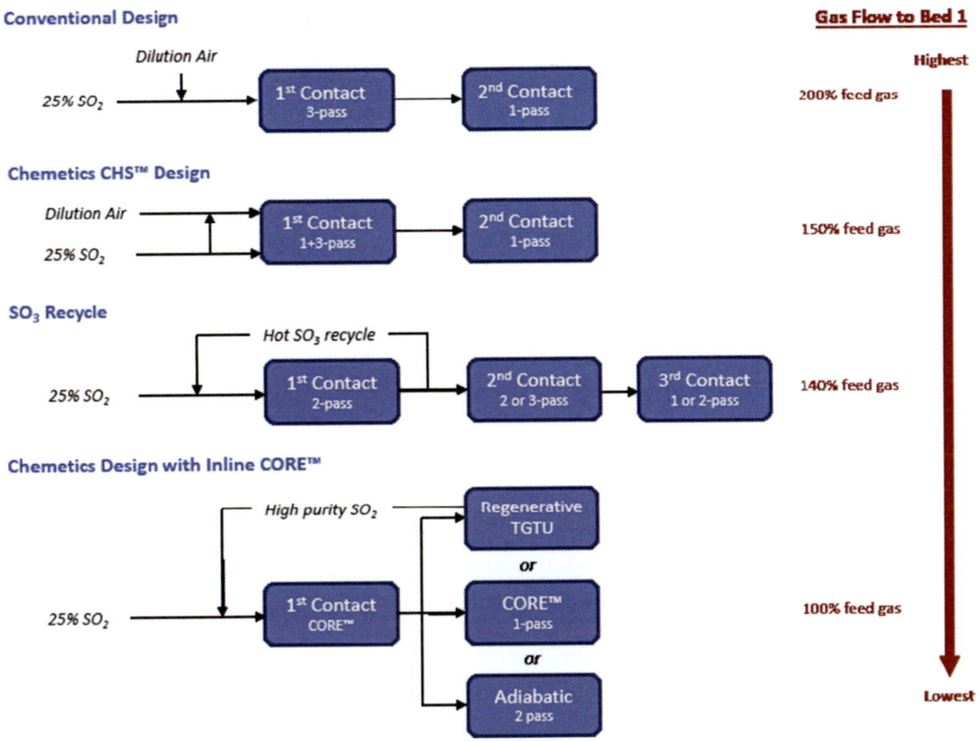

Figure 11.12 Comparison of methods for processing high-strength (25% SO_2) off-gas (Dijkstra, 2017). Illustration supplied courtesy of Chemetics Inc.

SO_3 absorption heat can be captured and used to generate steam or hot water using HRS (Heat Recovery System, MECS; Jung et al., 2010), HEROS (Heat RecOvery System; Outotec, 2014, 2016), or ALPHA (Dijkstra, 2019). HRS is the most common with ~90 installations worldwide. Low pressure steam (5–10 bar gauge, ~0.4 t steam/t H_2SO_4) is produced in a kettle tube boiler with hot, concentrated (~210°C, 99.6% H_2SO_4) sulfuric acid flowing through the tubes and the steam/water mixture on the shell side. Acid concentration and temperature control are critical in ensuring low corrosion rates in these systems. As a result, they are most suited for acid plants that treat feed gas with a steady flow and SO_2 concentration, like those from a flash furnace + flash converter. In addition, the site must have a use for the low pressure steam.

11.9 Alternative sulfur products

The SO_2 in Cu smelter gases is almost always captured as sulfuric acid. Other SO_2 capture products have been the following:

1. liquid SO_2
2. gypsum (Tian et al., 2018)
3. elemental sulfur

The processes for making these products are described by King and Forzatti (2009).

11.10 Summary

Modern smelting processes collect most of their SO_2 at sufficient strength for economic sulfuric acid manufacture. Most reverberatory smelting furnaces have been replaced. Peirce–Smith converting remains a problem for SO_2 collection, especially during charging and skimming when gas leaks into the workplace and at ground level around the smelter (Wu, 2019). Adoption of continuous converting processes such as Mitsubishi and Kennecott–Outotec flash converting will alleviate this problem.

This chapter has shown that most copper is extracted from sulfide minerals so that sulfur, in some form, is a byproduct of most copper extraction processes. The usual byproduct is sulfuric acid, made from the SO_2 produced during smelting and converting. Conventional sulfuric acid production entails the following:

1. cleaning and drying the furnace gases
2. catalytically oxidizing their SO_2 to SO_3 (with O_2 in the gas itself or in added air)
3. absorbing the resulting SO_3 into a 98.5% H_2SO_4–H_2O sulfuric acid solution.

The process is autothermal when the input gases contain about four or more vol.-% SO_2. Double absorption acid plants recover >99.7% of the input SO_2. SO_2 recovery can be increased even further by scrubbing the acid plant tail gas with basic solutions or by adding more catalyst beds with Cs-promoted catalyst.

Some modern smelting processes produce extrastrong SO_2 gases, 13+% SO_2. These strong gases tend to overheat during SO_2 oxidation, causing catalyst degradation and inefficient SO_2 conversion. This problem is leading to the development of catalysts which have low ignition temperatures and high degradation temperatures. Installation of preconverters, LUREC, or CORE are also used to handle high SO_2 strength feed gas.

The Peirce–Smith converter is the major environmental problem remaining in the Cu smelter. It tends to spill SO_2-bearing gas into the workplace and it produces gas discontinuously for the acid plant. Adoption of replacement converting processes began in the 1980s (Mitsubishi converter) and is continuing in the 2000s (Kennecott–Outotec flash converter, Ausmelt/ISASMELT converter, and Noranda converter).

References

Amiri, M., & Alihosseinpour, H. (2015). Identification of metallurgical sulfuric acid plants production line bottlenecks. In *Proceedings 33rd IIER international conference* (pp. 41–46).

Ante, A. (2010). Filsulfor and gypsulfor: Modern design concepts for weak acid treatment. In *Copper 2010, volume 2: Pyrometallurgy I* (pp. 601–614). Germany: Clausthal–Zellerfeld, GDMB.

Begg Cousland. (2018). *Mist elimination equipment for sulphuric acid production plants. (2018)* (Accessed 06.05.20) www.bcinsightsearch.com/userfiles/adverts/SID213212wpb1.pdf.

CCME (Canadian Council of Ministers of the Environment). (2019). *Canada's air* (Accessed 12.04.20) http://airquality-qualitedelair.ccme.ca/en/.

Chen, Y., Liao, T., Li, G., Chen, B., & Shi, X. (2012). Recovery of bismuth and arsenic from copper smelter flue dusts after copper and zinc extraction. *Minerals Engineering, 39*, 23–28.

Christensen, T. (2016). *Novel design of WSA technology for roaster/smelter applications* (Accessed 25.04.20) www.h2so4.com.br/downloads/RoundTable-Chile-2016/Novel+design+of+WSA+technology+for+roaster_smelter+applications.pdf.

Christensen, T. (2017). *Topsøe WSA — DC sulfuric acid technology; 4 plants now in successful operation* (Accessed 17.03.20) www.cobras2017.com/apresentações/topsoe1.pdf.

Clark, N. P., Ferraro, B. B., & Sturm, V. A. (November–December 2017). Safely increasing energy generation. *Sulphur, 373*, 3–7.

Conde, C. C., Taylor, B., & Sarma, S. (1999). Philippines Associated Smelting electrostatic precipitator upgrade. In D. B. George, W. J. Chen, P. J. Mackey, & A. J. Weddick (Eds.), *Copper 99–cobre 99, vol. V: Smelting operations and advances* (pp. 685–693). Warrendale, PA: TMS.

Council directive 1999/30/EC. (2010) (Accessed 12.04.20) https://eur-lex.europa.eu/legal-content/EN/TXT/?uri=celex:32008L0050.

Daum, K. H. (2000). Design of the world's largest metallurgical acid plant. In *Sulphur 2000 preprints* (pp. 325–338). London: British Sulphur.

Daum, K.-H. (2009). The LUREC® process — key to economic smelter acid plant operation. *Journal of the South African Institute of Mining and Metallurgy, 109*, 469–480.

Dijkstra, R. (2017). CORE™ — a new approach to sulphuric acid plant design. In *Presented at sulphur 2017*.

Dijkstra, R. (2019). Designing the world's largest energy recovery system. In *Sulphur + sulphuric acid preprints 2019*. London: CRU.

Dijkstra, R., Senyard, B., Shah, U., & Lee, H. (2017). Economical abatement of high-strength SO_2 off-gas from a smelter. *Journal of the South African Institute of Mining and Metallurgy, 117*, 1003–1007.

DKL Engineering. (2019). *Sulphuric acid on the web. Contact section — blowers* (Accessed 12.04.20) http://www.sulphuric-acid.com/techmanual/Contact/contact_blowers.htm.

DKL Engineering. (2016). *Sulphuric acid on the web. Strong acid system — acid cooling* (Accessed 06.05.20) http://www.sulphuric-acid.com/techmanual/Strong%20Acid/sa_acidcooling.htm.

Evans, C. M., Lawler, D. W., Lyne, E. G. C., & Drexler, D. J. (1998). Effluents, emissions and product quality. In *Sulphur 98 preprints* (Vol. 2, pp. 217–241). London: British Sulphur.

Guenkel, A. A., & Cameron, G. M. (2000). Packed towers in sulfuric acid plants — Review of current industry practice. In NA (Ed.), *Sulphur 2000 Preprints* (pp. 399–417). London: British Sulphur.

Halimi, J. (2020). *Anodic protection system (2020)* (Accessed 23.04.20) http://www.ssp-ir.com/En/Anodic.aspx.

Hands, R., & Connock, L. (March-April). Achieving lower SO_2 emissions from acid plants. *Sulphur, 327*, 28–38.

Hanekom, J. (2017). Sulphuric acid plant optimization and troubleshooting. *Journal of the South African Institute of Mining and Metallurgy, 117*, 1031–1034.

Hasselwander, K. (2009). Gas cleaning designs for smooth operations. *Sulphur, 321*(2), 36–43.

Hasselwander, K. (2019). *Gas cleaning plant designs for the smooth operation of metallurgical smelting facilities* (Accessed 04.04.20) https://www.outotec.com/products-and-services/newsletters/aor-newsletter/issue-2-2019/gas-cleaning-plant-designs-for-the-smooth-operation-of-metallurgical-smelting-facilities/.

Hattori, T., & Sunamoto, T. (2016). Heat recovery and energy saving in the acid plant at Tamano Smelter. In *Proceedings of copper 2016* (pp. 748–759). Tokyo: MMIJ.

Hoshi, M., Toda, K., Motomura, T., Takahashi, M., & Hirai, Y. (2010). Dryer fuel reduction and recent operation of the flash smelting furnace at Saganoseki Smelter & Refinery after the SPI project. In *Copper 2010, volume 2: Pyrometallurgy I* (pp. 779–791). Germany: Clausthal–Zellerfeld: GDMB.

Hultbom, K. G. (2003). Industrial proven methods for mercury removal from gases. In M. E. Schlesinger (Ed.), *EPD congress 2003* (pp. 147–156). Warrendale, PA: TMS.

Jones, D. M., Brandt, R., Nanda, A., & Piatkewicz, A. (2019). Environmental upgrade at the freeport McMoran Miami Smelter. In *Proceedings of the 58th conference of metallurgists hosting the international copper conference 2019*. Montreal: CIM (Article 595113).

Jung, K., Byun, G., & Shin, S. (2010). Profit enhancement through steam selling. In *Copper 2010, volume 2: Pyrometallurgy I* (pp. 831–838). Clausthal–Zellerfeld, Germany: GDMB.

Kamino, Y., Sato, S., Shimokawa, K., & Chida, H. (2016). Efficiency improvement at the sulfuric acid plant of the Saganoseki Smelter. In *Proceedings of copper 2016* (pp. 737–747). Tokyo: MMIJ.

King, M. J., & Forzatti, R. J. (2009). Sulphur based by-products from the non-ferrous metals industry. In J. Liu, J. Peacey, M. Barati, S. Kashani–Nejad, & B. Davis (Eds.), *Pyrometallurgy of nickel and cobalt 2009* (pp. 137–149). Montréal: CIM.

King, M. J., Moats, M. D., & Davenport, W. D. (2013). *Sulfuric acid manufacture: Analysis, control, and optimization* (2nd ed.). Oxford: Elsevier Science Press.

Köster, S. (2010). Waste heat boilers for copper smelting applications. In *Copper 2010, volume 2: Pyrometallurgy I* (pp. 879–891). Clausthal–Zellerfeld, Germany: GDMB.

Kruger, B. (2004). Recovery of SO_2 from low strength off-gases. In *International platinum conference 'platinum surges ahead'* (pp. 59–61). Johannesburg: SAIMM.

Kuang, J., Tu, F., Mao, G., & Liu, H. (2016). Numerical simulation of waste heat boiler for copper flash smelting. In *International conference on electronic information technology and intellectualization (ICEITI 2016)* (pp. 435–440). Lancaster, PA: DEStech Publ.

Kubo, N., Isshiki, Y., Satou, H., & Kurokawa, H. (2007). The acid plant expansion and energy saving at Toyo copper smelter. In A. E. M. Warner, C. J. Newman, A. Vahed, D. B. George, P. J. Mackey, & A. Warczok (Eds.), *Cu2007, volume III (book 2): The carlos díaz symposium on pyrometallurgy* (pp. 579–588). Montreal: CIM.

Liu, J., Baird, M. H. I., Kenny, P., Macnamara, B., & Vahed, A. (2006). Inco flash furnace froth column modifications. In F. Kongoli, & R. G. Reddy (Eds.), *Sohn international symposium on advanced processing of metals and materials, volume 8 – international symposium on sulfide smelting 2006* (pp. 291–302). Warrendale, PA: TMS.

Lossin, A., & Windhager, H. (1999). Improving the quality of smelter acid: The example of Norddeutsche Affinerie's "premium quality". In *Sulphur 99* (pp. 209–220). London: British Sulphur.

Louis, D. (2010). *Design considerations for sulphuric acid plants* (Accessed 23.04.20) www.sulphuric-acid.com/Sulphuric-Acid-on-the-Web/Copper-2010-DLouie=Design-Considerations.pdf.

Lüdtke, P., & Müller, H. (2003). Best available sulphuric acid technology for copper smelters: A state-of-the-art review. In C. Diaz, J. Kapusta, & C. Newman (Eds.), *Copper 2003–cobre 2003, volume IV–pyrometallurgy of copper (book 1): The herman schwarze symposium on copper pyrometallurgy* (pp. 543–554). Montreal: CIM.

Mars, P., & Maessen, J. G. H. (1964). The mechanism of the oxidation of sulphur dioxide on potassium–vanadium oxide catalysts. In W. M. H. Sachtler, G. C. A. Schuit, & P. Zweitering (Eds.), *3rd international Congress on catalysis* (Vol. 1, pp. 266–279). Amsterdam: North–Holland.

Mars, P., & Maessen, J. G. H. (1968). The mechanism and the kinetics of sulfur dioxide oxidation on catalysts containing vanadium and alkali oxides. *Journal of Catalysis, 10*, 1–12.

Ministerio Secretaría General de la Presdencia de la República de Chile. (2010). *Norma de calidad primaria de aire para dioxide de azufre (SO_2), D.S.N° 113/02* (Accessed 12.04.20) http://www.temasactuales.com/assets/pdf/gratis/CHLds113-02.pdf.

Ng'ambi, P., De Vries, D., Louie, D., & Hausler, T. (2016). Integrating smelter and sulfuric acid plant control at the Kansanshi Copper Smelter. In *Proceedings of copper 2016* (pp. 2308–2319). Tokyo: MMIJ.

Okanigbe, D. O., Popoola, A. P. I., & Adeleke, A. A. (2017). Characterization of copper smelter dust for copper recovery. *Procedia Manufacturing, 7*, 121–126.

Oshima, E., Igarashi, T., Nishikawa, M., & Kawasaki, M. (1997). Recent operation of the acid plant at Naoshima. In C. Diaz, I. Holubec, & C. G. Tan (Eds.), *Nickel–Cobalt 97, vol. III: Pyrometallurgical operations, environment, vessel integrity in high–intensity smelting and converting processes* (pp. 305–320). Montreal: CIM.

Outotec. (2014). *Improving efficiency and safety in metallurgical acid plants* (Accessed 02.04.20) mesaredondachile.com/pdf/2014/presentations/14-Outotec_Improving-efficiency-and-safety-in-metallurgical-acid-plants.pdf.

Outotec. (2016). *Outotec sulfuric acid technology* (Accessed 15.03.20) https://www.Outotec.com/globalassets/products/sulfuric-acid/ote_outotec-sulfuric-acid-technology-eng-web.pdf.

Parker, K. R. (1997). *Applied electrostatic precipitation*. London: Chapman and Hall.

Peippo, R., Holopainen, H., & Nokelainen, J. (1999). Copper smelter waste heat boiler technology for the next millennium. In D. B. George, W. J. Chen, P. J. Mackey, & A. J. Weddick (Eds.), *Copper 99—cobre 99, vol. V: Smelting operations and advances* (pp. 71—82). Warrendale, PA: TMS.

Revelli, E. (2019). *Reduce emission and increase production with Leap5™ catalyst* (Accessed 12.04.20) http://www.cobras2019.com/apresentacoes/topsoe.pdf.

Rios, G., Ruiz, I., Cruells, M., & Roca, A. (2019). Leaching of copper flash furnace dusts with weak acid bleed and arsenic precipitation as scorodite and arsenical jarosite. In *Proceedings of the 58th conference of metallurgists hosting the international copper conference 2019*. Montreal: CIM (Article 591168).

Rosenberg, H. (2006). Topsøe Wet Gas Sulphuric Acid (WSA) Technology — an attractive alternative for reduction of sulphur emissions from furnaces and converters. In *International platinum conference 'platinum surges ahead'* (pp. 191—198). Johannesburg: SAIMM, 191—8.

Roy, P., & Sardar, A. (2015). SO_2 emission control and finding a way out to produce sulphuric acid from industrial SO_2 emission. *Journal of Chemical Engineering & Process Technology, 6*, 230—236.

Shibata, T., & Oda, Y. (1990). Environmental protection for SO_2 gas at Tamano smelter. In *Sixth international flash smelting congress*.

Tian, H., Yuan, F., Xu, J., Chang, J., & Guo, Q. (2018). Performance evaluation of a gypsum—based desulfurizer for sulfur recovery from the smelter off—gas: Experimental analysis and thermodynamic performance. *Energy & Fuels, 32*, 2009—2018.

USEPA. (2020). *Table of historical sulfur dioxide national ambient air quality standards (NAAQS)* (Accessed 12.04.20) https://www.epa.gov/so2-pollution/table-historical-sulfur-dioxide-national-ambient-air-quality-standards-naaqs.

Veolia North America. (2020). *Sulfuric acid: Properties, usage, storage, handling* (Accessed 23.04.20) http://info.veolianorthamerica.com/hubfs/offers/info-sheets/industrial/sulfuric-acid-push-information.pdf.

Weidenbach, M., Dunn, G., & Teo, Y. Y. (2016). Removal of impurities from copper sulfide mineral concentrates. In *ALTA 2016 nickel—copper—cobalt proceedings* (pp. 1—17). Melbourne: ALTA Metallurgical Services.

Wu, L. (2019). The utilization of copper continuous smelting, converting, and fire refining processes in China. In *Proceedings of the 58th conference of metallurgists hosting the international copper conference 2019*. Montreal: CIM (Article 595879).

Suggested reading

Louie, D. K. (2008). *Handbook of sulphuric acid manufacturing* (2nd ed.). Richmond Hill, Canada: DKL Engineering.

Further reading

Australian Government Department of Sustainability, Environment, Water, Population and Communities. (2010). *National standards for criteria air pollutants in Australia* (Accessed 12.04.20) http://www.environment.gov.au/protection/publications/factsheet-national-standards-criteria-air-pollutants-australia.

Hay, S., Porretta, F., & Wiggins, B. (2003). Design and start-up of acid plant tail gas scrubber. In C. Diaz, J. Kapusta, & C. Newman (Eds.), *Copper 2003—cobre 2003, volume IV—pyrometallurgy of copper (book 1): The herman schwarze symposium on copper pyrometallurgy* (pp. 555—566). Montreal: CIM.

Ministry of the Environment, Government of Japan. (2020). *Environmental quality standards in Japan — air quality* (Accessed 12.04.20) http://www.env.go.jp/en/air/aq/aq.html.

Ministry of Environmental Protection (China). (2012). *Ambient air quality standards. (GB3095—2012)* (Accessed 26.04.20) https://perma.cc/8PHM-FPJ5.

CHAPTER 12

Fire refining (S and O removal) and anode casting

Virtually, all the molten copper produced by smelting/converting is subsequently electrorefined. It must, therefore, be suitable for casting into thin, strong, smooth anodes for interleaving with cathodes in electrorefining cells (Fig. 13.3). This requires that the copper be fire refined to remove most of its sulfur and oxygen.

The molten blister copper chemical quality will vary depending on the choice of batch or continuous converting. A batch converting process such as Peirce–Smith produces molten blister copper containing ~0.02% S and ~0.3% O. On the other hand, copper from single-step smelting and/or continuous converting produces molten blister copper containing up to 1% S and 0.2%–0.4% O. At these oxygen and sulfur levels, the elements would combine during solidification to form bubbles (*blisters*) of SO_2 in newly cast anodes, making them weak and bumpy. In stoichiometric terms, 0.01% dissolved sulfur and 0.01 dissolved oxygen in molten copper would combine to produce about 2 cm^3 of SO_2 (1083°C) per cm^3 of solidified copper.

Fire refining removes sulfur and oxygen from liquid blister copper by (a) air oxidation removal of sulfur as SO_2 down to ~0.003% S, and (b) hydrocarbon reduction removal of oxygen as CO and H_2O down to ~0.16% O. Sulfur and oxygen contents at the various stages of fire refining are summarized in Table 12.1.

In addition to the contained sulfur and oxygen, molten blister copper also contains minor elements such as arsenic, lead, antimony, bismuth, and precious metals. These elements will be partially removed during the oxidation and refining steps of the refining process and/or by the addition of suitable fluxes to eliminate them as oxides (in most cases) as part of a refining slag. In some cases, use of molten fluxes will be required to reduce levels of elements such as arsenic, lead, and antimony (Alvear et al., 1999). The degree of elimination of these elements will depend on the desired anode quality defined by the subsequent electrorefining process.

12.1 Industrial methods of fire refining

Fire refining is mostly carried out in rotary refining furnaces resembling Peirce–Smith converters (Fig. 12.1), or less often in hearth furnaces or tilting furnaces (Fig. 12.2). It is carried out at about 1200°C, which provides enough superheat for subsequent casting of anodes. The furnaces are heated by combusting hydrocarbon fuel throughout the process. About 1100 MJ of fuel are consumed per tonne of copper (Coursol et al., 2010).

Extractive Metallurgy of Copper, Sixth Edition
ISBN 978-0-12-821875-4, https://doi.org/10.1016/B978-0-12-821875-4.00012-2

Table 12.1 Sulfur and oxygen contents at various stages of fire refining.

Stage of process	Mass% S	Mass% O
Blister copper[a]	0.01–0.05 (Ramachandran et al., 2003)	0.1–0.8
After oxidation	0.002–0.005	0.3–1
After reduction (poling)	0.001–0.004	0.05–0.2
Cast anodes	0.003 ± 0.002 (Moats et al., 2007)	0.16 ± 0.04 (Moats et al., 2007)

[a]From Peirce–Smith and Hoboken converters. The "blister" copper from direct-to-copper smelting and continuous converting contains 0.2%–1% S and 0.2%–0.4% O.

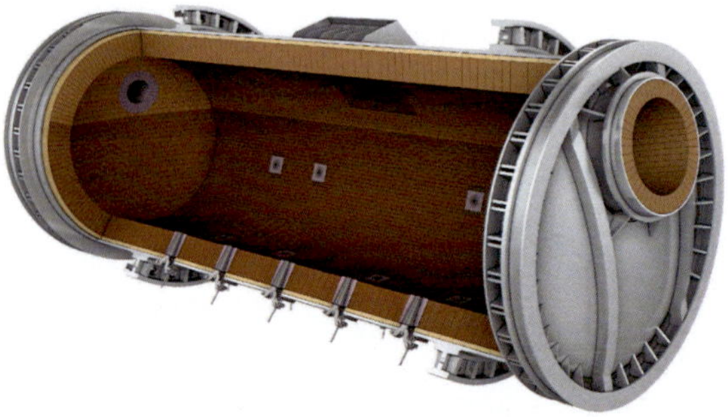

Figure 12.1 Rotary refining (anode) furnace, schematic view showing porous plugs on the furnace bottom, tuyeres position on the furnace shell, and gas uptake on the side wall. *(Photo courtesy of RHI-Magnesita.)*

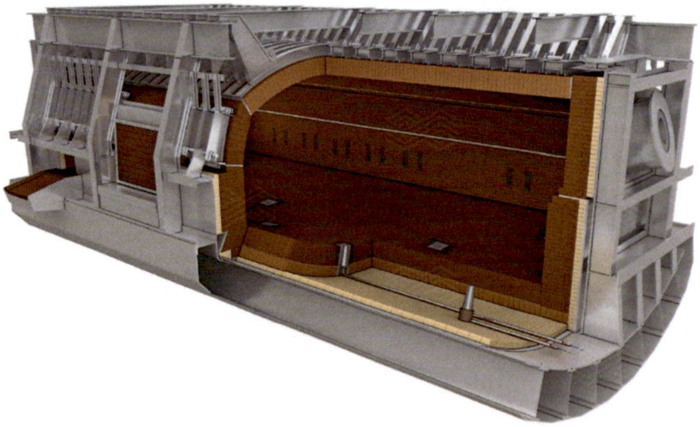

Figure 12.2 Tilting furnace schematics. Furnace has different position for melting and charging. *(Photo courtesy of RHI-Magnesita.)*

An important aspect to enhance removal of oxygen, sulfur, and minor elements is the degree of agitation that can be achieved within the molten bath. For this purpose, several means are used, such as tuyeres and porous plugs. In some cases, the input of nitrogen to enhance the desulfurization reaction and the overall desulfurization time has been successfully tested (Enriquez et al., 2018; Mokola et al., 2019).

12.1.1 Rotary furnace refining

Fig. 12.1 shows a rotary refining furnace. Furnaces are typically sized to hold between 250 and 400 tonnes. Typical dimensions are 4.2 m in diameter and 10 m in length with recent design increasing up to 4.9 m in diameter (George–Kennedy & Mackey, 2019). Air and hydrocarbon flowrates into refining furnaces are slow, to provide precise control of copper composition. Flows of injected air per tuyere will vary depending on the use of pure air or a mixture of air and hydrocarbons for deoxidation. Only ~2 tuyeres are used (Fig. 12.3, Table 12.2). Gas flowrates are 6–13 Nm^3/min per tuyere at 4–5 bar pressure.

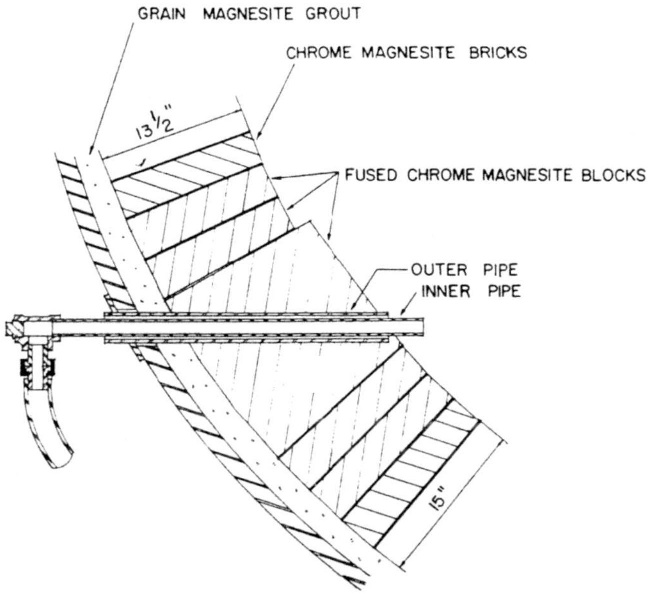

Figure 12.3 Detail of anode furnace tuyere. Note the two concentric pipes separated by castable refractory which permit easy replacement of the inside pipe as it wears back. The inside pipe protrudes into the molten copper to prevent seepage of gas back through the refractory wall of the furnace. *((McKerrow and Pannell, 1972) Gaseous deoxidation of anode copper at the Noranda smelter.* Canadian Metallurgical Quarterly, 11*, 629–633.)*

Table 12.2 Details of two rotary anode furnaces—one mold-on-wheel anode casting plant and one Contilanode anode casting plant.

Smelter	Aurubis AG, Hamburg	PT Smelting Co. Gresik, Indonesia
Nominal anode production tonnes/year	450,000	300,000
Maximum reported throughput	480,000	
Number of anode furnaces		
Total	2	3
Active	2	3
Furnace dimensions, m		
Diameter × length	4.25 × 10 (ID)	3.12 × 12.5 (ID)
Tuyeres		
Diameter, cm	0.8, 1, 1.2	
Number per furnace	2	2
Used during oxidation	2	2
Used during reduction	2	2
Reductant	Natural gas	Diesel oil
Production details		
Tap-to-tap duration, hours	9	11
Anode production, tonnes/cycle	270 (max 300)	400
Oxidation duration, hours	0.5	5
Air flowrate, Nm^3/min	6–7	50 air; 5 oxygen
Reduction duration, h	3	2
Reducing gas flowrate, Nm^3/min per tuyere	10	15 L per min
Scrap addition, tonnes/cycle	0	0–30
Anode casting		
Method	Mold on wheel	Contilanod
Number of wheels, m	1	
Diameter of wheels, m	12.8	
Number of molds per wheel	24	
Casting rate, tonnes/h	80–85	100
Automatic weighing	Yes	Yes
Anode mass, kg	408	400
Variation, kg	±4	±7

For the removal of impurities, tuyeres are also used for flux injection (Carrasco et al., 2003). Typical sodium carbonate fluxing requirements for removal of impurities have been discussed by Alvear et al., 1999. While Coursol et al. (2013) evaluated using molten calcium/sodium sulfates/arsenates salts for removal of arsenic. This last method was

Table 12.3 Details of Hazelett continuous anode casting plant at Gresik, Indonesia (Sato et al., 2010).

Smelter	**PT Smelting Co. Gresik, Indonesia**
Startup data	1998
Anode production tonnes/year	300,000
Casting machine size, m	
Length between molten copper entrance and solid copper exit	3.81
Band width (total)	1.65
Width of cast copper strip (between edge dams)	0.93
Length of lug	0.18
Thickness of cast strip	0.047
Thickness of lug	0.027
Band details	
Material	ASTM A607 Grade 45 steel
Life, tonnes of cast copper	1200
Lubrication	Silicone oil
Edge block details	
Material	Hardened bronze
Life, years	~3 years (~0.5 years for anode lug blocks)
Method of controlling copper level at caster entrance	Electromagnetic level indicator
Temperatures, °C	
Molten copper	1120–1150
Cast anode (leaving caster)	880–930
Casting details	
Casting rate, tonnes/h	100
Caster use, h/day	9
Method of cutting anodes from strip	Hydraulic shear
Anode details	
Mass, kg	400
Acceptable deviation	±7 kg
% Acceptable anodes	97

developed in the late 90 s at the Noranda Technology Centre and it was reassessed by the authors given the importance of safely slag arsenic, minimizing slag generation and copper losses achieving relatively high arsenic removal rates keeping the required arsenic content in the molten copper to ensure proper As/(Sb + As) molar rates for the electrorefining process.

Refining a 250 tonne charge of blister copper (0.03% S) takes up to 3 h: ~1 h for air injection (S removal) and ~2 h for hydrocarbon injection (O removal). High sulfur copper from direct-to-copper smelting and continuous converting takes up to 5 h to refine (Ramachandran et al., 2003). In the case of high sulfur blister copper, the refining process could take up to 24 h for a 400 tonne charge, including filling the furnace (Shibazaki, Shimizu & Oguma, 1991). Recent developments for reducing this time include innovative techniques such as coherent jet burners and use of agitation by porous plugs (Enriquez et al., 2018; Vukovic & Gamweger, 2014).

A typical sequence in rotary furnace refining is as follows:

(a) Molten copper is delivered by crane and ladle from converters to the anode furnace until it is filled to its capacity (200–400 tonnes),

(b) The accumulated charge is then desulfurized to ~0.003% S by blowing air into the molten copper,

(c) Addition of fluxes as required, depending on the targeted anode composition, followed by skimming, and

(d) The copper is deoxidized by blowing gaseous or liquid hydrocarbons into the molten copper bath.

Hydrocarbon blowing is terminated when the oxygen concentration has been lowered to ~0.16% O, as detected with disposable solid electrolyte probes (Heraeus–Electro-Nite, 2019) or by examination of copper test blocks. Copper with this oxygen content *sets* flat when it is cast into anodes.

12.1.2 Hearth furnace refining (Alarcon, 2005)

Although the rotary furnace dominates copper fire refining in primary smelters, secondary (scrap) smelters tend to use hearth-refining furnaces (Hanusch, 2010; Rinnhofer & Zulehner, 2005). They are better for melting scrap and other solid inputs. Sulfur is removed by reaction of the copper solids with an oxidizing flame above the bath and by injecting air through submerged tip steel lances and/or tuyeres. Deoxidation is also done by injecting hydrocarbons or hydrocarbons plus steam using the same type of injecting devices.

Some anode furnaces inject nitrogen into the molten copper through small holes in porous ceramic plugs in the bottom of the furnace (Filzwieser et al., 2003). The injected nitrogen stirs the molten copper and gives it uniform composition and temperature during fire refining and anode casting. This homogenization (a) minimizes solid slag buildups in otherwise cool parts of the anode furnace (Lee et al., 2003) and (b) produces uniform anode composition throughout an entire *cast* of anodes. Benefit (a) has been crucial in several smelters. Fig. 12.1 shows a typical arrangement of porous plugs in an anode furnace.

12.2 Chemistry of fire refining

Two chemical systems are involved in fire refining: (a) the Cu–O–S system (sulfur removal) and (b) the Cu–C–H–O system (oxygen removal).

12.2.1 Sulfur removal: the Cu–O–S system

The main reaction for removing sulfur from molten copper with air at 1200 °C is as follows:

$$[S] + O_2(g) \rightarrow SO_2(g) \qquad (12.1)$$

while oxygen dissolves in the copper by the reaction:

$$O_2\ (g) \rightarrow 2[O]\ (\text{in molten copper, } 1200°C) \qquad (12.2)$$

The equilibrium relationship between gaseous oxygen entering the molten copper and S in the molten copper is from Eq. (12.1):

$$K = p_{SO_2}/\left(\%SO_2 \times p_{SO_2}\right) \qquad (12.3)$$

where K is about 10^6 at 1200°C (Engh, 1992).

The large value of this equilibrium constant indicates that even at the end of desulfurization (~0.003 %S; pO_2 ~0.21 bar), SO_2 formation is strongly favored (i.e., $p_{SO2} >$ 1 bar) and S is still being eliminated. Also, oxygen is still dissolving.

12.2.2 Oxygen removal: the Cu–C–H–O system

The oxygen concentration in the newly desulfurized molten copper is ~0.3 mass % O. Most of this dissolved O would precipitate as solid Cu_2O inclusions during casting (Brandes & Brook, 1998), so it must be removed to a low level.

Copper oxide precipitation is minimized by removing most of the oxygen from the molten copper with injected gaseous or liquid hydrocarbons. Representative oxygen removal reactions are the following:

$$H_2C\ (s,l,g) + 2[O] \rightarrow H_2O\ (g) + CO\ (g) \qquad (12.4)$$

$$CO\ (g) + [O] \rightarrow CO_2\ (g) \qquad (12.5)$$

$$H_2\ (g) + [O] \rightarrow H_2O\ (g) \qquad (12.6)$$

where [O] denotes dissolved oxygen.

12.3 Choice of hydrocarbon for deoxidation

The universal choice for removing S from copper is air. Many different hydrocarbons are used for O removal, but natural gas, oil, liquid petroleum gas, and propane/butane are favored (Ramachandran et al., 2003).

Gas and liquid hydrocarbons are injected into the copper through the same tuyeres used for air injection. Natural gas is blown in directly, sometimes with steam (Talvensaari, 2015). Liquid petroleum gas, propane, and butane are blown in after vaporization. Oil is atomized and blown in with steam.

Oxygen removal typically requires ~5 kg of gas or liquid hydrocarbons per tonne of copper (Ramachandran et al., 2003). This is about twice the stoichiometric requirement, assuming that the products of the reaction are CO and H_2O.

12.4 Minor metals removal

Chapters 4 through 9 indicate that significant fractions of the impurities entering a smelter end up in the metallic copper. The fire refining procedures described above do not remove these impurities to a significant extent. The impurities report mostly to the anodes.

As long as impurity levels in the anodes are not excessive, electrorefining and electrolyte purification keep the impurities in the cathode copper product at low levels. With excessively impure blister copper, however, it can be advantageous to eliminate a portion of the impurities during fire refining (Carrasco et al., 2003; Larouche et al., 2003). The process entails adding appropriate fluxes during the oxidation stage of fire refining. The flux may be blown into the copper through the refining furnace tuyeres or it may be added prior to charging the copper to the furnace.

Antimony and arsenic are removed from molten copper in the anode furnace by fluxing with alkaline fluxes such Na_2CO_3, $CaCO_3$, and CaO (Alvear et al., 1999; Carrasco et al., 2003; Larouche et al., 2003). Removals up to 90% are obtained. Lead, on the other hand, is removed from molten anode furnace copper by adding a combination of SiO_2 flux and fayalite slag. This was practiced extensively by the Timmins smelter, now shut (Newman et al., 1991), and Caletones (Carrasco et al., 2003). About 75% removal was obtained.

12.4.1 Fundamentals of minor element removal

Most of the minor metals contained in molten copper have a high oxygen affinity and therefore are more stable in their oxide form during the oxidation process. Addition of appropriate fluxes allows the formation of complex oxides as part of molten slag structures. Removal of these molten slags is accomplished by skimming from the surface of the molten bath.

The oxidation of an impurity X in molten copper can be represented by the following chemical reaction:

$$\text{X(in copper)} + (\upsilon/4)\text{O}_2 = \text{XO}_{\upsilon/2} \tag{12.7}$$

The distribution ratio of X, L_X, which is defined by (%X in the slag)/[%X in metal], can be evaluated from the following expression:

$$L_X \equiv \frac{(\%X \text{ in slag})}{[\%X \text{ in metal}]} \equiv \frac{K_{12.7} \cdot (n_T) \cdot \gamma_X \cdot p_{O_2}^{\nu/4}}{[n_T] \cdot \gamma_{XO_{\nu/2}}} \tag{12.8}$$

where $K_{(12.7)}$ is the equilibrium constant of the oxidation reaction of the impurity X, defined by Eq. (12.7), $[n_T]$ and (n_T) are the molar amount of 100 g metal and slag, respectively, γ_X is the Raoultian activity coefficient of component i, and pO_2 is the standardized oxygen partial pressure at the slag–metal interface.

As discussed by (Fujisawa, Alvear Flores, Shimpo, & Yamauchi, 1997), to achieve a high impurity removal, a high distribution ratio is required. Higher values of $K_{(13.7)}$, pO_2, and γ_X as well as lower values of $\gamma_{XO_{v/4}}$ are, therefore, preferable.

The oxidizing refining ability of minor elements can be defined by the following expression derived from the oxidation reaction of an element X in molten copper (Alvear et al., 1999):

$$A_X = K_{(13.7)} \cdot \gamma_X \cdot p_{O_2}^{v/4} \tag{12.9}$$

where $K_{(12.7)}$ is the equilibrium constant of the oxidation reaction of the impurity X, defined by Eq. (12.7), γ_X is the Raoultian activity coefficient of component X, and pO_2 is the standardized oxygen partial pressure at the slag–metal interface. Based on Eq. (12.9), the refining ability of minor metals was calculated under conditions of oxygen saturation and is presented in Fig. 12.4. A large oxidizing refining ability index represents a more impurity oxide and therefore a smaller deportment of that particular element to the molten copper phase.

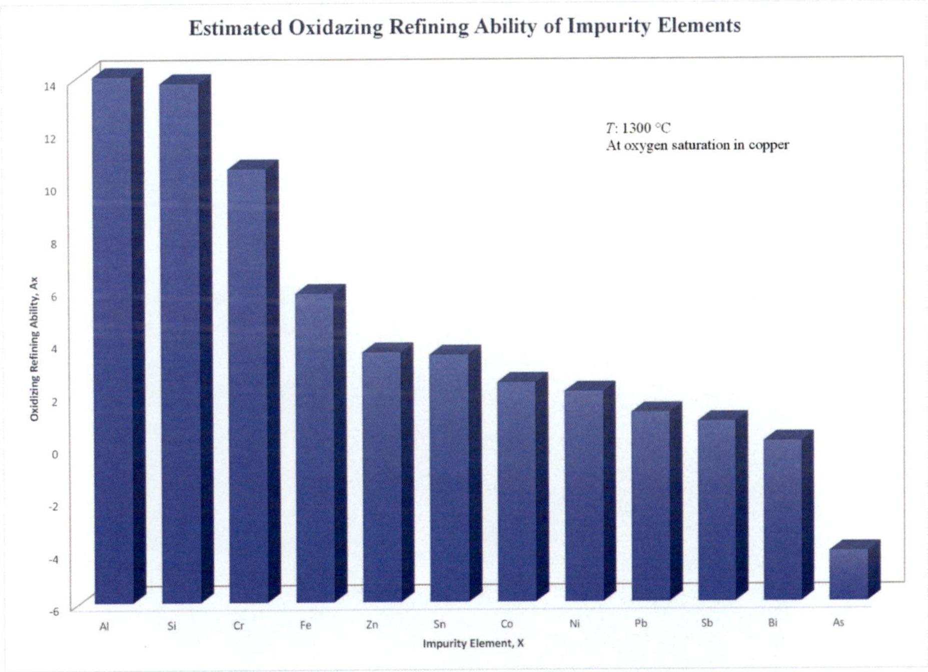

Figure 12.4 Estimated oxidizing refining ability index, A_X, for impurities in molten copper at 1300°C, at oxygen saturation conditions. *(Figure modified from Alvear (1997).)*

12.5 Casting anodes

The final product of fire refining is molten copper, ~0.003% S, 0.16% O, 1200°C, ready for casting as anodes. Virtually, all copper anodes are cast in open anode shaped impressions on the top of flat copper molds. 16–32 such molds are placed on a large horizontal rotating wheel (Fig. 12.5, Table 12.2). The wheel is rotated to bring a mold under the copper stream from the anode furnace where it rests while the anode is being poured (Figs. 12.5 and 12.6 and Fig 1.8 (Chapter 1)) . When the anode impression is full, the

Figure 12.5 Anode casting wheel. The molten copper is poured at 10 o'clock on the wheel. The solidified anodes are lifted out of the molds at 7 o'clock on the wheel. The mass of copper in the ladles is sensed by load cells. The sensors automatically control the mass of each copper pour without interrupting copper flow from the anode furnace. The anode molds are copper, usually cast at the smelter, occasionally machined into a cast copper block. *(Photograph courtesy of Miguel Palacios, Atlantic Copper, Huelva, Spain.)*

Figure 12.6 Copper anodes cast from a casting wheel. *(Photo courtesy of Mount Isa Copper Smelter. © Glencore Australia.)*

wheel is rotated to bring a new mold into casting position and so on. Spillage of copper between the molds during rotation is avoided by placing one or two tiltable ladles between the refining furnace and casting wheel. Most casting wheels operate automatically, but with devoted human supervision.

The newly poured anodes are cooled by spraying water on the tops and bottoms of the molds while the wheel rotates (Edens & Hannemann, 2005). They are stripped from their molds (usually by an automatic raising pin and lifting machine) after a 3/4 rotation. The empty molds are then sprayed with a barite water wash (30 vol.-% barite, 70% water) to prevent sticking of the next anode (Wenzl et al., 2007b). Excessive levels of barite will negatively impact the quality of the anode and the further processing of anode slimes from the electrorefining process.

Casting rates are ~50–100 tonnes of anodes per hour (Ramachandran et al., 2003). The limitation is the rate at which heat can be extracted from the solidifying/cooling anodes. The flow of copper from the refining furnace is adjusted to match the casting rate by rotating the taphole up or down (rotary furnace) or by blocking or opening a tapping notch (hearth furnace). New anode casting installations cast their anodes in pairs to speed up casting rates (Alarcon, 2005; Outotec, 2013).

12.5.1 Anode molds

The molds into which anodes are cast are always copper. They are usually made by (a) casting molten anode copper (occasionally cathode copper; Wenzl et al., 2007b) into large 1.2 m × 1.2 m open molds, then (b) pressing an anode-shaped die into the surface of the molten copper. The mold and die are both steel. The die is water cooled.

Several methods have been tested and reported to ensure large mold life (Edens & Hannemann, 2005). According to the authors, a major step in increasing copper anode mold life was the development of milled molds of continuous cast copper using phosphorus deoxidized copper.

Machining gives a smooth, crack-free surface to the anode shape, considerably increasing the number of anodes that can be cast before the mold surface deteriorates. In all cases, the anode molds shall be perfectly leveled on the casting wheel, to ensure that the anode faces are exactly parallel.

12.5.2 Anode uniformity

The most important aspect of anode casting, besides flat surfaces, is uniformity of thickness. This uniformity ensures that all the anodes in an electrorefining cell reach the end of their useful life at the same time. Automatic control of the mass of each pour of copper (i.e., the mass and thickness of each anode) is now used in most plants (Moats et al., 2007).

The usual practice is to sense the mass of metal poured from a tiltable ladle, using load cells in the ladle supports as sensors. Anode mass is normally 360−410 kg. Anode-to-anode mass variation in a smelter or refinery is ±2−4 kg with automatic weight control.

Recent anode designs have incorporated (a) knife-edged lugs which make the anode hang vertically in the electrolytic cell and (b) thin tops where the anode is not submerged (i.e., where it remains undissolved during refining). The latter feature decreases the amount of undissolved anode scrap which must be recycled when the anode is electorefined.

12.5.3 Anode preparation

Anode flatness, uniformity, and verticality are critical in obtaining good electrorefinery performance. For this reason, most refineries treat their anodes in an automated anode preparation machine to improve flatness and verticality (Outotec, 2018). The machine

(a) weighs the anodes and directs underweight and overweight anodes to remelting
(b) straightens the lugs and machines a knife edge on each lug (for vertical hanging)
(c) presses the anodes flat with a *full body press*
(d) loads the anodes in a spaced rack for dropping into an electrorefining cell.

Inclusion of these anode preparation steps has resulted in increased refining rates, improved cathode purities, and decreased electrorefining energy consumption.

12.6 Continuous anode casting (Hazelett, 2019)

A schematic view of the Hazelett continuous casting of anodes is shown in Fig. 12.7. The technology is being used by several smelters/refineries. The advantages of the Hazelett system over mold-on-wheel casting are uniformity of anode product and a high degree of mechanization/automation.

Fig. 12.8 shows the Hazelett process schematics. Copper is poured at a controlled rate (30−100 tonnes per hour) from a ladle into the gap between two moving water-cooled low-carbon steel belts. The product is an anode thickness continuous strip of copper moving at 4−6 m/min (Fig. 12.8, Table 12.3). The thickness of the strip is controlled by adjusting the gap between the belts. The width of the strip is determined by adjusting the distance between bronze or stainless steel edge blocks which move at the same speed as the steel belts.

Recent Hazelett Contilanod casting machines have periodic machined edge blocks into which copper flows to form anode support lugs (Fig. 12.7). The lug shape is machined half anode thickness in the top of these specialized blocks. The blocks are machined at a 5 degrees angle to give a knife-edge support lug. Identical positioning of the lug blocks on opposite sides of the strip is obtained by heating or cooling the dam blocks between the specialized lug blocks.

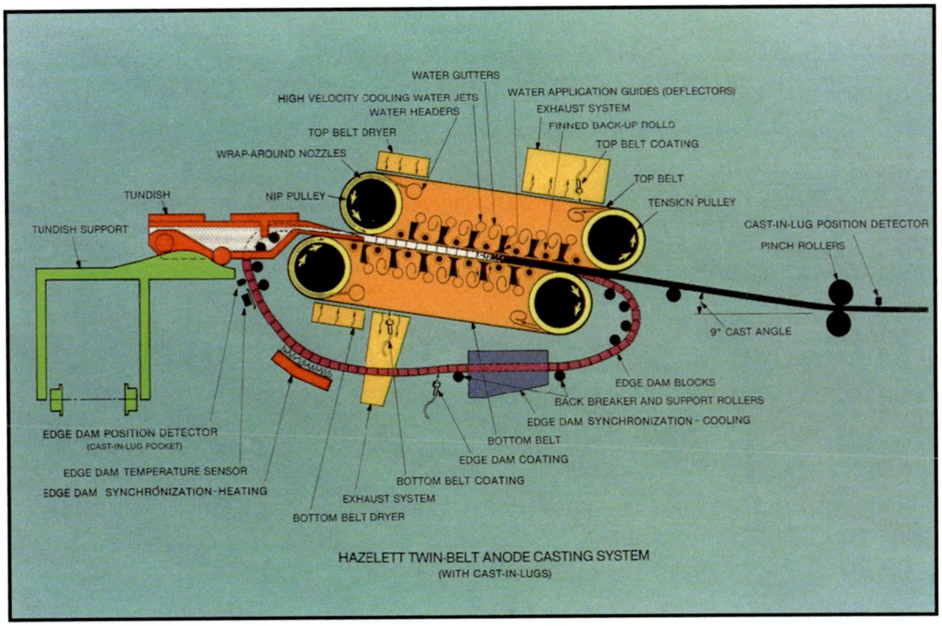

Figure 12.7 Hazelett twin-belt casting machine for continuously casting copper anode strip. The molten copper is contained in the casting machine by means of moving "dam blocks" on both sides of the machine. The anode strip is ~0.047 m thick and ~1 m wide. *(Schematic courtesy of David Hazelett and SMS group.)*

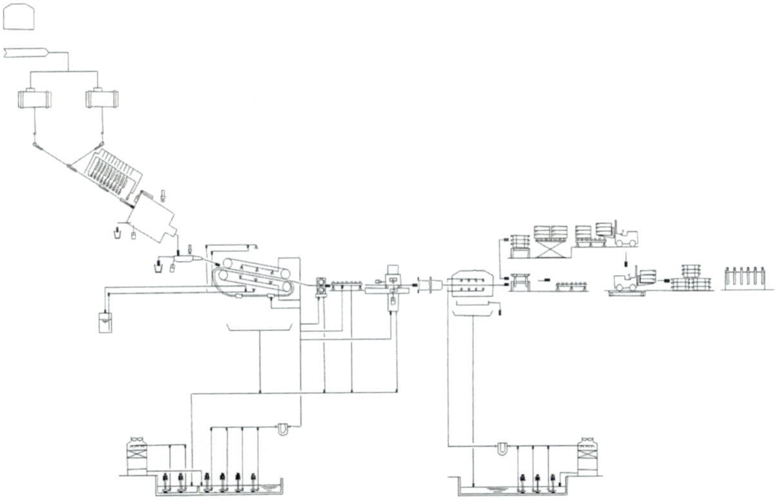

Figure 12.8 Process schematics for the Hazelett—cast copper strip process. *(Schematic courtesy of D. Hazelett and SMS group.)*

The caster produces a copper strip with regularly spaced anode lugs. Individual anodes are produced from this strip by a traveling hydraulic shear. Details of the operation are given by Hazelett (2019). Adoption of the process seems to have stopped, probably because copper fire refining (hence anode casting) is not usually not completely continuous.

12.7 New anodes from rejects and anode scrap

Smelters and refineries reject 2%—3% of their new anodes because of physical defects or incorrect masses. They also produce 12%—20% undissolved anode scrap after a completed electrorefining cycle (Moats et al., 2007). These two materials are remelted and cast into fresh anodes for feeding back to the electrorefinery. The postrefining scrap is thoroughly washed before remelting.

The reject and scrap anodes are often melted in the smelter Peirce—Smith converters. There is, however, an increasing tendency to melt them in Asarco-type shaft furnaces (Chapter 19) in the electrorefinery itself. The Asarco shaft furnace is fast and energy efficient for this purpose. Sulfur and oxygen concentrations in the product copper are kept at normal anode levels by using low sulfur fuel and by adjusting the O_2/fuel ratio in the Asarco furnace burners.

Antimony and arsenic are removed from molten copper in the anode furnace by fluxing with alkaline fluxes such Na_2CO_3, $CaCO_3$, and CaO (Carrasco et al., 2003; Larouche et al., 2003). Removals up to 90% are obtained. Lead, on the other hand, is removed from molten anode furnace copper by adding SiO_2 flux. This was practiced extensively by the Timmins smelter for anode production, now shut (Newman et al., 1991). Caletones smelter also used this technique to produce pyrorefined copper (Carrasco et al., 2003). About 75% removal was obtained.

12.8 Summary

The final step in Cu pyrometallurgical processing is casting of flat, thin anodes for electrorefining. The anodes must be strong and smooth surfaced for efficient electrorefining—bubbles or blisters of SO_2 are unacceptable.

Blister formation is prevented by removing sulfur and oxygen from molten copper by air oxidation then hydrocarbon reduction. The air and hydrocarbons are usually injected into the molten copper via one or two submerged tuyeres in a rotary anode furnace. Minor elements can be removed to the required standards defined by each particular tank house operation, by the addition of appropriate fluxes.

Anodes are virtually always cast in open molds on a large horizontal rotating wheel. Uniformity of anode mass is critical for efficient electrorefining so most smelters automatically weigh the amount of copper that is poured into each anode mold. The molds are also carefully leveled on the wheel to give parallel anode faces.

The cast anodes are often straightened and flattened in automated anode preparation machines. Their lugs may also be machined to a knife edge. Uniform, straight, flat, vertically hung anodes give pure cathodes, high productivity, and efficient energy use in the electrorefinery.

References

Alarcon, V. R. (2005). Anode casting operation, process improvements and anode quality at the Ilo refinery. In A. Ross, T. Warner, & K. Scholey (Eds.), *Converter and fire refining practice* (pp. 133—137). Warrendale, PA: TMS.

Alvear Flores, G. R. F., Tanaka, A., Fukuyama, H., Fujisawa, T., & Yamauchi, Ch (1999). Application of Na_2CO_3 flux to the removal of impurity elements from molten copper. *Shigen to Sozai, 115*, 89—96.

Brandes, E. A., & Brook, G. B. (1998). *Smithells metals reference book* (7th ed.). Oxford: Butterworth—Heinmann.

Carrasco, C., Figueroa, F., Lopez, J., & Alvear, G. (2003). Optimization of future of copper pyrometallurgical refining at Codelco's El Teniente Division. In C. Díaz, J. Kapusta, & C. Newman (Eds.), *Pyrometallurgy of copper (Book 1) the Hermann Schwarze symposium on copper pyrometallurgy: Vol. IV. Copper 2003—cobre 2003* (pp. 405—414). Montreal: CIM.

Coursol, P., Mackey, P. J., & Diaz, C. M. (2010). Energy consumption in copper sulphide smelting. In , *Pyrometallurgy: Vol. II. Copper 2010*. Germany: Clausthal—Zellerfeld. GDMB 649—658.

Coursol, P., Stubina, N., Mackey, P. J., & Zamalloa, M. (2013). Arsenic removal from blister copper — a re-evaluation of the use of $CaSO_4$, CaO and Na_2SO_4 fluxes. In R. Bassa, R. Parra, A. Luraschi, & S. A. Demetrio (Eds.), *Nickolas Themelis symposium on pyrometallurgy and process engineering: Vol. III. Copper 2013* (pp. 1005—1021). Santiago: IIMCH.

Edens, T., & Hannemann, D. (2005). New casting moulds for anode copper. In A. Ross, T. Warner, & K. Scholey (Eds.), *Converter and fire refining practice* (pp. 159—165). Warrendale, PA: TMS.

Engh, T. A. (1992). *Principles of metal refining*. Oxford: Oxford University Press.

Enriquez, J., Walton, R., Deneys, A., Chan, A., Bielec, B., & Kilchyk, V. (2018). Reducing refining cycle times to extend anode furnace campaign life at Kennecott Copper. In B. Davis, et al. (Eds.), *Extraction 2018* (pp. 519—531). Warrendale, PA: TMS.

Filzwieser, A., Wallner, S., Caulfield, K., & Rigby, A. J. (2003). The COP KIN system part II: Performance and benefits — a world wide overview. In C. Díaz, J. Kapusta, & C. Newman (Eds.), *Pyrometallurgy of copper (Book 1) the Hermann Schwarze symposium on copper pyrometallurgy: Vol. IV. Copper 2003—cobre 2003* (pp. 415—425). Montreal: CIM.

Fujisawa, T., Alvear Flores, G. R. F., Shimpo, Y., & Yamauchi, Ch (1997). Thermodynamics of Cu_2O-based slags treatment for recycling of copper scraps. In T. Nakamura, & M. Maeda (Eds.), *Tokyo symposium on recycling and treatment of metals* (pp. 15—21). Tokyo: Center for Eco-Mining.

George—Kennedy, D., & Mackey, P. J. (2019). Conversion and refining. In R. C. Duune, K. S. Komar, & C. A. Young (Eds.), *SME minerals processing & extractive metallurgy handbook* (pp. 1489—1505). Littleton, CO: SME.

Hanusch, B. (2010). New highly efficient rotary furnace for environmentally friendly refining process. In , *Pyrometallurgy I: Vol. 2. Copper 2010* (pp. 721—730). Clausthal—Zellerfeld, Germany: GDMB.

Hazelett. (2019). *Copper bar casting technology*. www.hazelett.com/copper-bar-technology. (Accessed 22 September 2020).

Heraeus Electro-Nite. (2019). *Oxygen activity control in copper.* https://www.heraeus.com/en/hen/products_and_solutions_hen/copper/oxygen_activity_control/oxygen_activity_control.html. (Accessed 22 September 2020).

Larouche, P., Harris, R., & Wraith, A. (2003). Removal technologies for minor elements in copper smelting. In C. Díaz, J. Kapusta, & C. Newman (Eds.), *Pyrometallurgy of copper (Book 1) the Hermann Schwarze symposium on copper pyrometallurgy: Vol. IV. Copper 2003—cobre 2003* (pp. 385—404). Montreal: CIM.

Lee, S.-S., Kim, B.-S., & Choi, S.-R. (2003). Application of the porous plug system in the anode furnace at Onsan smelter. In F. Kongoli, K. Itagaki, C. Yamaguchi, & H. Y. Sohn (Eds.), *High—temperature metals production: Vol. II. Yazawa international symposium* (pp. 447—457). Warrendale, PA: TMS.

McKerrow, G. C., & Pannell, D. G. (1972). Gaseous deoxidation of anode copper at the Noranda smelter. *Canadian Metallurgical Quarterly, 11*, 629—633.

Moats, M., Davenport, W., Demetrio, S., Robinson, T., & Karcas, G. (2007). Electrolytic copper refining — 2007 world tankhouse operating data. In G. E. Houlachi, J. D. Edwards, & T. G. Robinson (Eds.), *Copper electrorefining and electrowinning: Vol. V. Cu 2007* (pp. 195—241). Montreal: CIM.

Mokola, E., Bwalya, C., Musonda, K., Makayi, A., & Hanschar, L. (2019). Anode furnace nitrogen gas purging system at Kansanshi Copper Smelter. In *Proceedings of the 58th conference of metallurgists hosting the international copper conference 2019 (Article 593111).*

Newman, C. J., MacFarlane, G., Molnar, K., & Storey, A. G. (1991). The Kidd Creek copper smelter — an update on plant performance. In C. Díaz, C. Landolt, A. Luraschi, & C. J. Newman (Eds.), *Pyrometallurgy of copper: vol. IV. Copper 91—cobre 91* (pp. 65—80). New York, NY: Pergamon Press.

Outotec. (2013). *Outotec anode casting.* www.youtube.com/watch?v=dB6epzqEbqM. (Accessed 22 September 2020).

Outotec. (2018). *Outotec® anode preparation machine.* https://www.outotec.com/products-and-services/technologies/hydrometallurgy/anode-preparation-machine/. (Accessed 22 September 2020).

Ramachandran, V., Diaz, C., Eltringham, T., Jiang, C. Y., Lehner, T., Mackey, P. J., Newman, C. J., & Tarasov, A. (2003). Primary copper production — a survey of operating world copper smelters. In C. Díaz, J. Kapusta, & C. Newman (Eds.), *Pyrometallurgy of copper (Book 1) the Hermann Schwarze symposium on copper pyrometallurgy: Vol. IV. Copper 2003—cobre 2003* (pp. 3—119). Montreal: CIM.

Rinnhofer, H., & Zulehner, U. (2005). Gas—fired furnaces for copper and copper alloys. In A. von Starck, A. Mühlbauer, & C. Kramer (Eds.), *Handbook of thermoprocessing technologies: Fundamentals, processes, components, safety* (pp. 358—364). Essen: Vulkan—Verlag.

Sato, H., Adji, D. S., Prayoga, A., & Bouman, T. S. (2010). PT. Smelting, Gresik smelter and refinery. In , *Pyrometallurgy II: Vol. 3. Copper 2010* (pp. 1143—1153). Clausthal—Zellerfeld, Germany: GDMB.

Shibasaki, T., Shimizu, T., & Oguma, N. (1991). Analysis of process dynamics and improvements on actual operation on anode furnace. In T. Lehner, P. J. Koros, & V. Ramachandran (Eds.), *International symposium on injection in process metallurgy* (pp. 265—276). Warrendale, PA: TMS.

Talvensaari, H. (2015). *Reduced operational costs and increased efficiency through new developments in fire refining.* www.outotec.com/products-and-services/newsletters/smelting-newsletter/smelting-issue-3–2015/new-developments-in-fire-refining/. (Accessed 22 September 2020).

Vukovic, G., & Gamweger, K. (2014). A new bottom gas purging system for stationary and tilting anode furnaces. In T. Jiang, et al. (Eds.), *5th International symposium on high temperature metallurgical processing* (pp. 25—31). Warrendale, PA: TMS.

Wenzl, C., Filzwieser, A., & Antrekowitsch, H. (2007b). Review of anode casting — part II: Physical anode quality. *Erzmetall, 60*(2), 83—88.

Suggested reading

Antrekowitsch, H., Wenzl, C., Filzwieser, I., & Offenthaler, D. (2005). Pyrometallurgical refining of copper in an anode furnace. In A. Ross, T. Warner, & K. Scholey (Eds.), *Converter and fire refining practice* (pp. 191—202). Warrendale, PA: TMS.

Gamweger, K. (2010). Introduction of a slide gate system for copper anode furnaces. In , *Pyrometallurgy I: Vol. 2. Copper 2010* (pp. 713—720). Clausthal—Zellerfeld, Germany: GDMB.

Pagador, R. U., Malazarte, M. C., & Gonzales, T. W. (2005). Current practices at the converter and anode furnace operations of PASAR. In A. Ross, T. Warner, & K. Scholey (Eds.), *Converter and fire refining practice* (pp. 89–96). Warrendale, PA: TMS.

Potesser, M., Holleis, B., Demuth, M., Spoljaric, D., & Zauner, J. (2010). Customized burner concepts for the copper industry. In , *Pyrometallurgy I: Vol. 2. Copper 2010* (pp. 1051–1062). Clausthal–Zellerfeld, Germany: GDMB.

Rigby, A. J., & Wiessler, M. P. (2007). Interactive 3D modeling of the refractory linings of vessels for copper production. In A. E. M. Warner, et al. (Eds.), *(Book 2): The Carlos Díaz symposium on pyrometallurgy: Vol. III. Cu 2007* (pp. 661–666). Montreal: CIM.

Ross, A. G., Warner, T., & Scholey, K. (Eds.). (2005). *Converter and fire refining practices*. Warrendale, PA: TMS.

Wenzl, C., Filzwieser, A., & Antrekowitsch, H. (2007a). Review of anode casting – part I: Chemical anode quality. *Erzmetall, 60*(2), 77–83.

Yamauchi, Ch (2003). Application of sodium carbonate slag to copper refining. In F. Kongoli, K. Itagaki, C. Yamaguchi, & H. Y. Sohn (Eds.), *High–temperature metals production: Vol. I. Yazawa international symposium* (pp. 3–23). Warrendale, PA: TMS.

CHAPTER 13

Electrolytic refining

Almost all copper is treated by an electrolytic process during its production from ore. It is either electrorefined from impure copper anodes or electrowon from leach/solvent extraction (SX) solutions. Considerable copper scrap is also electrorefined. This chapter describes electrorefining. Electrowinning is discussed in Chapter 16. Approximately 75% of world cathode copper is produced by electrorefining (International Copper Study Group, 2017).

Electrorefining entails

(a) electrochemically dissolving copper from impure copper anodes into an electrolyte containing $CuSO_4$ and H_2SO_4;

(b) selectively electroplating pure copper from this electrolyte *without the anode impurities.*

Electrorefining serves two purposes:

(a) it produces copper essentially free of impurities;

(b) it separates valuable impurities (e.g., gold and silver) from copper for recovery as byproducts.

Copper anodes with a typical purity of 98.5%–99.5% Cu are electrorefined to produce cathodes with a purity of >99.997% Cu. Electrorefined copper, melted and cast, contains less than 20 parts per million (ppm) impurities; the oxygen content is controlled at 0.018%–0.025%.

Table 13.1 presents industrial ranges of copper anode and cathode compositions and the purity requirements for copper cathode to be sold as Grade A quality on the London Metal Exchange and New York Mercantile Exchange (Comex) (LME, 2021). Figs. 13.1 and 13.2 show a typical flow sheet and a modern industrial electrorefining plant, respectively.

13.1 The electrorefining process

An electrical potential is applied between a copper anode and a metal cathode in an electrolyte containing $CuSO_4$ and H_2SO_4. The following processes occur:

(a) Copper is electrochemically dissolved from the anode into the electrolyte, producing copper cations plus electrons:

$$Cu^0_{anode} \rightarrow Cu^{2+} + 2e^- \quad E^0 = -0.34\ V \tag{13.1}$$

(b) The electrons produced by Reaction (13.1) are conducted toward the cathode through the external circuit and power supply.

Extractive Metallurgy of Copper, Sixth Edition
ISBN 978-0-12-821875-4, https://doi.org/10.1016/B978-0-12-821875-4.00005-5

Table 13.1 Industrial ranges of typical copper anode and cathode compositions (Moats et al., 2019) and specification for Grade A copper cathode (LME, 2021).

	Typical industrial ranges of values		LME Grade A specification
Element	**Anodes (%)**	**Cathodes (%)**	**Copper cathode (%, maximum)**
Cu	98.2–99.8	Up to 99.998	—
Ag	0.01–0.75	0.0005–0.0025	0.0025
As	Up to 0.25	<0.0005	0.0005[a]
Au	Up to 0.03	Trace	—
Bi	Up to 0.06	<0.0003	0.00020[b]
Fe	0.001–0.03	<0.0003	0.0010[c]
O	0.035–0.35	Not determined	—
Ni	0.003–0.6	<0.00001–0.002	[c]
Pb	0.001–0.9	<0.00001–0.001	0.0005
S	0.001–0.018	0.0002–0.013	0.0015[d]
Sb	Up to 0.13	<0.0005	0.0004[a]
Se	0.002–0.12	<0.0002	0.00020[b]
Te	0.001–0.065	<0.0002	0.00020[b]

[a](As + Cd + Cr + P + Sb) maximum = 0.0015%.
[b](Bi + Se + Te) maximum = 0.0003%, of which (Se + Te) maximum = 0.00030%.
[c](Co + Fe + Ni + Si + Sn + Zn) maximum = 0.0020%.
[d]Determined on an as-cast sample.

(c) The Cu^{2+} cations in the electrolyte migrate to the cathode by convection and diffusion.

(d) The electrons and Cu^{2+} ions recombine at the cathode surface to form copper metal (without the anode impurities):

$$Cu^{2+} + 2e^{-} \rightarrow Cu^{0}_{cathode} \quad E^{0} = +0.34\ V \tag{13.2}$$

Overall copper electrorefining is the sum of Reactions (13.1) and (13.2):

$$Cu^{0}_{impure} \rightarrow Cu^{0}_{pure} \quad E^{0} = 0.0\ V \tag{13.3}$$

In practice, resistance to current flow must be overcome by applying a potential between the anode and cathode. Small overvoltages must also be applied to plate copper on the cathode (~0.05 V) and dissolve copper from the anode (~0.1 V). There are also ohmic drops across the electrolyte (0.11–0.13 V) and the electrical contacts (0.04–0.08 V) (Nicol, 2006). Applied industrial anode–cathode potentials are typically about 0.3 V (Moats et al., 2019).

The impurities in the anodes do not dissolve or are not plated at the cathode, so the purity of the copper plated onto the cathodes is much higher than that of the anode starting material.

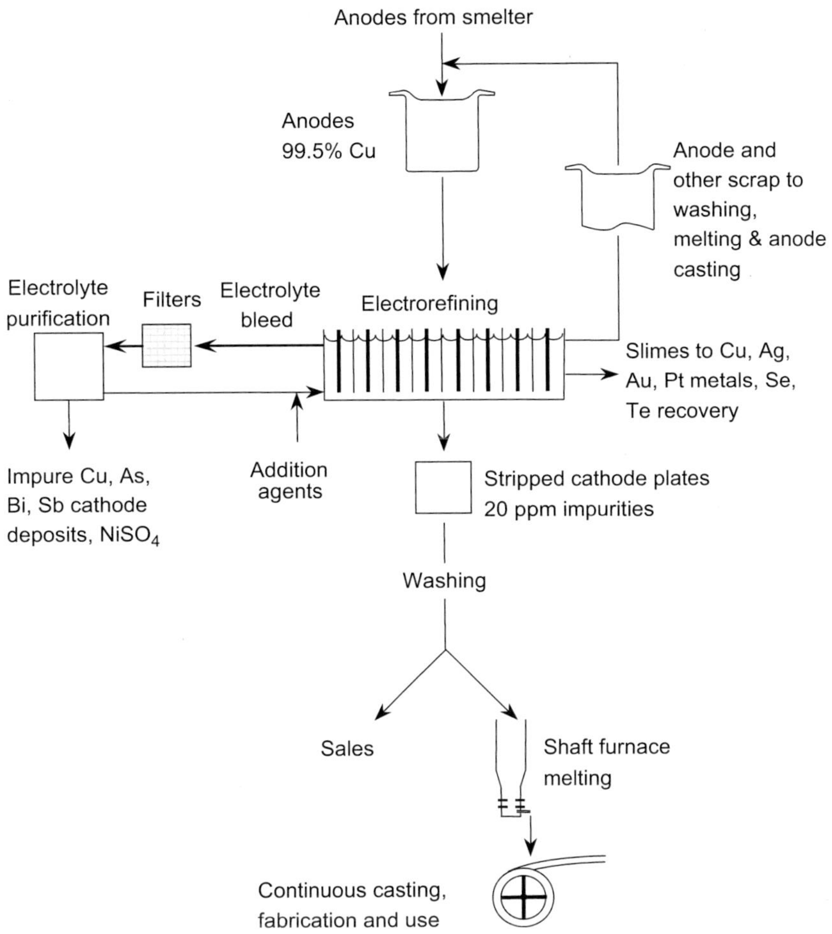

Figure 13.1 Copper electrorefinery flow sheet. Impure copper anodes are electrolytically dissolved in a $CuSO_4$—H_2SO_4 electrolyte and pure copper is plated from solution as cathodes. The electrolyte purification circuit treats a small fraction of the electrolyte (Section 13.9). The remainder is recirculated directly to refining (after reagent additions and heating).

13.2 Chemistry of electrorefining and behavior of anode impurities

The principal impurities in copper anodes are Ag, As, Au, Bi, Co, Fe, Ni, O, Pb, S, Sb, Se, and Te (Table 13.1) (Noguchi et al., 1992). As lower grade and more complex copper ores are processed, the complexity of the anode composition also increases (Alvear Flores et al., 2020). Similarly, reprocessing and recycling of copper scrap contribute to the increasing variety and concentrations of impurity elements that can be present. These elements must be prevented from entering the cathode copper. Their behavior is governed by their position in the electrochemical series (Table 13.2). At the anode, elements with

Figure 13.2 A modern copper electrorefinery in Australia. Note the overhead crane used for moving anodes and cathodes to and from the cells. *(Courtesy of Glencore Technology.)*

Table 13.2 Standard electrochemical potentials of elements in copper electrorefining (25°C, unit thermodynamic activity).

Electrochemical reaction	Standard reduction potential (25°C), V
$Au^{3+} + 3e \rightarrow Au^0$	1.50
$O_2 + 2H^+ + 2e \rightarrow H_2O$	1.229
$Pt^{2+} + 2e \rightarrow Pt^0$	1.2
$Pd^{2+} + 2e \rightarrow Pd^0$	0.83
$Ag^+ + e \rightarrow Ag^0$	0.800
$Fe^{3+} + e \rightarrow Fe^{2+}$	0.771
$H_2SeO_3 + 4H^+ + 4e \rightarrow Se^0 + 3H_2O$	0.74
$TeO_4^- + 8H^+ + 7e \rightarrow Te^0 + 4H_2O$	0.472
$\mathbf{Cu^{2+} + 2e \rightarrow Cu^0}$	**0.337**
$BiO^+ + 2H^+ + 3e \rightarrow Bi^0 + H_2O$	0.32
$HAsO_2 + 3H^+ + 3e \rightarrow As^0 + 2H_2O$	0.25
$SbO^+ + 2H^+ + 3e \rightarrow Sb^0 + H_2O$	0.21
$2H^+ + 2e \rightarrow H_2$	0.000[a]
$Pb^{2+} + e \rightarrow Pb^0$	−0.126
$Sn^{2+} + 2e \rightarrow Sn^0$	−0.136
$Ni^{2+} + 2e \rightarrow Ni^0$	−0.250
$Co^{2+} + 2e \rightarrow Co^0$	−0.277
$Fe^{2+} + 2e \rightarrow Fe^0$	−0.440
$Zn^{2+} + 2e \rightarrow Zn^0$	−0.763

[a](pH = 0; pH_2 = 1 atm).

Table 13.3 Deportment of elements present in anode to the slimes and the electrolyte.

Element	Deportment to slimes (%)	Deportment to electrolyte (%)
Cu	<0.2	>99.8
Au	100	0
Ag	>99	<1
Se	98	2
Te	98	2
Pb	98	2
Bi	60[a]	40
Sb	60[b]	40
As	25[c]	75
S	1	99
Ni	1	99
Co	1	99
Fe	0	100
Zn	00	100

[a]with 0.1% Pb in anode.
[b]with 0.1% As, Bi, Pb, and Sb (each) in anode.
[c]with 0.1% As in anode.
From Larouche, P. (2001). *Minor elements in copper smelting and refining*. M. Eng. Thesis. Montreal, Canada: McGill University.

less positive reduction potentials than Cu dissolve under the applied potential; elements with more positive reduction potentials remain in solid form. At the cathode, elements with more positive reduction potentials deposit preferentially, while elements with more negative potentials will remain in solution. The relative deportments of the various impurity elements present in the anode are shown in Table 13.3.

13.2.1 Au, Ag, and platinum-group metals

Gold and the platinum-group metals (PGMs: Pt, Pd, Rh, Ir, Ru, Os) do not dissolve in sulfate electrolyte, so they cannot plate at the cathode. They form solid *slimes* that adhere to the anode surface or fall to the bottom of the electrolytic cell. These slimes are periodically collected and processed for byproduct recovery (Chen & Dutrizac, 2004; Chen & Dutrizac, 2007; Hait et al., 2009).

Silver is often present with Cu in recycled electronic scrap, so increasing amounts are present in cast anodes. Ag dissolves to a small extent in the electrolyte and is more noble than Cu (Table 13.2), so any Ag^+ present in solution will be reduced and codeposited at the cathode. Cathode copper typically contains 8–10 ppm (g/t) silver. Fortunately, silver is a rather benign impurity in copper.

13.2.2 Se and Te

Selenium and tellurium are present in anodes mainly as compounds with copper and silver and do not dissolve in $CuSO_4-H_2SO_4$ electrolyte. They enter the slimes in these bound forms, e.g., Cu_2(Se,Te), Ag_2(Se,Te), CuAg(Se,Te) (Artzer et al., 2018; Chen & Dutrizac, 2003; Chen & Dutrizac, 2007; Ramírez, Ríos, & Martín, 2007).

13.2.3 Pb and Sn

Lead and tin are present in electronic solder, so are increasingly present in copper anodes as a consequence of increased recycling of electronic scrap. Lead is present in the anode in solid solution with copper and as a secondary complex oxide phase along grain boundaries. Most lead dissolves from the anode and forms $PbSO_4$ by precipitation. Tin forms SnO_2 on the anode. Both of these compounds are sparingly soluble and report to the slimes.

13.2.4 As, Bi, Co, Fe, Ni, S, and Sb

As, Bi, Co, Fe, Ni, S, and Sb dissolve extensively in the electrolyte; however, Cu plates at a lower applied potential than these elements (Table 13.2) so they remain in the electrolyte. Their presence in cathode copper is due to entrapment of electrolyte or slime particles (Atenas & Muñoz, 2007; Beauchemin et al., 2008; Chen & Dutrizac, 2000; Möller, Friedrich, & Bayanmunkh, 2010; Rodrigues Bravo, 2007; Santos Moraes, 2007). To avoid excessive buildup of these elements in the electrolyte and contamination of the cathodes, they are continuously removed from a bleed of the electrolyte stream (Fig. 13.1).

Of particular importance are As, Sb, and Bi. These elements affect anode passivation and cathode quality (see Section 13.6.3, Moats et al., 2016; Wesstrom, 2014). The presence of arsenic in the anode inhibits passivation and prevents the formation of floating slimes. Sb and Bi are precipitated into the slimes layer by maintaining high levels of As in the electrolyte. Most refineries intentionally keep the As levels high, sometimes employing *anode doping* to add As to the anode, to maintain a molar ratio of $As/(Sb + Bi) > 2$ in the anode to assist in controlling Sb and Bi (Moats et al., 2019). As can also be added directly to the electrolyte (Xiao et al., 2012). Doping of Pb into the anode to maintain a Pb/As ratio >1.5 assists in agglomerating anode slimes, causing them to sink to the bottom of the cells and preventing cathode contamination (Zeng et al., 2016).

Ni at concentrations greater than 2000 ppm can form nickel oxide particles within the anode. Ni can also form an oxide phase called *kupferglimmer* ($Cu_3Ni_{2-x}SbO_{6-x}$, where $x = 0.1-0.2$) at high concentrations and in the presence of Sb and/or Sn. NiO and kupferglimmer do not dissolve rapidly and report to the slimes. High concentrations of Ni in the electrolyte cause passivation and detrimental affect current efficiency (Sado, Ueno, Watanabe, & Takebayashi, 2019; Tetsuka & Okamoto, 2019).

13.2.5 O

Oxygen occurs in the anode as various oxides, with Cu_2O being the most prevalent. Cu_2O dissolves chemically (not electrochemically) due to the high acidity of the electrolyte. The dissolution rate of an anode is therefore greater than would be predicted from Faraday's Law (see Section 16.3). This results in a buildup of copper in the electrolyte and requires a bleed of electrolyte for decopperization in liberator cells (Section 13.9).

13.2.6 Slimes

The impurity elements that do not dissolve in the electrolyte, along with ~2% of dissolved copper that does not plate, form anode slimes that drop to the bottom of the cell. The quantity, morphology, pore structure, and adhesion properties of the slimes can affect mass-transfer processes at the anode–electrolyte interface by retarding diffusion of Cu^{2+} ions near the anode surface (Moats & Hiskey, 2010). This can cause precipitation of $CuSO_4 \cdot 5H_2O$ on the anode surface and its passivation. Cathode copper becomes contaminated by entrapment of slime particles in the depositing copper as the deposit grows. Recovery of slimes from the cells and their treatment for recovery of precious metals and other specialty metals (such as Bi, Te, and Se) can significantly influence the profitability of a refinery (Section 13.10).

13.3 Equipment

Industrial electrorefining uses large (~1 × 1 m) thin (40–50 mm) anodes and thin (1–10 mm) cathodes interleaved approximately 50 mm apart in a cell filled with electrolyte (Fig. 13.3). The anodes in the cell are all at the same potential; the cathodes are all at a lower potential. Power is provided by a rectifier. Anodes and cathodes are evenly spaced along the cell to ensure an even distribution of current to all electrodes.

The process is continuous until electrodes are loaded or unloaded from the cells. Electrolyte containing $CuSO_4$ and H_2SO_4 continuously enters at the bottom end of each cell. It leaves the cells (slightly less pure) by continuously overflowing the opposite end of the cell into an electrolyte collection system. Anodes continuously dissolve and pure copper continuously plates on the cathodes.

13.3.1 Anodes

The impure copper anodes are cast at a smelter or at the refinery itself (Fig. 13.4A), as described in Chapter 12. They are typically 4–5 cm thick, with a mass of 300–400 kg (Fig. 13.4B). The starting masses of the anodes should be very similar so that they all dissolve at the same rate. The anodes slowly become thinner as the copper dissolves into the electrolyte. Once they reach 15%–20% of their original mass (after a typical electrorefining time of 21 days), they are removed before they break up and fall into the cell. This anode scrap is washed free of slimes, dried, remelted, and cast into new anodes.

Anode casting and preparation are critical to ensuring that they dissolve evenly and that cathodes of optimum purity are produced. Anode preparation machines are now widely used in the industry. These allow cast anodes to be weighed, straightened, and the lugs machined automatically to ensure good electrical contact with the busbars.

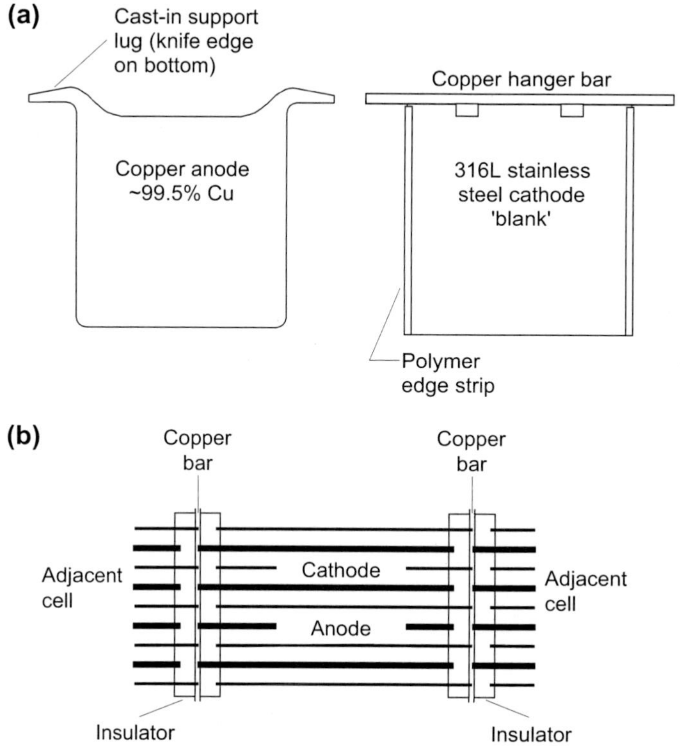

Figure 13.3 (a) Copper anode and stainless-steel cathode. The cathode is about 1 m^2 in area. The anode is slightly smaller. (b) Schematic of electrorefining circuitry. Current flows between the anodes and cathodes through the electrolyte.

13.3.2 Cathodes

Modern refineries use stainless-steel *blanks* as the starting cathode. This is a stainless steel sheet welded to copper support bars (Fig. 13.4C). Copper is electrodeposited onto these cathodes for 7—10 days. The copper-plated cathodes are then removed from the cell and replaced with fresh steel blanks.

The copper-plated cathodes are washed with hot water to remove electrolyte. The copper deposits (30—80 kg on each side of the blank) are machine-stripped from the stainless steel. These are packed, strapped, and sent to market (Fig. 13.5) or melted and cast into wire, rod, or other final product. The stripped blanks are washed and returned to the refining cells. It may be necessary to polish the stainless steel surface to maintain its smoothness and avoid plated copper sticking onto rough patches.

The blanks are usually flat, cold- and bright-rolled 316L stainless steel, ~3 mm thick. Electrodeposited copper attaches quite firmly to this surface so that it does not accidently detach during refining. The vertical edges of the blanks are covered with tight-fitting

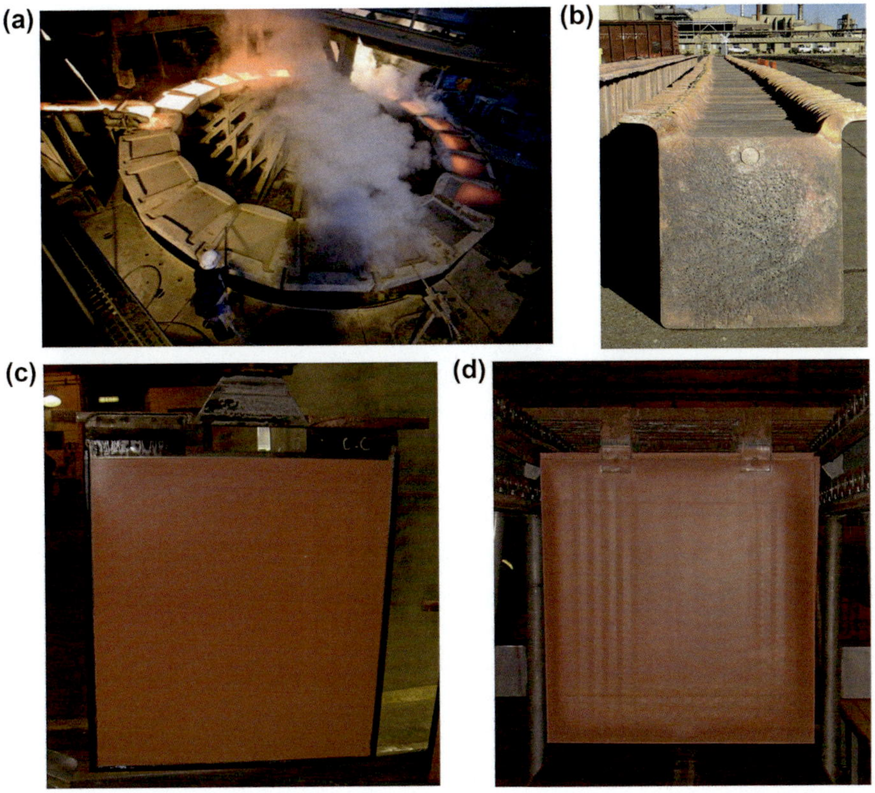

Figure 13.4 (a) Anode casting wheel showing mold shape . (b) Cast anodes ready for electrorefining. Electrorefined copper cathodes plated onto (c) stainless-steel blanks and (d) copper starter sheets. Note the plastic edge strips on the permanent cathode and the copper hangers on the starter sheet. *((A) Photograph courtesy of Palabora Mining Company. (B) Photograph courtesy of M. Moats, Missouri University of Science and Technology. (C and D) Photographs courtesy of T. Robinson, Metso Outotec.)*

Figure 13.5 Electrorefined copper cathodes ready for shipping. *(Courtesy of Glencore Technology.)*

polymer *edge strips* (see Fig. 13.4C) that prevent copper from depositing completely around the cathode and allow removal of the electrorefined copper plates from the stainless steel. The base of the blank is machined to a sharp-edged V groove to allow for easy detachment of the plated copper.

Older refineries use thin copper "*starter sheet*" cathodes, hung by starter-sheet loops on copper support bars. These are thin sheets of copper that have been plated onto titanium or stainless steel, stripped, flattened, and then hanger bars are attached (Fig. 13.4D). These cathodes have higher mechanical stresses than stainless-steel blanks and have a tendency to warp. They are often embossed to improve rigidity (Nicol, 2006). They need to be removed after about 2 days of plating to be straightened in presses and then returned to the cells. Use of these cathodes therefore makes refinery management more difficult. They also have a tendency to more frequent short-circuiting, which reduces cathode quality.

Many refineries (especially in Europe, North America, and Asia where labor costs are high) have switched from this older technology to stainless-steel blanks. In 2019, approximately 70% of refineries used *permanent cathode technology* (Moats et al., 2019): Glencore Technology (the IsaKidd process) and Multotec Outotec supply this refinery tankhouse technology (Pranowo, Aslin, Eriksson, & Heferen, 2019). A lower average labor requirement per tonne of cathode is reported for tankhouses using permanent cathode technology, compared with those that use starter sheets (Moats et al., 2019). This technology also allows use of higher current densities (~320 A/m^2, compared with about 280 A/m^2 for starter sheets), higher current efficiency (96.2% vs. 95.3%) (Moats et al., 2019), as well as better cathode quality, higher productivity, and lower operational costs.

13.3.3 Cells

Industrial refining cells are typically about 6 m long. They are wide and deep enough (~1.1 × 1.3 m) to accommodate the anodes and cathodes with 0.1–0.2 m underneath. Modern cells are designed with adequate height between the bottom of the electrodes and the cell floor to minimize contamination of the cathode from slimes that fall to the base of the cell. Each cell typically contains about 60 anode–cathode pairs connected in parallel.

Modern cells are made of precast polymer concrete. These are usually cast with built-in structural supports, electrolyte distributors, and drains, which are advantageous for fitting them into the tankhouse infrastructure.

13.3.4 Electrical components

The cells are connected electrically in series to form sections of 20–40 cells. Each section can be electrically isolated for inserting and removing anodes and cathodes and for cleaning and maintenance. The number of cells in each section is chosen to maximize the efficiency of these operations.

The electrical connection between cells is made by connecting the cathodes of one cell to the anodes of the adjacent cell. The connection is made by seating the cathodes of one cell and the anodes of the next cell on a common copper *distributor bar* or *busbar* (Fig. 13.3). Good contacts between the anodes, cathodes, and distributor bar are essential to minimize energy loss and ensure uniform current distribution to all electrodes.

Electrorefining requires direct voltage and current. These are obtained by converting commercial alternating current to direct current at the refinery. Silicon-controlled rectifiers are used.

13.4 Typical refining cycle

Production electrorefining begins by installing anodes and cathodes into the empty cells of a freshly cleaned section of the refinery. They are precisely spaced in a rack and brought to each cell by crane. This process is usually completely automated. The cells are then filled with electrolyte and connected to the power supply. Electrolyte flows continuously in and out of the cells. The anodes begin to dissolve and pure copper begins to plate on the cathodes.

Each anode is electrorefined until about 80%–85% of the mass has dissolved. Two or three cathode plating cycles are produced from a single anode. Each copper cathode typically weighs 50–160 kg. This multicathode plating cycle ensures that cathodes do not grow too close to the slime-covered anodes and become contaminated. Copper-loaded cathodes are removed from the cells after 7–10 days of plating and a new rack of stainless-steel blanks is inserted.

The copper-plated cathodes are removed and washed to remove electrolyte and slimes. The plated copper is then machine-stripped from the stainless-steel blanks, sampled, and stacked for shipping, typically in batches of 2.5 t. Cathodes produced from starter sheets are handled similarly but do not require stripping. A robotic stripping machine in a modern refinery is shown in Fig. 13.6.

Figure 13.6 Robotic cathode stripping machine. *(Courtesy of Glencore Technology.)*

Electrolyte is then drained from the cell (through an elevated standpipe), the anodes and cell walls are hosed down with water, and the slimes are drained from the bottom of the cell. The drained electrolyte is sent to filtration and storage prior to being returned to the cells for the next anode refining cycle. The slimes are sent to a Cu and byproduct metal recovery plant (Section 13.10).

The cells are inspected regularly during refining to locate short-circuited anode—cathode pairs. The inspection is done using infrared scanners, gauss meters, and cell voltage—monitoring systems to locate "hot" electrodes.

Short circuits are caused by nonvertical electrodes, bent cathodes, or nodular cathode growths between anodes and cathodes. Shorts waste electrical current and lead to impure copper, due to settling of slimes on nodules and nonvertical cathode surfaces. Shorts are eliminated by straightening the electrodes and removing the nodules (see Section 13.7).

13.5 Electrolyte

Copper refining electrolytes typically contain 40—50 g/L Cu, 170—200 g/L H_2SO_4, 10—20 g/L Ni, up to 20 g/L As, and various impurities (Table 13.4). The nature and concentration of the impurities can vary quite widely, depending on the anode composition, which, in turn, depends on the feed to the smelter (Table 13.1).

Conductivity of the electrolyte increases with increasing H_2SO_4 concentration, but decreases with increasing Cu and Ni concentrations (Price & Davenport, 1981). High conductivity improves energy efficiency, but higher acid concentration also increases corrosion in the tankhouse. Too high a Cu concentration may lead to passivation of the anodes (see Section 13.6.3) and increase viscosity of the electrolyte. This, in turn, can increase the solids in suspension and consequently reduce cathode purity due to incorporation of impurity particles into the copper cathode as it plates.

Table 13.4 Compositions of copper refining electrolytes (Moats et al., 2019). Impurity levels can be removed to lower levels than shown, but at extra cost. Each refinery chooses its impurity levels to give high-purity marketable cathode copper at minimum cost.

Component	Concentration
Cu	36—50 (typical 45—50) g/L
H_2SO_4	160—200 (typical 150—180) g/L
Cl	0.01—0.06 g/L
As	2—12 g/L
Bi	0.0—0.5 g/L
Fe	0.1—2 g/L
Ni	0.8—26 g/L
Sb	0.05—0.5 g/L
Protein colloids (glue)	30—140 g/t Cu cathode
Thiourea	20—170 g/t Cu cathode
Avitone	0—40 g/t Cu cathode

The incoming electrolyte is heated by steam to 60−65°C to improve conductivity and mass transfer: it leaves the cell about 2°C cooler (Moats et al., 2019). Electrolyte is circulated through each cell at ~1.2 m^3/h. At this flowrate, the electrolyte in each cell is completely replaced within a few hours. A steady electrolyte circulation is essential to:

(a) bring warm purified electrolyte into the cell;
(b) ensure uniform concentrations of Cu^{2+} and leveling and grain-refining agents across all cathode surfaces;
(c) remove dissolved impurities from the cell.

Modern developments in electrolyte flow arrangements ensure that the above requirements are met, even when employing high current densities, by minimizing the diffusion boundary layer at the electrode surfaces. Operation above 400 A/m^2 without loss of cathode quality or current efficiency is now routinely achieved (Wenzl, Filzwieser, Filzwieser, & Anzinger, 2010).

13.5.1 Electrolyte additives

Electrorefining practice is trending to increasingly higher current densities (Section 13.8) to maximize production from existing refineries; however, as current density is increased, the cathode deposits become rougher with plating time and uneven surfaces are created. Impurities from the electrolyte and slimes produced at the anode adsorb onto the deposit and become encapsulated in the deposit as it grows. Under these conditions, passivation of the anodes is promoted (see Section 13.6.3).

To counteract these effects, deposition of smooth, dense, pure copper is promoted by adding leveling and grain-refining agents to the electrolyte (De Maere & Winand, 1995; Hiskey, 2019; Wang & O'Keefe, 1984). Without these, the cathode deposits would be dendritic and soft, and have lower purity.

13.5.1.1 Leveling agents

The principal *leveling agents* are protein colloid "bone glues" (Fig. 13.7A). All copper refineries use these glues, added at a concentration of 50−140 g per tonne of cathode copper (Moats et al., 2019). The glues consist of large protein molecules (MW 10 000−30 000)

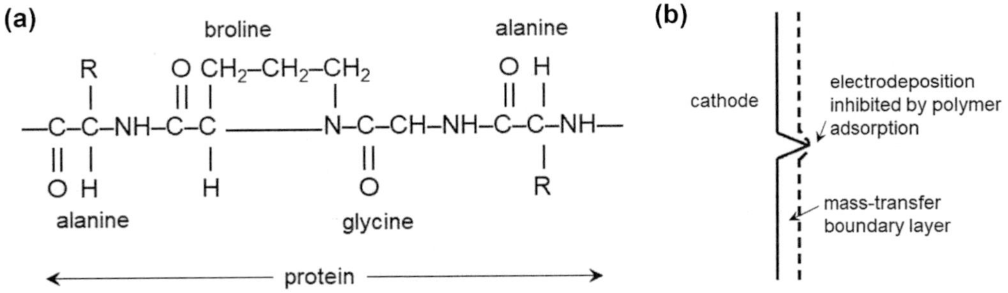

Figure 13.7 (a) Structure of glue molecules, comprising a series of amino acids linked together into a large protein structure, and (b) mechanism of action of glue as a leveling agent.

that form large cations in the electrolyte. The glue exerts its leveling action by electrodepositing large protein molecules at the negatively charged tips of rapidly growing copper grains (Fig. 13.7B). This creates an electrically resistant barrier at the tips of the protruding crystals and encourages further copper deposition sideways and in the "valleys" between the protrusions. The net result is dense and level growth of the copper crystals on the cathode surface.

A practical issue associated with tankhouse management is that glue decomposes within 1.5−2 h at high temperatures and high acidity, and its degradation is catalyzed by H_2SO_4. It is therefore essential to constantly monitor the glue concentration in the electrolyte and ensure that it is maintained at optimum levels. Excess glue leads to rough, striated, and brittle deposits; too low a glue concentration causes nodulation (Fig. 13.8). The CollaMat system is widely used for measuring glue concentrations (Stantke, 2002).

13.5.1.2 Grain-refining agents

The principal *grain-refining agents* are thiourea and chloride. Avitone, a sulfonated petroleum liquid, is sometimes used with thiourea as a grain refiner.

Thiourea ($(NH_2)_2C = S$) acts as a grain refiner by promoting the formation of new copper nuclei and inhibiting the growth of existing crystals. Thiourea molecules adsorb on the cathode surface and prevent crystal growth. The mechanism of action of thiourea is believed to occur via the reduction of Cu^{2+} to Cu^{+} at the cathode surface. Thiourea forms a cationic complex with Cu^{+} on the cathode surface. This promotes copper growth with fine-grained crystal morphology (Fig. 13.9). Thiourea is added to the electrolyte at 60−80 g/t cathode copper. If the concentration is too high, its adsorption on the cathode surface can lead to sulfur contamination of the cathode. Thiourea concentrations can be measured using the Reatrol system (Conard et al., 1990).

(a) (b) (c)

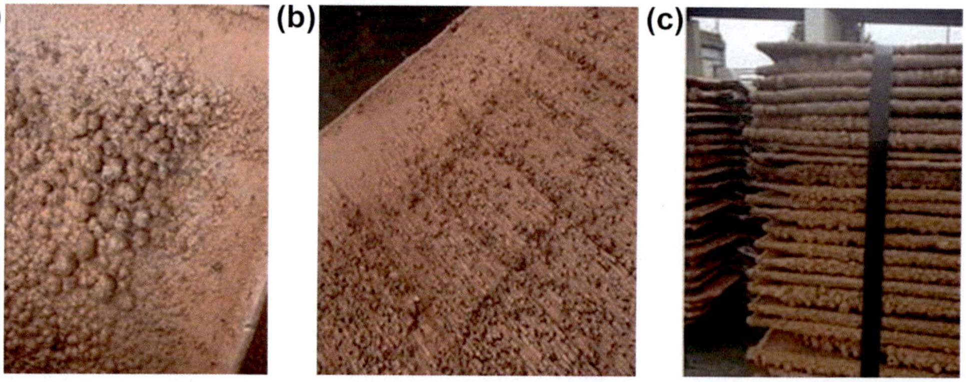

Figure 13.8 Poor quality of electrorefined cathode showing (a) nodules on cathode surface, (b) entrainment of anode slimes in the deposit and *"roping"* or striations on the cathode and (c) dendrite growth on the edges of the cathode. *(Courtesy of T. A. Mulare, University of Pretoria.)*

(a)

(b)

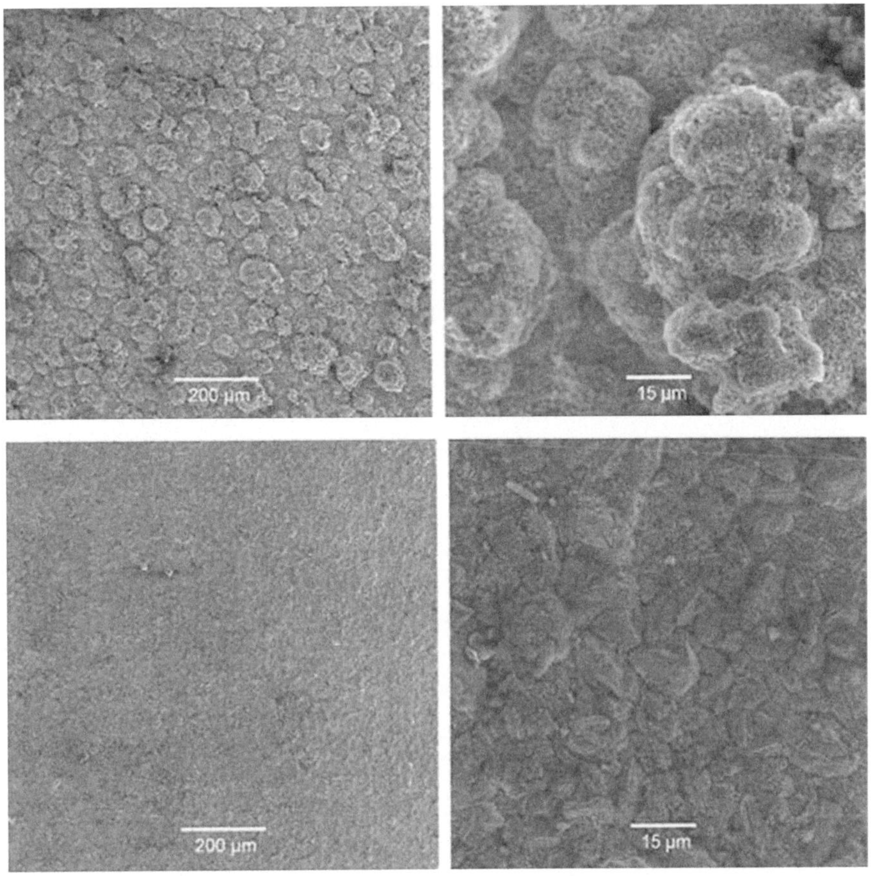

Figure 13.9 Scanning electron microscopy images of (a) dendritic cathode morphology, showing the rough surface created at high current density without the use of grain refiners and (b) the desired, compact cathode morphology with a smooth, dense, homogenous, and fine-grained surface. *(Courtesy of T. A. Muhlare, University of Pretoria.)*

Chloride (20–50 mg/L in electrolyte or 70–100 g/t Cu, added as HCl or NaCl) improves and brightens cathode deposits when combined with organic additives. It is believed to coadsorb with thiourea (Nakano, Oue, Fukushima, & Kobayashi, 2010; Tantavichet et al., 2009). Chloride also acts as a grain-refining agent. Too high a Cl concentration (>50 mg/L) will promote pitting corrosion of the stainless-steel cathode blanks.

Avitone is a proprietary hydrocarbon-based sodium sulfonate surfactant. It is used in combination with glue at 10–30 g/t Cu to give a dense, smooth deposit on the cathode. It also acts as a detergent, wetting oily patches on the anodes and cathodes. Too little Avitone can cause slimes to adhere to the surface of the cathode, leading to entrapment of slimes and contamination of the cathode.

13.5.2 Electrolyte temperature

Electrolyte is heated to ~65°C using steam. The beneficial effects of heating include
(a) increasing the solubility of $CuSO_4 \cdot 5H_2O$, preventing it from crystallizing on the anode (see Section 13.6.2);
(b) lowering the electrolyte density and viscosity (Price & Davenport, 1981), which reduces migration of slimes in the cell;
(c) increasing the rate of the electrochemical reactions (Eqs. 13.1 and 13.2).

Too high a temperature leads to excessive evaporation and energy consumption, and can increase the degradation rate of organic additives.

13.5.3 Electrolyte filtration

Insoluble solid particles present in the electrolyte can cause problems in electrorefining. Solids can originate from breakdown of the anodes, from anode slimes, or from the mold coating used when casting the anodes (see Chapter 20). At high viscosity, insoluble particles are suspended in the electrolyte. Solids that come into contact with the growing crystal surface can serve as substrates for nucleation. This enhances the local growth rate and grain roughening starts, which can initiate the growth of dendrites. Adequate filtration of the recirculating electrolyte is critical.

13.6 Maximizing cathode copper purity

The principal technical objective of electrorefining is to produce high-purity cathode copper. The main factors influencing the purity of cathode copper and minimizing contamination by impurities are
(a) chemical conditions, particularly electrolyte composition, clarity, leveling and grain-refining agent concentrations, temperature, and circulation rate;
(b) the physical arrangement of the anodes and cathodes in the electrolytic cells;
(c) electrical conditions, particularly current density;
(d) thorough washing of cathodes after electrorefining and their removal from the cells;
(e) controlling the impurity levels in the electrolyte by bleeding electrolyte from the refinery and removing the impurities (see Section 13.9).

13.6.1 Physical factors affecting cathode purity

Slimes particles, with their high concentrations of impurities, are kept away from the cathodes by ensuring a smooth flow of electrolyte so that slimes are not transported from the anodes and cell bottoms to the cathodes: turbulence in the electrolyte should be avoided. Contamination is also avoided by filtering the electrolyte (especially that from cell cleaning) before it is recycled to electrorefining (see Fig. 13.1).

13.6.2 Chemical factors affecting cathode purity

The chemical conditions that lead to highest-purity cathode copper are

(a) constant availability of high Cu^{2+} electrolyte at the cathode surface;
(b) constant availability of appropriate concentrations of leveling and grain-refining agents;
(c) uniform 65°C electrolyte temperature;
(d) absence of slimes particles in the electrolyte at the cathode faces;
(e) controlled concentrations of dissolved impurities in the electrolyte;
(f) enough As in the electrolyte to prevent the formation of floating slimes and promote precipitation of Sb and Bi in the slimes layer.

A high Cu^{2+} concentration (40–50 g/L) in the electrolyte and steady recirculation of the electrolyte through the cells ensures constant availability of Cu^{2+} ions over the cathode faces. This allows a constant rate of plating, which gives uniformity of crystal grain size in the cathode. This is helped by maintaining constant concentrations of leveling and grain-refining agents in the electrolyte (Section 13.5.1).

13.6.3 Electrical factors affecting cathode purity

The main electrical factor affecting cathode purity is cathode current density, *i.e.*, the rate at which electricity is passed through the cathodes, measured in amperes per square meter of cathode area (A/m^2).

High current densities give rapid copper plating, but can cause growth of nodules and protrusions. These can entrap slimes in the cathodes, lowering cathode purity, and can lead to short circuiting. Excessive current density may cause anodes to *passivate* by producing Cu^{2+} ions at the anode surface faster than they can convect away. The net result is a high concentration of Cu^{2+} at the anode surface and precipitation of a coherent $CuSO_4 \cdot 5H_2O$ layer on the anode (Chen & Dutrizac, 1991; Moats & Hiskey, 2010). This $CuSO_4 \cdot 5H_2O$ layer isolates and electrically insulates the anode from the electrolyte and blocks further oxidation of Cu to Cu^{2+}. The likelihood of passivation increases with increasing Cu^{2+}, acid, and sulfate concentrations in the electrolyte (Moats & Hiskey, 2007). High levels of nickel in the anode can also cause passivation through the formation of a similar $NiSO_4.mH_2O$ layer (Doucet & Stafiej, 2007). The problem is exacerbated if the impurities in the anode form a coherent slimes layer (Moats & Hiskey, 2006). It is well known that anodes that have high levels of Sb, Sn, and Pb are also readily passivated (Atenas & Muñoz, 2007; Mubarok et al., 2007a,b), while electrolyte additives can also play some role (Moats & Hiskey, 2000; Nakano, Oue, Fukushima, & Kobayashi, 2010; Tantavichet et al., 2009). Arsenic is the only element known to inhibit passivation (Moats & Hiskey, 2006).

Passivation can usually be avoided by operating at current densities below 300 A/m^2, depending on the impurities in the anode, and by maintaining the As concentration above 300 ppm in the anode. Increasing the chloride content of the electrolyte can also be helpful (Moats & Hiskey, 2000). Warm electrolyte (with its higher $CuSO_4 \cdot 5H_2O$ solubility) also helps. Automatic temperature control in each cell is now widely practiced (Rantala, Larinkari, & Menese, 2009). Each refinery must balance these competing economic factors.

13.7 Minimizing energy consumption and maximizing current efficiency

Cathode current efficiencies in modern copper electrorefineries range from ~93% to >98%. The unused current is wasted as anode-to-cathode short-circuits (~3%), stray current to ground (~1%), and reoxidation of cathode copper by O_2 and Fe^{3+} (~1%).

Short-circuiting is caused by a cathode touching an anode. This is avoided by precise, vertical electrode placement and controlled additions of leveling and grain-refining agents to the electrolyte. Its effect is minimized by locating and immediately breaking cathode—anode contacts whenever they occur.

Stray current loss is largely due to current flow to ground via spilled electrolyte. It is minimized by good housekeeping around the refinery.

Copper plated on the cathode can be reoxidised back into the electrolyte by the presence of oxidizing agents, such as oxygen or Fe(III). The reactions are (see Table 13.2)

$$Cu^0 + O_2 + 2H^+ + 2e \rightarrow Cu^{2+} + H_2O \tag{13.4}$$

$$Cu^0 + 2Fe^{3+} \rightarrow Cu^{2+} + 2Fe^{2+} \tag{13.5}$$

This current wastage can be avoided by minimizing oxygen absorption in the electrolyte, by avoiding electrolyte turbulence, and keeping electrolyte flow as smooth and quiet as possible, while still maintaining adequate hydrodynamics for mass transfer. The presence of Fe in the electrolyte should be avoided where possible.

Other housekeeping factors that can have a significant impact on current efficiency include

(a) ensuring even distribution of current to all electrodes;
(b) ensuring that electrical contacts are clean;
(c) ensuring that electrodes are equally spaced and hanging vertically.

The electrical power consumption of an electrorefinery, defined as

$$P = \frac{\text{Total electrical energy consumed in the refinery (kWh)}}{\text{Total mass of cathode copper produced (t)}} \tag{13.6}$$

is 300—400 kWh/t Cu. Power consumption is minimized by maximizing current efficiency and by maintaining good electrical connections throughout the refinery.

Hydrocarbon fuel is also used in an electrorefinery—mainly for heating electrolyte and melting anode scrap. Electrolyte heating energy is minimized by insulating the cells, tanks, and pipes and by covering the cells with canvas or plastic sheeting to avoid evaporation and heat loss. Anode scrap melting energy is minimized by minimizing scrap production, *i.e.*, by casting thick, equal mass anodes, and by equalizing current between all anodes and cathodes. Employing an energy-efficient melting furnace will also assist in limiting overall power consumption.

13.8 Treatment of electrolyte bleed

Refineries are increasingly producing saleable and valuable byproducts from the treatment of their bleed streams and anode slimes (Wesstrom, Olave, & Ramirez, 2019).

Soluble anode impurities continuously dissolve into the electrolyte (Section 13.2). To prevent their accumulation, these are continuously removed from an electrolyte bleed stream. As, Bi, Co, Fe, Ni, and Sb are the main impurities removed this way. Some 1%–2% of the Cu that dissolves from the anodes is not plated on the cathodes (see Section 13.2.6) and is also removed from the electrolyte bleed stream. About 0.1–0.2 m^3 of electrolyte is bled and purified per tonne of product copper.

The impurities and Cu are usually removed in three sequential steps (Hoffman, 2004; Ríos, Ramírez, & Arbizu, 2010; Stantke & Leuprecht, 2010; Stoyanova, Nedeleva, Sraev, Gerov, & Stoyanova, 2016):

(a) Copper is electrowon using Pb–Sn–Ca anodes and stainless steel cathodes in a series of *liberator cells*, in which Cu in the electrolyte is sequentially reduced from ~45 g/L to < 1 g/L;

(b) Water is evaporated from the Cu-depleted electrolyte and $NiSO_4$ crystals are precipitated from the concentrated solution. In some operations, $CuSO_4$ may also be recovered by evaporation and crystallization of hydrated $CuSO_4$. Vacuum and centrifuge technologies may be employed;

(c) As, Bi, and Sb are electrowon from the Cu-depleted electrolyte into an impure Cu–As–Bi–Sb cathode deposit or *sludge* that can be sent for toll refining for recovery of the contained value.

The remaining concentrated acid (~100 g/L H_2SO_4) is returned to electrolyte storage to maintain the acid balance in the refinery. A small portion is neutralized or sold to prevent a gradual buildup of Ca, K, Mg, and Na ions in the electrolyte.

More sophisticated and specific technologies are now used for individual and bulk recovery of impurities, such as As, Bi, Co, and Sb, as well as for recovery of Cu, Ni, and H_2SO_4. Solvent extraction (Agrawal et al, 2008, 2009; Nagai, 1997; Stevens & Gottliebsen, 2000; Szymanowski, 1998), ion exchange (Arroyo et al., 2017; Nagai, 1997; Sado, Ueno, Watanabe, & Takebayashi, 2019; Shaw, 2019; Sole et al., 2017), molecular recognition technology (MRT) (Sharma et al., 2019), crystallization (Sado, Ueno, Watanabe, & Takebayashi, 2019), and electrowinning (Agrawal, Bagchi, et al., 2007; Agrawal, Kumari, et al., 2007; Jin et al., 2015; Lundström et al., 2019) are reported.

13.9 Treatment of slimes

Treatment of anode slimes, particularly to recover precious and specialty metals for additional revenue, is receiving increasing attention as a refinery process (Mahmoudi et al., 2020; Xing et al., 2020). Treatment of more complex ores and a greater tonnage and variety of copper-containing scrap in recent years has resulted in greater amounts and varied compositions of anode contaminants.

The recovery process depends on the chemical and mineralogical composition of the slimes. Base metals are typically removed first in a pretreatment process. In pyrometallurgical pretreatment, elements such as Ni, Sn, and Zn form alloys with precious metals at ~ 1000°C and can be separated from the slag (Safarzadeh et al., 2018). Hydrometallurgical methods involve an upfront leaching step to dissolve the base metals and species such as Te and Se. The precious metal-enriched material is then typically leached under oxidizing conditions to dissolve the contained Au, Ag, and PGMs using lixiviants such as cyanide, thiourea, thiosulfate, and halide systems. The precious metals are recovered from solution using conventional technologies, such as solvent extraction, ion exchange, adsorption, and electrowinning (Elkayar & Kryst, 2019; Komori, Ito, Okada, & Iwahori, 2010; Nagai et al., 2017; Wang & Kim, 2018; Xing et al., 2020; Xing & Lee, 2019).

13.10 Industrial electrorefining

Table 13.5 describes the operating conditions of seven modern electrorefineries (Moats et al., 2019). These refineries have been selected to show the diversity of operating conditions, from one of the largest in the world (Jinlong Copper, China: 450,000 t/a production) to one of the oldest refineries in the world that is still operating (the Pan Pacific Copper refinery at Saganoseki, Japan: commissioned in 1916). A relatively new refinery is Onsan #1 in South Korea (commissioned in 2017). Xianguang #2 tankhouse (China) uses the highest current density (420 A/m^2), while Aurubis Lünen (Germany) treats one of the lowest-quality feedstocks to produce high-purity cathode. More detailed information about these and other major refineries is provided by Moats et al., 2019.

13.11 Recent developments and emerging trends in copper electrorefining

Developments and trends in electrorefining relate to maximizing production, safety, and product purity by upgrading of existing infrastructure and equipment, while minimizing energy consumption and ensuring sustainability of the industry. Artificial intelligence and machine learning are increasingly employed for managing and interrogating data, assisting in decision-making, and improving process control. There is increasing

Table 13.5 Selected industrial copper electrorefining data.

Refinery	Jinlong Copper Tongling China	Xianguang Copper Tankhouse #2 China	Boliden Harjavalta Oy Pori Finland	Aurubis Lünen Germany	Pan Pacific Copper Saganoseki Japan	LS-Nikko Copper Onsan #1 S. Korea	Uralelektromed Copper Verkhnyaya Pyshma Russia
Startup date	**1997**	**2011**	**1941**	**1994**	**1916**	**2017**	**2012/2018**
Capacity cathode Cu (kt/a)	**400**	**300**	**153**	**220**	**235**	**440**	**320**
Copper source (excluding scrap)							
Primary (%)		100	82	100	82	66.8	
Secondary (%)			18		18	33.2	
Electrolytic cells							
Number (total)	1440	720	782	924	768	1392	888
Material	Cement concrete	Concrete	Polymer concrete	Concrete	Polymer concrete/concrete	Polymer concrete	Polymer concrete
Lining material	FRP	Fiberglass		Lead	PVC (concrete)	None	None
Length × width × depth (m)	6070 × 1400 × 1720	5840× 1170× (1400/1600)	3580 × 1050 × (1240–1330)	4000 × 1000 × 1400	5190 × 1109 × 1495	4970 × 1066 × (1460–1317)	6550 × 1152 × (1335–1500)
Anodes/cathodes per cell	56/55, 51/50	55/54	35/34	39/38	50/49	50/49	63/62
Anodes							
Cu (%)	99.3–99.6	98.95	99	98.4	99.37	99.36	99.37
Ag (ppm)	180–350		680	1000	712	762	650
As (ppm)	1500–2000	3000	1700	1000	940	1567	750
Au (ppm)			23	10	67	56.61	22
Bi (ppm)	200–300	1100	328	120	130	184	25
Fe(ppm)	20–50	30	60	10	10	45	20
Ni (ppm)	500–1000	100	5000	6000	1640	2169	950
O (ppm)	1500–2000	1700	2100	1500	1320	1370	1500

Continued

Table 13.5 Selected industrial copper electrorefining data.—cont'd

Refinery	Jinlong Copper Tongling China	Xianguang Copper Tankhouse #2 China	Boliden Harjavalta Oy Pori Finland	Aurubis Lünen Germany	Pan Pacific Copper Saganoseki Japan	LS-Nikko Copper Onsan #1 S. Korea	Uralelektromed Copper Verkhnyaya Pyshma Russia
Pb (ppm)	300–800	3500	343	3000	120	355	800
S (ppm)	40–60	50	10	20	–	10	40
Sb (ppm)	150–300	300	300	1100	220	337	1000
Se (ppm)	150–300	–	532	50	460	570	600
Te (ppm)	30–60	–	138	50	70	163	180
Length × width × thickness (mm)	1089 × 948 × 45	1000 × 960 × 46	940 × 940 × 46	1045 × 846 × 50	1040 × 915 × 47	1183 × 1170 × 51	997 × 914 × 50
Initial mass of anode (kg)	378	385	350	400	375	422	392
Center-line spacing (mm)	100–105	100	98	98	100	98	100
Life (d)	18	15	19	21	18	21	19
Scrap after refining (%)	14.5	14	20	10.5	16.4	13	14.1
Anode slimes (kg/t anode)	4.0–4.5	10	7	15	3.4	3.6	7
Cathodes							
Type	Starter sheet and stainless steel	Stainless steel	Stainless steel	Stainless steel	Stainless steel	Stainless steel	Stainless steel
Plating time (d)	9	5	6–6–7	7	9.0	7	110
Mass Cu plated (total) (kg)	160	110	80–140	2 × 60	75	122	0.6
Cu (%)	99.99	99.997	99.998	99.99	99.99	99.9972	>99.99
Ag (ppm)	5–10	4–7	12	10	14	11.73	8
As (ppm)	0.3	<0.5	0.4	<1	<0.1	0.22	0.3
Bi (ppm)	<1.0	<0.5	0.2	<1	<0.1	0.16	0.3
Fe (ppm)	<1.0	<1	0.3	<1	<0.1	0.21	1
Ni (ppm)	<1.0	<0.5	1	<1	<0.1	0.27	1.3
S (ppm)	<4	2–4	4	4	5.47	10.1	4.5
Sb (ppm)	<1.0	<1	0.3	<1	<0.1	0.1	1
Se (ppm)	<1.0	<0.5	<0.3	<1		0.14	0.6
Te (ppm)	<1.0	<1.5	0.3	<1		0.18	0.6

Electrolyte							
Cu (g/L)	45	45–55	55	45	46.7	43–47	50
H_2SO_4 (g/L)	190	160–200	170	170	168.9	170–180	160
As (g/L)	10	5–15	13	10	3.4	5.7	5
Bi (g/L)	0.05	0.1–0.4	0.45	0	0	0.19	
Cl (g/L)	0.04	0.03–0.07	0.06	0.06	0.06	0.04–0.05	0.05
Fe (g/L)	1	1.0–1.5		0.2	0.42	0.17	0.2
Ni (g/L)	12	8–12	15	20	14.1	16.0	7.5
Sb (g/L)	0.05	0.1–0.3	0.32	0.4	0.2	0.27	0.5
Inlet temperature (°C)	64	64	65	63	66	65	65
Outlet temperature (°C)	62	64	65–68	62	67	63–64	65
Bone glue (g/t Cu cathode)	50	50–100	100	70	75.7	94	95
Thiourea (g/t Cu cathode)	80	50–100	60	70	170.7	130	65
Avitone (g/t Cu cathode)	20						
HCl/NaCl (g/t Cu cathode)	To 0.04 g/L	30–70	60	60 mg/L HCl	91	507	100
Power and energy							
Cathode current density (A/m^2)	310/290	420	310	335	322	323	330
Cathode current efficiency (%)	99	≥99%	88	97	97.1	96.0	98
Cell voltage (mV)	295	400	300–500	360	370	370	370
Power (kWh/t Cu) (plating only)	320	490	445		351	325	375

From Moats, M., Filzwieser, A., Wang, S., Davenport, W. G., Siegmund, A., & Robinson, T. (2019). Global survey of copper electrorefining: 2019 world tankhouse operating data. In: *Copper 2019 (Paper 594940)*. Montreal: Canadian Institute of Mining, Metallurgy and Petroleum.

mechanization and automation throughout refineries, including the use of robots, sophisticated crane electrode-handling systems, anode preparation machines, and cathode handling machines (Larinkari, Nord, & Schmidt, 2019; Morris & Madariaga, 2019).

To lower the carbon footprint of an operation, there is a drive to reduced specific energy consumption by lowering voltage drop across the electrolyte, such as by reducing anode–cathode spacing by increasing anode thickness (mass) or changing the anode shape. A related trend to increase production from an existing infrastructure is the use of a greater proportion of the anode in each cycle and steadily increasing current densities. Many tankhouses now routinely operate at 350 A/m^2 (Moats et al., 2019; Takenaka, Yoneyama, Matsubara, Yamaguchi, & Takebayashi, 2019). Sophisticated electrolyte-flow technologies and acid-mist management systems allow operation at current density as high as 420 A/m^2 (the highest in the industry in 2019) without compromising current efficiency, cathode quality, or operator health (Filzwieser, Hanel, & Filzwieser, 2019; Penna, Villaseca, & Villarroel, 2019).

Other trends in operational practice include (Moats et al., 2019)

- continued adoption of permanent cathode technology;
- better control of anode casting to improve uniformity in size and mass;
- increasing implementation of polymer concrete cells;
- parallel electrolyte flow (METTOP-BRX technology; Filzwieser, Hanel, & Filzwieser, 2019) to allow operation above 400 A/m^2;
- increasing use of short-circuit detection systems and voltage control monitoring;
- use of section switches to electrically isolate cells for harvesting and maintenance.

Treatment of increasingly complex smelter feeds (both ores and secondary materials) is increasing the variety and concentrations of impurity elements deporting to the anodes and the need to recover valuable components from the anode slimes and spent electrolyte. Many of these impurities are considered critical elements (Sb, Se, Te) or are a source of environmental concern (As, Bi, Pb, Cd). These need to be recovered for sale and reuse or disposed of in a stable and sustainable manner.

Increasingly sophisticated chemistry and technologies continue to evolve to ensure high-quality cathode.

13.12 Summary

This chapter has shown that electrolytic refining is the principal method of mass-producing high purity (>99.97%) copper. The other is electrowinning (Chapter 16). Copper from electrorefining, after melting and casting, contains less than 20 ppm impurities, plus oxygen, which is controlled at 0.018%–0.025%.

Electrorefining entails electrochemically dissolving copper from impure copper anodes into an electrolyte containing $CuSO_4$ and H_2SO_4, and then electrochemically depositing pure copper from the electrolyte onto stainless steel or copper cathodes. The process is continuous.

Insoluble impurities in the anode adhere to the anode or fall to the bottom of the refining cell. They are removed and processed for byproduct recovery. Soluble impurities leave the cell in continuously flowing electrolyte. These are removed from an electrolyte bleed stream.

Production of high-purity cathode copper is achieved by precise control of chemical, physical, and electrical tankhouse parameters. These include

(a) evenly spaced, flat, vertical anodes and cathodes;

(b) precise control of temperature, electrolyte composition, supply of leveling and grain-refining agents constant, and flow distribution throughout each cell;

(c) provision of constant controlled current and current distribution to all electrodes and management of short-circuits and other current losses.

Important modern trends are to treat anodes with more complex impurity suites at higher current density, while maximizing cathode purity, productivity, and operator safety and minimizing power consumption, waste production, and environmental damage.

References

Agrawal, A., Bagchi, D., Kumari, S., Kumar, V., & Pandey, B. D. (2007). Recovery of nickel powder from copper bleed electrolyte of an Indian copper smelter by electrolysis. *Powder Technology, 177*, 133–139.

Agrawal, A., Kumari, S., Bagchi, D., Kumar, V., & Pandey, B. D. (2007). Recovery of copper powder from copper bleed electrolyte of an Indian copper smelter by electrolysis. *Minerals Engineering, 20*, 95–97.

Agrawal, A., Manoj, M. K., Kumari, S., Bagchi, D., Kumar, V., & Pandey, B. D. (2008). Extractive separation of copper and nickel from copper bleed stream by solvent extraction route. *Minerals Engineering, 21*, 1126–1130.

Agrawal, A., Bagchi, D., Kumari, S., & Pandey, B. D. (2009). An overview of process options and behavioral aspects of the copper values recovered from copper bleed stream of a copper smelter developed at the National Metallurgical Laboratory. *Mineral Processing and Extractive Metallurgy Review, 30*, 136–162.

Alvear Flores, G., Risopatron, C., & Pease, J. (2020). Processing of complex materials in the copper industry: Challenges and opportunities ahead. *JOM Journal of the Minerals, Metals & Materials Society*. https://doi.org/10.1007/211837-020-04255-9

Arroyo, F., Almansa, A. R., Ruiz, I., González, I., Ríos, G., Pereira, C. F., & Vilches, L. F. (2017). Optimizing operating conditions in an ion-exchange column treatment applied to the removal of Sb and Bi impurities from an electrolyte of a copper electro-refining plant. *Hydrometallurgy, 171*. https://doi.org/10.1016/j.hydromet.2017.06.009

Artzer, A., Moats, M., & Bender, J. (2018). Removal of antimony and bismuth from copper electrorefining electrolyte: Part I—a review. *JOM Journal of the Minerals, Metals and Materials Society, 70*(7). https://doi.org/10.1007/s11837-018-3075-x

Atenas, A. C., & Muñoz, P. A. (2007). Electrorefining high level arsenic cast anode. In G. E. Houlachi, J. D. Edwards, & T. G. Robinson (Eds.), *Copper 2007, vol. V, Copper electrorefining and electrowinning* (pp. 3–12). Montreal: Canadian Institute of Mining, Metallurgy and Petroleum.

Beauchemin, S., Chen, T. T., & Dutrizac, J. E. (2008). Behaviour of antimony and bismuth in copper electrorefining circuits. *Canadian Metallurgical Quarterly, 47*(1), 9–26.

Chen, T. T., & Dutrizac, J. E. (1991). A mineralogical study of anode passivation in copper electrorefining. In W. C. Cooper, D. J. Kemp, G. E. Lagos, & K. G. Tan (Eds.), *Copper 91, vol. III, Hydrometallurgy and electrometallurgy* (pp. 369–389). New York: Pergamon Press.

Chen, T. T., & Dutrizac, J. E. (2000). A mineralogical overview of the behaviour of nickel during copper electrorefining. *Metallurgical Transactions B, 21B*(2), 229–328.

Chen, T. T., & Dutrizac, J. E. (2003). The behaviour of tellurium during the decopperizing of copper refinery anode slimes. In J. E. Dutrizac, & C. G. Clements (Eds.), *Copper 2003, vol. V, Copper electrorefining and electrowinning* (pp. 287–307). Montreal: Canadian Institute of Mining, Metallurgy and Petroleum.

Chen, T. T., & Dutrizac, J. E. (2004). Gold in the electrorefining of copper and the decopperizing of copper anode slimes. *JOM Journal of the Minerals, Metals and Materials Society, 56*(8), 48–52.

Chen, T. T., & Dutrizac, J. E. (2007). Mineralogical characterization of a conventional copper refinery anode slimes treatment circuit. In G. E. Houlachi, J. D. Edwards, & T. G. Robinson (Eds.), *Copper 2007, vol. V, Copper electrorefining and electrowinning* (pp. 159–172). Montreal: Canadian Institute of Mining, Metallurgy and Petroleum.

Conard, B. R., Rogers, B., Brisebois, R., & Smith, C. (1990). Inco copper refinery addition agent monitoring using cyclic voltammetry. In P. L. Claessens, & G. B. Harris (Eds.), *Electrometallurgical plant practice* (pp. 195–209). Warrendale, PA: The Metals Society.

De Maere, C., & Winand, R. (1995). Study of the influence of additives in copper electrorefining, simulating industrial conditions. In W. C. Cooper, D. B. Dreisinger, J. E. Dutrizac, H. Hein, & G. Ugarte (Eds.), *Copper 95, vol. III, Electrorefining and hydrometallurgy of copper* (pp. 267–286). Montreal: Canadian Institute of Mining, Metallurgy and Petroleum.

Doucet, M., & Stafiej, J. (2007). Processing high nickel slimes at CCR refinery. In G. E. Houlachi, J. D. Edwards, & T. G. Robinson (Eds.), *Copper 2007, vol. V, Copper electrorefining and electrowinning* (pp. 173–182). Montreal: Canadian Institute of Mining, Metallurgy and Petroleum.

Elkayar, K., & Kryst, K. (2019). Utilizing ion exchange for impurity control in copper electrolyte. In *Copper 2019 (paper 592548)*. Montreal: Canadian Institute of Mining, Metallurgy and Petroleum.

Filzwieser, A., Hanel, M. B., & Filzwieser, I. (2019). Current density increase within the last 150 years. In *Copper 2019 (paper 597216)*. Montreal: Canadian Institute of Mining, Metallurgy and Petroleum.

Hait, J., Jana, R. K., & Sanyai, S. K. (2009). Processing of copper electrorefining anode slimes: A review. *Mineral Processing and Extractive Metallurgy, 118*(4), 240–252.

Hiskey, J. B. (2019). The historical development of electrolyte additives and their specific role and influence on cathode quality. In *Copper 2019 (paper 599392)*. Montreal: Canadian Institute of Mining, Metallurgy and Petroleum.

Hoffman, J. E. (2004). The purification of copper refinery electrolyte. *JOM Journal of the Minerals, Metals and Materials Society, 56*(7), 30–33.

International Copper Study Group. (2017). *Directory of copper mines and plants up to 2020*. Portugal: Lisbon. Available from: http://www.icsg.org.

Jin, W., Laforest, P. I., Luyima, A., Read, W., Navarro, L. G., & Moats, M. (2015). Electrolytic recovery of bismuth and copper as powder from acidic sulfate effluents using an emew® cell. *RSC Advances, 5*(62). https://doi.org/10.1039/C5RA08318D

Komori, K., Ito, S., Okada, S., & Iwahori, S. (2010). Hydrometallurgical process of precious metals in Naoshima smelter & refinery. In *Copper 2010, vol. 4, Electrowinning and –refining* (pp. 1403–1411). Clausthal-Zellerfeld: Gesellschaft der Metallurgen und Bergleute.

Larinkari, M., Nord, P., & Schmidt, M. (2019). Latest development in cathode stripping machines. In *Copper 2019 (paper 594055)*. Montreal: Canadian Institute of Mining, Metallurgy and Petroleum.

Larouche, P. (2001). *Minor elements in copper smelting and refining*. M. Eng. Thesis. Montreal, Canada: McGill University.

LME. (2021). *London metal exchange: Special contract rules for copper grade A*. Available from: https://www.lme.com/en-GB/Metals/Non-ferrous/Copper/Physical.

Lundström, M., Hannula, P., Yliniemi, K., Wilson, B. P., Barranco, V., Janas, D., & Hubin, A. (2019). Solution purification of copper electrorefining electrolyte – a novel way to recover precious metals. In *Copper 2019 (paper 588813)*. Montreal: Canadian Institute of Mining, Metallurgy and Petroleum.

Mahmoudi, A., Shakibania, S., Mokmeli, M., & Rashchi, F. (2020). Tellurium, from copper anode slime to high purity product: A review paper. *Metallurgical and Materials Transactions B, 51B*(12), 2555–2575.

Moats, M., Filzwieser, A., Wang, S., Davenport, W. G., Siegmund, A., & Robinson, T. (2019). Global survey of copper electrorefining: 2019 world tankhouse operating data. In *Copper 2019 (paper 594940)*. Montreal: Canadian Institute of Mining, Metallurgy and Petroleum.

Moats, M. S., & Hiskey, J. B. (2000). The role of electrolyte additives on passivation behaviour during copper electrorefining. *Canadian Metallurgical Quarterly, 39*(3), 297–306.

Moats, M. S., & Hiskey, J. B. (2006). The effect of anode composition on passivation of commercial copper electrorefining anodes. In K. Kongoli, & R. G. Reddy (Eds.), *Sohn international symposium. Advanced processing of metals and materials, vol. 6. New, improved and existing technologies: Aqueous processing and electrochemical technologies* (pp. 507–518). Warrendale, PA: The Minerals, Metals and Materials Society.

Moats, M. S., & Hiskey, J. B. (2007). The effect of electrolyte composition on passivation of commercial electrorefining anodes. In G. E. Houlachi, J. D. Edwards, & T. G. Robinson (Eds.), *Copper 2007, vol. V, Copper electrorefining and electrowinning* (pp. 47–58). Montreal: Canadian Institute of Mining, Metallurgy and Petroleum.

Moats, M., Aslin, N., Pranowo, A., & Alvear, G. (2016). Behaviour and benefits of arsenic in copper electrorefining. *CIM Journal,* 7(3). https://doi.org/10.15834/cimj.2016.16

Moats, M. S., & Hiskey, J. B. (2010). How anodes passivate in copper electrorefining. In *Copper 2010, vol. 4, Electrowinning and –refining* (pp. 1463–1482). Clausthal-Zellerfeld: Gesellschaft für Bergbau, Metallurgie, Rohstoff- und Unwelttechnik).

Möller, C. A., Friedrich, B., & Bayanmunkh, M. (2010). Effect of As, Sb, Bi and oxygen in copper anodes during electrorefining. In *Copper 2010, vol. 4, Electrowinning and –refining* (pp. 1495–1506). Clausthal-Zellerfeld: Gesellschaft für Bergbau, Metallurgie, Rohstoff- und Unwelttechnik.

Morris, T. T., & Madariaga, R. (2019). A contemporary approach to an age-old process, using robotics to strip starter sheets for use in a conventional copper refinery. In *Copper 2019 (paper 595521)*. Montreal: Canadian Institute of Mining, Metallurgy and Petroleum.

Mubarok, A., Antrekowitsch, H., & Mori, G. (2007a). Passivation behaviour of copper anodes with various chemical compositions. In P. Aryalebechi (Ed.), *Materials processing fundamentals* (pp. 225–234). Warrendale, PA: The Minerals, Metals and Materials Society.

Mubarok, A., Antrekowitsch, H., Mori, G., Lossin, A., & Leuprecht, G. (2007b). Problems in the electrolysis of copper anodes with high contents of nickel, antimony, tin and lead. In G. E. Houlachi, J. D. Edwards, & T. G. Robinson (Eds.), *Copper 2007, vol. V, Copper electrorefining and electrowinning* (pp. 59–76). Montreal: Canadian Institute of Mining, Metallurgy and Petroleum.

Nagai, T. (1997). Purification of copper electrolyte by solvent extraction and ion-exchange techniques. *Mineral Processing and Extractive Metallurgy Review, 17*(4), 143–168.

Nagai, H., Shibata, E., & Nakamura, T. (2017). Development of methods for concentration and dissolution of Rh and Ru from copper slime. *Hydrometallurgy, 169*, 282–289.

Nakano, H., Oue, S., Fukushima, H., & Kobayashi, S. (2010). Synergistic effects of thiourea, polymer additives and chloride ions on copper electrorefining. In *Copper 2010, vol. 4, Electrowinning and –refining* (pp. 1531–1544). Clausthal-Zellerfeld: Gesellschaft für Bergbau, Metallurgie, Rohstoff- und Unwelttechnik.

Nicol, M. J. (2006). *Electrowinning and electrorefining of metals*. Short course. Perth, Australia: Murdoch University.

Noguchi, F., Iida, N., Nakamura, T., & Ueda, Y. (1992). Behavior of anode impurities in copper electrorefining. *Metallurgical Review of the Mining and Metallurgy Institute of Japan, 8*, 83–98.

Penna, F., Villaseca, C., & Villarroel, R. (2019). Applications of Lateralvex® and Lateralflow® technologies in Minera Antucoya, an AMSA tank-house in Chile. In *Copper 2019 (paper 591456)*. Montreal: Canadian Institute of Mining, Metallurgy and Petroleum.

Pranowo, A., Aslin, N. J., Eriksson, P. O., & Heferen, G. J. (2019). Stainless steel "permanent" cathode plates – fixed assets of consumables?. In *Copper 2019 (paper 590965)*. Montreal: Canadian Institute of Mining, Metallurgy and Petroleum.

Price, D. C., & Davenport, W. G. (1981). Physico-chemical properties of copper electrorefining and electrowinning electrolytes. *Metallurgical Transactions B, 12B*, 639–643.

Ramírez, R., Ríos, G., & Martín, J. (2007). Anode slime leaching and tellurium removal at Atlantic Copper refinery. In G. E. Houlachi, J. D. Edwards, & T. G. Robinson (Eds.), *Copper 2007, vol. V, Copper electrorefining and electrowinning* (pp. 125–138). Montreal: Canadian Institute of Mining, Metallurgy and Petroleum.

Rantala, A., Larinkari, M., & Menese, T. (2009). Advanced wireless monitoring and integrated information management for improving electrorefinery performance. In *Hydrocopper 2009* (pp. 102–113). Santiago: Gecamin.

Ríos, G., Ramírez, R., & Arbizu, C. (2010). Management of electrolyte impurities at the Atlantic copper refinery. In *Copper 2010, vol. 4, Electrowinning and –refining* (pp. 1599–1616). Clausthal-Zellerfeld: Gesellschaft für Bergbau, Metallurgie, Rohstoff- und Unwelttechnik.

Rodrigues Bravo, J. L. (2007). Studies on bismuth content variation in Caraiba Metais' electrolytic copper. In G. E. Houlachi, J. D. Edwards, & T. G. Robinson (Eds.), *Copper 2007, vol. V, Copper electrorefining and electrowinning* (pp. 149–158). Montreal: Canadian Institute of Mining, Metallurgy and Petroleum.

Santos Moraes, I. M. (2007). Upsets on antimony contents in electrolytic copper produced at Caraiba Metais. In G. E. Houlachi, J. D. Edwards, & T. G. Robinson (Eds.), *Copper 2007, vol. V, Copper electrorefining and electrowinning* (pp. 139–148). Montreal: Canadian Institute of Mining, Metallurgy and Petroleum.

Sado, A., Ueno, A., Watanabe, K., & Takebayashi, K. (2019). Recent operational improvements at Saganoseki electrorefinery. In *Copper 2019 (paper 592998)*. Montreal: Canadian Institute of Mining, Metallurgy and Petroleum.

Safarzadeh, M. S., Horton, M., & Rythoven, A. D. V. (2018). Review of recovery of platinum group metals from copper leach residues and other resources. *Mineral Processing and Extractive Metallurgy Review, 39*, 1–17.

Sharma, A., Rao, B. V., Patel, N. R., Adiga, R., Izatt, N. E., Bruening, R. L., ... Izatt, S. R. (2019). Copper refining electrolyte purification using SuperLig® molecular recognition technology for bismuth removal – the experience at Birla Copper. In *Copper 2019 (paper 587553)*. Montreal: Canadian Institute of Mining, Metallurgy and Petroleum.

Shaw, D. R. (2019). Ion exchange in hydrometallurgy. In B. Wassink, & E. Asselin (Eds.), *Copper hydrometallurgy: Principles and practice* (pp. 189–212). Montreal: Canadian Institute of Mining, Metallurgy and Petroleum.

Sole, K. C., Mooiman, M. B., & Hardwick, E. (2017). Ion exchange in hydrometallurgical processing: An overview and selected applications. *Separation and Purification Reviews, 47*(2), 159–178.

Stantke, P. (2002). Using CollaMat to measure glue in copper electrolyte. *JOM Journal of the Metals Society, 54*(4), 19–22.

Stantke, P., & Leuprecht, G. (2010). Process change for the treatment of spent electrolyte at Aurubis Hamburg. In *Copper 2010, vol. 4, Electrowinning and –refining* (pp. 1663–1674). Clausthal-Zellerfeld: Gesellschaft für Bergbau, Metallurgie, Rohstoff- und Unwelttechnik.

Stevens, G. W., & Gottliebsen, K. (2000). The recovery of sulphuric acid from copper tankhouse electrolyte bleeds—a case study on the development of solvent extraction processes. In *Minprex 2000 international congress on mineral processing and extractive metallurgy* (pp. 57–64). Victoria: Australasian Institute of Mining and Metallurgy.

Stoyanova, V., Nedeleva, K., Sraev, A., Gerov, K., & Stoyanova, V. (2016). Impurities management in Aurubis Bulgaria refinery through flexible operation control. In *Copper 2016 (paper EL1-5)*. Tokyo: Mining and Materials Processing Institute of Japan.

Szymanowski, J. (1998). Removal of toxic elements from copper electrolyte by solvent extraction. *Mineral Processing and Extractive Metallurgy Review, 18*(3–4), 389–418.

Takenaka, K., Yoneyama, T., Matsubara, S., Yamaguchi, Y., & Takebayashi, M. (2019). High current density operation at Toyo tankhouse. In *Copper 2019 (paper 606676)*. Montreal: Canadian Institute of Mining, Metallurgy and Petroleum.

Tantavichet, N., Damronglerd, S., & Chailapakul, O. (2009). Influence of the interaction between chloride and thiourea on copper electrodeposition. *Electrochimica Acta, 55*, 240–249.

Wang, C.-T., & O'Keefe, T. J. (1984). The influence of additives and their interactions on copper electrorefining. In P. E. Richardson, S. Srinivasan, & R. Woods (Eds.), *International symposium on electrochemistry in mineral and metal processing* (Vol. 84–10, pp. 655–670). Pennington, NJ: The Electrochemical Society.

Tetsuka, D., & Okamoto, H. (2019). Effect of antimony, nickel and sulfuric acid in copper electrorefining. In *Copper 2019 (paper 593000)*. Montreal: Canadian Institute of Mining, Metallurgy and Petroleum.

Wang, S., & Kim, D. (2018). *Process for the recovery of gold from anode slimes*. US Patent No. US 2018/0021743 A1.
Wenzl, C., Filzwieser, I., Filzwieser, A., & Anzinger, A. (2010). Newest developments using the METTOP-BRX technology. In *Copper 2010, vol. 4, Electrowinning and —refining* (pp. 1713—1722). Clausthal-Zellerfeld: Gesellschaft für Bergbau, Metallurgie, Rohstoff- und Unwelttechnik.
Wesstrom, B. (2014). The effects of high antimony in electrolyte. In *Bill Davenport honorary symposium, conference of metallurgists COM 2014*. Montreal: Canadian Institute of Mining, Metallurgy and Petroleum.
Wesstrom, B. C., Olave, B., & Ramirez, A. (2019). Freeport McMoRan El Paso refinery electrolyte management. In *Copper 2019 (paper 595381)*. Montreal: Canadian Institute of Mining, Metallurgy and Petroleum.
Xiao, F., Mao, J., Cao, D., Shen, X., & Volinsky, A. A. (2012). The role of trivalent arsenic in removal of antimony and bismuth impurities from copper electrolytes. *Hydrometallurgy, 125—126*, 76—80.
Xing, W. D., & Lee, M. S. (2019). Development of a hydrometallurgical process for the recovery of gold and silver powders from anode slime containing copper, nickel, tin, and zinc. *Gold Bulletin, 52*, 69—77.
Xing, W. D., Sohn, S. H., & Lee, M. S. (2020). A review on the recovery of noble metals from anode slimes. *Mineral Processing and Extractive Metallurgy Review, 41*(2), 130—143.
Zeng, W., Wang, S., & Free, M. L. (2016). Experimental studies of the effects of anode composition and process parameters on anode slimes adhesion and cathode copper purity by performing electrorefining in a pilot-scale cell. *Metallurgical and Materials Transactions B, 47*, 3178—3191.

Suggested reading

Aromaa, J., Forsén, O., Lundström, & Latostenmaa, P. (2019). One hundred years of electrolytic copper production in Finland. In *Copper 2019 (paper 582049)*. Montreal: Canadian Institute of Mining, Metallurgy and Petroleum.
Baranek, W., Chmielarz, A., & Śmieszek, Z. (2007). Development of electrorefining processes in the Polish copper industry. In G. E. Houlachi, J. D. Edwards, & T. G. Robinson (Eds.), *Copper 2007, vol. V, Copper electrorefining and electrowinning* (pp. 335—344). Montreal: Canadian Institute of Mining, Metallurgy and Petroleum.
Culver, W. W., Mackey, P. J., Wesstrom, B., & Wraith, A. E. (2019). The first electrolytic copper refinery in the USA at the Chemical Copper Co. Phoenixville, PA — history revisited. In *Copper 2019 (paper 588113)*. Montreal: Canadian Institute of Mining, Metallurgy and Petroleum.
Djurov, I. (2010). The new ISA 2000 refinery of Aurubis in Pirdop. In , *4. Copper 2010, Electrowinning and —refining* (pp. 1307—1325). Clausthal-Zellerfeld: Gesellschaft der Metallurgen und Bergleute.
Dufresne, R. P. (2019). Technological advances in the electrolytic cell, maintenance and replacement. In *Copper 2019 (paper 582911)*. Montreal: Canadian Institute of Mining, Metallurgy and Petroleum.
Ikemoto, T., Yamashita, T., & Inoue, O. (2019). Electrolytic refining of high Ni copper anode at the Tankhouse in Naoshima smelter and refinery. In *Copper 2019 (paper 594797)*. Montreal: Canadian Institute of Mining, Metallurgy and Petroleum.
Nakayama, S., Fujihara, M., Narita, M., & Saski, Y. (2019). Recent improvements at Tamano refinery. In *Copper 2019 (paper 593955)*. Montreal: Canadian Institute of Mining, Metallurgy and Petroleum.
Okabe, S., & Mackey, P. J. (2019). Pioneer days of electrolytic copper refining in Japan: From the first copper refinery up to the present day. In *Copper 2019 (paper 577350)*. Montreal: Canadian Institute of Mining, Metallurgy and Petroleum.
Tan, K., & Hu, C. (2019). History of China's electrolytic copper production. In *Copper 2019 (paper 599250)*. Montreal: Canadian Institute of Mining, Metallurgy and Petroleum.
Wraith, A. E., Mackey, P. J., & Protheroe Jones, J. (2019). Origins of electrorefining: Birth of the technology and the world's first commercial electrorefinery. In *Copper 2019 (paper 605404)*. Montreal: Canadian Institute of Mining, Metallurgy and Petroleum.

CHAPTER 14

Hydrometallurgical copper extraction: introduction and leaching

Previous chapters describe the processes of concentration, pyrometallurgy, and electrorefining that convert copper sulfide ores and concentrates into high-purity electrorefined copper. These processes account for about 80% of primary copper production.

The remaining 20% of primary copper production comes from *hydrometallurgical* processing of mainly copper oxide and chalcocite ores. This copper is recovered by leaching (this chapter), followed by solvent extraction (SX) (Chapter 15) and electrowinning (EW) (Chapter 16). The final product is electrowon cathode copper, which is equal to or often greater in purity than electrorefined copper.

In 2020, about 4.2 million tonnes per year of cathode copper were produced using hydrometallurgical processes.

14.1 Copper recovery by hydrometallurgical flowsheets

Modern hydrometallurgical flowsheets all comprise three essential steps for the recovery of copper: leaching, followed by SX and EW.

Leaching involves dissolving Cu^{2+} (or Cu^{+}) from copper-containing minerals into an aqueous H_2SO_4 solution, known as the *lixiviant,* to produce a *pregnant leach solution* (PLS). In addition to copper, the PLS will also contain other impurity species, such as Fe, Al, Co, Mn, Zn, Mg, and Ca that may be present in the ore and are leached with the copper. The leach *residue* (solids remaining after leaching) contains *gangue* or "waste" minerals, such as alumina, silica, and insoluble iron oxides/hydroxides/sulfates. The gangue is disposed of in tailings dams or dumps. The PLS is fed to the SX circuit.

SX treats the impure PLS to *purify* and *upgrade* the solution to produce a high-purity copper electrolyte suitable for EW of copper metal. In the *extraction* step, copper is selectively loaded into an organic solvent. This contains an extractant that reacts selectively with copper over other metal cations present in the PLS. The copper-barren aqueous *raffinate* leaving the extraction circuit is higher in acid and is returned to the leach circuit as the lixiviant. In the *strip* step, copper is stripped from the *loaded organic* solvent into the *advance electrolyte* from which copper is electrowon. The *stripped organic* solvent is recycled to the extraction step.

Extractive Metallurgy of Copper, Sixth Edition
ISBN 978-0-12-821875-4, https://doi.org/10.1016/B978-0-12-821875-4.00003-1

In *electrowinning*, Cu^{2+} in the purified advance electrolyte from SX is reduced to copper metal at the cathode by the application of a DC electrical current. Sulfuric acid, produced at the anode of the EW cell, is returned to the SX circuit in the copper-depleted *spent electrolyte* to strip more copper from the loaded organic solvent.

The integration of these three hydrometallurgical process steps is illustrated in Fig. 14.1.

14.2 Chemistry of the leaching of copper minerals

Leaching involves solubilizing the valuable components of an ore in an aqueous solution. In the case of copper minerals, sulfuric acid is almost always used as the leaching medium. Oxide copper minerals dissolve readily in acid; some sulfides, such as chalcocite and covellite, also dissolve under atmospheric conditions. More refractory sulfides (chalcopyrite and bornite) usually require high temperatures and pressures to break down the crystal lattices to release copper from the minerals into solution.

14.2.1 Leaching of copper oxide minerals

Fig. 14.2 shows the Pourbaix (potential pH) diagram for the Cu—O—S system. Copper can be dissolved as Cu^{2+} under mildly acidic conditions (<pH 5). Examples of some copper minerals that leach in this manner are as follows:

$$\text{Tenorite}: CuO + H_2SO_4 \rightarrow Cu^{2+} + SO_4^{2-} + H_2O \quad (14.1)$$

$$\text{Azurite}: 2CuCO_3 \cdot Cu(OH)_2 + 3H_2SO_4 \rightarrow 3Cu^{2+} + 3SO_4^{2-} + 2CO_2 + 4H_2O \quad (14.2)$$

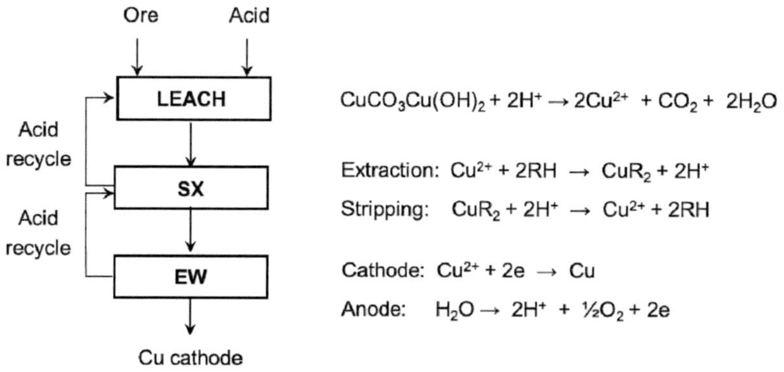

Figure 14.1 Simplified hydrometallurgical flowsheet, showing the acid balance between leaching, solvent extraction, and electrowinning (Sole, 2008). The leaching reaction is shown for malachite, but similar reactions can be written for other minerals. RH is the organic extractant used in solvent extraction. Solvent extraction and electrowinning are described in Chapters 15 and 16, respectively.

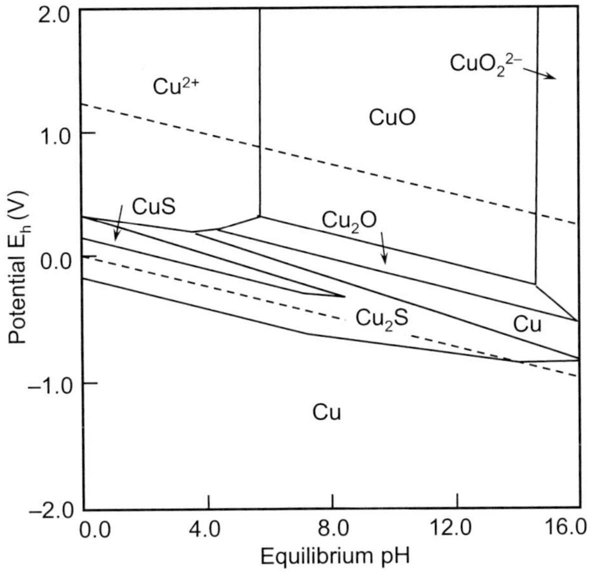

Figure 14.2 Pourbaix diagram of the Cu—S—O—H_2O system at 25°C. [Cu] = [S] = 10^{-4} *M* (House, 1987). Minerals such as CuO can be leached by lowering the pH (adding acid). Sulfides and native copper require an oxidizing agent and acid.

$$\text{Malachite: } CuCO_3 \cdot Cu(OH)_2 + 2H_2SO_4 \rightarrow 2Cu^{2+} + 2SO_4^{2-} + CO_2 + 3H_2O \tag{14.3}$$

$$\text{Cuprite: } Cu_2O + H_2SO_4 \rightarrow Cu^{2+} + SO_4^{2-} + Cu + H_2O \tag{14.4}$$

$$\text{Chrysocolla: } CuO \cdot SiO_2 + H_2SO_4 \rightarrow Cu^{2+} + SO_4^{2-} + SiO_2 \cdot nH_2O + (3-n)H_2O \tag{14.5}$$

14.2.2 Leaching of copper sulfide minerals

Elemental copper and copper sulfide minerals require the use of acid and an oxidizing agent to break the mineral lattice and release Cu^{2+} into solution.

Elemental copper found in nature can be leached by either oxygen or ferric ion:

$$\text{Metallic copper: } Cu + \tfrac{1}{2}O_2 + H_2SO_4 \rightarrow Cu^{2+} + SO_4^{2-} + H_2O \tag{14.6a}$$

$$Cu + 2Fe^{3+} + 3SO_4^{2-} \rightarrow Cu^{2+} + 2Fe^{2+} + 3SO_4^{2-} \tag{14.6b}$$

All copper sulfides require the presence of Fe^{3+} and O_2 as oxidizing agents for leaching to occur. The copper sulfide is oxidized by Fe^{3+}. The resulting Fe^{2+} is reoxidized to Fe^{3+} by O_2. The Fe(II)/Fe(III) redox couple acts in a catalytic manner in these reactions (shown for elevated temperature reaction conditions where the sulfur-containing product is sulfate, rather than the elemental sulfur that forms under ambient conditions):

$$\text{Chalcocite: (i)} \quad Cu_2S + Fe_2(SO_4)_3 \rightarrow Cu^{2+} + SO_4^{2-} + CuS + 2FeSO_4 \tag{14.7a}$$

$$\text{(ii)} \quad 4FeSO_4 + O_2 + 2H_2SO_4 \rightarrow 2Fe_2(SO_4)_3 + 2H_2O \tag{14.7b}$$

$$\text{(overall)} \quad Cu_2S + {}^1\!/\!_2O_2 + H_2SO_4 \rightarrow Cu^{2+} + SO_4^{2-} + CuS + H_2O \tag{14.7c}$$

$$\text{Covellite: (i)} \quad CuS + Fe_2(SO_4)_3 \rightarrow Cu^{2+} + SO_4^{2-} + 2FeSO_4 + S \tag{14.8a}$$

$$\text{(ii)} \quad 4FeSO_4 + O_2 + 2H_2SO_4 \rightarrow 2Fe_2(SO_4)_3 + 2H_2O \tag{14.8b}$$

$$\text{(overall)} \quad CuS + {}^1\!/\!_2O_2 + H_2SO_4 \rightarrow Cu^{2+} + SO_4^{2-} + S + H_2O \tag{14.8c}$$

Under some conditions (especially biological leaching), elemental sulfur that forms as a reaction product can be converted to sulfuric acid:

$$\text{Sulfur: } 2S + 3O_2 + 2H_2O \rightarrow 2H_2SO_4 \tag{14.9}$$

A Pourbaix diagram for the Cu—Fe—O—S system is given in Fig. 14.3. Very refractory minerals, such as chalcopyrite, require high temperatures and pressures to enable leaching to take place at economically viable rates (Section 14.8). The leaching reactions are complex and vary depending on the specific experimental conditions, including the sample mineralogy, gangue minerals, temperature, acid concentration, oxidation—reduction potential, and the presence of ligands (Asselin, 2019; Hiskey, 1993; Lazarro & Nicol, 2003; Nicol & Lazarro, 2003; Yue & Asselin, 2014). One typical series of reactions that occur under oxidizing conditions and at high temperatures is shown below:

$$\text{Chalcopyrite: (i)} \quad 2CuFeS_2 + 4Fe_2(SO_4)_3 \rightarrow 2Cu^{2+} + 2SO_4^{2-} + 10FeSO_4 + 4S \tag{14.10a}$$

$$\text{(ii)} \quad 4FeSO_4 + O_2 + 2H_2SO_4 \rightarrow 2Fe_2(SO_4)_3 + 2H_2O \tag{14.10b}$$

$$\text{(iii)} \quad 2S + 3O_2 + 2H_2O \rightarrow 2H_2SO_4 \tag{14.10c}$$

$$\text{(overall)} \quad 4CuFeS_2 + 17O_2 + 4H_2O \rightarrow 4Cu^{2+} + 4SO_4^{2-} + 2Fe_2O_3 + 4H_2SO_4 \tag{14.10d}$$

14.3 Leaching methods

The main methods employed for the leaching of copper minerals are as follows:

- *percolation* leaching processes, which include heap, dump, and vat leaching;
- *agitation* leaching, which takes place in a series of stirred tanks at close to ambient temperature and pressure;

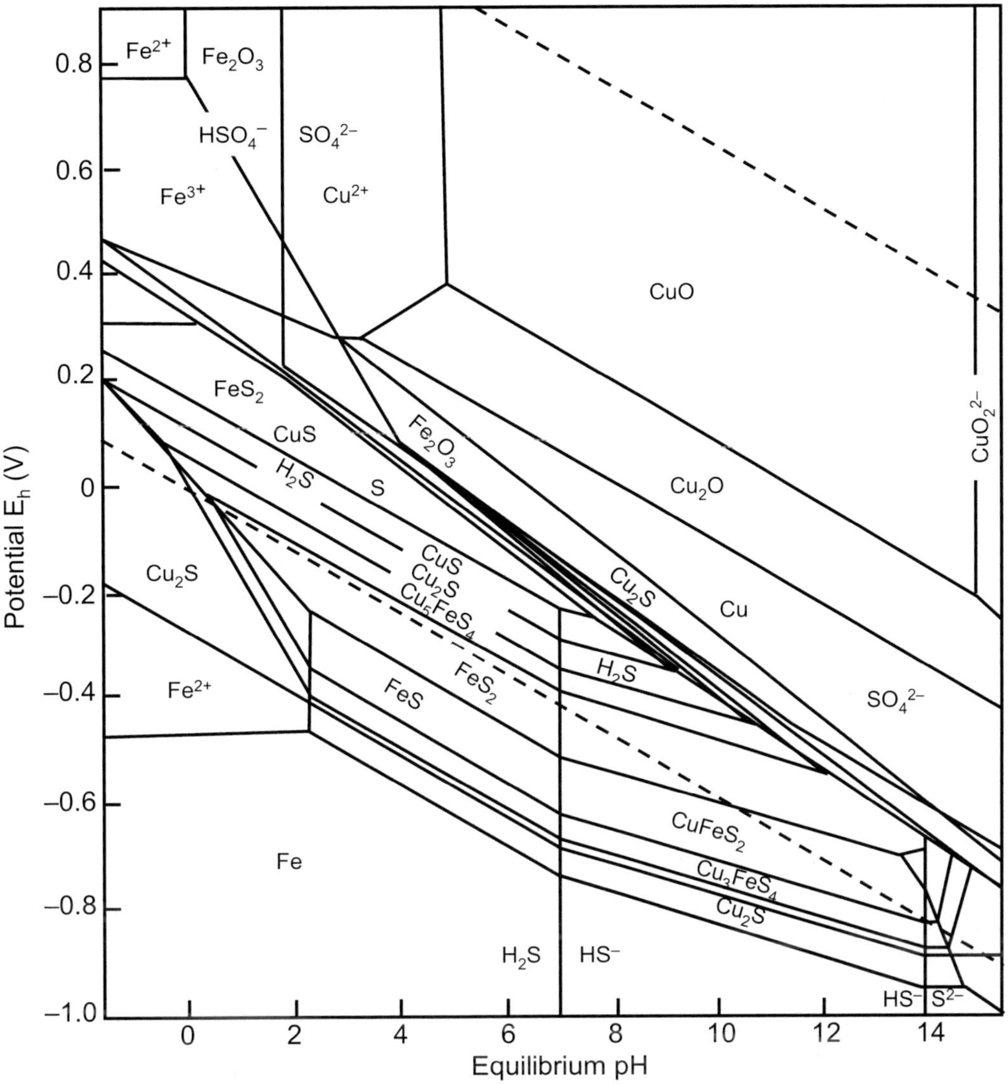

Figure 14.3 Pourbaix diagram of the Cu—Fe—S—O—H_2O system at 25°C. [Cu] = 0.01 *M*; [Fe] = [S] = 0.1 *M* (Hayes, 1993, pp. 235—251; Peters, 1976). Most minerals require the application of both an acidic environment and an oxidizing agent to leach copper as Cu^{2+}.

- *pressure oxidation* (POx) leaching, which uses high temperatures and pressures; and
- *in situ* leaching (ISL), in which the minerals are leached as they naturally occur in the Earth's crust, i.e., without the need to mine the ore.

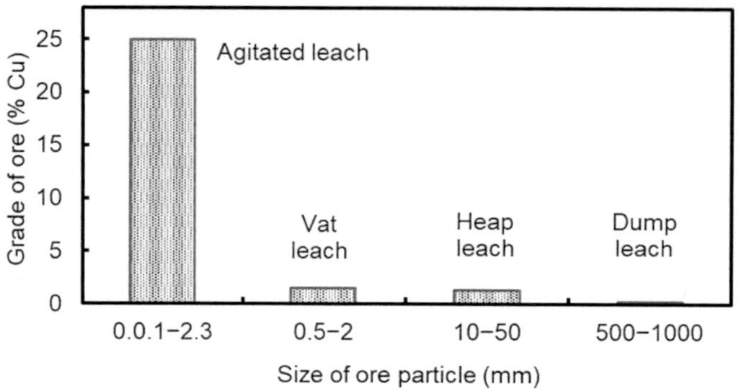

Figure 14.4 Relationship between copper grade and particle size of ore for different methods of atmospheric leaching.

The choice of leaching method depends on the copper grade of the ore, the mineralogy of the copper-bearing ore, and the particle size to which is it reduced (Fig. 14.4). The topography of the mine site, geographical location of the ore body, climatic conditions, and prevailing economic conditions also influence this decision.

14.4 Heap leaching

Heap leaching is the most important method of hydrometallurgical copper extraction. Heap leaching is used for treating oxide and lower-grade secondary sulfide ores that contain up to ~2% Cu. The ore is crushed to a uniform particle size (typically 12–50 mm), often agglomerated, and then stacked on large flat-topped heaps in a controlled manner. Heaps are typically ~7 m high and have a surface area of 0.01–1 km^2. The heap has an impermeable pad at the base. The heap is irrigated with a sulfuric acid solution under normal atmospheric conditions. The solution is distributed evenly over the surface of the ore mass. It trickles through the ore under the influence of gravity and copper dissolves into solution by Reactions (14.1) through (14.8). The copper-containing PLS is collected on a sloped impermeable liner at the base of the heap and directed to a PLS storage pond. Copper is recovered from the PLS by SX (Chapter 15) and EW (Chapter 16) to produce metallic copper. The acid-rich raffinate from SX is returned to the heap for further leaching. A recirculating *intermediate leach solution* (ILS) system is employed to manage the copper concentration of the PLS that is fed forward to SX. Fig. 14.5 shows the flow configuration and solution compositions for a typical heap-leach flowsheet.

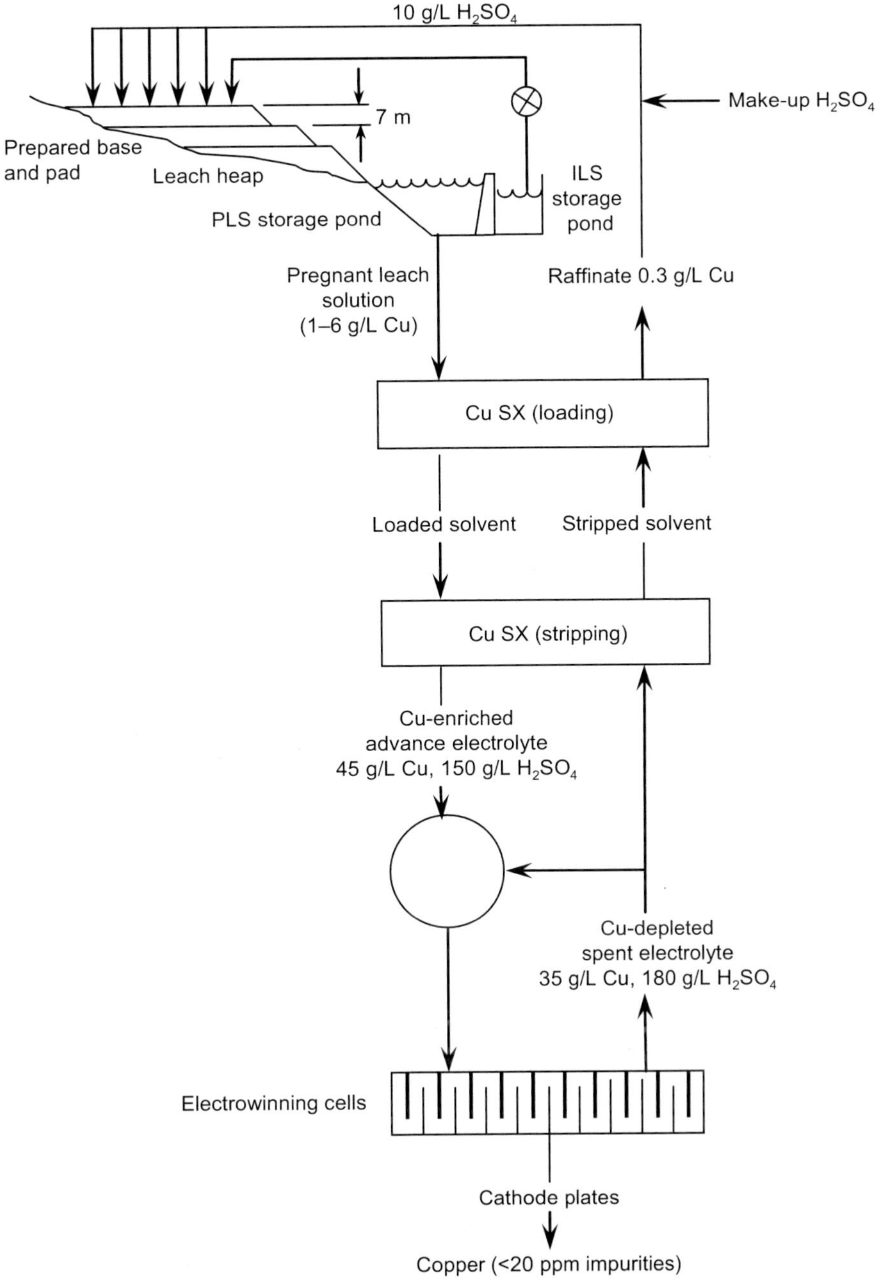

Figure 14.5 Copper heap leach/solvent extraction/electrowinning flowsheet.

14.4.1 Chemistry of heap leaching

The oxide ores listed in Table 14.1 and secondary sulfides, such as chalcocite, are readily leached. Bornite, covellite, and native copper are also leached under biological oxidizing conditions. Chalcopyrite is not leached to any significant extent under the relatively mild conditions of heap leaching (Section 14.8).

As discussed in Section 14.2, nonsulfide copper minerals are leached directly by H_2SO_4 according to reactions such as follows:

$$CuO + H_2SO_4 \rightarrow Cu^{2+} + SO_4^{2-} + H_2O \quad (14.1b)$$

Leaching of sulfide minerals requires an oxidizing agent as well as H_2SO_4. The oxidizing agent is usually dissolved O_2 from air, such as follows:

$$Cu_2S + \frac{1}{2}O_2 + H_2SO_4 \rightarrow CuS + CuSO_4 + H_2O \quad (14.7d)$$

Heap leaching of sulfides is assisted by microorganisms. Bacteria, which occur naturally in the ore, act as a catalyst for the leaching reaction. The bacterial action increases the reaction rate so that the time frame required for leaching of these minerals becomes economic (Section 14.4.1.2).

Table 14.1 Copper minerals normally found in leach heaps.

Type	Common minerals	Mineral formula
Secondary minerals		
Carbonates	Azurite	$2CuCO_3 \cdot Cu(OH)_2$
	Malachite	$CuCO_3 \cdot Cu(OH)_2$
Hydroxychlorides	Atacamite	$Cu_2Cl(OH)_3$
Hydroxysilicates	Chrysocolla	$CuO \cdot SiO_2 \cdot 2H_2O$
Native copper	Metal	Cu^0
Oxides	Cuprite	Cu_2O
	Tenorite	CuO
Sulfates	Antlerite	$CuSO_4 \cdot 2Cu(OH)_2$
	Brochantite	$CuSO_4 \cdot 3Cu(OH)_2$
Supergene sulfides	Chalcocite	Cu_2S
	Covellite	CuS
	Bornite	Cu_5FeS_4
Primary sulfide minerals		
Hypogene sulfides	Chalcopyrite	$CuFeS_2$
	Enargite	Cu_3AsS_4
	Pyrite (source of Fe^{2+}, Fe^{3+}, and H_2SO_4)	FeS_2

Carbonates, oxides, hydroxychlorides, hydroxysilicates, and sulfates are generally referred to as "oxide" ores. They leach quickly. Chalcocite also leaches reasonably rapidly, while bornite and covellite leach slowly. Chalcopyrite is not easily leached.

14.4.1.1 Oxidation by Fe^{3+}

Reaction (14.7c) represents the overall sulfide leaching reaction; however, the ferric ion, Fe^{3+}, is necessary for rapid leaching. Iron-containing minerals in the ore, such as pyrite (FeS_2), are oxidized by Fe^{3+} ions in the presence of *sulfur-oxidizing bacteria*, releasing Fe^{2+} ions. The Fe^{2+} ions are rapidly reoxidized to Fe^{3+} by oxygen and catalyzed by *ferrous-oxidizing bacteria* on the surface of the pyrite to maintain a high redox potential. Pyrite is not only a source of Fe^{3+}, but is also an important source of acid (Eq. 14.11a). Copper sulfides (and native copper) are leached by Fe^{3+}, also assisted by direct bacterial action:

$$2FeS_2 + 7O_2 + 2H_2O \rightarrow 2Fe^{2+} + 2SO_4^{2-} + 2H_2SO_4 \quad (14.11a)$$

$$O_2 + 4Fe^{2+} + 4SO_4^{2-} + 2H_2SO_4 \rightarrow 4Fe^{3+} + 6SO_4^{2-} + 2H_2O \quad (14.11b)$$

$$Cu_2S + 10Fe^{3+} + 15SO_4^{2-} + 4H_2O \rightarrow 2Cu^{2+} + 10Fe^{2+} + 12SO_4^{2-} + 4H_2SO_4 \quad (14.11c)$$

The Fe^{2+} ions produced by Reaction (14.11c) are then reoxidized by Reaction (14.11b) and the process becomes cyclic. Some direct oxidation (Eq. 14.7c) may also occur. PLS from heap leaching typically contains 1–5 g/L Fe.

14.4.1.2 Bacterial action

The above reactions can proceed without bacterial action, but they are accelerated by orders of magnitude by the enzyme catalysts produced by bacteria. The catalytic actions are most commonly attributed to *Acidothiobacillus ferrooxidans*, *Leptospirillum ferriphilumooxidans*, and *Acidothiobacillus thiooxidans*. At moderate temperatures (40–45°C), *Acidothiobacillus caldus* prevails, while at elevated temperature (>60°C), *Sulfolobus metallicus* and *Metallosphaera* spp are the dominant bioleaching strains (Brierley, 2016; Brierley & Brierley, 2013; Guézennec, 2019; Watling, 2006). The bacteria are rod shaped, 0.5 μm × 2 μm long, or spherical and ~1 μm diameter (Fig. 14.6).

Bacteria are indigenous to both oxide and sulfide ores and their surrounding aqueous environment. They are present in leach heaps in the order of 10^{12} bacteria per tonne of ore (Brierley & Brierley, 2019). Like all bacteria, they adapt readily to changes in their environment.

The following conditions promote optimum activity of the bacteria:

(a) lixiviant pH between 1.5 and 6 (optimum ~ pH 2);

(b) temperature between 5 and 45°C (optimum ~30°C; heat is often generated in leach heaps and dumps by exothermic sulfide oxidation reactions);

(c) an adequate O_2 supply, often obtained by gently blowing air through perforated pipes beneath sulfide ore leach heaps (Connelly, 2019, 2020);

(d) no organics in lixiviant or heap, so adequate removal of organic entrainment from the SX raffinate is essential (Chapter 15).

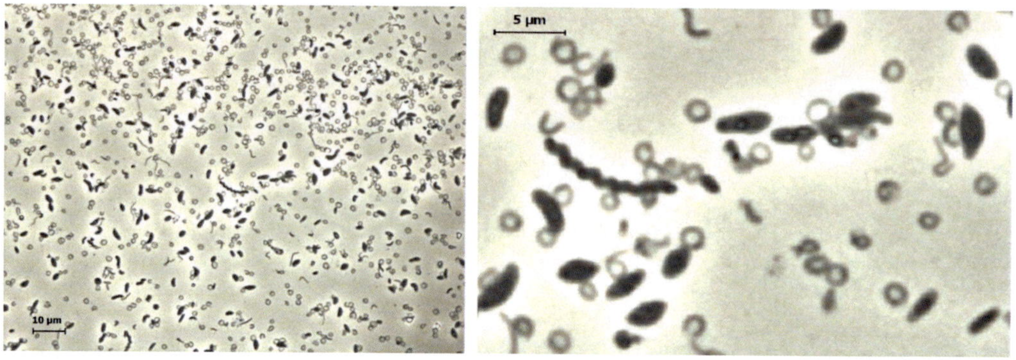

Figure 14.6 Naturally occurring mixed bacterial culture found in copper sulfide heap leaching. Note the presence of both rod-shaped and round bacteria. *(Courtesy of F. Perrot, CSIRO Australian Minerals Resource Centre.)*

Bacteria may also need small amounts of minor nutrients, such as NH_4^+ and PO_4^{3-}, for cell growth (Brierley & Brierley, 2019). Once leaching has started, the bacteria adapt so that the nutrients they need are provided by minerals in the ore: nitrogen is generally sourced from blasting chemicals used in the mine and phosphorus from the leaching of phosphates in the ore. If the ore is phosphorus-deficient, phosphate may be added to the leach solution.

Bacteria are sensitive to the presence of chloride, nitrate, and total dissolved solids (TDS) in the leach solution. Where possible, it is preferable to use high-quality water for make-up of the lixiviant, because this enhances the rate of leaching. Bacteria are, however, very adaptable: some Chilean operations have chloride contents up to 40 g/L; some Australian operations have up to 10 g/L TDS in the make-up water; and it is not unusual to have 200 g/L total sulfate content in heap leach solutions.

14.4.1.3 Effect of chloride

Chloride can occur in heap-leach solutions, either by dissolution of chloride-containing minerals, such as atacamite ($Cu_2Cl(OH)_3$), or by addition of NaCl in the agglomeration step. Chloride ions form stable complexes with both Cu^+ and Cu^{2+}. The Cu(I)/Cu(II) couple can act as a more favorable redox catalyst than the Fe(II)/Fe(III) couple in the leaching of sulfide minerals (Nicol and Akilan, 2016; Nicol & Basson, 2017; Nicol et al., 2010; Velásquez-Yévenes, Nicol, & Miki, 2010; Velásquez-Yévenes, & Miki, & Nicol, 2010). Regeneration of Cu^+ by dissolved oxygen occurs rapidly at ambient temperature (Petersen & van Staden, 2019):

$$4Cu^+ + O_2 + 4H^+ \rightarrow 4Cu^{2+} + 2H_2O \tag{14.12}$$

This concept is applied to heap leaching of sulfide materials at selected sites in Chile.

14.4.1.4 Rate of leaching

Oxide minerals leach relatively rapidly under heap leach conditions, achieving copper recoveries of 75%–95% within 30–100 days (Petersen & van Staden, 2019). Leaching of secondary sulfides is much slower (Dixon & Liu, 2019). Depending on the mineralogy of a heap, leach times may vary from ~90 days to 3 years to yield copper recoveries of 70%–90% (Ghorbani et al., 2016). Chalcopyrite hardly reacts at all in heap leaching.

Heap leaching is diffusion controlled. Long leach times are due to geotechnical characteristics of the ore, especially the presence of clays and micas (Miller, 2003). Kinetics are also influenced by porosity of the heap and access of the lixiviant and oxygen (Petersen & van Staden, 2019).

14.4.2 Industrial heap leaching

Table 14.2 provides details of selected heap leach operations in different locations.

14.4.2.1 Construction of a heap

Leach heaps can be either *multilift permanent* heaps (Fig. 14.7A) or *on/off* (*dynamic*) heaps (Fig. 14.7B). *Multilift heaps* consist of an initial lift built on an impermeable surface. After this lift has been leached, subsequent lifts are built on top of the first lift. Each lift is 5–8 m high; the total heap height can be up to 200 m (Connelly & West, 2009). *Dynamic heaps* consist of a single lift (7–8 m high) built on an impermeable surface. The leached material is removed after leaching and replaced by a new lift of fresh ore. Dynamic pads are the most economic installation for large-tonnage operations (5–7 million t/a). When the ore contains clay and will not support equipment to stack over the leached lift, a dynamic pad is required.

Multilift heaps have the advantages that (Breitenbach, 2005; Iasillo & Schlitt, 1999; Scheffel, 2002) (i) the ore needs to be moved only once (onto the heap) and (ii) lixiviant can flow through all the lifts until leaching is moved to another area, permitting additional recovery of copper from slower-leaching minerals in the lower lifts (More often, however, an intermediate liner is placed on top of the leached ore to limit the acid consumed by the gangue in the lower lift). The disadvantages of a multilift heap are that construction of a heap, which may ultimately become 200 m high, requires a strong impermeable base and versatile heap-building equipment. The initial base area must be large enough to hold all of the ore that will be mined and leached during the life of the operation. There can be significant copper inventory locked up in the heap for extended periods of time. Control of the internal solution head and geotechnical stability of the heap can also be challenging (Miller, 2003).

Dynamic heaps have the advantages that they are simple to construct, it is easier to control the leaching at optimum conditions (Section 14.4.2.7), their base need not be as strong as those needed for multilift heaps, and the aeration and pipework can be maintained when ore is emptied from the pad. The ability to regularly remove tailings samples

Table 14.2 Selected industrial heap leach data.

Operation	Cerro Verde, Peru	El Abra, Chile	Mantoverde, Chile	Morenci, USA	Spence, Chile
Start-up date	1977	1996	1995	1987	2006
Cathode capacity (t/a)	90,000	194,000	65,000	366,000	200,000
H_2SO_4 consumption (t/t Cu)	1.5	3–4	12.2		3.3
Ore feed to leach					
Predominant Cu minerals	Chalcocite, covellite, chalcopyrite	Chrysocolla	Brochantite, chrysocolla	Chalcocite, chrysocolla	Atacamite, chalcocite
Average Cu grade (%)	0.64	0.63	0.61	0.26	1.18, 1.13
Average leachable Cu (%)	0.54	0.55	0.50	0.23	0.88
Leachable Cu recovered to PLS (%)	85	78	82.35	53	70–75
Ore preparation					
Crush (Y/N)	Y	Y	Y	Y	Y
Agglomeration in rotating drum (Y/N)	Y	Y	Y	Y	Y
Ore size on heap (mm)	10%–15% +9.5	80% <15	95% −15	80% −12 mm crushed, 300 mm ROM	70%–75% 12.5, 95%–19

Heap					
Permanent or on−off?	Permanent	On−off	On−off	Permanent	On−off
Building machinery	Conveyer stacking	Conveyer stacking	Conveyer stacking	Conveyer stacking	Conveyer stacking
Area under leach (m^2)	1,000,000	30,000	549,000	1,300,000	500,000 and 1,800,000
# Cells under leach	45	80	7	44	16
Typical lift height (m)	7	8	5	4.5	10
Aeration pipes beneath heap (Y/N)	Y	N	N	Y	Y
Liner material	HDPE	HDPE	HDPE	Compacted soil liner	HDPE
Liner thickness (mm)	1.5	1.5	1.5		1.5
Acid cure (Y/N)					
On heap or during agglomeration?	Y - rotating drum	Y - rotating drum	Y - rotating drum	Y - rotating drum	Y - rotating drum
H_2SO_4 in cure solution (g/L)	Conc. H_2SO_4	Conc. H_2SO_4	Conc. H_2SO_4	Conc. H_2SO_4	Conc. H_2SO_4
H_2SO_4 addition (kg/t ore)	3.5	15	20−32	2−3	19 and 7
Cure time before leaching (d)	1	3−4	1	3	1

Continued

Table 14.2 Selected industrial heap leach data.—cont'd

Operation	Cerro Verde, Peru	El Abra, Chile	Mantoverde, Chile	Morenci, USA	Spence, Chile
Lixiviant					
Source of lixiviant	Raffinate or recirculated PLS	Raffinate + H_2SO_4	Raffinate + ILS + H_2SO_4	ILS + H_2SO_4	Raffinate + ILS
H_2SO_4 concentration (g/L)	4–8	7–10	6–10	5–8	13 and 8
Cu concentration (g/L)	0.3	0.2–0.3	2–3 and 0.4–0.6	0.3	0.3–0.5
Application rate (m^3/h per m^2 area)	0.012–0.065	0.08–0.12	0.12 and 0.01	0.006	0.014
Irrigation system	Drippers, wobblers	Emitters	Drippers, wobblers	Mainly emitters	Drippers
Distance apart along header pipe (m)	0.2	0.46	0.55	0.45	0.6
Distance apart between header pipes (m)	0.4	0.46	0.6	0.45	0.6
Leach cycle sequence	Acid cure in rotating drum, build cured ore into heap, leach 90 days, then 15 days on, 15 days off	Acid cure in rotating drum, build cured ore into heap, rest 3–4 days, leach 120 days	Acid cure in rotating drum, build cured ore in heap, leach 140 d	Agglomerate, cure fines in rotating drum, stack in 7 m lifts, leach for 120 days, rest 15 days, leach 15 days, rest 15 days, leach 15 days	Acid cure in rotating drum, build cured ore into heaps, rest 20 days, leach 262 days (oxide ore) and 536 days (sulfide ore)

PLS					
H_2SO_4 concentration (g/L)	3–4	3–4	<1	2–3	2–4
Cu concentration (g/L)	3	3	3–5	2–4	5–6 and 3–4
Temperature (°C)	19–25	15–20	18–20	32	18–20
Flowrate from leach (m^3/h)	2500	10,000	1480–1500	20,000	1600 and 1200

Figure 14.7 Heap leaching methods. (a) Multilift heap leach operation in Chile, showing the PLS and ILS collection ponds in the foreground. (b) Building a dynamic heap under subarctic conditions in Finland. (c) Impermeable plastic liner and drainage pipes laid down in preparation for the construction of a heap. (d) Plastic-lined trenches for the collection of PLS and ILS at the base of a heap. (e) Irrigation of the surface of a heap using wobblers. Some undesirable pooling of lixiviant on the surface of the heap is evident. *((a) Courtesy of Miller Metallurgical Services; (b and c) Courtesy of J.W. Mann.)*

also means that metal accounting is more reliable (Pyper et al., 2019). Their main disadvantages are that the ore must be moved twice (on and off the pad) and the associated costs of residue disposal. Dynamic heaps are usually the preferred choice for ore grades above 1% Cu (Scheffel, 2002).

14.4.2.2 Impermeable base

Leach heaps are built on an impermeable base. This permits complete collection of the leached Cu^{2+} in the PLS and prevents solution penetration into the underlying environment and potential contamination of ground water.

The base typically consists of a synthetic plastic liner or geomembrane sheet (such as high- or low-density polyethylene, polyvinyl chloride, or chlorinated polyvinyl chloride), 1–2 mm thick, with a layer of rolled clay or earth (0.1–0.5 m thick) beneath, and a 0.5 m layer of finely crushed and screened rock (<2 cm diameter) above it. Perforated PLS collection pipes and aeration pipes are placed on top of this layer (Fig. 14.7C and D). The base is sloped to direct the PLS to a collection pond. The slope should be less than 5% (5 m drop in 100 m horizontal) to avoid slippage of the heap on the plastic liner (Breitenbach, 2005). Considerable care is taken to avoid puncturing the plastic liner during construction to avoid loss of copper-containing PLS and environmental contamination.

14.4.2.3 Pretreatment of the ore

The simplest method of leaching involves placement of run-of-mine (ROM) ore directly onto the leach pad. This is the cheapest method, but it gives the slowest and least-efficient recovery of copper to the PLS.

More often, ROM ore is placed on the heap and then pretreated by trickling strong H_2SO_4 solution through the heap. This *acid cure* rapidly dissolves Cu^{2+} from readily soluble oxide minerals and acidifies the heap, thereby preventing precipitation of ferric sulfate during subsequent leaching. Typically, 10–20 kg H_2SO_4 per tonne of ore is applied to the heap over a period of ~10 days (Iasillo & Schlitt, 1999).

The rate of copper extraction and overall copper recovery decreases with decreasing particle size of the ore because the local saturation increases (Miller, 2008). Many heap leach operations crush their ROM ore to 1–5 cm pieces: crushing to less than 1 cm does not further improve extraction, while crushing to below 0.5 cm can adversely decrease heap permeability.

In most modern heap leach operations, *agglomeration* is used as an intermediate step between crushing and stacking (Bouffard, 2005, 2008; Velarde, 2005). The crushed ore (10–50 mm particle size) is agglomerated with concentrated H_2SO_4 and SX raffinate in revolving drums (typically 3 m diameter, 9 m long, sloped at ~6 degrees) (Fig. 14.8).

(a)

(b)

Figure 14.8 (a) Rotating drum used for the agglomeration of ore, usually with the addition of acid, before stacking into heaps for leaching. (b) Agglomerated ore on heap.

Agglomeration allows uniform application of acid to all surfaces of the mineral particles to cure the ore and improve the leach kinetics. The fines created during crushing are aggregated into larger particles so that they do not reduce the permeability of the heap.

When secondary sulfides are present in the mineralogy, a *ferric cure* may be used in conjunction with the acid cure. An acid-containing solution with >3 g/L Fe^{3+} may be added either directly to the heap or during the agglomeration process. An appropriate bacterial culture and its nutrients or chloride may also be added to the agglomeration process.

The agglomerated material is then placed on the leach heaps.

14.4.2.4 Stacking of ore on heap

Heaps of ore to be leached are stacked on the impermeable base by dumping from trucks or by stacking with a wheel loader or mobile conveyor. Trucks have the advantage of simplicity, but they tend to compact the heap as they run over it. This reduces permeability, slows leach kinetics, and limits oxygen circulation in the heap. Mobile conveyors avoid this problem and are now widely used (Figs. 14.9 and 14.10).

The heap height is a compromise to maximize copper recovery at the lowest cost (Connelly, 2020). A lower heap height can require greater capital due to the greater area under leach, increased pumping requirements from more irrigation, extra drainage piping, and a lower PLS grade. In contrast, a lower heap height can produce better overall copper recovery, may not require forced aeration, and is unlikely to suffer permeability problems from compaction at the base of the heap.

14.4.2.5 Aeration of heap

As discussed in Section 14.2, the leaching of sulfides requires the presence of oxygen (reactions 14.7—14.10). Oxygen also enhances bacterial activity (Watling, 2006). Oxygen is introduced to a heap by placing perforated plastic pipes about 1 m above the heap base and blowing air gently and uniformly upward through the heap (Lupo, 2010). Forced air supply is also used when pyrite oxidation is the main source of solution heating in colder climates.

The additional cost of providing oxygen to the heap is more than offset by the resulting increase in copper recovery and enhanced kinetics when used for leachable copper grades above 1%. Smaller and lower-grade heaps rely on natural air convection for oxygen supply.

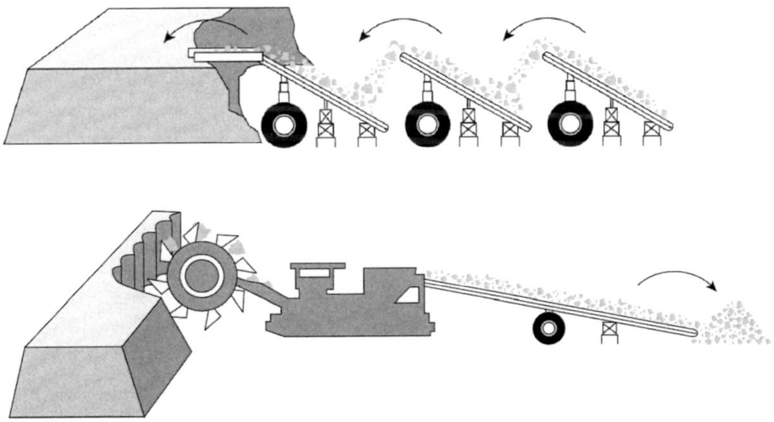

Figure 14.9 Diagrammatic representation of the loading of a heap using mobile conveyer belts. Unloading of a heap on completion of leaching is undertaken using bucket-wheel reclaimer.

(a)

(b)

(c)

(d)

Figure 14.10 Modern mechanized equipment for the loading and unloading of on–off heap leach pads. (a) TNT Super Portable mobile conveyer ~70 m long at an Arizona heap leach site. (b, c) Fixed-pad, multiple-lift heap stacking system at El Abra in Chile, which processes 8550 t/h ore using Super Portable Technology: 23 portable horizontal feed conveyers (each 85 m long), a 90 m long horizontal conveyor, and 75 m radial stacker. (d) Bucket-wheel reclaimer recovering an on–off heap after leaching. *(Photographs courtesy of Terra Nova Technologies and Miller Metallurgical Services.)*

14.4.2.6 Irrigation of heap

The lixiviant is typically the copper-depleted raffinate returning from SX (Fig. 14.5). This has a typical composition of 0.5 g/L Cu and 5–10 g/L H_2SO_4. Additional sulfuric acid may be added to the raffinate before it is recycled to the heap. Acid in excess of the stoichiometric requirement is required due to the leaching of gangue minerals, such as those containing Fe, Co, Mn, Al, Ca, Mg, Na, K, P, and carbonates (Salgado, Pérez, & Alvayai, 2009; Seyedbagheri, van Staden, & McLaren, 2009). Water may be added to replace evaporative and ore hold-up losses.

The lixiviant is added via an equi-spaced network of plastic pipes laid on top of the heap (Fig. 14.7E). *Sprinklers, wobblers*, and *drip emitters* are used to distribute the lixiviant (Dixon et al., 2012). Sprinklers and wobblers distribute solution evenly over large areas. Drip emitters require more maintenance to avoid blockages, but they avoid excessive evaporation and cooling losses.

The lixiviant addition rate typically ranges from about 8 to 12 L/h per m^2 of heap surface, but can be as high as 20 L/h/m^2 (Pyper et al., 2019). This rate prevents accumulation of lixiviant on the heap surface and allows free movement of air in the heap. The optimum irrigation rate can be determined from air and water conductivity tests that are available for use in heap design (Guzman, Scheffel, & Flaherty, 2008, Guzman, Scheffel, Ahlborn, Ramos, & Flaherty, 2007; Petersen & van Staden, 2019).

The lixiviant is generally at ambient temperature. In cold regions, it may be heated to enhance the kinetics of Cu extraction (Saario & Riekkola-Vanhanen, 2011). The leaching reactions are exothermic, so the heap will increase marginally in temperature as the leach liquor percolates through the heap; however, significant heat is lost by evaporation from wetted surfaces and in the air flow though the heap. The acid concentration and flowrate of the lixiviant and the rate of aeration of the heap all need to be balanced to ensure that the heap does not get too hot (which will cause the bacteria to die) or too cold (which slows down the rate of leaching) (Dixon, 2000; Schlitt, 2006).

14.4.2.7 Optimum leaching conditions

The following conditions promote optimum leaching of oxide and sulfide material in heaps:

(a) uniform heaps of optimally sized, cured, and agglomerated ore that maintain permeability throughout their life;

(b) leach conditions that maximize bacterial activity (30°C, pH 1.5–2.0, 5–10 g/L H_2SO_4);

(c) uniform lixiviant application (~ 10 L/h/m^2) onto the heap surface without pooling;

(d) well-designed impervious heap base, sloping less than 5%, with an efficient solution collection system;

(e) adequate heap temperature, provided in cold regions by heating raffinate, insulating pipes, and covering heaps with plastic mesh or sheet;

(f) a controlled, uniform air supply (~ 10 L/min/m^2 of heap surface), if used, blown in from perforated pipes beneath the heap;

(g) heap height that is appropriate to the geomechanical characteristics and mineralogy of the ore, to ensure unsaturated conditions and that sufficient oxygen and acid are available throughout the active leaching region.

14.4.2.8 Collection of PLS

The PLS exiting the base of a heap typically contains 1–6 g/L Cu^{2+}. The solution flows by gravity down plastic drainpipes (~10 cm diameter) on the sloping heap base to a collection trench (Fig. 14.7D). The pipes have slits 2 mm wide, 20 mm long through which the PLS enters. The pipes are spaced 2–4 m apart, at about 45 degrees across the slope. High-grade PLS flows to a PLS pond, from where it is sent to the SX circuit by gravity or pumping. Low-grade ILS is reacidified and recycled back to the heap for further copper leaching and PLS grade management (Fig. 14.5).

14.5 Dump leaching

Dump leaching is very similar to heap leaching: the main differences between these two leaching methods are the copper grade and particle size of the ore. Dump leaching is used for treating low-grade ROM ore, typically 0.1%–0.4% Cu oxide or sulfide materials. It is practiced mainly in North America and Chile. It is an inexpensive method of treating mined material that would otherwise be considered as waste. Industrial details of a Chilean dump leach operation are presented in Table 14.3.

Table 14.3 Industrial dump leach data.

Operation	Los Bronces, Chile
Start-up date	1998
Cathode capacity (t/a)	40,000
Ore feed to leach	
Predominant Cu minerals	Chalcocite 13%, chalcopyrite >70%
Average Cu grade (%)	0.35
Average leachable Cu (%)	0.07
Leachable Cu recovered to PLS (%)	33
Ore preparation	
Crush (Y/N)	N
Agglomeration in rotating drum (Y/N)	N
Ore size on heap (mm)	ROM, up to 500 mm
Heap	
Permanent or on–off?	Permanent
Building machinery	Trucks
Area under leach (Ha)	15–25
Typical lift height (m)	60
Aeration pipes beneath heap (Y/N)	N
Liner material	N

Table 14.3 Industrial dump leach data.—cont'd

Operation	Los Bronces, Chile
Acid cure (Y/N)	
On heap or during agglomeration?	N
H_2SO_4 in cure solution (g/L)	N
H_2SO_4 addition (kg/t ore)	N
Cure time before leaching (d)	N
Lixiviant	
Source of lixiviant	Raffinate + H_2SO_4
H_2SO_4 concentration (g/L)	4–5
Cu concentration (g/L)	0.1
Application rate (m^3/h per m^2 area)	3.4–8.5
Irrigation system	Drippers
Distance apart along header pipe (m)	0.5
Distance apart between header pipes (m)	0.2
Leach cycle sequence	Continuous leach for 360 d
PLS	
H_2SO_4 concentration (g/L)	0.5–1
Cu concentration (g/L)	2–3
Temperature (°C)	10–18
Flowrate from leach (m^3/h)	1600–1900

ROM ore is piled into dumps that may be as high as 200 m (Fig. 14.11). The material is not crushed or agglomerated: rock sizes range up to ~500 mm. Leaching, using acidified raffinate, can take several years. The eventual copper recovery can vary from 30% to 75%, depending on the mineralogy and the hydraulic and geomechanical characteristics of the ore.

14.6 Vat leaching

Vat leaching is carried out in large reaction vessels (*vats*). It is employed for low-grade (0.5%–1% Cu) oxide and carbonate minerals where the leach kinetics are fast and a relatively short contact time with the leach liquor is required (Schlitt & Johnston, 2010). Sulfides cannot be leached because of the lack of aeration in the vats. Table 14.4 illustrates typical commercial vat leaching operating conditions.

The ore is crushed to 5–10 mm. After acid curing, 4000–13,000 t ore is loaded into large square vats (20–30 m sides, 5–8 m deep) (Fig. 14.12). Vats are filled with ore from conveyer trolleys that run on rails along the length of the train of vats. The vats are typically constructed of lined, reinforced concrete with a raised, slatted floor, supported by concrete foundations, through which lixiviant is introduced.

(a) (b)

Figure 14.11 Dump leaching at Los Bronces, a mountainous site in Chile. (a) The natural topography has been used to contain the dump leach material. The leach solution flows by gravity through the dumps and is collected in PLS ponds at the base of the valley. (b) Close-up view showing the dripper pipes laid on the surface of the dump. *((A) Courtesy of Anglo American.)*

Table 14.4 Industrial data for vat leach operations. Mantos Blancos closed in 2020; Mina Justa was commissioned in 2021 and is the only copper vat leach in the world.

Operation	Mantos Blancos, Chile	Mina Justa, Peru
Start-up date	1961	2021
Cathode capacity (t/a)	40,000	58,000
H_2SO_4 consumption (kg/t ore)	12–20	31–36
Ore to leach		
Cu mineralogy	Atacamite, chrysocolla, cuprite	Oxides
Total Cu in ore (%)	0.5–1	0.56
Acid-soluble Cu (%)	60–70	82
Vats		
No of vats	12	15
Length and width of vat (m)	20 × 20	41.6 × 36
Height of vat (m)	7.5	7
Mass of dry solids in vat (kt)	4.4	12.8

Table 14.4 Industrial data for vat leach operations. Mantos Blancos closed in 2020; Mina Justa was commissioned in 2021 and is the only copper vat leach in the world.—cont'd

Operation	Mantos Blancos, Chile	Mina Justa, Peru
Residence time of ore in vat (h)	80	12 each, 144 total
Moisture content of tailings (%)	11	9–13
Ore preparation		
Crush	Y	Y
Agglomeration in rotating drum	Y	15 s retention time, 11% target moisture
Particle size to leach (mm)	30% +6.25 mm	−8 mm
Acid cure		
In vat or during agglomeration	During agglomeration	In agglomeration drum
H_2SO_4 addition (kg/t ore)	10–12	15
Cure time before leaching (h)	8 max.	–
Leach cycle sequence	Four vats leaching PLS, four vats leaching ILS, one being loaded, one being unloaded, one being drained, one being repaired	Four vats leaching for high-grade (HG) SX, eight vats for low-grade (LG) SX, one on ore filling, one on draining, one surge
Lixiviant		
Source of lixiviant	ILS followed by raffinate	Raffinate from HG and LG SX
H_2SO_4 concentration (g/L)	ILS 6–8; raffinate 9–10	HG 10; LG 6
Cu concentration (g/L)	ILS 1–3; raffinate 0.2–0.5	0.15–0.85
PLS		
H_2SO_4 concentration (g/L)	4–7	1.5
Cu concentration (g/L)	4–6	HG 4–8; LG 1–2
Total flow from leach (m^3/h)	600	900

Raffinate from SX is used as the lixiviant. The lixiviant is fed upflow into each vat until solution starts to overflow from the top of the vat. Vat leaching is operated in flooded fashion, using a flowrate of about 375 $L/h/m^2$. This keeps fine particles in suspension and prevents them from compacting and plugging at the bottom of the vat. The leach liquor moves from one vat to the next in a countercurrent manner so that fresh

Figure 14.12 Vat leaching at Mantos Blancos, Chile (closed 2020). Approximately 40,000 t/a Cu were produced from leaching in 12 vats. Mina Justa, Peru, commissioned a vat leaching operation in 2021, which is the only site in the world to use this leaching technology. *(Courtesy of Anglo American.)*

lixiviant (which contains the highest acid concentration) is contacted with ore that has been leaching the longest (i.e., only the most refractory minerals remain unleached). After an extended period of time (4–20 days) of recirculating the lixiviant through the ore in the vat, the copper-containing PLS is drained from the vats and sent for copper recovery.

To avoid soluble copper losses, the wet leached solids are water washed several times. The copper-containing spent wash water is combined with the PLS. The solids are removed from the vat using a bucket wheel or clam shell excavator that moves on a gantry adjacent to the vat train, loaded onto trucks, and sent to the tailings' disposal area.

Because of the high ore:lixiviant ratio in the vats, it is possible to achieve very high copper concentration in the PLS (~50 g/L Cu). The PLS from vat leaching will usually be blended with PLS from other leaching processes to provide a more dilute feed to SX. Alternatively, the high-grade solution can be fed to a direct EW circuit (see Section 16.9.1).

Vat leaching can also be configured as a continuous process (Mackie & Trask, 2009). Flowrates up to 30,000 L/h/m^2 enable good bed fluidization and rapid kinetics of leaching. For readily leached ores, the residence time on a heap leach is 120–500 days; this can be reduced to 24–72 h in a continuous vat operation.

14.7 Agitation leaching

Agitation leaching is used for minerals that have rapid leach kinetics. Leaching is carried out in a series of stirred tanks. Copper recovery can approach 100% within a few hours.

14.7.1 Oxide minerals

Agitation leaching is used for oxide minerals that have rapid leach kinetics, such as carbonates, and reasonable grade (0.8%—5% Cu). It can be applied to ores or to milled tailings from other leach circuits. This method is widely used in the African Copperbelt (Reolon, Gazis, & Amos, 2009; Roux, van Rooyen, Minnaar, Robles, & Cronje, 2010).

Following crushing and milling, the material is leached in stirred tanks. Finely ground ore is kept in suspension in the lixiviant by agitating the slurry mechanically or pneumatically. The fine particle size offers a large surface area that exposures more of the mineral to the leach solution, leading to fast kinetics and high copper recoveries. The slurry reports to solid—liquid separation, usually a countercurrent decantation (CCD) circuit, in which some further leaching can take place. The recovery of acid-soluble copper is usually above 98%, generating a PLS that contains 3—12 g/L Cu, although can range as high as 40 g/L (Sole & Tinkler, 2016). Ores high in chrysocolla need longer leach times because this mineral has slower leach kinetics.

The single largest copper loss in many of these plants is due to low recovery of PLS from the leach residue due to poor washing. Efficient operation of the CCD circuits is essential to maximize copper recovery and maintain the water balance in the circuit. These circuits can also have high operating costs due to the need for fine milling and increased acid requirements for the fine ore, followed by the subsequent need to neutralize the tailings for disposal. Various flowsheet configurations have been proposed to minimize these disadvantages (Roux, van Rooyen, Minnaar, Robles, & Cronje, 2010).

14.7.2 Sulfide minerals

Secondary sulfides of reasonable grade (>6% Cu) can also be treated using agitation leaching. The method is similar to that described for oxide leaching, but elevated temperature and an oxidizing agent are required (see Section 14.2.2). Most operations use air (oxygen), but plants in Australia, Laos, and Spain use Fe^{3+} as the oxidizing agent (Fleury, Delgado, & Collao, 2010).

The PLS generated from sulfide leaching is usually more concentrated than that of oxide leaching, typically 12—25 g/L Cu. Copper recovery varies from 95% to 98%; leach times range from less than 24 h to several days.

The Albion process is the first commercial atmospheric leaching process that is able to treat chalcopyrite (Stieper, 2018; Voigt, Stieper, & Hourn, 2019). The chalcopyrite concentrate is very finely ground to <10 μm. The reaction is accelerated by sparging of hypersonic oxygen into the leach tanks and operating above 90°C. The leach time is about 24 h.

Table 14.5 Industrial data for selected copper agitation leach plants.

Operation	Kansanshi, Zambia	Kamoto Copper, DRC	Metalkol, DRC	Sable Zinc[†] Kabwe, Zambia	Las Cruces, Spain
Start-up date	2004	2008	2018	2017	2009
Cathode Cu (t/a)	51,738	300,000 (design)	105,000	8000	72,000 (design)
H_2SO_4 consumption (t/t Cu)	24.3		2–2.3		20
Feed to leach					
Predominant Cu minerals	Malachite, chrysocolla, Cu–Fe silicate, Cu–Fe oxide, chalcopyrite, secondary Cu sulfide	Major malachite; minor pseudomalachite, chrysocolla, kolwezite, cuprite, chalcocite, copper-bearing clays	Malachite, chrysocolla, secondary sulfides (mainly chalcocite and covellite)	Concentrate of chalcopyrite, bornite, some chalcocite	65%–75% pyrite, 6% digenite and chalcocite, minor covellite, chalcopyrite, galena, and blende
Average Cu grade (%)*	TCu 0.72 ASCu 0.47	TCu 3.66 ASCu 3.11	TCu 1.2–1.46	TCu 28–30 ASCu 2.1	6.22
Average S grade (%)	–	–	–	19	41
Ore size to leach (μm)	80–150	80% −106	80% 100–250	80% 12–14	100% −150
Ore fed to leach (t/h)	700–1200	1000–1200	700–1000	0.6	
Agitation tank					
No of tanks	8 (train 1) 5 (train 2)	12 (2 × 6)	5	2 (active)	8
Tank height (m)	12.3–17.2 (train 1) 17.2 (train 2)	12.25	12.23	5	
Tank diameter (m)	9–12.6 (train 1) 12.6 (train 2)	11	11	5	

Impeller type	Mixtec	Afromix	Mixtec	Dual-blade hydrofoil	
Impeller power (kW)	90–200	160	90	4–20	
Oxygen addition (Y/N)	N	N	N	Y (hypersonic sparge)	
Solids' density (%)	30–35	35	32	44–46	
Solids' residence time (h)	~5.5	2.8	2.0	36	
Lixiviant					
Source	Raffinate, conc. H_2SO_4	HG raffinate, conc. H_2SO_4, acidified raffinate	HG raffinate, conc. H_2SO_4	Raffinate	Ferric sulfate, raffinate
H_2SO_4 concentration (g/L)	150–250	80–100	150–200	4	~60
Cu concentration (g/L)	0.5–1.0 and 1.5–6 (if PLS is recycled)	0.7–1.2	7.7–9.6		~10
Fe concentration (g/L)	3–7	2–2.5	2–3	8–10	25 Fe^{2+}, 25 Fe^{3+}
Temperature (°C)	25–35	Ambient	35	94	90
Feed rate to leach (m^3/h)	1100–1700	2545	2300		
PLS					
H_2SO_4 concentration (g/L)	8–12	2–5	6–8	<15	
Cu concentration (g/L)	1.5–6	13–18	6–9.5	30–40	43
Fe concentration (g/L)	3–7	2–2.5	3		50
Temperature (°C)	40–50	45	45	94–98	
Flowrate from leach (m^3/h)	1930–2900	2900	1600		313

Continued

Table 14.5 Industrial data for selected copper agitation leach plants.—cont'd

Operation	Kansanshi, Zambia	Kamoto Copper, DRC	Metalkol, DRC	Sable Zinc[†] Kabwe, Zambia	Las Cruces, Spain
Leach residue					
Predominant mineralogy	Major chalcopyrite, minor covellite, chrysocolla	Covellite, minor malachite, pseudomalachite			
Cu in residue (%)	0.1–0.4	0.2–0.3	0.2–0.36	0.4–0.6	
Leach efficiency					
Cu extraction (leach only) (%)	50	88	75	>99	92.5 (target)
Cu extraction (leach + CCDs) (%)	72	92–95	93–96		—

*TCu: total copper; ASCu: Acid-soluble Cu.
[†]Sable plant is the first commercial demonstration of atmospheric leaching of chalcopyrite using the Albion process (Stieper, 2018; Voigt, Stieper, & Hourn, 2019).

Table 14.5 shows typical operating data for several plants where agitation leaching is practiced.

The main advantages of heap or dump leaching compared with milling of the ore followed by agitation leaching are as follows (Pyper et al., 2019):

- lower capital costs;
- lower operating costs;
- no need for a subsequent solid/liquid step;
- no tailings disposal.

In contrast, the advantages of agitation leaching include the following:

- higher metal recovery;
- much shorter process time;
- rapid response to process changes;
- opportunity for byproducts recovery, such as cobalt or gold (Sole et al., 2018).

14.8 Pressure oxidation leaching

POx leaching uses high temperatures and pressures to promote the leaching of refractory minerals, such as chalcopyrite. When the chalcopyrite copper grade is too low or too variable for a pyrometallurgical process, pressure leaching can be used to treat the concentrate product from flotation (in contrast to the other leaching methods that process whole ore) (Frischmuth et al., 2019).

Temperature affects both the thermodynamics and kinetics of leaching. An increase in temperature and pressure promotes leaching because it permits a much higher concentration of oxygen to be dissolved in solution and allows faster kinetics for reactions that are slow or thermodynamically unfavorable at atmospheric conditions.

Elevated temperature—pressure leaching is carried out in an autoclave. The pressure inside the leaching vessel is increased by pressurizing the autoclave using oxygen; the temperature is typically increased by steam injection on start-up. The autoclave is usually horizontal, with a series of compartments though which the lixiviant and solids move in a cocurrent manner (Fig. 14.13). The material to be leached is ground to a fine particle size (typically ~10 μm) to allow maximum exposure of the lixiviant to the mineral surfaces.

Medium-temperature medium-pressure and high-temperature high-pressure processes have been developed to commercial scale.

14.8.1 High-temperature high-pressure oxidation leaching

In high-temperature and -pressure leaching (200—230°C, ~3000 kPa), leaching is rapid and >99% of the copper is typically dissolved. All of the sulfide sulfur in the concentrate is converted to H_2SO_4, which consumes oxygen. Unless the resulting acid can be used for heap or agitation leaching in other parts of the flowsheet, this acid needs to be neutralized ahead of downstream processing. Iron in the concentrate reports to the leach residue

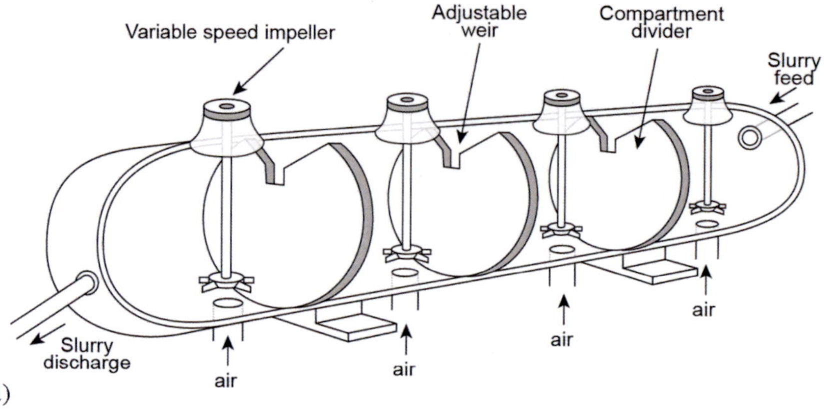

(a)

(b)

Figure 14.13 (a) Schematic representation of a four-compartment horizontal autoclave (baffles inside each compartment not shown). (b) Autoclave installation at Kansanshi, Zambia. *(Courtesy of W. Kuazi.)*

as hematite (Fe_2O_3), which is environmentally stable and easily filterable. In the presence of sufficient dissolved oxygen, the leach reaction can be represented as follows (Green et al., 2018; McDonald & Muir, 2007a):

$$4CuFeS_2 + 17O_2 + 4H_2O \rightarrow 4Cu^{2+} + 4SO_4^{2-} + 2Fe_2O_3 + 4H_2SO_4 \qquad (14.12)$$

High-temperature and -pressure of copper concentrates was practiced at a demonstration scale at Bagdad, Arizona (16,000 t/a Cu cathode) (Green et al., 2018). Commercial pressure leach facilities operate at Kansanshi, Zambia (30,000 t/a Cu) and for the treatment of a copper-containing pyrite concentrate at Sepon, Laos (80,000 t/a Cu) (Bell, Hoey, & Lui, 2010; Keokhounsy, Moore, & Liu, 2006; Sherrit, Pavlides, & Keekes, 2005). Details of these plants are given in Table 14.6.

Table 14.6 Industrial data for three copper high-temperature and -pressure oxidation leach plants (2021).

Operation	Kansanshi, Zambia	Morenci, USA	Sepon, Laos
Start-up date	2007	2015	2005
Cathode capacity (t/a)	30,202 design, 15,484 actual 2020	50,000	80,000
H_2SO_4 consumption (kg/t ore)		190–520	–
Ore/concentrate feed to autoclave			
Source of solids' feed to autoclave	Concentrate	Concentrate	Pyrite concentrate
Cu mineralogy	70%–80% chalcopyrite, 3%–10% chalcocite	Chalcopyrite	76% pyrite, refractory sulfides including covellite, enargite, and chalcopyrite
Cu in ore/concentrate (%)	23–27	35–40	2–3
S in ore/concentrate (%)	28–35	30–31	40–45
Particle size feed to leach (μm)	P_{80} 75 μm	P_{80} 7 μm, P_{98} 15 μm	P_{80} 125 μm
Solids' feedrate to leach (t/h)	12–15	15–22	10
Autoclave			
No of autoclaves	2	2	1
No of compartments per autoclave	5		4
Length (m)	26.8		14
Diameter (m)	3.35		4.2
Shell material	Carbon steel		Mild steel
Lining material	Lead layer, three layers of refractory bricks		Glass fiber-reinforced furan membrane, three layers of refractory bricks
Impeller type	4 × 8-blade (comp 1), 4 × 4-blade (comp 2–5)		8-blade (comp 1 and 2) and 4-blade (comp 3 and 4) Rushton turbines
Impeller power (kW)	4 × 55 kW (comp 1), 4 × 45 kW (comp 2–5)		110

Continued

Table 14.6 Industrial data for three copper high-temperature and -pressure oxidation leach plants (2021).—cont'd

Operation	Kansanshi, Zambia	Morenci, USA	Sepon, Laos
Impeller specific power (kW/m^3)	0.211		3–4
Temperature of operation (°C)	220	200	220
Pressure of operation (kPa)	3000	1000–3200	2800–3000
Oxygen overpressure (kPa)	800	690–1380	600–800
Oxygen grade (%)	>96		98
Solids' density (%)	50–55 (slurry feed to autoclave, excludes raffinate for cooling)		45
Residence time of solids in leach (min)	98	90	60
Temperature control mechanism	Raffinate injection for cooling (exothermic reaction)		Raffinate injection for cooling (exothermic reaction)
Lixiviant (solution feed to autoclave)			
Source of lixiviant	High-grade raffinate + H_2SO_4 make-up		Raffinate
H_2SO_4 concentration (g/L)	70–75		30
Cu concentration (g/L)	<1.3		1.0–1.5
Fe concentration (g/L)	<3		30–35
Temperature (°C)	<40		Ambient
Flowrate to leach (m^3/h)	75–90		110
PLS (discharge from autoclave)			
H_2SO_4 concentration (g/L)	56–62	20.1	60
Cu concentration (g/L)	10–15	26	3

Table 14.6 Industrial data for three copper high-temperature and -pressure oxidation leach plants (2021).—cont'd

Operation	Kansanshi, Zambia	Morenci, USA	Sepon, Laos
Fe concentration (g/L)	<17	0.4	40
Temperature (°C)	95		100–102 after flash
Flowrate from leach (m^3/h)	80–110 per clave	250	120–140
Total Cu in solids feed recovered to PLS (%)	99	98.5	97
Residue			
Predominant mineralogy of residue	Basic ferric sulfate	Basic ferric sulfate	Basic ferric sulfate
Cu in residue (%)	<0.3	<0.3	0.05

Kansanshi and Morenci leach chalcopyrite concentrate directly. Sepon used pressure leaching primarily to produce ferric ion, which was then used to leach chalcocite (the plant is now converted for gold production). The Morenci plant was converted from a medium-temperature pressure leach process in 2015 (Green et al., 2018).

14.8.2 Medium-temperature medium-pressure oxidation leaching

Medium-temperature and -pressure processes (140–165°C, 1000–1200 kPa) require fine grinding of the feed material (80%–10 μm or finer) and often rely on the addition of chloride to attain reasonable kinetics of leaching (for example, see McDonald & Muir, 2007b; Nicol et al., 2010; Senanayake, 2009; Velásquez-Yévenes, Nicol, & Miki, 2010; Yoo et al., 2010). Sulfides are converted to a mixture of H_2SO_4 and elemental sulfur:

$$4CuFeS_2 + 5O_2 + 4H_2SO_4 \rightarrow 4Cu^{2+} + 4SO_4^{2-} + 2Fe_2O_3 + 8S + 4H_2O \quad (14.13)$$

$$4CuFeS_2 + 17O_2 + 2H_2SO_4 \rightarrow 4Cu^{2+} + 10SO_4^{2-} + 4Fe^{3+} + 2H_2O \quad (14.14)$$

The melting point of sulfur is 119°C, so it is critical to avoid coating of the surface of the chalcopyrite particles by molten sulfur and to avoid agglomeration of the finely ground solids. Surfactants, such as calcium lignosulfate, and other additives are used (Steyl, 2004).

Depending on the temperature and free acid present in solution, iron will be rejected as basic ferric sulfate ($Fe(OH)SO_4$) or jarosite ($Fe_3(OH)_6(SO_4)_2 \cdot 2H_2O$); hematite can form at higher temperatures. It is desirable to produce an insoluble iron product because residual Fe^{3+} in solution can be transferred through SX and negatively affect current efficiency in EW (Chapter 16).

Medium-temperature and -pressure processes have been proven at a demonstration scale at Bagdad, Arizona (14,000 t/a Cu) (Marsden et al., 2007c) and at UHC Carajas, Brazil (10,000 t/a Cu) (Defreyne et al., 2008; Hayton, Defreyne, & Murray, 2009; Jones, Xavier, Gonçalves, Costa, & Torres, 2006; Omena et al., 2007). The first commercial implementation of this technology ran at Morenci, Arizona (67,000 t/a Cu) from 2007 to 2014 (Freeport McMoRan, 2008; Marsden et al., 2007d).

14.9 In situ leaching

A less-common technology is ISL, also known as *solution mining*. ISL of copper was reportedly practiced by the Romans and in China prior to 200 AD (Hiskey et al., 2019; Kuhn & Alley, 2019). Oxide minerals are most suited to ISL because their leaching does not require the presence of an oxidant. Sulfide ores can be leached by adding ferric sulfate or oxygen to the leach solution; naturally occurring sulfide-reducing bacteria can assist the leaching kinetics.

ISL is typically applied to a rock mass that is still "in place," i.e., as it occurs in nature in its undisturbed geological setting. ISL can also be used to recover copper from old mine workings, such as pit slopes, block caving subsidence zones, and underground stopes (Hiskey et al., 2019). The sulfuric acid lixiviant is introduced through injection wells, percolates through the ore, and dissolves readily soluble minerals. The PLS is pumped out and processed on the surface for copper recovery (Gunnison Copper, 2021).

Because the leach liquors are acidic, there are environmental concerns for potential contamination of groundwater and aquifers, so great care must be taken to ensure appropriate geological conditions for this type of leaching.

14.10 Hydrometallurgical processing of chalcopyrite concentrates

Chalcopyrite is essentially unleached under the mild oxidizing conditions of heap leaching (very slow leach kinetics). It can, however, be leached under oxidizing conditions at elevated temperatures (Section 14.8). This has led to extensive study of leaching of chalcopyrite *concentrates* as an alternative to smelting for copper recovery. Some economic and process drivers for treating chalcopyrite concentrates hydrometallurgically, rather than by smelting, include the following (Baxter, 2016; Dreisinger, 2019; Green et al., 2018):

(a) the relatively high and cyclically variable cost of external smelting and refining charges;

(b) limited availability of smelting capacity in some locations;

(c) high capital cost of installing new smelting capacity;
(d) increasingly stringent environmental legislation, particularly avoiding the generation of gaseous effluents such as SO_2;
(e) the ability to treat high-impurity concentrates, particularly those that contain elements that are deleterious in electrorefining, such as As, Sb, and F;
(f) recovery of valuable byproducts, such as cobalt and precious metals;
(g) the ability to treat lower-grade concentrates on site and increasing overall recovery from the ore body;
(h) the construction of small leach plants at mine sites, rather than shipping concentrate to large distant smelters, thereby eliminating freight and transport costs;
(i) cost-effective generation of sulfuric acid at mine sites for use in heap and agitation leaching;
(j) overall lower cost production of copper (if credit can be obtained for the acid production).

A variety of approaches have been taken to finding an economically viable process for the hydrometallurgical treatment of refractory copper sulfides (Baxter, 2019; Dreisinger, 2019). Sulfate-based processes include bacterially assisted processes (Batty & Rorke, 2006; Gericke et al., 2010; Pradhan et al., 2008; Watling, 2015) and low-, medium-, and high-temperature pressure leaching (Marsden et al., 2007a—d; Evans & Johnson, 1999; Defreyne et al., 2008; Dixon, Baxter, & Sylwestrzak, 2007; Hayton, Defreyne, & Murray, 2009; Marsden et al., 2007a—d; Miller, Sylwestrzak, & Baxter, 2008). Halide- (Hyvärinen & Hämäläinen, 2005; Miura et al., 2016; Moyes, Sammut, & Houllis, 2002), ammonia- (Arbiter & McNulty, 1999; Duyvesteyn & Sabacky, 1995; Moyo & Petersen, 2019), and nitrate-based (Anderson, Fayran, & Twidwell, 2010) processes have also received attention.

14.11 Future developments

Heap leaching of copper ores is widely practiced, particularly in North and South America, for the leaching of acid-soluble oxide minerals and secondary sulfides, such as chalcocite. The contribution to hydrometallurgical copper production by heap leaching is expected to continue, because this technology is relatively simple, robust, flexible, and low cost.

Improvements in heap leaching efficiency continue through optimization of crushing, acid curing, agglomeration, understanding solution and air conductivities, heap construction, aeration, lixiviant composition and application rate, bacterial activity, and temperature. Significant advances in materials-handling equipment that can assist in rapidly and uniformly stacking and reclaiming on—off heaps and minimize labor requirements continue to be brought to the industry (see, for example, Terra Nova Technologies, 2021). Continuous improvement in heap leaching aims to maximize the rate of leaching and overall copper recovery. Advances in heap leaching of secondary sulfides focus on the use of forced aeration to supply adequate oxygen requirements throughout the heap, addition of chloride to catalyze the leach kinetics, and bio-augmentation of sulfur-reducing bacteria.

Development of commercially viable technologies for the hydrometallurgical processing of chalcopyrite remains of continuing research interest. Considerable efforts continue to seek better understanding of the chemistry and mechanisms of leaching and the behavior of reaction products and byproducts (see, for example, Cordoba et al., 2008a-d; Hiroyoshi et al., 2001, 2004, 2008; Klauber, 2008; Nicol et al., 2010, 2016, 2017; Watling, 2013). Heap leaching of primary sulfides is becoming a reality. El Abra (Chile) has implemented the Sulfolix process to treat high-bornite material by heap leaching augmented by addition of biomass and aeration: sulfur-reducing bacteria are generated in bioreactors and the biomass solution is supplied directly to the heaps. A small-scale atmospheric leaching plant to treat chalcopyrite concentrate has been commercialized in Zambia using the Albion process (Stieper, 2018; Voigt, Stieper, & Hourn, 2019). The Rapid Oxidative Leach process offers potential for atmospheric leaching of chalcopyrite using mechanical—chemical activation of the ore (Chaiko, Baczek, Rocks, Walters, & Klepper, 2015; Chaiko et al., 2015; Mulligan et al., 2018).

POx leaching of refractory sulfides has not achieved wide commercial acceptance and has only been applied at a few operations. This technology may in future offer niche potential for treating high-impurity concentrates that cannot be processed using pyrometallurgical techniques (Baxter, 2016).

ISL may, in future, offer a relatively low-cost option for the recovery of copper under appropriate geotechnical and mineralogical conditions (Seaman, Vollert, & O'Callaghan, 2019; Sinclair & Thompson, 2015).

14.12 Summary

Hydrometallurgical extraction accounts for about 4.5 million tonnes of cathode copper per year (about 20% of total primary copper production). Most of this is produced by heap leaching.

Heap leaching consists of trickling H_2SO_4-containing lixiviant uniformly through flat-surfaced heaps of crushed ore, agglomerated, or ROM ore. Oxide ores are leached by H_2SO_4 without the need for oxidation. Chalcocite and covellite need the presence of oxygen and Fe^{3+} to be oxidized and leached by H_2SO_4-containing solutions.

Leaching of sulfide ores at economic rates is made possible by indigenous bacteria that increase the mineral leaching kinetics by several orders of magnitude. The rate-determining kinetics is diffusion in the saturated zones of the ore. Bacterial activity is maximized at pH 1.5–2, a temperature of ~30°C, and with an adequate O_2 supply.

Heap leaching generates a PLS containing 1–6 g/L Cu^{2+}. The PLS is collected on a sloping impervious plastic liner beneath the leach heaps and sent to SX and EW for copper production (Chapters 15 and 16). The Cu^{2+}-depleted raffinate from SX is recycled to leaching, with the acid strength usually increased by the addition of concentrated H_2SO_4.

The rate of leaching and the overall copper recovery are maximized by optimizing the conditions for diffusion control: crush size, acid curing, agglomeration, heap permeability, lixiviant composition, aeration, and bacterial activity.

Low-grade ores are treated by dump leaching, while vat leaching can be used to treat higher-grade oxide ore. Agitation leaching in stirred tanks is applied to both oxide and sulfide materials: the PLS concentrations can be up to 25 g/L Cu. Agitation leaching is a more expensive technology but can give very high copper recoveries. It is mainly applied on the African Copperbelt.

Most copper minerals are amenable to leaching under atmospheric conditions. A critical exception is chalcopyrite ($CuFeS_2$), which does not dissolve under percolation leach conditions. It can be leached under oxidizing conditions at elevated temperatures, either using the assistance of thermophilic microbes or high temperatures and pressures. Several process routes for the hydrometallurgical treatment of chalcopyrite have been proposed and are the topics of ongoing research, but few have yet been successfully commercialized. Total POx (220°C, 3000 kPa) circuits to treat chalcopyrite-containing concentrates operate in Laos, Zambia, and the United States of America.

Table 14.7 summarizes the key differences between the main leaching methodologies.

Table 14.7 Comparison of different leaching methods for hydrometallurgical recovery of copper.

Parameter	Dump	Heap	Vat	Agitation	Agitation	Pressure
Material treated	ROM, oxide, or secondary sulfide	Oxides, secondary sulfides, tailings, usually milled, agglomerated, and acid cured, bacteria assisted	Oxides, secondary sulfides	Oxides and tailings	Sulfides	Primary and secondary sulfides, refractory flotation concentrates
Cu grade (%)	0.1–0.4	0.3–2.3	>0.8	0.8–5.0	Up to 30	Up to 25
Particle size (mm)	Up to 1000	12–50	0.5–2	0.1–0.2	0.01–0.2	0.01–0.75
Leach time	Years	Months to years	Weeks	Days	Days	Hours
Cu in PLS (g/L)	0.5–3	1.5–8	6–40	2–30	12–50	25–80
Cu recovery (%)	35–75	Up to 90%	>90	85–100	95–99	>96

References

Anderson, C. G., Fayran, T. S., & Twidwell, L. G. (2010). Industrial NSC pressure oxidation of combined copper and molybdenum concentrates. In *Copper 2010* (Vol. 7, pp. 2589–2620). Clausthal-Zellerfeld-feld: Gesellschaft für Bergbau, Metallurgie, Rohstoff- und Umwelttechnik.

Arbiter, N., & McNulty, T. (1999). Ammonia leaching of copper sulfide concentrates. In S. K. Young, D. B. Dreisinger, R. P. Hackl, & D. G. Dixon (Eds.), *Copper 99: vol. IV. Hydrometallurgy of copper* (pp. 197–212). Warrendale, PA: The Minerals, Metals and Materials Society.

Asselin, E. (2019). Thermodynamics of copper leaching. In B. Wassink, & E. Asselin (Eds.), *Copper hydrometallurgy: Principles and practices* (pp. 1–23). Montreal: Canadian Institute of Mining, Metallurgy and Petroleum.

Batty, J. D., & Rorke, G. V. (2006). Development and commercial demonstration of the BioCOP™ thermophile process. *Hydrometallurgy, 83*(1–4), 83–89.

Baxter, K. (2016). Are we any closer to hydromet overtaking smelting for copper sulphide concentrates? In *ALTA 2016 nickel cobalt copper conference* (pp. 1–32). Melbourne: ALTA Metallurgical Services.

Baxter, K. (2019). Copper concentrate leaching: Fundamentals and practical aspects. In B. Wassink, & E. Asselin (Eds.), *Copper hydrometallurgy: Principles and practices* (pp. 79–110). Montreal: Canadian Institute of Mining, Metallurgy and Petroleum.

Bell, M., Hoey, M., & Lui, M. (2010). *Design, construction and commissioning of the Sepon POX II circuit. ALTA nickel cobalt copper 2010.* Melbourne: ALTA Metallurgical Services.

Bouffard, S. C. (2005). Review of agglomeration practice and fundamentals in heap leaching. *Mineral Processing and Extractive Metallurgy Review, 26*, 233–294.

Bouffard, S. C. (2008). Agglomeration for heap leaching: Equipment design, agglomerate quality control, and impact on the heap leach process. *Minerals Engineering, 21*(15), 1115–1125.

Breitenbach, A. J. (2005). *Heap leach design and construction practices in the 21st century*. Available from: http://ore-max.com/pdfs/resources/heap_leach_pad_design_and_construction_practices_in_the_21st_century.pdf Accessed April 2021.

Brierley, C. L. (2016). Biological processing of sulfidic ores and concentrates—Integrating innovations. In V. Lakshamanan, R. Roy, & V. Ramachandran (Eds.), *Innovative process development in metallurgical industry: Concept to commission* (pp. 109–135). Berlin: Springer.

Brierley, C. L., & Brierley, J. A. (2013). Progress in bioleaching: Part B Application of microbial processes by the minerals industry. *Applied Microbiology and Biotechnology, 97*, 7545–7552.

Brierley, J. A., & Brierley, C. L. (2019). Bioleaching. In R. C. Dunne, S. S. Kowatra, & C. A. Young (Eds.), *SME handbook of minerals processing and extractive metallurgy* (pp. 1267–1281). Littleton, CO: Society for Mining, Metallurgy and Exploration.

Chaiko, D., Baczek, F., Rocks, S. S., Walters, T., & Klepper, R. (2015). The FLSmidth® rapid oxidative leach (ROL) process. Part I: Mechano-chemical process for chalcopyrite. In *54th Annual conference of metallurgists*. Montreal: Canadian Institue of Mining, Metallurgy and Petroleum.

Chaiko, D., Rocks, S. S., Walters, T., Asihene, S., Eyzaguirre, C., Klepper, R., … McMahon, G. (2015). The FLSmidth® rapid oxidative leach (ROL) process. Part II: A new chemical activation process for chalcopyrite. In *54th Annual conference of metallurgists*. Montreal: Canadian Institute of Mining, Metallurgy and Petroleum.

Connelly, D. (2019). Technology options and innovation for copper heap leaching. In *Copper 2019 (Paper No. 566372)*. Montreal: Canadian Institute of Mining, Metallurgy and Petroleum.

Connelly, D. (2020). Technology options and innovation for copper heap leaching. *Mine and Quarry World, 17*(5), 46–50.

Connelly, D., & West, J. (2009). Trips and traps for copper heap leaching. In *ALTA copper 2009*. Melbourne: ALTA Metallurgical Services.

Cordoba, E. M., Muñoz, J. A., Blázquez, M. L., González, F., & Ballester, A. (2008a). Leaching of chalcopyrite with ferric ion. Part I: General aspects. *Hydrometallurgy, 93*(3–4), 81–87.

Cordoba, E. M., Muñoz, J. A., Blázquez, M. L., González, F., & Ballester, A. (2008b). Leaching of chalcopyrite with ferric ion. Part II: Effect of redox potential. *Hydrometallurgy, 93*(3–4), 88–96.

Cordoba, E. M., Muñoz, J. A., Blázquez, M. L., González, F., & Ballester, A. (2008c). Leaching of chalcopyrite with ferric ion. Part III: Effect of redox potential on the silver-catalyzed process. *Hydrometallurgy, 93*(3–4), 97–105.

Cordoba, E. M., Muñoz, J. A., Blázquez, M. L., González, F., & Ballester, A. (2008d). Leaching of chalcopyrite with ferric ion. Part IV: The role of redox potential in the presence of mesophilic and thermophilic bacteria. *Hydrometallurgy, 93*(3–4), 106–115.

Defreyne, J., Brace, T., Miller, C., Omena, A., Matos, M., & Cobral, T. (2008). Commissioning UHC: A Vale copper refinery based on CESL technology. In C. A. Young, P. R. Taylor, C. G. Anderson, & Y. Choi (Eds.), *Hydrometallurgy 2008* (pp. 357–366). Littleton, CO: Society of Mining, Metallurgy and Exploration.

Dixon, D. G. (2000). Analysis of heat conservation during copper sulfide heap leaching. *Hydrometallurgy, 58*, 27–41.

Dixon, D. G., Baxter, K., & Sylwestrzak, L. (2007). Galvanox™ treatment of copper concentrates. In *ALTA copper 2007*. Melbourne: ALTA Metallurgical Services.

Dixon, D. G., & Liu, W. (2019). Copper sulfide leaching kinetics – fundamentals and case studies. In B. Wassink, & E. Asselin (Eds.), *Copper hydrometallurgy: Principles and practices* (pp. 24–51). Montreal: Canadian Institute of Mining, Metallurgy and Petroleum.

Dixon, S. N., Shearer, M. N., & Steele, D. J. (2012). *Comparison of spray and drip leaching systems*. Available from: http://ore-max.com/pdfs/resources/comparison_of_spray_and_drip_leaching_systems.pdf Accessed April 2021.

Dreisinger, D. (2019). New developments in hydrometallurgical copper ore and concentrate leaching. In *Copper 2019 (Paper No. 595433)*. Montreal: Canadian Institute of Mining, Metallurgy and Petroleum.

Duyvesteyn, W. P. C., & Sabacky, B. J. (1995). Ammonia leach process for Escondida concentrates. *Transactions of the Institute of Mining and Metallurgy, 104*, C125–C140.

Evans, H. A., & Johnson, G. (1999). *Activox technology for treatment of copper-gold sulphide concentrates. Oretest colloquium '99*. Victoria: Australian Institute of Mining and Metallurgy.

Fleury, F., Delgado, E., & Collao, N. (2010). A new technology for processing hydrometallurgical copper ore: Cobre Las Cruces Project. *Copper 2010, 5*, 1871–1898. Clausthal-Zellerfeldfeld: Gesellschaft für Bergbau, Metallurgie, Rohstoff- und Umwelttechnik.

Freeport McMoRan. (2008). Construction and start-up of the Freeport-McMoRan concentrate leach plant at Morenci, AZ. In *ALTA copper 2008*. Melbourne: ALTA Metallurgical Services.

Gericke, M., Govender, Y., & Pinches, A. (2010). Tank bioleaching of low-grade chalcopyrite concentrates using redox control. *Hydrometallurgy, 104*(3–4), 414–419.

Ghorbani, Y., Franzidis, J.-P., & Petersen, J. (2016). Heap leaching technology—current status, innovations, and future directions: A review. *Minerals Processing and Extractive Metallurgy Review, 37*(2), 73–119.

Green, C., Robertson, J., & Marsden, J. O. (2018). Pressure leaching of copper concentrates at Morenci, Arizona – 10 years of experience. *Minerals and Metallurgical Processing, 35*(3), 109–116.

Gunnison Copper. (2021). *In-situ copper recovery*. Available from: https://gunnisoncopper.com/index.php/about/in-situ-copper-recovery Accessed April 2021.

Guzman, A., Scheffel, R., Ahlborn, G., Ramos, S., & Flaherty, S. (2007). Geochemical profiling of a sulfide leaching operation: The rest of the story. In G. E. Houlachi, J. D. Edwards, & T. G. Robinson (Eds.), *Vol. VI. Copper 2007*. Montreal: Canadian Institute of Mining, Metallurgy and Petroleum.

Guzman, A., Scheffel, R., & Flaherty, S. (2008). The fundamentals of physical characterisation of ore for leach. In C. A. Young (Ed.), *Hydrometallurgy 2008* (pp. 937–966). Littleton, CO: Society for Mining, Metallurgy and Exploration.

Hayes, P. C. (1993). *Process principles in minerals and materials production*. Sherwood, Queensland: Hayes Publishing Co.

Hayton, N., Defreyne, J., & Murray, K. (2009). UHC copper refinery: An update on the Vale project based on CESL technology. In *HydroCopper 2009* (pp. 114–123). Santiago: Gecamin.

Hiroyoshi, N., Kitagawa, H., & Tsunekawa, M. (2008). Effect of solution composition on the optimum redox potential for chalcopyrite leaching in sulfuric acid solutions. *Hydrometallurgy, 91*, 144–149.

Hiroyoshi, N., Kuroiwa, S., Miki, H., Tsunekawa, M., & Hirajima, T. (2004). Synergistic effect of cupric and ferrous ions on active-passive behaviour in anodic dissolution of chalcopyrite in sulfuric acid solutions. *Hydrometallurgy, 74*, 103–116.

Hiroyoshi, N., Miki, H., Hirajima, T., & Tsunekawa, M. (2001). Enhancement of chalcopyrite leaching by ferrous ions in acidic ferric sulfate solutions. *Hydrometallurgy, 60*, 185–197.

Hiskey, J. B. (1993). Chalcopyrite semiconductor electrochemistry and dissolution. In R. G. Reddy, & R. N. Weizenbach (Eds.), *vol. I. Extractive metallurgy of copper, nickel and cobalt (the Paul E. Queneau international symposium)* (pp. 949–969). Warrendale, PA: The Minerals, Metals and Materials Society.

Hiskey, J. B., Mansanti, J. G., & McNulty, T. (2019). Solution mining and in situ leaching. In R. C. Dunne, S. S. Kowatra, & C. A. Young (Eds.), *SME handbook of minerals processing and extractive metallurgy* (pp. 1191–1206). Littleton, CO: Society for Mining, Metallurgy and Exploration.

House, C. I. (1987). Potential-pH diagrams and their application to hydrometallurgical systems. In G. A. Davies (Ed.), *Separation processes in hydrometallurgy* (pp. 3–19). Chichester: Ellis Horwood.

Hyvärinen, O., & Hämäläinen, M. (2005). Hydrocopper – a new technology producing copper directly from concentrate. *Hydrometallurgy, 77*, 61–65.

Iasillo, E., & Schlitt, W. J. (1999). Practical aspects associated with evaluation of a copper heap leach project. In G. V. Jergensen, II (Ed.), *Copper leaching, solvent extraction, and electrowinning technology* (pp. 123–138). Littleton, CO: Society for Mining, Metallurgy and Exploration.

Jones, D., Xavier, F., Gonçalves, L. R., Costa, R., & Torres, V. (2006). CESL process - semi-industrial operation at CVRD Sossego plant. In *ALTA copper 2006*. Melbourne: ALTA Metallurgical Services.

Keokhounsy, S., Moore, T., & Liu, M. (2006). Hydromet at Sepon. *ALTA copper 2006*. Melbourne: ALTA Metallurgical Services.

Klauber, C. (2008). A critical review of the surface chemistry of acidic ferric sulphate dissolution of chalcopyrite with regards to hindered dissolution. *International Journal of Minerals Processing, 86*, 1–17.

Kuhn, M. C., & Alley, R. D. (2019). Copper hydrometallurgy. In R. C. Dunne, S. S. Kowatra, & C. A. Young (Eds.), *SME handbook of minerals processing and extractive metallurgy* (pp. 1643–1651). Littleton, CO: Society for Mining, Metallurgy and Exploration.

Lazarro, I., & Nicol, M. (2003). The mechanism of the dissolution and passivation of chalcopyrite: An electrochemical study. In C. A. Young, A. M. Alfantazi, C. G. Anderson, D. B. Dreisinger, B. Harris, & A. James (Eds.), *Hydrometallurgy 2003: Vol. 1. Leaching and solution purification* (pp. 405–417). Warrendale, PA: The Minerals, Metals and Materials Society.

Lupo, J. (2010). Assessment of natural and forced air in ore heaps. In *Third international workshop on process hydrometallurgy, Hydroprocess 2010*. Santiago: Gecamin.

Mackie, D., & Trask, F. (2009). Continuous vat leaching – first copper pilot trials. In *ALTA copper 2009*. Melbourne: ALTA Metallurgical Services.

Marsden, J. O., Wilmot, J. C., & Hazen, N. (2007a). Medium-temperature pressure leaching of copper concentrates—Part I: Chemistry and initial process development. *Minerals and Metallurgical Processing, 24*(6), 193–204.

Marsden, J. O., Wilmot, J. C., & Hazen, N. (2007b). Medium-temperature pressure leaching of copper concentrates—Part II: Development of direct electrowinning and an acid-autogenous process. *Minerals and Metallurgical Processing, 24*(6), 205–217.

Marsden, J. O., Wilmot, J. C., & Mathern, D. R. (2007c). Medium-temperature pressure leaching of copper concentrates—Part III: Commercial demonstration at Bagdad, Arizona. *Minerals and Metallurgical Processing, 24*(6), 218–225.

Marsden, J. O., Wilmot, J. C., & Smith, R. J. (2007d). Medium-temperature pressure leaching of copper concentrates—Part IV: Application at Morenci, Arizona. *Minerals and Metallurgical Processing, 24*(6), 226–236.

McDonald, R. G., & Muir, D. M. (2007a). Pressure oxidation leaching of chalcopyrite. Part I. Comparison of high and low temperature reaction kinetics and products. *Hydrometallurgy, 86*(3–4), 191–205.

McDonald, R. G., & Muir, D. M. (2007b). Pressure oxidation leaching of chalcopyrite: Part II: Comparison of medium temperature kinetics and products and effect of chloride ion. *Hydrometallurgy, 86*(3–4), 206–220.

Miller, G. (2003). Ore geo-technical effects on copper heap leach kinetics. In C. A. Young, A. M. Alfantazi, C. G. Anderson, D. B. Dreisinger, B. Harris, & A. James (Eds.), *Hydrometallurgy 2003: vol. 1. Leaching and solution purification* (pp. 329–342). Warrendale, PA: The Minerals, Metals and Materials Society.

Miller, G. (2008). The effective particle size in heap leaching. In *ALTA copper 2008*. Melbourne: ALTA Metallurgical Services.

Miller, K. J., Sylwestrzak, L. A., & Baxter, K. G. (2008). Treatment of copper sulphide deposits—evaluation of a GalvanoxTM versus Sepon circuit configuration. In *ALTA copper 2008*. Melbourne: ALTA Metallurgical Services.

Miura, A., Manabe, M., Sato, K., Kuwano, K., Kawashiro, S., & Ohata, A. (2016). Leaching of low-grade primary copper sulfide ore with JX iodine process. In *Copper 2016 (paper hy5-1)*. Tokyo: Mining and Materials Processing Institute of Japan.

Moyes, J., Sammut, D., & Houllis, F. (2002). The Intec copper process: Superior and sustainable metals production. *ALTA 2002 copper-7 forum*. Melbourne: ALTA Metallurgical Services.

Moyo, T., & Petersen, J. (2019). The hydrometallurgy of chalcopyrite in ammoniacal solutions. In *Copper 2019 (paper No. 594898)*. Montreal: Canadian Institute of Mining, Metallurgy and Petroleum.

Mulligan, M., Chaiko, D., Baczek, F., Rocks, S., Eyzaguirre, C., Dickinson, C., & Klepper, R. (2018). The FLSmidth® rapid oxidative leach (ROL) process: A mechano-chemical approach and industrial applications for rapid metal sulphide dissolution. *Journal of the Southern African Institute of Mining and Metallurgy, 117*(8), 741–747.

Nicol, M. J. (2017). The anodic behaviour of chalcopyrite in chloride solutions: Overall features and comparison with sulfate solutions. *Hydrometallurgy, 169*, 321–329.

Nicol, M. J., & Akilan, C. (2016). Kinetics of the oxidation of iron(II) by oxygen and hydrogen peroxide in concentrated chloride solutions – a re-evaluation and comparison with the oxidation of copper(I). *Hydrometallurgy, 166*, 123–129.

Nicol, M., & Basson, P. (2017). The anodic behaviour of covellite in chloride solutions. *Hydrometallurgy, 172*, 60–68.

Nicol, M., & Lazarro, I. (2003). The role of non-oxidative processes in the leaching of chalcopyrite. In P. A. Riveros, D. Dixon, D. B. Dreisinger, & J. Menacho (Eds.), *Copper 2003* (vol. VI, pp. 367–381). Montreal: Canadian Institute of Mining, Metallurgy and Petroleum (Book 1).

Nicol, M., Miki, H., & Velásquez-Yévenes, L. (2010). The dissolution of chalcopyrite in chloride solutions: Part 3. Mechanisms. *Hydrometallurgy, 103*(1–4), 86–95.

Nicol, M. J., Zhang, S., & Tjandrawan, V. (2016). Cathodic reduction of iron(III) and copper(II) on various sulfide minerals in chloride solutions. *Hydrometallurgy, 166*, 113–122.

Omena, A., Geraldo, J., Lima, L., & Cabral, T. (2007). Technical feasibility study to produce cathodes in the UHC Carajas plant by the hydrometallurgical CESL route. In *HydroCopper 2007* (pp. 67–73). Santiago: Gecamin.

Peters, E. (1976). Direct leaching of sulfides: Chemistry and applications. *Metallurgical Transactions B, 7*, 505–517.

Petersen, J., & van Staden, P. (2019). Copper heap leaching: Process, principles, and practical considerations. In B. Wassink, & E. Asselin (Eds.), *Copper hydrometallurgy: Principles and practices* (pp. 52–78). Montreal: Canadian Institute of Mining, Metallurgy and Petroleum.

Pradhan, N., Nathsarma, K. C., Srinivasa Rao, K., Sukla, L. B., & Mishra, B. K. (2008). Heap bioleaching of chalcopyrite: A review. *Minerals Engineering, 21*(5), 355–365.

Pyper, R., Seal, T., Uhrie, J. L., & Miller, G. C. (2019). Heap and dump leaching. In R. C. Dunne, S. S. Kowatra, & C. A. Young (Eds.), *SME handbook of minerals processing and extractive metallurgy* (pp. 1207–1224). Littleton, CO: Society for Mining, Metallurgy and Exploration.

Reolon, M., Gazis, T., & Amos, S. (2009). Optimizing acid utilisation and metal recovery in African Cu/Co flowsheets. In *Hydrometallurgy 2009* (pp. 1–20). Johannesburg: South African Institute of Mining and Metallurgy.

Roux, L., van Rooyen, M., Minnaar, E., Robles, E., & Cronje, I. (2010). Copper oxide agitated leaching on the African Copperbelt and possible heap leach application. *Copper 2010, 5*, 1999–2016. Clausthal-Zellerfeldfeld: Gesellschaft für Bergbau, Metallurgie, Rohstoff- und Umwelttechnik.

Saario, P., & Riekkola-Vanhanen, M. (2011). Talvivaara bio heap leaching process. In *Percolation leaching: The status globally and in Southern Africa, Symposium series 269* (pp. 97–109). Johannesburg: South African Institute of Mining and Metallurgy.

Salgado, C., Pérez, C., & Alvayai, C. (2009). Acid management in heap leaching. Are we doing it right?. In *HydroCopper 2009* (pp. 68–76). Santiago: Gecamin.

Scheffel, R. E. (2002). Copper heap leach design and practice. In A. L. Mular, D. N. Halbe, & D. J. Barratt (Eds.), *Mineral processing plant design, practice, and control* (Vol. 2, pp. 1571–1605). Littleton, CO: Society for Mining, Metallurgy and Exploration.

Schlitt, J., & Johnston, A. (2010). The Marcobre vat leach system: A new look at an old process. *Copper 2010, 5*, 2039–2057. Clausthal-Zellerfeldfeld: Gesellschaft für Bergbau, Metallurgie, Rohstoff- und Umwelttechnik.

Schlitt, W. J. (2006). The history of forced aeration in copper sulfide leaching. In *SME annual meeting (preprint 06–019)*. Littleton, CO: Society for Mining, Metallurgy & Exploration.

Seaman, B., Vollert, L., & O'Callaghan, J. (2019). In-situ recovery in hard rock applications: Idealistic notion or realistic future processing option?. In *Copper 2019 (paper No. 590358)*. Montreal: Canadian Institute of Mining, Metallurgy and Petroleum.

Senanayake, G. (2009). A review of chloride-assisted copper sulfide leaching by oxygenated sulfuric acid and mechanistic considerations. *Hydrometallurgy, 98*, 21–32.

Seyedbagheri, A., van Staden, P., & McLaren, C. (2009). A study of acid-gangue reactions in heap leach operations. In *HydroCopper 2009* (pp. 68–76). Santiago: Gecamin.

Sherrit, R., Pavlides, A. G., & Keekes, B. L. (2005). Design and commissioning of the Sepon copper pressure oxidation circuit. In *Metplant 2005*. Victoria: Australasian Institute of Mining and Metallurgy.

Sinclair, L., & Thompson, J. (2015). In situ leaching of copper: Challenges and future prospects. *Hydrometallurgy, 157*, 306–324.

Sole, K. C. (2008). Solvent extraction in the hydrometallurgical processing and purification of metals: Process design and selected applications. In M. Aguilar, & J. L. Cortina (Eds.), *Solvent extraction and liquid membranes: Fundamentals and applications in new materials* (pp. 141–200). New York: Taylor and Francis.

Sole, K. C., Parker, J., Cole, P. M., & Mooiman, M. B. (2018). Flowsheet options for cobalt recovery from African copper–cobalt hydrometallurgy circuits. *Minerals Processing and Extractive Metallurgy Review, 40*(3), 194–206.

Sole, K. C., & Tinkler, O. S. (2016). Copper solvent extraction: Status, operating practices and challenges in the African Copper Belt. *Journal of the Southern African Institute of Mining and Metallurgy, 116*(6), 553–560.

Steyl, J. D. T. (2004). The effect of surfactants on the behaviour of sulphur in the oxidation of chalcopyrite at medium temperature. In M. J. Collins, & V. G. Papangelakis (Eds.), *Pressure hydrometallurgy 2004 international conference on the use of pressure vessels for metals extraction and recovery* (pp. 101–117). Montreal: Canadian Institute of Mining, Metallurgy and Petroleum.

Stieper, G. (2018). First chalcopyrite copper concentrate leaching using Albion processTM technology. In *Hydroprocess 2018*. Santiago: Gecamin.

Terra Nova Technologies. (2021). *Projects*. Available from: https://www.tntinc.com/about/projects Accessed April 2021.

Velarde, G. (2005). Agglomeration control for heap leaching processes. *Mineral Processing and Extractive Metallurgy Review, 26*, 219–231.

Velásquez-Yévenes, L., Nicol, M., & Miki, H. (2010a). The dissolution of chalcopyrite in chloride solutions: Part 1. The effect of solution potential. *Hydrometallurgy, 103*(1–4), 108–113.

Velásquez-Yévenes, L., Miki, H., & Nicol, M. (2010b). The dissolution of chalcopyrite in chloride solutions: Part 2: Effect of various parameters on the rate. *Hydrometallurgy, 103*(1–4), 80–85.

Voigt, P., Stieper, G., & Hourn, M. (2019). First commercialization of the Albion processTM for copper. In *Copper 2019 (paper 576516)*. Montreal: Canadian Institute of Mining, Metallurgy and Petroleum.

Watling, H. R. (2006). The bioleaching of sulphide minerals with emphasis on copper sulphides—a review. *Hydrometallurgy, 84*(1–2), 81–108.

Watling, H. R. (2013). Chalcopyrite hydrometallurgy at atmospheric pressures: I. Review of acid sulfate, sulfate–chloride and sulfate–nitrate process options. *Hydrometallurgy, 140*, 163–180.

Watling, H. R. (2015). Review of biohydrometallurgical metals extraction from polymetallic mineral resources. *Minerals, 5*, 1–60.

Yoo, K., Kim, S.-K., Lee, J.-C., Ito, M., Tsunekawa, M., & Hiroyoshi, N. (2010). Effect of chloride ions on leaching rate of chalcopyrite. *Minerals Engineering, 23*(6), 471–477.

Yue, G., & Asselin, E. (2014). Kinetics of ferric ion reduction on chalcopyrite and its influence on leaching up to 150°C. *Electrochimica Acta, 146*, 307–321.

Suggested reading

Frischmuth, R., Krumins, T., Pearson, M., & Fraser, K. S. (2019). Pressure leaching and oxidation. In R. C. Dunne, S. S. Kowatra, & C. A. Young (Eds.), *SME handbook of minerals processing and extractive metallurgy* (pp. 1239–1265). Littleton, CO: Society for Mining, Metallurgy and Exploration.

Guézennec, A. G. (2019). Bioleaching of copper ores and concentrates: Fundamentals and practical aspects. In B. Wassink, & E. Asselin (Eds.), *Copper hydrometallurgy: Principles and practices* (pp. 111–132). Montreal: Canadian Institute of Mining, Metallurgy and Petroleum.

Habashi, F. (2009). Kinetics and mechanisms of leaching copper minerals. In *HydroCopper 2009* (pp. 176–193). Santiago: Gecamin.

John, L. W. (2011). The art of heap leaching—the fundamentals. In *Percolation leaching: The status globally and in Southern Africa. Symposium series 269* (pp. 17–42). Johannesburg: South African Institute of Mining and Metallurgy.

Kyle, J. I., Maxwell, D., & Alexander, B. L. (2011). In-situ techniques of solution mining. In P. Darling (Ed.), *SME mining engineering handbook* (pp. 1103–1119). Littleton, CO: Society for Mining, Metallurgy and Exploration.

Marsden, J. O., & Botz, M. M. (2017). Heap leach modelling – a review of approaches to metal production forecasting. *Minerals and Metallurgical Processing, 34*(2), 53–64.

Petersen, J. (2019). Heap leaching of copper ores – state of the science. In *Copper 2019 (paper 597297)*. Montreal: Canadian Institute of Mining, Metallurgy and Petroleum.

Scheffel, R. E., Guzmán, A., & Dreier, J. E. (2016). Development of metallurgy guidelines for copper heap leach. *Minerals and Metallurgical Processing: Special Issue on Process Development Testing, 33*(4), 187–199.

Schippers, A., Hedrich, S., Vasters, J., Dribe, M., Sand, W., & Willschier, S. (2014). Biomining: Metal recovery from ores with microorganisms. In A. Schippers, F. Glombitza, & W. Sand (Eds.), *Advances in chemical engineering/biotechnology – geobiotechnology I: Metal-related issues* (pp. 1–47). Berlin: Springer-Verlag.

Taylor, A. (2020). Heap leaching and its application to copper, gold, uranium and nickel ores. In *ALTA short course*. Melbourne: ALTA Metallurgical Services.

Vera, M., Schippers, A., & Sand, W. (2013). Progress in bioleaching: Fundamentals and mechanisms of bacterial metal sulfide oxidation. Part A. *Applied Microbiology and Biotechnology, 97*, 7529–7541.

CHAPTER 15

Solvent extraction

Solvent extraction (SX) serves the purpose of *purifying* and *upgrading* the pregnant leach solution (PLS) produced by the leaching operation (Chapter 14) to give an electrolyte from which high-quality copper cathode can be electrowon (Chapter 16).

Most leach liquors, irrespective of the leaching process by which they are produced, are too dilute in copper (1−10 g/L Cu), too impure (containing 1−20 g/L Fe, as well as other deleterious species, such as Mn, chlorides, or nitrates), and not sufficiently conductive (generally having a low acid content of 10 g/L H_2SO_4 to pH 2.5) for direct electrodeposition of high-purity cathode copper. Electrowinning (EW) from these solutions would give soft, powdery, and impure copper deposits.

Industrial EW requires pure copper-rich electrolytes (45−55 g/L Cu) with high conductivity. This high concentration of copper ensures that (i) Cu^{2+} ions are always available for plating at the cathode surface at current densities that allow economical rates of plating, being readily renewed by mass transfer and (ii) gives smooth, dense, high-purity, readily marketable cathode copper. The high conductivity of the electrolyte is provided by a high acid concentration, typically 175−190 g/L H_2SO_4.

SX provides the means for producing pure, concentrated Cu electrolytes from dilute, impure leach liquors. It is a crucial step in the production of ~4.64 million tonnes of copper cathode per year (International Copper Study Group, 2018). The combined technology of SXEW is important in enabling lower-grade materials to be economically treated for recovery of copper by hydrometallurgical techniques.

15.1 The solvent extraction process

SX flowsheets used in purifying of copper leach liquors comprise two essential continuous processes—*extraction* and *stripping* (Fig. 15.1):

(a) In the *extraction* process, the PLS (feed to the SX process) (containing 1−20 g/L Cu^{2+}, 0.5−15 g/L H_2SO_4) is contacted with an organic phase containing an organic *extractant* that reacts selectively with Cu. The extractant complexes with Cu^{2+}, resulting in transfer of the metal ion from the aqueous phase into the organic phase, leaving all other impurity species present in the PLS in the aqueous phase.

(b) The organic phase, now containing Cu (known as the *loaded organic* phase), and the aqueous phase, now depleted of Cu (known as the *raffinate*), are separated by gravity.

(c) The raffinate is recycled back to the leach circuit where the acid generated in the extraction process can be used (see Section 15.2).

Extractive Metallurgy of Copper, Sixth Edition
ISBN 978-0-12-821875-4, https://doi.org/10.1016/B978-0-12-821875-4.00020-1

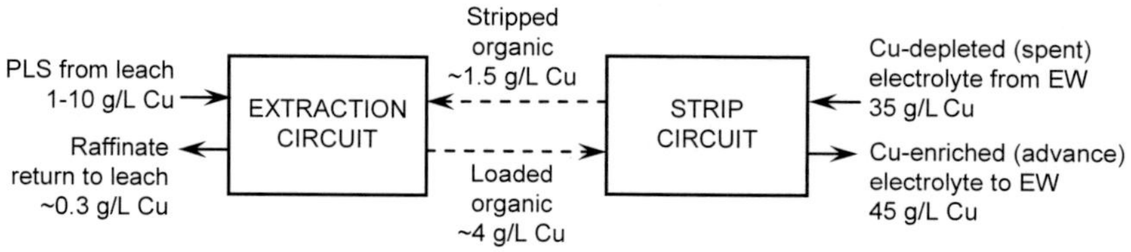

Figure 15.1 Schematic copper SX process. The inputs are pregnant leach solution (PLS) and Cu-depleted (spent) electrolyte. The products are Cu-enriched (advance) electrolyte and low-Cu raffinate. Aqueous streams are shown as *solids lines*; organic flows are indicated by *dotted lines*. The extraction and strip circuits may each contain several stages, through which the aqueous and organic streams flow in counter-current directions (see Section 15.5). The PLS can contain a wide range of Cu concentrations; the SX output, however, delivers a consistent electrolyte composition to the EW circuit. Values shown in this diagram are indicative of Chilean operating conditions.

(d) In the *stripping* process, the Cu-loaded organic phase is contacted with a strong acid *spent* (or *lean*, *barren*, *weak*) *electrolyte* (175–190 g/L H_2SO_4) from the EW circuit, which strips Cu from the organic into the electrolyte.

(e) The Cu-depleted organic phase (known as the *stripped* or *barren organic*) and the Cu-enriched aqueous phase (known as the *advance*, *rich*, *loaded*, or *pregnant electrolyte*) are separated by gravity.

(f) The stripped organic phase is recycled back to the extraction circuit for renewed contact with fresh PLS.

(g) The advance electrolyte is sent to EW where Cu^{2+} is reduced and electrodeposited on the cathode as pure metallic copper.

15.2 Chemistry of copper solvent extraction

The organic extractant removes Cu^{2+} from the PLS by the following extraction reaction:

$$\underset{\text{stripped organic}}{2\,RH_{(org)}} + \underset{\text{PLS }(1-20\text{ g/L Cu})}{Cu^{2+}_{(aq)} + SO_4^{2-}{}_{(aq)}} \rightarrow \underset{\text{loaded organic}}{R_2Cu_{(org)}} + \underset{\text{raffinate }(0.05-1\text{ g/L Cu})}{2\,H^{+}_{(aq)} + SO_4^{2-}{}_{(aq)}} \quad (15.1)$$

where RH represents the extractant (see Section 15.3).

From the Law of Mass Action (Le Chatelier's Principle), loading of the extractant with Cu^{2+} is favored by a low acid (H^+) concentration in the aqueous phase. The dilute H_2SO_4 concentration of the PLS (pH 1.5–2.5) permits extraction of Cu^{2+} from the aqueous phase into the organic phase. Two protons (H^+) are generated for each Cu^{2+} that is extracted, so the extraction reaction produces acid.

The organic and aqueous phases are then separated by gravity. The raffinate, which is depleted in Cu^{2+} but enriched in H^+, is returned to the leach circuit to dissolve more copper. The copper-loaded organic phase is sent forward to the strip circuit where copper is stripped back into an aqueous phase using the high acid content of Cu^{2+}-depleted spent electrolyte. The strip reaction is the reverse of Eq. (15.1):

$$\underset{\substack{\text{high acid, Cu}-\\ \text{depleted electrolyte}\\ (\sim 180\text{ g/L }H_2SO_4,\\ \sim 35\text{ g/L Cu})}}{2\,H^+_{(aq)} + SO_4^{2-}{}_{(aq)}} + \underset{\substack{\text{loaded}\\ \text{organic}\\ \text{extractant}}}{R_2Cu_{(org)}} \rightarrow \underset{\substack{\text{stripped}\\ \text{organic}\\ \text{extractant}}}{2\,RH_{(org)}} + \underset{\substack{\text{Cu}-\text{replenished}\\ \text{electrolyte}\\ (\sim 165\text{ g/L }H_2SO_4,\\ \sim 45\text{ g/L Cu})}}{Cu^{2+}_{(aq)} + SO_4^{2-}{}_{(aq)}} \quad (15.2)$$

The equilibrium of the strip reaction is pushed to the right by the high acid concentration of the spent electrolyte. Cu is stripped from the organic extractant, which regenerates the extractant for recycle back to the extraction circuit. The electrolyte is enriched to the required high Cu^{2+} concentration necessary for EW.

SX is a reversible equilibrium-controlled process, which operates on a "pH swing." Extraction is favored by low aqueous-phase acidity and stripping is favored when the aqueous phase contains a high acid concentration.

15.3 Composition of the organic phase

The organic phase comprises two essential components:

(a) the *extractant* is the active molecule that complexes with Cu^{2+}, enabling it to be transferred from the aqueous to the organic phase;

(b) the *diluent* is an inert hydrocarbon carrier for the extractant that is added to reduce the viscosity of the organic phase (so that it can be easily mixed with the aqueous phases to allow mass transfer to occur) and the density of the organic phase (to allow for gravity separation of the two phases).

The organic phase should be immiscible with $CuSO_4-H_2SO_4$ solutions and fluid enough (viscosity of 2–10 cP) for pumping, continuous mixing, and gravity separation.

15.3.1 Extractants

Modern copper extractants are based on the oxime functionality. These molecules complex with the Cu^{2+} ion in a process known as chelation ("chelate" coming from the Greek word for "claw") (Fig. 15.2). It is the nature of this chelating reaction that allows these extractants to react very selectively with Cu, rejecting other species in the PLS to the raffinate.

There are four main classes of modern copper extractants (Table 15.1): *ketoximes*, *modified aldoximes*, *modified aldoxime–ketoxime mixtures*, and *aldoxime–ketoxime mixtures*. The selection of the extractant that is most appropriate for a given application depends on the properties of the PLS (including the pH, Cu tenor, and nature of impurities),

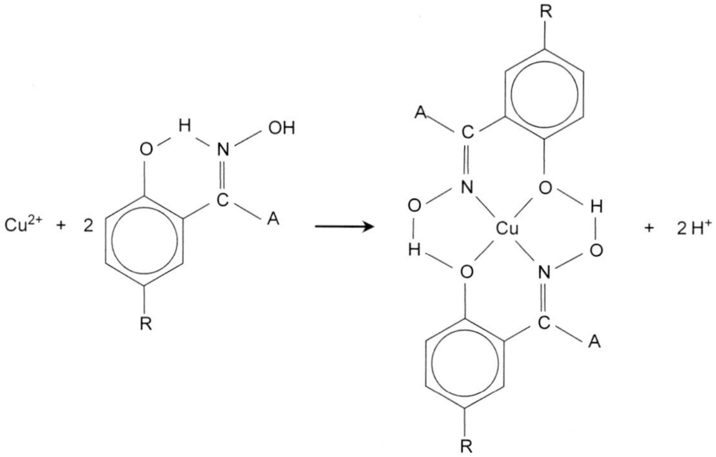

Figure 15.2 Structure of oxime extractant and Cu—oxime complex. The copper complex is formed from reaction of a Cu^{2+} cation with two oxime molecules, releasing two protons (Eq. 15.1). For aldoximes: $R = C_9H_{19}$ or $C_{12}H_{25}$, A = H; for ketoximes: $R = C_9H_{19}$, A = CH_3.

Table 15.1 Properties of modern copper extractants.

Property	Modified aldoximes[a]	Aldoxime—ketoxime mixtures[b]	Modified aldoxime—ketoxime mixtures[b]	Ketoximes
Extractive strength	Strong/ customized	Strong/customized	Customized	Weak
Stripping ability	Customized	Customized	Very good	Excellent
Cu/Fe selectivity	Excellent	Moderate	Good	Low
Kinetics of extraction	Very fast	Fast	Fast	Slower
Rate of phase separation	Very fast	Very fast	Very fast	Very fast
Chemical stability	Very good	Very good	Very good	Excellent
Crud generation	Dependent on PLS and operation			
Entrainment losses	Dependent on operation and diluent			

[a] Esters are most commonly employed as the modifier. The strength of the extractant depends on the relative aldoxime: modifier ratio.
[b]Aldoxime—ketoxime extractants are customized by their relative quantities: the greater the proportion of aldoxime, the stronger is the extractant, *i.e.*, the more Cu it will extract at lower pH values.
Adapted from Sole, K. C. (2008). Solvent extraction in the hydrometallurgical processing and purification of metals: Process design and selected applications. In M. Aguilar, M., & J. L. Cortina (Eds.), *Solvent extraction and liquid membranes: Fundamentals and applications in new materials* (pp. 141—200). New York: Taylor and Francis.

the characteristics of the advance electrolyte required, the number of extraction and strip stages available, and, often, on the prevailing economic conditions (which may influence whether a particular plant is trying to achieve maximum copper recovery from the PLS or maximum copper throughput) (Kordosky, 2008). Extractants are typically employed at concentrations ranging from 5 to 40 vol.% in the diluent. The extractant concentration chosen depends on the copper tenor of the PLS (see Section 15.7).

Ketoximes are relatively weak, low-selectivity extractants, but have high chemical stability. They are therefore seldom used on their own, but are favored for niche applications, such as the treatment of leach solutions that contain species like nitrates (common at some plants in the Atacama Desert of Chile), which are strong oxidizing agents and would otherwise rapidly degrade the extractant (Hurtado-Guzmán & Menacho, 2003; Virnig, Eyzaguirre, Jo, & Calderon, 2003).

Aldoximes are considered strong extractants because they are able to extract copper from high-tenor and low pH solutions; however, they can only be stripped by very high acid concentrations (+225 g/L H_2SO_4). Such high acid degrades the extractant and is too corrosive for industrial EW. Aldoximes are therefore used in mixtures with ketoximes and/or modifiers, such as highly branched alcohols or esters (Maes, Tinkler, Moore, & Mejías, 2003). Modified aldoximes can be stripped at an acid strength compatible with that of copper EW electrolytes and can be tailored to the desired extractant strength. They exhibit better selectivity for Cu over Fe than aldoxime—ketoxime mixtures and are often favored when the PLS has a high iron content. Aldoxime—ketoxime mixtures allow lower acid concentrations to be used for stripping and are appropriate when PLS Cu tenors are lower and there is a low concentration of Fe in the PLS. Mixtures of modified aldoximes and ketoximes have properties intermediate between modified aldoximes and aldoxime—ketoxime mixtures.

15.3.2 Diluents

The selection of the diluent in which the extractant is dissolved depends on local availability, the concentration of extractant employed, and safety, health, and environmental considerations. Diluents are moderately refined, high-flashpoint petroleum distillation byproducts (purified kerosene) from the petrochemical industry. They range in aromaticity from aliphatic (fully saturated hydrocarbon chains) to partially aromatic (containing unsaturated ring structures). Some of the advantages and disadvantages of choosing an aliphatic over partially aromatic diluent for Cu SX are given in Table 15.2.

Modern copper plants are concerned with minimizing their environmental impact and carbon footprint, and ensuring safe working conditions for their operators (Hopkins, 2005; Miller, 2005; Smith, 2009). For these reasons, a diluent with the lowest possible aromaticity that still gives favorable chemical and physical behavior is usually selected. Diluents with <20% aromatic content are typically preferred, because these provide

Table 15.2 Comparative properties of aliphatic and partially aromatic diluents.

Advantages of aliphatic diluents
• higher flash point assists in minimizing fire risk • lower volatility reduces evaporation losses and improves operator environment • lower rate of oxidative degradation • lower carbon footprint
Disadvantages of aliphatic diluents
• increased viscosity, especially at lower temperatures • can reduce solubilization of complexes, especially at high extractant concentrations • reduced solubility of impurities, which may increase crud formation • less selective for Fe transfer by diluent • slower rate of phase separation • higher organic-in-aqueous entrainment, leading to higher extractant and diluent consumptions • May have poorer crud management

the best compromise of properties: good solvation of the copper—extractant complex in the organic phase, good phase separation from the aqueous phases, and acceptable health and safety characteristics.

15.4 Equipment

Copper SX is carried out in mixer-settlers (Fig. 15.3). While details of the equipment design vary (see, for example, Currie & Miller, 2018; Hursi & Saario, 2018; Mallory, Kehn, Iniguez, & Filgueras, 2016; Taylor, 2007), the essential functions always consist of the following steps:

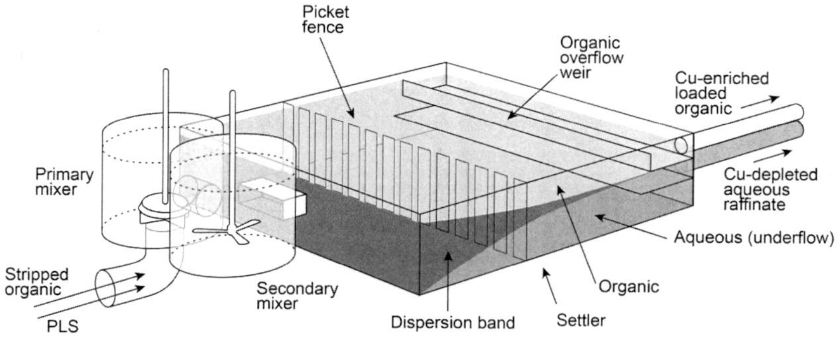

Figure 15.3 Conventional mixer-setter used for copper SX. Note the two mixing compartments, the large settler, and the organic overflow/aqueous underflow system. Flow is distributed evenly across the settler by passing the emulsion from the mixer first through a chevron (not shown) and then through one or more picket fences.

(a) the aqueous and organic phases are pumped into a mixer at predetermined rates;
(b) the two phases are mixed with an impeller in two or three mixing stages;
(c) the aqueous–organic mixture overflows from the mixer through flow distributors into a flat settler where the phases separate by gravity as they move along the length of the settler based on the difference in their specific gravities (organic and aqueous phases have specific gravities of approximately 0.85 and 1.1, respectively);
(d) the organic phase overflows and the aqueous phase underflows a weir at the far end of the settler, leading to separation of the two phases.

Flowrates of aqueous and organic phases through mixer-settlers can range from 2 to 4000 m^3/h.

15.4.1 Mixer designs

The *mixer* is designed to create a well-mixed aqueous–organic dispersion to ensure good mass transfer for the extraction and stripping reactions. The drop-size distribution should be as narrow as possible, with minimum air entrainment, which can stabilize crud (see Section 15.8.3), and without forming a stable emulsion (which inhibits subsequent phase separation).

Modern mixers consist of two or three mixing chambers. The smaller first chamber employs a pumping impeller (~0.02 kWh/m^3) because it is required to *pump* the two phases into the mixer from an adjacent mixer-settler or a holding tank. Subsequent mixers are larger and operate more gently (~0.01 kWh/m^3), serving to *maximize mass transfer*. The extraction and stripping kinetics of copper SX systems are relatively slow: mixer aqueous–organic contact times of 120–180 s are required for the chemistry to approach equilibrium. Entrainment of very fine droplets is avoided by using low tip-speed (<350 m/min) impellers. Modern mixer designs, illustrated in Fig. 15.4, enable good mass transfer while minimizing the formation of fine drops (Hursi & Saario, 2018; Kokotov et al., 2010).

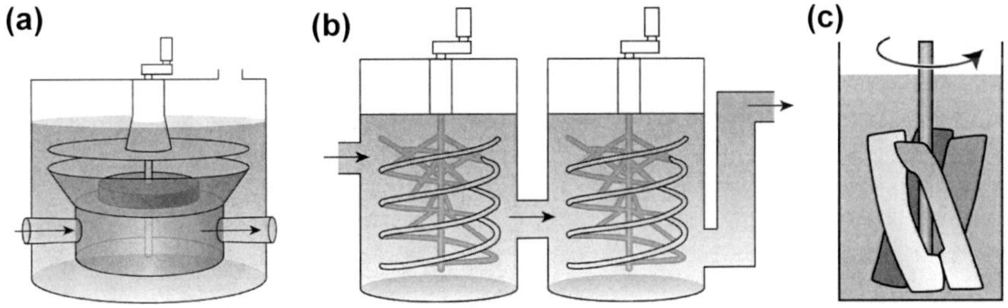

Figure 15.4 Modern mixer designs. The pumping and mixing functions are separated in Metso Outotec technology, using first (a) a dispersion overflow pump (DOP) mixer, able to pump large flow volumes into the mixer, which achieves high mass-transfer efficiency while controlling the size of droplets formed, followed by (b) Spirok mixers which allow gentle mixing of the emulsion and minimize crud formation (Pekkala, Laitala, Nyman, Ekma, & Weatherseed, 2007). In the Tenova turbulent settling design (c), homogeneous mixing with a narrow drop size distribution enhances the subsequent settling step and minimizes entrainment and crud losses (Kokotov et al., 2010).

15.4.2 Settler designs

The *settler* is designed to separate the dispersion into separate aqueous and organic layers. When the aqueous—organic dispersion overflows from the mixer, it passes through one or more flow distributors to give smooth, uniform forward flow that allows the aqueous and organic phases to separate across the full width of the settler (Taylor, 2007)

Large modern settlers are often square in plan and can vary in length up to 30 m, depending on the PLS flowrate. Settlers are usually wide to minimize linear velocity at low organic depths and to minimize the crest height of the organic phase overflowing the weirs. Settlers have a typical depth of 1 m. The aqueous phase is ~0.5 m deep, while the organic phase is ~0.3 m deep. The vertical position of the aqueous—organic interface is controlled by an adjustable weir. Conventional copper SX settlers are usually designed for a settling rate of 4—5 m^3/m^2h and a maximum linear velocity of each phase of 0.06 m/s. This means that residence times of each phase in the settlers are 10—20 min, which is sufficient to achieve complete phase separation.

The main function of the settler is to ensure smooth horizontal flow and clean separation (no entrainment) of the two phases, so considerable effort has gone into various designs to promote coalescence and phase disengagement. The most common of these are flow distributors, known as *chevrons* or *picket fences*, which can be installed at various positions along the length of the settler (Fig. 15.5A). They assist in slowing down the velocity of the emulsion, which backs up behind the installation and the emulsion itself acts as a coalescence medium. Construction using a hydrophobic or hydrophilic material can assist with coalescence of the dispersed phase. Picket fences can be prone to mechanical blockages if solids enter the settler or precipitation occurs.

Alternative configurations include the use of *reverse—flow settlers* (Fig. 15.5C), in which the mixer overflow emulsion is directed into a launder running along the side of a settler, before passing through an angled chevron and then settling in a direction back toward the mixer (Tenova Advanced Technologies, 2018). The *deep dispersion gate* (DDG) and *vertical smooth flow* (VSFX) are other patented designs that promote smooth flow in the settler, allowing additional time for the coalescence process (Fig. 15.5D, (Hursi, Saario, & Weatherseed, 2018; Metso Outotec, 2020).

15.5 Circuit configurations

15.5.1 Series circuit

The simplest Cu SX circuit typically has two extraction stages in *series*, with counter-current flow of the aqueous and organic phases, and a single strip stage (Fig. 15.6A). The two extraction stages typically transfer ~90% of Cu present in the PLS onto the extractant: the remaining Cu in the raffinate is not lost—it merely circulates around the leach circuit. The single strip stage strips 50%—65% of the Cu from the loaded organic into the advance electrolyte (depending on the extractant and operating conditions): the

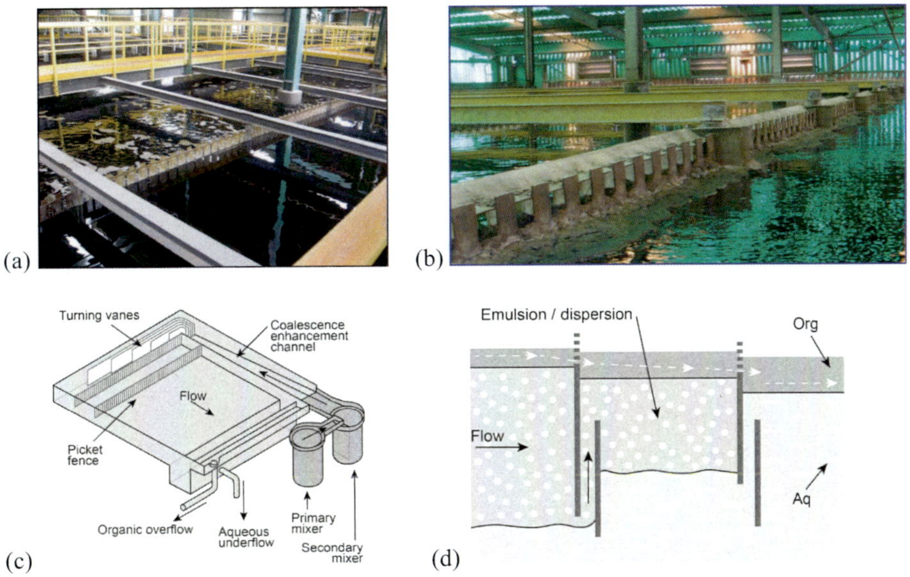

Figure 15.5 Modern settler designs include features that enhance droplet coalescence and phase separation from the emulsion overflowing from the mixer. Picket fences across a settler restrict the movement of (a) turbulence and bubbles and (b) crud through the circuit. (c) In reverse-flow settlers, the emulsion is directed along a launder on the side of the settler, then is further slowed by turning vanes before settling uniformly across the settler in a reverse direction. (d) Deep dispersion gate (DDG) system for slowing flow and assisting coalescence (Pekkala, Laitala, Nyman, Ekma, & Weatherseed, 2007).

remainder circulates around the SX circuit. This 2E−1S configuration tends to predominate in the industry, although 2E−2S circuits are common in Chile and 3E−2S circuits are widely found in Africa.

The efficiency of Cu transfer can be increased by adding extraction and strip stages to the circuit, increasing the extractant concentration, or changing the relative flowrates of the organic and aqueous flowrates (the *O/A ratio*). Depending on the characteristics of the PLS (pH, Cu tenor, and impurities), up to three extraction stages and two strip stages may be employed.

In plants that have very large PLS flowrates, the PLS may be split into several *trains*. Each train is processed in an identical SX circuit, capable of treating 500−4000 m^3/h. Each train may transfer 20−380 t Cu from PLS to electrolyte per day, depending on the Cu content and flowrate of the PLS. Modern engineering practice often chooses this flow arrangement because it allows for minimizing production losses that may occur should one circuit become nonfunctional in case of fire or other damage because additional flow can then be routed through the remaining trains (Hopkins, 2005).

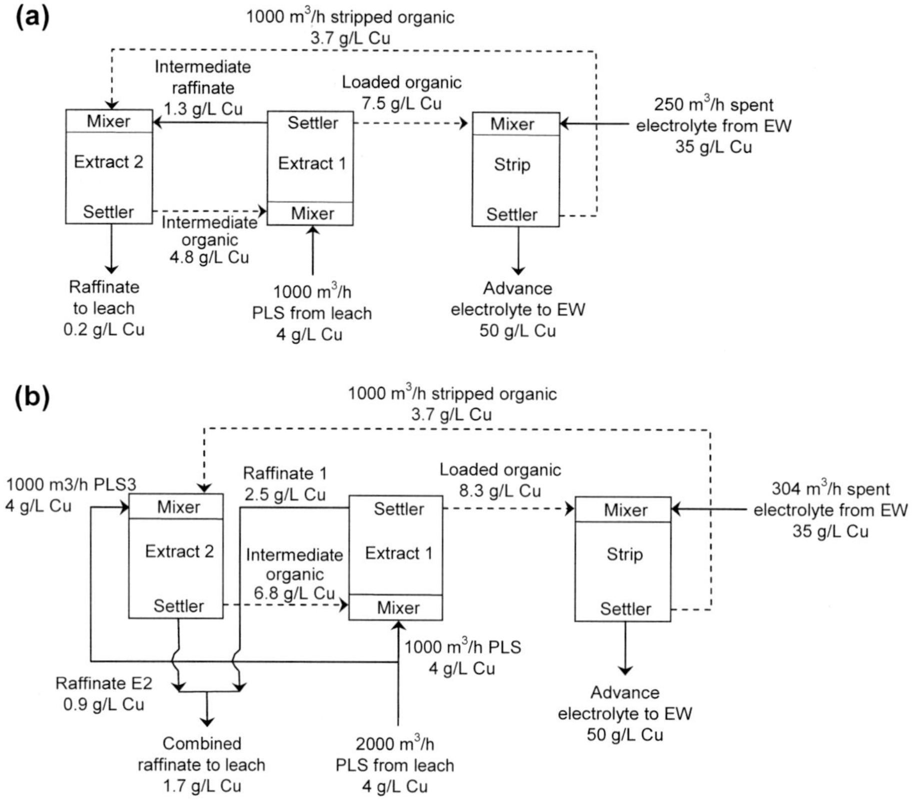

Figure 15.6 Plan views of (a) series and (b) parallel SX circuits with two extraction mixer-settlers and one strip mixer-settler. Industrial mixer-settlers are built close together to minimize plant area and flow distances. This is also the reason that most plants are arranged in a "head-to-tail" arrangement, as shown. Note the increased copper production that can result from a series–parallel arrangement, albeit with a higher raffinate concentration (lower overall copper recovery from PLS). Details of the flowrates and Cu concentrations are given in Tables 15.3 and 15.4.

15.5.2 Parallel and series–parallel circuits

To increase copper production, a *parallel* or *series–parallel* flow arrangement may be used instead of series flow in the extraction circuit. In the example shown in Fig. 15.6B, PLS is fed in parallel into the mixers of both extraction stages, while the flow of the organic phase remains the same as in the series circuit.

This arrangement allows the same circuit to process double the PLS flowrate, provided that the extractant concentration is increased to transfer the additional copper. The overall rate of copper transfer from the PLS to the EW circuit can therefore be increased. Introducing a parallel flow in a circuit is a low-cost way of increasing the productivity of an existing series plant. A plant will often be converted from a series to parallel or series—parallel flow configuration toward the end of the plant life. When the ore grades are reduced, the Cu concentration in the leach liquor decreases but the leach volumes remain the same, so copper production can be maintained by increasing the PLS throughout (Nisbett et al., 2008, Nisbett, Peabody, & Crane, 2009).

The disadvantage of this circuit is that overall copper recoveries are reduced (*i.e.*, raffinate Cu tenors increase) due to lower efficiency of extracting Cu from the PLS into the organic than using a series circuit. This is because (i) the parallel portion of the PLS goes through only one extraction mixer-settler (rather than two) before returning to leach and (ii) the series portion of the PLS is contacted with organic phase that is already partially loaded with Cu in the second extraction stage, so extraction efficiency in the first extraction stage is reduced.

15.5.3 Inclusion of a wash stage

In some circuits, a *wash* stage is included is included between the extraction and strip circuits. The purpose of this circuit is to "*wash out*" impurities that are physically entrained in the loaded organic phase to avoid these being transferred to the electrolyte. The wash liquor is typically dilute H_2SO_4 (pH 2). The spent wash liquor is returned to the extraction circuit to ensure that no copper is lost.

Examples of impurities that may be carried across are chlorides and nitrates (anionic species that are not chemically complexed by the extractant), as well as Fe, Mn, and Co that may be present in the PLS (Bednarski & McCallum, 2016). Many plants in the Atacama Desert of Chile, where the PLS contains significant quantities of chlorides and nitrates, use a wash stage (Hein, 2005).

15.6 Quantitative design of a series circuit

This section describes how a preliminary design of a series SX circuit is carried out. It is based on the design criteria in Table 15.3 and the circuit illustrated in Fig. 15.6A.

Table 15.3 Preliminary specifications for the design of a Cu SX plant.

No of extraction stages	2
No of strip stages	1
PLS composition	4 g/L Cu, pH 2
PLS flowrate to be processed	1000 m^3/h
Specified O/A ratio in extraction circuit	1/1 (m^3/h organic flow/m^3/h aqueous flow)[a]
Composition of spent electrolyte	35 g/L Cu, 180 g/L H_2SO_4
Composition of advance electrolyte required	50 g/L Cu

[a] An O/A ratio of ~1 permits good mass transfer in the mixer (see Section 15.9.3).

15.6.1 Determination of extractant concentration required

As a starting point, one "rule of thumb" used in the industry is that for each 1 g/L Cu extraction required, the concentration of extractant required will be approximately 4 vol.%. Therefore, for a PLS containing 4 g/L Cu, an extractant concentration of 16 vol.% is required. The SX plant requires 1000 m^3/h of 16 vol.% extractant diluted in an appropriate diluent (see Section 15.3.2) to be pumped into the extraction circuit.

This calculation is the first step in selecting the composition of the organic phase for a proposed SX circuit. The chosen organic must then be tested with actual leach and EW solutions to ensure suitability for the proposed operation.

15.6.2 Determination of extraction and stripping isotherms

Experimental laboratory test work is carried out to determine the equilibrium loading and stripping characteristics of the selected organic phase with the PLS and electrolyte compositions. An extraction isotherm is measured by contacting the PLS and stripped organic phase batchwise at different O/A ratios, allowing each system to reach equilibrium. The Cu concentrations in each phase are analyzed. By plotting the Cu in the aqueous phase against the Cu in the organic phase, the maximum loading of Cu into that particular organic phase can be determined (Fig. 15.7A). Stripping isotherms can be similarly determined using loaded organic and spent electrolyte as the input streams.

15.6.3 Determination of extraction efficiency

Using the experimental extraction isotherm and the specified extraction O/A, a *McCabe—Thiele construction* can be employed to determine the raffinate concentration that will be produced from two stages of counter-current contact of PLS and stripped organic. The graphical output is seen in Fig. 15.7A. This shows that, for the operating conditions specified in Table 15.3, the raffinate will contain 0.2 g/L Cu, which corresponds to an extraction efficiency of 94.8%. This calculation also shows that the maximum loading of the extractant is 7.5 g/L Cu under these conditions.

15.6.4 Determination of equilibrium stripped organic Cu concentration

The high H_2SO_4 concentration in spent electrolyte (180 g/L) strips Cu from the loaded organic into the electrolyte (Eq. 15.2). From the strip isotherm (Fig. 15.7A), it is determined that the 50 g/L Cu electrolyte specified for return to the EW tankhouse will be at equilibrium with ~3.8 g/L Cu in the stripped organic phase.

15.6.5 Transfer of Cu extraction into organic phase

The overall mass flow rate for the extraction of Cu into the organic phase is given by the following mass balance:

$$\text{Cu extraction (kg/h Cu)} = \text{PLS flowrate}\left(m^3/h\right) \times (\text{Cu in PLS} - \text{Cu in raffinate})\,(g/L) \tag{15.3}$$

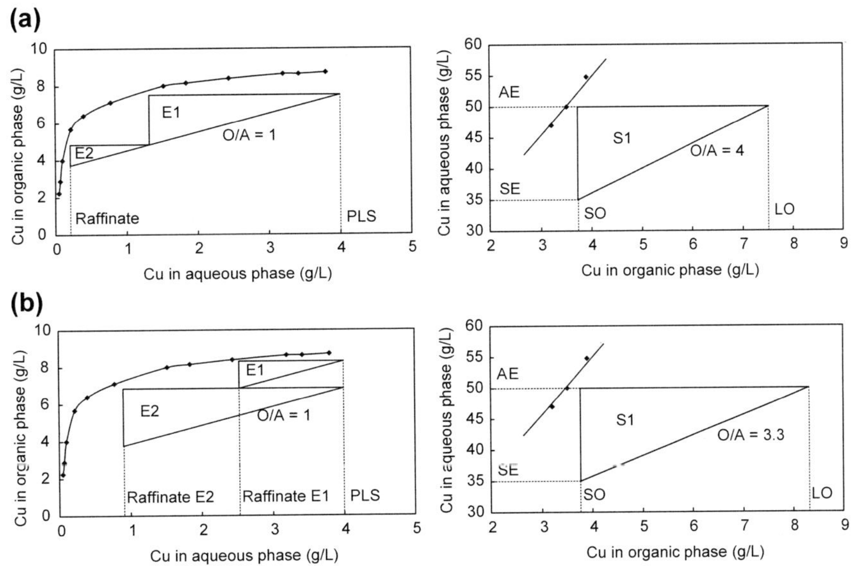

Figure 15.7 Extraction and stripping isotherms for (a) series and (b) parallel circuits shown in Fig. 15.6. McCabe–Thiele constructions are used to determine the Cu concentrations of the raffinate and stripped organic (SO). *AE*, advance electrolyte; *LO*, loaded organic; *SE*, spent electrolyte. Note that a *stage efficiency* of 90% has been assumed for both extraction and stripping. This means that mixing is only 90% efficient, as is common in practice. For quantitative mixing, the left-hand sides of the McCabe–Thiele horizontal construction lines would coincide with the equilibrium isotherm lines.

In this case (and converting to appropriate units)

$$\text{Cu extraction (kg/h Cu)}$$
$$= 1000\ \text{m}^3/\text{h PLS} \times (4.0\ \text{g/L Cu in PLS} - 0.2\ \text{g/L Cu in raffinate}) = 3800\ \text{kg/h Cu}.$$

This is also the overall mass flow rate at which metallic copper will have to be plated in the EW tankhouse. This allows the design engineer to calculate the cathode area and current density required for the proposed EW plant (see Chapter 16).

15.6.6 Determination of electrolyte flowrate required to strip Cu transferred

Section 15.6.5 indicates that the EW plant must plate 3800 kg/h metallic copper. For the strip circuit, this means that 3800 kg/h Cu must be transferred from the loaded organic phase to the electrolyte. By mass balance and using the specified compositions of spent and advance electrolyte (Table 15.3), the rate at which electrolyte must flow into and out of the strip mixer-settler can be calculated:

$$\begin{matrix}\text{Rate of Cu} \\ \text{plating in EW} \\ (\text{kg/h Cu})\end{matrix} = \begin{matrix}\text{electrolyte} \\ \text{flowrate}\left(\text{m}^3/\text{h}\right)\end{matrix} \times \begin{pmatrix}\text{Cu in electrolyte} & - & \text{Cu in electrolyte} \\ \text{leaving SX(g/L Cu)} & & \text{entering SX(g/L Cu)}\end{pmatrix} \tag{15.4}$$

The electrolyte flowrate required is therefore the following:

$$\text{electrolyte flowrate} = \frac{3800\ \text{kg/h Cu}}{(50 - 35)\ \text{kg/m}^3\text{Cu}} = 253\ \text{m}^3/\text{h}.$$

In other words, the O/A in the strip circuit will be 1000 m^3/h (organic flowrate)/ 253 m^3/h (aqueous flowrate) = 4.

Fig. 15.6A summarizes the flows and Cu concentrations in the newly designed plant.

15.6.7 Alternative approach

In the previous example, the relative flowrates of the aqueous and organic phases were specified (O/A ratio) and the extraction efficiency that could be achieved using these operating parameters was calculated from the McCabe—Thiele construction. An alternative approach to the circuit design is to specify the extraction efficiency required, say 97%, and then calculate what O/A is required to achieve this value. It may be necessary to add a third extraction stage to achieve the specified extraction efficiency. Alternatively, by using two strip stages (instead of one), the stripping will be more efficient, so the Cu concentration on the loaded organic phase will be lower and greater transfer of Cu will be possible.

Sophisticated software is today available that is able to quickly and accurately model such scenarios. This allows a circuit to be designed and tested for robustness using a computer model, thereby minimizing the experimental work that is required to optimize a circuit design for a particular project (Du Preez & Taute, 2015; BASF, 2020; Solvay, 2020).

15.7 Quantitative comparison of series and series—parallel circuits

The influence of the circuit flow configuration on SX performance is illustrated by comparing the series circuit discussed in Section 15.7 with an equivalent series—parallel circuit (Fig. 15.6B). All parameters remain the same as indicated in Table 15.3: the only difference is that PLS is fed to both extraction stages.

The extraction and stripping isotherms are shown in Fig. 15.7B and the performance results are shown in Table 15.4. The series—parallel circuit allows a higher rate of Cu

Table 15.4 Comparison of the performance of series and parallel flow configurations using two extraction stages and one strip stage. The operating conditions are given in Table 15.3, the circuits are illustrated in Fig. 15.6, and the isotherms in Fig. 15.7.

	Series	Parallel
Cu in PLS (g/L)	4	4
Extractant concentration (vol.%)	16	16
Cu in raffinate E2 (g/L)	0.2	0.9
Cu in raffinate E1 (g/L)	—	2.5
Cu in total (combined) raffinate (g/L)	0.2	1.7
Cu on loaded organic (g/L)	7.5	8.3
Volume of PLS processed (m^3/h)	1000	2000
Cu recovery from PLS (%)	94.8	57.1
Cu transferred to strip (kg/h)	3800	4580

throughput (*i.e.*, increases overall Cu production), but at the expense of lower overall Cu recovery from the PLS. A series circuit will usually be converted to a series–parallel circuit when the Cu tenor of the PLS drops significantly (see Section 15.5.2).

15.8 Minimizing impurity transfer and maximizing electrolyte purity

To produce high-purity copper cathode in EW (Chapter 16), the advance electrolyte produced by the SX step must minimize contamination from impurities. An efficient extractant must therefore transfer Cu from the PLS to the electrolyte while not transferring impurities, particularly Fe, Mn, and Cl.

There are three mechanisms by which impurities can be carried forward to the electrolyte:

(a) *chemical transfer*, in which cationic impurity species are coextracted by the extractant along with Cu^{2+};

(b) *physical transfer*, in which finely divided droplets of aqueous phase are not completely separated from the organic phase after mixing and are transferred to the strip circuit in the loaded organic phase;

(c) *crud* formation, in which unwanted solids move through the circuit and can transfer adsorbed impurities.

15.8.1 Coextraction of impurities

Modern oxime extractants, particularly modified aldoximes, are very selective for Cu over other cations (see Fig. 15.8). Fe(III) is the only cation typically present in a leach solution that can complex to a limited extent with the extractant. Coextraction of Fe can be minimized by appropriate control of pH and by operating at a high loading of the extractant by Cu, which means that Fe is "pushed off" the organic phase due to lack of capacity (Tinkler, Cronje, Soto, Delvallee, & Hangoma, 2009). It should be

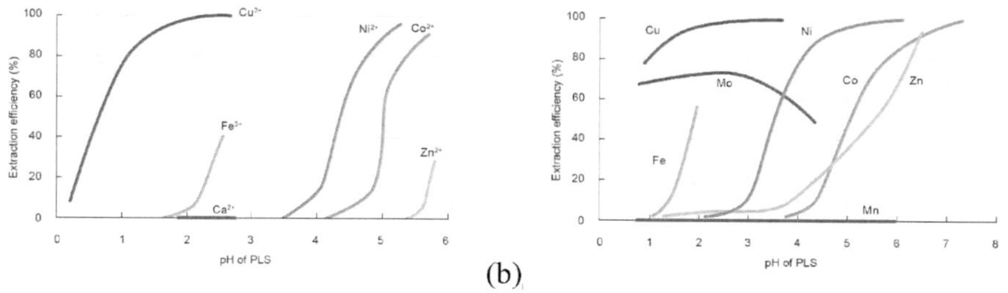

(a) (b)

Figure 15.8 Effect of pH on the selectivity of extraction of metal cations by (a) ketoxime and (b) aldoxime extractants. *(Data courtesy of BASF. (2020).* Redbook mining solutions. *http://www.basf.com. (Accessed 13 March 20)).*

noted, however, that the presence of up to 2 g/L Fe in the electrolyte is beneficial to the EW process (see Section 16.5.2.3).

15.8.2 Transfer of impurities to electrolyte by entrainment

Both cationic and anionic impurities can be transferred from the PLS to the advance electrolyte as dispersed droplets in the loaded organic phase. This is known as *aqueous-in-organic entrainment*. This can cause both beneficial and deleterious elements to deport to the electrolyte (Cole et al., 2016).

Anions present in the PLS are not chemically complexed by the cation exchanger employed as the extractant, but can be transferred through the SX circuit by entrainment. Chloride can have very detrimental effects in EW (see Section 16.5.2.6), causing pitting corrosion of stainless-steel cathode blanks and the evolution of corrosive Cl_2 gas. Nitrates are a strong oxidizing agent that can degrade the SX organic phase and cause the deposition of nodules in EW. Mn(II) that reports to EW can be oxidized to higher valence states, where it can cause loss of current efficiency and degradation of the organic phase in SX (Bednarksi, Soderstrom, & Wiggett, 2011).

Transfer of Fe from the PLS to the electrolyte is beneficial in small amounts (see Section 16.5.2.3) and particularly if high concentrations of Mn are present (Abbey & Moats, 2018; Miller, 2011). Deportment of Co from the leach liquor to EW is also helpful in reducing operating costs, because $CoSO_4$ needs to be added to the electrolyte to reduce anode oxidation (see Section 16.4.6.2).

In many of the African Copper belt operations, silica is present in high quantities in the ores and deports to the PLS, where it creates significant operating problems in SX (Sole & Tinkler, 2016). It can transfer through the circuit in colloidal form, cause crud formation, and cause severe difficulties in holding phase continuity and achieving phase separation. Approaches to dealing with silica include use of an upfront diluent wash (Alexander, van der Merwe, Lumbule, & Kgomo, 2018), changing phase continuity (see Section 15.13.2, Kashala et al., 2018), and use of various additives

(Bender, Catling, Milligan, & Delgado, 2018; Brown et al., 2018; Sole, Crundwell, Dlamini, & Kruger, 2018).

Physical entrainment of the dispersed phase in the continuous phase and its transfer through the SX circuit can be minimized by the following:

(a) appropriate choice of diluent to promote acceptable phase disengagement and to minimize transfer of neutral species by solvation by the diluent;
(b) choice of a low-density, low-viscosity extractant, particularly for circuits in which a high extractant concentration is required;
(c) avoiding overmixing and air entrainment into the mixing boxes (see Section 15.4.1);
(d) installing picket fences or other features in the settler to promote phase disengagement (see Section 15.4.2);
(e) ensuring adequate residence time for phase disengagement in the settlers;
(f) contacting the loaded organic phase in a holding tank with a hydrophobic media (such as polymer scrap) that will assist coalescence of the aqueous droplets;
(g) "washing out" droplets of PLS from the loaded organic in a *wash* stage (see Section 15.5.3).

15.8.3 Crud

Almost all SX circuits experience a phenomenon known as *crud* formation. Crud is a solid-stabilized emulsion that is formed from various contaminants in the SX circuit. These can include solids in the PLS (colloidal silica from dust or from silica present in the PLS), vegetation, mold, organic degradation precipitates, carryover of organic constituents from up- and down-stream operations, and air entrainment (Mbao et al., 2018; Sole, 2008).

Crud usually forms a layer at the aqueous—organic interface of the settler or floats on the surface of the organic phase. Small amounts of interfacial crud can be beneficial to phase separation, by acting as a coalescence medium. Large amounts can occupy a significant volume of the settler and may cause deterioration of the rate of phase separation. Floating crud is more harmful, because this can move from one stage to another with the organic phase, thereby transferring adsorbed and entrained impurities through the circuit to the electrolyte.

Crud formation is minimized by minimizing the amount of solids entering the SX circuit in the PLS. This is done by allowing the PLS to settle in ponds and by directing rain runoff away from PLS collection areas. Agitation and pressure leaching solutions can be processed through clarification equipment, such as countercurrent decantation thickeners followed by pinned-bed clarifiers, ahead of the SX circuit. Minimizing air entrainment to the mixers and dust egress to the circuit (by installed roofs over the SX mixer-settlers) and careful selection of extractant and diluent for a particular application can be helpful in mitigating the detrimental effects of crud (Sole, 2008). Additives that reduce crud formation can also be employed (Mbao et al., 2018).

(a) (b)

Figure 15.9 (a) Photograph showing measurement of interfacial crud accumulation between the aqueous and organic layers in an operating plant. (b) Vacuuming of crud from the bottom of the settler using Crud Forx. *(Courtesy of ProProcess).*

Crud is removed from the settler by regular vacuum pumping (Fig. 15.9, Megaw, Muzadi, Kashala, & Sanchez, 2016). In extreme cases, it may be necessary to periodically empty the settler and pump the crud from the settler floor. Crud is usually separated from its host organic by filtering on a plate-and-frame or vacuum-drum filter. A centrifuge can also be used (Hartmann, Horbach, & Kramer, 2010; Mukutuma, Mbao, Schwartz, Chisakuta, & Feather, 2007). The liquid portion can be washed out with diluent, organic, or water in a reslurry tank. The crud solids are usually sent to the leach heaps and the organic component is returned to the SX circuit after clay treatment (see Section 15.9.1, Cole, Zambra, Bednarski, Thomas, & Soderstrom, 2015).

15.9 Operational considerations

15.9.1 Stability of operation

Industrial SX circuits are easily controlled and forgiving, allowing a consistent electrolyte composition to be transferred to EW.

Consider, for example, how the circuit examined in Section 15.6 responds to an increase in Cu concentration in PLS (which would happen if more easily leached ore is encountered in the mine). Suppose that the PLS tenor increases from the 4 g/L Cu in Fig. 15.6A to 4.4 g/L Cu. Using the extraction isotherm of Fig. 15.7A with the new PLS composition, it can be shown that (Fig. 15.10A)

(a) the raffinate will now contain 0.3 g/L Cu rather than 0.2 g/L Cu;
(b) the overall extraction efficiency decreases from 95% to 93%;
(b) but 4090 kg/h Cu will be transferred to the loaded organic phase instead of 3800 kg Cu.

The resulting flows and Cu concentrations are shown in Fig. 15.7B. Of course, the rate at which copper is plated in EW will also have to increase to 4090 kg/h Cu. This can be done by increasing current density and/or by bringing unused cells into operation.

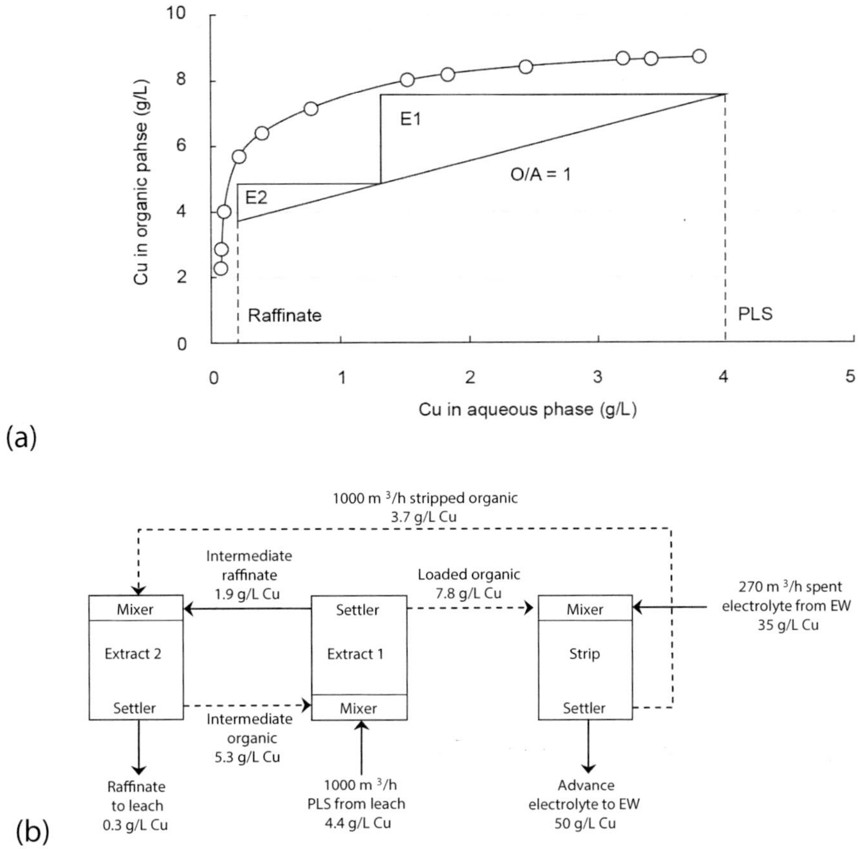

Figure 15.10 The SX circuit of Fig. 15.6A has been perturbed by receiving a PLS containing 4.4 g/L Cu instead of 4 g/L Cu. It is assumed that Cu EW rate has been increased (by increasing current density) to match the rate at which Cu is being transferred from PLS to electrolyte. (a) Modified McCabe–Thiele construction for the extraction isotherm shown in Fig. 15.7A for higher Cu concentration in PLS (4.4 g/L Cu). The construction for a PLS containing 4 g/L Cu is shown in dotted lines. (b) Modified concentrations and solution flows for the new PLS. Note that the only operating variable that has to be changed is the electrolyte recycle flowrate.

15.9.2 Phase continuity

When two phases are mixed, the emulsion that forms will be either *aqueous continuous*, containing organic droplets (the *dispersed* phase) in an aqueous matrix (the *continuous* phase), or it may be *organic continuous*, with aqueous droplets dispersed in an organic phase (Sole, 2008).

Copper SX plants usually operate with organic-continuous mixing. The rate of phase separation is usually faster from an organic-continuous emulsion. Organic continuity also ensures compact crud formation and minimizes crud movement through the circuit.

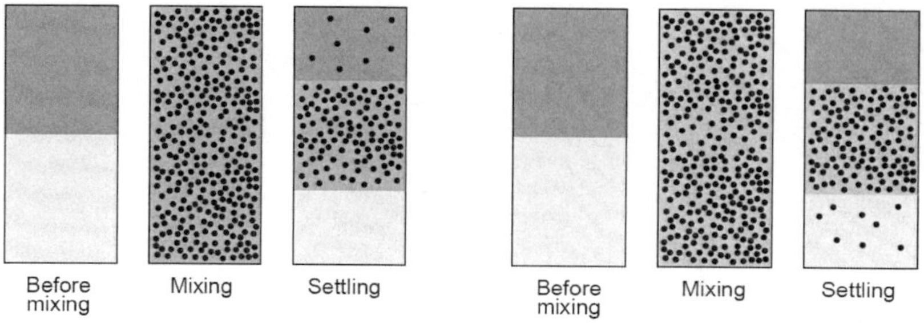

Figure 15.11 The effect of mixing in (a) organic- and (b) aqueous-phase continuity on the physical transfer of (a) aqueous droplets to the settled organic phase or (b) organic droplets to the settled aqueous phase. *(Adapted from Cramer, K., & Tinkler, O. (2019). Copper solvent extraction: Practical issues and troubleshooting. In B. Wassink, & E. Asselin (Eds.),* Copper hydrometallurgy: Principles and practice *(pp. 132–147). Montreal: Canadian Institute of Mining, Metallurgy and Petroleum).*

As shown in Fig. 15.11 (Cramer & Tinkler, 2019), the aqueous phase that settles out after organic-continuous mixing should contain no organic droplets, so losses to aqueous phases that leave the circuit (raffinate and advance electrolyte) are minimized. In certain instances, however, it may be more appropriate to operate specific stages in aqueous continuity.

15.9.3 Organic health, losses, and recovery

Ensuring ongoing health of the organic phase is critical to successful SX operation. The conditions of the organic phase affect both physical and metallurgical performance: poor organic quality can compromise overall copper recovery, decrease extraction and strip kinetics and Cu/Fe selectivity, increase phase disengagement times and entrainment, and cause crud problems (Cole, Zambra, Bednarski, Thomas, & Soderstrom, 2015). These issues will have detrimental effects on operating costs, reagent consumption, electrolyte purity, and the quality of the copper cathode.

One of the main operating costs of an SX plant is the replacement of extractant and diluent. The main cause of diluent loss is evaporation. Both extractant and diluent are lost to crud (Section 15.8.3) and by entrainment in the raffinate and advance electrolyte (Section 15.8.2). Both extractant and diluent can be destroyed by hydrolysis and chemical degradation. Appropriate choice of diluent and a roof over settlers can minimize evaporative losses (Bishop et al., 1999). Losses to crud can be prevented by minimizing crud formation in the circuit and recovering the contained organic phase (Section 15.9.3). Organic entrainment losses can be minimized or recovered by passing the aqueous stream through an aftersettler, or through a coalescer, flotation cell, activated carbon column, or dual media or CoMatrix filter. These types of equipment all allow additional settling time and promote coalescence, so that the organic phase can be recovered and returned to the SX circuit (Greene & Rubio, 2010).

The host organic is regenerated by mixing it with acid-activated bentonite clay or zeolite then filtering (Cole, Zambra, Bednarski, Thomas, & Soderstrom, 2015; Quilodrán, Soderstrom, & Zambra, 2015). This process also removes degradation products (formed by oxidative degradation of the extractant and diluent) from the organic phase, which can also be harmful to phase separation and extraction kinetics. The organic is then returned to the SX circuit.

15.10 Industrial solvent extraction plants

SX plants are designed to match the production rate at which Cu is leached in the preceding leach operation. They vary in capacity from 5 to 400 kt Cu per year (Table 2.7). Table 15.5 gives operational details of five SX plants. These include plants in which the leach liquors are generated by dump and run-of-mine leaching, heap leaching, and pressure leaching. The range of operating conditions is evident, demonstrating the versatility of this technology. Further details are given in Sole, Taute, Tinkler, Steeples, & Zarate, 2019.

15.11 Safety in solvent extraction plants

SX plants contain large volumes of combustible organic components, so the risk of a catastrophic fire is always present. Extreme care must be taken in both the design and operation of SX circuits to minimize and mitigate this hazard (Hopkins, 2005; Miller, 2005; Smith, 2009). In particular, the potential for the formation and occurrence of organic aerosols and mist droplets must be paid particular attention, because these can create conditions for static electricity.

Some design issues to consider include the following:

- Splitting the PLS into parallel trains that are spaced sufficiently far apart to avoid loss of the entire SX circuit in the event of an incident;
- Use conductive piping and vessels, minimize pipe velocities and pipe lengths, and ensure that pipes always run full;
- Minimize the energy input by use of low-turbulence mixer designs and low-pressure organic pump;
- All tanks should have bottom entry.

Operational good practices include the following:

- Avoiding potential for static electricity occurrence;
- Avoiding aerosol formation, particularly "waterfall" effects in launders and overflows;
- SX area should be fenced and access restricted;
- Only intrinsically safe radios, tools, and instrumentation should be used and hot-work permits required;
- Control spillage and regularly remove organic from aqueous tanks and ponds;
- Ensure that safe crud-removal procedures are in place;
- Use appropriate personal protective equipment to ensure that operators are always electrically grounded to earth.

Table 15.5 Selected industrial copper solvent-extraction plant data (2019). Additional details of other major SX plant operations are given by Sole, Taute, Tinkler, Steeples, & Zarate, 2019.

Operation	Lomas Bayas, Chile	Radomiro Tomić, Chile	Cobre Las Cruces, Spain	Kinsevere, DRC	Kansanshi, Zambia
Startup date	1998		2018	2018	2018
Nameplate capacity (kt Cu/a)	94	320	72	60	60
SX plant					
Number of trains	1	4	1	2	2
Series extraction stages/train	2	2	3	2	2
Parallel extraction stages/train	4	0	2	0	1
Strip stages/train	1	1	2	2	1
Wash stages/train	1	1	1	0	0
Mixers					
Design	Fluor Daniel	Outokumpu	Outotec	Lightnin	Mixtec
Material	Stainless steel 254 SMO	Vinyl resin FRP	Fiberglass	316L stainless steel	316L stainless steel
Primary impeller	Closed turbine	DOP	DOP	Agitator mixer with electric motor	Single top or top and bottom shrouded radial blade
Secondary impeller	Helical	Spirok	Helical spirok system	Agitator turbine mixer	Varying pitch angle, chord 3 blade (radial) axial upflow turbine
Mixer box shape	Square, cylindrical	Cylindrical	Square		Cylindrical
Extraction residence time (s)	43	180	125	100–120	120
Strip residence time (s)	43	120	94	85–90	120

Settlers					
Length × width × depth (m)	18 × 26 × 1.2 (7) 21 × 27 × 1.2 (2) 34 × 17 × 1.2 (1)	28.5 × 28.5 × 1	18.7 × 20.3 × 2	21 × 16.2	16.5 × 25 × 1.1
Picket fences per settler	2	3	4	3	2
Extraction settling velocity (m^3/m^2h)	4.5	4	3.37	4.3	4.6
Strip settling velocity (m^3/m^2h)	4	2.5	3.37	3.4	4.5
Material	Concrete lined with HDPE	Structural cement, HDPE	Concrete/ fiberglass	316L stainless steel	Concrete/fiberglass
Active height of settler (mm)	0.88		1.8	1	0.9
Organic depth (mm)	0.31	0.3	0.5	0.3	0.25
Organic phase					
Extractant	LIX 84IC/NR20	Acorga M5774	LIX 973NSC-LV	Acorga OPT5540	Acorga OPT5510/ LIX984NC
Diluent	Solvex Minería	Escaid 110	Escaid 110	ShellSol 2325	Shellsol 2325, Fluiden HF 1826
[Extractant] (vol.%)	27	18–20	32	20–25 (LG) 25–30 (HG)	22
Flowrate per train (m^3/h)	1200	1600	1150/140	780–820	1200
Extractant consumption (kg/t Cu)	7	2.09	2.8	3	>3.5
Diluent consumption (kg/t Cu)	30	8.5	3	12	29

Continued

Table 15.5 Selected industrial copper solvent-extraction plant data (2019). Additional details of other major SX plant operations are given by Sole, Taute, Tinkler, Steeples, & Zarate, 2019.—cont'd

Operation	Lomas Bayas, Chile	Radomiro Tomić, Chile	Cobre Las Cruces, Spain	Kinsevere, DRC	Kansanshi, Zambia
PLS					
Source	Heaps, ROM	Heaps	Atmospheric agitation leach	Agitated leach, CCD thickeners	Agitated leach, CCD thickeners
Series flowrate/train (m^3/h)	1000	1600	280	640 (HG) 630 (LG)	650
Parallel flowrate/train (m^3/h)	950	—	150		650
Cu (g/L)	1.75–1.90	4.5–5.5	38.0	10 (HG) 6–8 (LG)	2.4 (HG) 0.9 (LG)
Fe (g/L)	13	7.4	30.0	0.4–0.6	3.5
Co (g/L)	—	—	0.125	1.0–1.5	0.2
Mn (g/L)	4	1.6	0.192	1.5–2.0	0.35
Cl (g/L)	6000	9.1	1500	50–80	0.02
pH	2.3	1.3	0.8	1.5–2.0	1.3
Total suspended solids (mg/L)	5		100	50–150	30–300
Raffinate					
Cu (g/L)	0.3	0.35–0.45	11	0.6 (HG) 0.3 (LG)	0.07
H_2SO_4 (g/L)	4	6–8	60	18–20 (HG) 14–16 (LG)	15
Entrained organic (mg/L)	30	35	20	—	—

Spent electrolyte					
Cu (g/L)	43	35	38	35	35–38
H_2SO_4 (g/L)	195	195	185	180–195	185–190
Co (mg/L)	180	150–180	190	600–800	100–120
Flowrate/train (m^3/h)	850	550–620	690	320 (HG) 220 (LG)	300
Advance electrolyte					
Cu (g/L)	45	55	48	50–55	50–55
Entrained organic (mg/L)	3	6	10	–	–
Performance					
Extraction efficiency (%)	90	92–94	93	94–95	85–95
Cu transfer (g/L/vol.% extractant)	0.31	0.22–0.25	0.45	0.20–0.25	~0.15

15.12 Current and future developments

15.12.1 Extractants

With the development of flowsheets that are able to treat sulfide materials using hydrometallurgical techniques, the use of pressure oxidation for leaching is increasing. The need to process these high-temperature liquors has led to the recent introduction of an SX reagent that does not employ oxime functionality and is stable at elevated temperatures (Bender, Yañez, Copp, & McDonald, 2018).

Another application in which reagent degradation must be avoided is the processing of high-nitrate (up to 40 g/L) liquors, such as in some Chilean operations (Virnig, Eyzaguirre, Jo, & Calderon, 2003). Nitration-resistant extractant formulations are available for use at such sites (Yáñez, Soto, Soderstrom, & Bednarski, 2009). At sites that contain high levels of Mn in the leach liquor, oxidation-resistant formulations can be used to reduce the risk of degradation that could result from permanganate formation in EW owing to entrainment of PLS (Cramer & Tinkler, 2019).

15.12.2 Equipment

The enormous capital cost of building a new plant can be offset to an extent by shortening the project timeline from conception to installation. Many new plants are today being built in countries or locations that are difficult to access and/or have poor local engineering infrastructure. For these reasons, it is desirable to carry out as much of the manufacturing as possible prior to delivery of the equipment to site. This has been very successfully achieved by modularizing the design of mixer-settlers, so that they can be readily transported in standard container sizes (Fig. 15.12). Almost all equipment construction can be done in the country of origin under the direct supervision of the manufacturer and the final product can be readily installed on site in a very short time. This approach leads to lower engineering and construction costs, faster project delivery, and improved consistency, quality, and safety of the installed equipment (Hursi & Saario, 2018; Hursi et al., 2018).

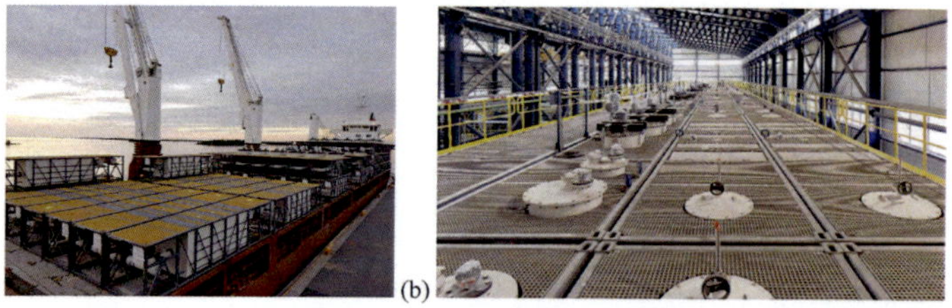
(a) (b)

Figure 15.12 VSFX modular mixer-settler units (a) awaiting shipping and (b) installed for commercial operation. *(Courtesy of Metso Outotec).*

15.13 Summary

Impure leach solutions containing relatively low concentrations of copper are processed to high copper concentration, pure EW electrolytes by SX. It is a crucial step in producing high-purity electrowon copper from leached ores.

The SX process consists of extracting Cu^{2+} from aqueous PLS into a copper-specific organic extractant, separating the aqueous and organic phases by gravity, and then stripping Cu^{2+} from the organic extractant into high-H_2SO_4 electrolyte returned from the EW circuit. The process is continuous and is carried out in large mixer-settlers.

The SX process is equilibrium-controlled. Loading of Cu into the organic extractant is favored by the low acid concentration (0.5−5 g/L H_2SO_4) of the PLS. Stripping of Cu from the organic extractant is favored by the high acid concentration (180−190 g/L H_2SO_4) of the electrolyte.

The copper-specific organic extractants used in the copper industry are modified aldoximes, ketoximes, and modified aldoxime−ketoxime and aldoxime−ketoxime mixtures. The relative strength of the extractant can be tailored for the particular application. The extractant is dissolved in a kerosene-type diluent to a concentration appropriate to the Cu tenor of the PLS (typically 5−40 vol.%).

About 4.64 million tonnes of Cu per year are treated by SX followed by EW. Copper production by SXEW continues to grow as increasing volumes of ore and concentrates are processed by leaching technologies.

References

Abbey, C., & Moats, M. (2018). Pb−Ca−Sn anode potential as a function of cobalt, iron and manganese in synthetic sulfuric acid electrolytes. In *Extraction 2018* (pp. 1535−1546). Montreal: Canadian Institute of Mining, Metallurgy and Petroleum.

Alexander, D., van der Merwe, C., Lumbule, R., & Kgomo, J. (2018). Innovative process design for oxide ores in the Democratic Republic of the Congo. In *Copper cobalt Africa* (pp. 225−238). Johannesburg: Southern African Institute of Mining and Metallurgy.

BASF. (2020). *Redbook mining solutions*. http://www.basf.com (Accessed 13 October 2020).

Bednarksi, A., Soderstrom, M., & Wiggett, S. (2011). Oxidation in Cu SX processes. In *ALTA copper 2011*. Melbourne: ALTA Metallurgical Services.

Bednarski, A., & McCallum, T. (2016). SX impurity transfer and wash stage considerations. In *SME annual meeting 2016*. Littleton, CO: Society for Mining, Metallurgy and Exploration.

Bender, J., Catling, M., Milligan, J., & Delgado, B. (2018). The impact of polymeric clarifying agents on the phase disengagement and kinetics of solvent extraction. In *ALTA nickel cobalt copper 2018* (pp. 267−274). Melbourne: ALTA Metallurgical Services.

Bender, J., Yañez, H., Copp, R., & McDonald, T. (2018). New generation of highly selective and degradation resistant reagents for copper extraction. In *HydroCopper 2018*. Santiago: Gecamin.

Bishop, M. D., Gray, L. A., Greene, M. G., Bauer, K., Young, T. L., May, J., ... Amerson-Treat, I. (1999). Investigation of evaporative losses in solvent extraction circuits. In S. K. Young, D. B. Dreisinger, R. P. Hackl, & D. G. Dixon (Eds.), *Copper 99, vol. IV, Hydrometallurgy of copper* (pp. 277–289). Warrendale, PA: The Minerals, Metals and Materials Society.

Brown, L., Lin, J., Paliza, L., Rees, K., Thiel, N., Boskovic, S., & Coveney, A. (2018). Ameliorating the impact of colloidal silica on copper SX – the MMG Sepon example. In *ALTA nickel cobalt copper 2018* (pp. 253–266). Melbourne: ALTA Metallurgical Services.

Cole, P., Bednarski, L., Thomas, D., Muteba, D., Banza, G., & Soderstrom, M. (2016). Understanding aqueous-in-organic entrainment in copper solvent extraction. *Journal of the Southern African Institute of Mining and Metallurgy, 116*, 525–531.

Cole, P., Zambra, R., Bednarski, T., Thomas, L., & Soderstrom, M. (2015). A holistic approach to organic quality in copper solvent extraction. In *ALTA nickel cobalt copper 2015*. Melbourne: ALTA Metallurgical Services.

Cramer, K., & Tinkler, O. (2019). Copper solvent extraction: Practical issues and troubleshooting. In B. Wassink, & E. Asselin (Eds.), *Copper hydrometallurgy: Principles and practice* (pp. 132–147). Montreal: Canadian Institute of Mining, Metallurgy and Petroleum.

Currie, B., & Miller, G. (2018). Design and material selection considerations for solvent extraction and electrowinning plants. In *ALTA nickel cobalt copper 2018* (pp. 283–308). Melbourne: ALTA Metallurgical Services.

Du Preez, N. P., & Taute, J. J. (2015). Evaluation of copper solvent-extraction circuit data and performance. In *Copper cobalt Africa 2015* (pp. 291–302). Johannesburg: Southern African Institute of Mining and Metallurgy.

Greene, W., & Rubio, D. (2010). Aqueous coalescer for electrolyte and raffinate entrained organic removal. In *Hydroprocess 2010* (pp. 90–92). Santiago: Gecamin.

Hartmann, T., Horbach, U., & Kramer, J. (2010). Copper crud treatment, concentration-dependent pond depth adjustment for decanter centrifuges, DControl®. In *Hydroprocess 2010* (pp. 82–83). Santiago: Gecamin.

Hein, H. (2005). The importance of a wash stage in copper solvent extraction. In J. M. Menacho, & J. Casas de Prada (Eds.), *HydroCopper 2005* (pp. 425–436). Santiago: Universidad de Chile.

Hopkins, W. (2005). Fire hazards in SX plant design-an update. In *ALTA SX/IX world summit on SX fire protection*. Melbourne: ALTA Metallurgical Services.

Hursi, T., & Saario, R. (2018). Outotec's VSF®X implementation to provide modular solvent extraction technology. In *Extraction 2018* (pp. 1987–1995). Warrendale, PA: The Minerals, Metals and Materials Society.

Hursi, T, Saario, R., & Weatherseed, M. (2018). Development, design and implementation of Outotec's VSF® X solvent extraction technology. In *ALTA nickel cobalt copper 2018* (pp. 275–282). Melbourne: ALTA Metallurgical Services.

Hurtado-Guzmán, C., & Menacho, J. M. (2003). Oxime degradation chemistry in copper solvent extraction plants. In P. A. Rivieros, D. G. Dixon, D. B. Dreisinger, & J. H. Menacho (Eds.), *Copper-Cobre 2003. Hydrometallurgy of copper: Modelling, impurity control and solvent extraction* (Vol. I, Book 2, pp. 719–734). Montreal: Canadian Institute of Mining, Metallurgy and Petroleum.

International Copper Study Group. (2018). *Directory of copper mines and plants 2015 to 2020*. Lisbon: International Copper Study Group. http://www.icsg.org (Accessed 13 October 20).

Kashala, A., Mitshabu, G., Cheng, Y., Bradford, L., Modi, A., & Tshisand, P. (2018). Management of mixing continuity in a solvent-extraction plant with a leach solution of high silica at Ruashi Mining. In *Copper cobalt Africa* (pp. 303–310). Johannesburg: Southern African Institute of Mining and Metallurgy.

Kokotov, Y., Braginsky, L., Shteinman, D., Slonim, E., Barfield, V., & Grinbaum, B. (2010). Turbulent settling (TS) technology for solvent extraction. In *ALTA copper 2010*. Melbourne: ALTA Metallurgical Services.

Kordosky, G. A. (2008). Development of solvent-extraction processes for metal recovery—finding the best fit between the metallurgy and the reagent. In B. A. Moyer (Ed.), *Solvent extraction: Fundamentals to industrial applications. International solvent extraction conference ISEC 2008* (Vol. 1, pp. 3–16). Montreal: Canadian Institute of Mining, Metallurgy and Petroleum.

Maes, C., Tinkler, O., Moore, T., & Mejías, J. (2003). The evolution of modified aldoxime copper extractants. In , *Copper 2003: Vol. I, Book 2. Hydrometallurgy of copper: Modelling, impurity control and solvent extraction* (pp. 753–760). Montreal: Canadian Institute of Mining, Metallurgy and Petroleum.

Mallory, K., Kehn, R., Iniguez, A., & Filgueras, M. (2016). The effect of impeller design on phase separation efficiency for solvent extraction auxiliary tanks. In *ALTA nickel cobalt copper 2016* (pp. 290–300). Melbourne: ALTA Metallurgical Services.

Mbao, B., Cohen, L., Fischmann, A., Moya, L., Moser, M., & Tinkler, O. (2018). Additive to mitigate crud formation in solvent extraction operations. In *Copper cobalt Africa* (pp. 321–330). Johannesburg: Southern African Institute of Mining and Metallurgy.

Megaw, D., Muzadi, P., Kashala, A., & Sanchez, A. R. (2016). Crud Forx™ advance in solvent extraction crud management technology. In *ALTA nickel cobalt copper 2016* (pp. 275–289). Melbourne: ALTA Metallurgical Services.

Metso Outotec. (2020). *VSF®X plant*. None. https://www.outotec.com/globalassets/products/solvent-extraction/ote_outotec_vsf_x_plant_eng_web.pdf (Accessed 13 October 20).

Miller, G. (2005). Engineering design for lowering fire risk. In *ALTA SX/IX world summit on SX fire protection*. Melbourne: ALTA Metallurgical Services.

Miller, G. (2011). Methods of managing manganese effects on copper solvent extraction plant operations. *Solvent Extraction and Ion Exchange, 29*(5–6), 837–853.

Mukutuma, A., Mbao, B., Schwartz, N., Chisakuta, G., & Feather, A. (2007). A case study on the operation of a Flottweg Tricanter® centrifuge for solvent-extraction crud treatment at Bwana Mkubwa, Ndola, Zambia. In *Africa's base metals resurgence* (pp. 393–405). Johannesburg: Southern African Institute of Mining and Metallurgy.

Nisbett, A., Peabody, S., & Crane, P. (2009). Concepts for developing high performing copper solvent extraction circuits. In *ALTA copper 2009*. Mebourne: ALTA Metallurgical Services.

Nisbett, A., Peabody, S., & Kordosky, G. A. (2008). Developing high-performing copper solvent-extraction circuits: Strategies, concepts, and implementation. In B. A. Moyer (Ed.), *Solvent extraction: Fundamentals to industrial applications. International solvent extraction conference ISEC 2008* (Vol. 1, pp. 93–100). Montreal: Canadian Institute of Mining, Metallurgy and Petroleum.

Pekkala, P., Laitala, H., Nyman, B., Ekma, E., & Weatherseed, M. (2007). Performance highlights of a modern VSF™ SX plant. In *ALTA copper 2007*. Melbourne: ALTA Metallurgical Services.

Quilodrán, A., Soderstrom, M., & Zambra, R. (2015). Organic phase quality and optimization of the clay treatment process. In T. Graber, M. E. Taboada, & F. Valenzuela (Eds.), *HydroProcess 2015*. Santiago: Gecamin. Santiago: Gecamin.

Smith, B. (2009). Fires in SX plants – recent trends in fire prevention and fire protection. In *ALTA copper 2009*. Melbourne: ALTA Metallurgical Services.

Sole, K. C. (2008). Solvent extraction in the hydrometallurgical processing and purification of metals: Process design and selected applications. In M. Aguilar, & J. L. Cortina (Eds.), *Solvent extraction and liquid membranes: Fundamentals and applications in new materials* (pp. 141–200). New York: Taylor and Francis.

Sole, K. C., Crundwell, F. K., Dlamini, N., & Kruger, G. (2018). Silica mitigation in copper solvent-extraction circuits. In *Copper cobalt Africa* (pp. 331–342). Johannesburg: Southern African Institute of Mining and Metallurgy.

Sole, K. C., Taute, J. J., Tinkler, O., Steeples, J., & Zarate, G. (2019). Global survey of copper solvent extraction: 2018 operating data and practice. In *Copper-cobre 2019, paper 581100*. Montreal: Canadian Institute of Mining, Metallurgy and Petroleum.

Sole, K. C., & Tinkler, O. S. (2016). Copper solvent extraction: Status, operating practices and challenges in the African Copper Belt. *Journal of the Southern African Institute of Mining and Metallurgy, 116*(6), 553–560.

Solvay. (2020). *Minchem®/MEUM™*. https://www.solvay.com/en/brands/minchem-meum (Accessed 13 October 20).

Taylor, A. (2007). Review of mixer-settler types and other possible contactors for copper SX. In *ALTA copper 2007*. Melbourne: ALTA Metallurgical Services.

Tenova Advanced Technologies. (2018). *Solvent extraction plant*. https://www.youtube.com/watch?v=9miKFqfvXhI (Accessed 13 October 20).

Tinkler, O. S., Cronje, I., Soto, A., Delvallee, F., & Hangoma, M. (2009). The ACORGA OPT-series: Industrial performance vs. aldoxime:ketoxime reagents. In *ALTA copper 2009*. Melbourne: ALTA Metallurgical Services.

Virnig, M., Eyzaguirre, D., Jo, M., & Calderon, J. (2003). Effects of nitrates on copper SX circuits: A case study. In , *Copper 2003: Vol. I, Book 2. Hydrometallurgy of copper: Modelling, impurity control and solvent extraction* (pp. 795–810). Montreal: Canadian Institute of Mining, Metallurgy and Petroleum.

Yáñez, H., Soto, A., Soderstrom, M., & Bednarski, T. (2009). Nitration in copper SX? Cytec Acorga provides a new reagent. In E. Domic, & J. Casas (Eds.), *HydroCopper 2009* (pp. 332–341). Santiago: Gecamin.

Suggested reading

Blass, E. (2004). Engineering design and calculation of extractors for liquid-liquid systems. In J. Rydberg, M. Cox, C. Musikas, & G. R. Choppin (Eds.), *Solvent extraction principles and practice* (2nd ed., pp. 367–414). New York: Marcel Dekker.

Cox, M. (2004). Solvent extraction in hydrometallurgy. In J. Rydberg, M. Cox, C. Musikas, & G. R. Choppin (Eds.), *Solvent extraction principles and practice* (2nd ed., pp. 455–506). New York: Marcel Dekker.

Godfrey, J. C., & Slater, M. J. (Eds.). (1994). *Liquid-liquid extraction equipment*. Chichester, UK: John Wiley & Sons.

Han, K. N., & Soderstrom, M. (2019). Solvent extraction. In *SME mineral processing and extractive metallurgy handbook* (pp. 1348–1358). Littleton, CO: Society of Mining, Metallurgy and Exploration.

Laitala, H., Hakkarainen, J., & Rodríguez, C. (2009). Plant design and operational aspects in solvent extraction impurity control. In *HydroCopper 2009* (pp. 311–317). Santiago: Gecamin.

Ritcey, G. M. (2006). *Solvent extraction: Principles and applications to process metallurgy* (2nd ed., Vols. 1 & 2). Ottawa: Gordon M. Ritcey and Associates.

CHAPTER 16

Electrowinning

Chapters 14 and 15 show how leaching followed by solvent extraction (SX) produces a high-purity $CuSO_4-H_2SO_4$ electrolyte containing ~45 g/L Cu^{2+}. This chapter explains how the Cu^{2+} ions in this electrolyte are reduced and electrowon as pure metallic copper in the form of cathodes.

The purity specifications of the copper cathode are dictated by the various international metal exchanges. Table 13.1 shows the purity requirements for copper cathode sold to the London Metal Exchange (LME) and New York Mercantile Exchange (Comex) (LME, 2020).

16.1 The electrowinning process

The electrowinning (EW) process entails the following:

(a) immersing metal *cathodes* and inert (but conductive) *anodes* into a purified electrolyte containing $CuSO_4$ and H_2SO_4;

(b) applying an electrical current from an external source, such as a rectifier, which causes current to flow through the electrolyte between the cathodes and anodes;

(c) plating pure metallic copper from the electrolyte onto the cathodes using the energy provided by the electrical current to drive the reduction of the Cu^{2+} ions to Cu^0 metal.

The cathodes are usually stainless-steel blanks. The anodes are usually rolled Pb alloy sheets; titanium with a precious metal oxide coating is also used in some modern operations. Copper is electroplated onto the cathodes for 4–7 days, after which the plated copper is machine-stripped from the stainless-steel cathode blanks, washed, packaged, and sold.

Almost 4 million tonnes of copper are electrowon per year (International Copper Study Group, 2019). Almost all of this is produced by leach–SX–EW, although a small proportion is by direct EW from a leach solution.

16.2 Chemistry of copper electrowinning

The EW cathode reaction is the same as for electrorefining, i.e.:

$$Cu^{2+} + 2e \rightarrow Cu^0 \quad E^0 = +0.34\ V \tag{16.1}$$

Extractive Metallurgy of Copper, Sixth Edition
ISBN 978-0-12-821875-4, https://doi.org/10.1016/B978-0-12-821875-4.00007-9

The anode reaction is, however, completely different. Water is decomposed at the inert anode to form oxygen gas and release protons:

$$H_2O \rightarrow H^+ + OH^- \rightarrow \tfrac{1}{2}O_2 + 2H^+ + 2e \quad E^0 = -1.23\ V \tag{16.2}$$

The overall EW reaction is the sum of Reactions (16.1) and (16.2) plus sulfate ions:

$$Cu^{2+} + SO_4^{2-} + H_2O \rightarrow Cu^0 + \tfrac{1}{2}O_2 + 2H^+ + SO_4^{2-} \quad E^0 = -0.89\ V \tag{16.3}$$

The EW products are as follows:

(a) pure copper metal at the cathode;
(b) oxygen gas at the anode;
(c) regenerated sulfuric acid in the electrolyte.

The copper is stripped from the cathode blank, washed, packed, and strapped into bundles weighing 1–3 t, and sent to market. The oxygen bubbles burst at the surface of the electrolyte and are released to the atmosphere. The acid is returned to SX in Cu-depleted spent electrolyte where the free H^+ ions are used to strip Cu from the loaded organic phase into the advance electrolyte (Eq. 15.2).

16.3 Electrical requirements

The electrical potential or cell voltage needed for EW is ~2.0 V, compared with ~0.3 V for electrorefining (Chapter 13). As illustrated in Fig. 16.1 (Nicol, 2006), the voltage components comprise the following:

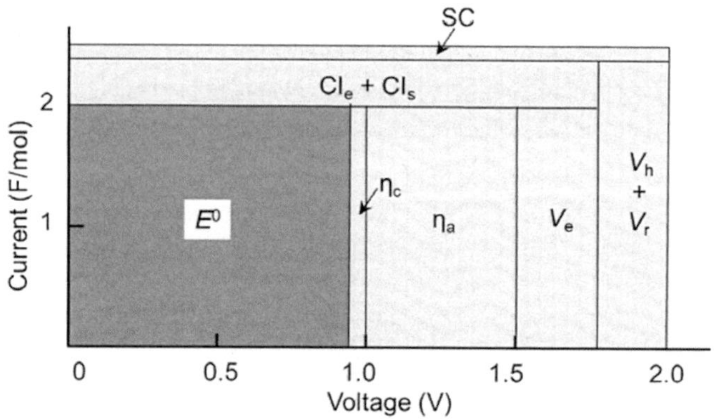

Figure 16.1 Current and voltage contributions to power consumption in a copper electrowinning cell (see text for symbols). Although the theoretical power requirement is ~750 kWh/t Cu (dark-shaded area), the actual power requirement is approximately 2000 kWh/t (dark- + light-shaded areas), which represents an energy efficiency of about 30%. It is therefore extremely important to minimize the energy influences of side reactions, shorts, and stray currents. *(From Nicol, M. (2006).* Electrowinning and electrorefining of metals. Short course. *Murdoch University, Perth.)*

- theoretical voltage for Reaction (16.3), E^0 0.89 V.
- overvoltage for copper deposition at cathode, $\eta_c \sim 0.05-0.1$ V.
- overvoltage for oxygen evolution at anode, $\eta_a \sim 0.5$ V.
- Ohmic potential drop across electrolyte, $V_s \sim 0.25-0.3$ V
- Ohmic potential drops across cell hardware and rectifier, $V_h + V_r \sim 0.3$ V

The current requirement for EW comprises the following:

- stoichiometric current requirement for Eq. (16.1), nF
- current inefficiencies due to side reactions, CI_e
- current inefficiencies due to shorts, CI_s
- stray currents in the tankhouse, SC

The power consumption, *P*, is the product of current and voltage (Eq. 16.4), which is given by the total area under the graph in Fig. 16.1:

$$P = VI. \tag{16.4}$$

The rate of plating of Cu from solution is given by Faraday's Law:

$$m = MIt\xi/nF, \tag{16.5}$$

where *m* is the mass of Cu plated (g);

M is the molar mass of Cu (63.54 g/mol);

I is the current passed (A);

t is the time for which the current is passed (s);

ξ is the current efficiency (i.e., the fraction of the total current used in producing Cu) (see Section 16.7);

n is the number of electrons involved in the plating of Cu (equal to 2 from Eq. 16.1);

F is the Faraday constant (96 485 C/mol = 96 485 As/mol).

The theoretical power consumption for plating 1 t of copper is therefore

$$P = \frac{(1.23 - 0.34)\ \text{V} \times 2/\text{mol} \times 96\ 485\ \text{As}}{63.54\ \text{g/mol}} = 2703\ \text{Ws/g} = 754\ \text{kWh/t Cu}$$

In practice, the energy requirement for copper EW is ~2000 kWh/t. This is considerably greater than that for copper electrorefining (200–300 kWh/t Cu). This difference is due to the large voltage requirements for EW, as well as current losses. The energy efficiency of Cu EW is therefore only about 30%.

16.4 Equipment

16.4.1 Cathodes

Most copper EW tankhouses employ reuseable 316L stainless-steel cathode blanks, which are virtually identical to electrorefining cathodes. LDX duplex and superior grades of stainless steels are also used for cathode blanks. Using a blank material onto which

Figure 16.2 Stainless-steel cathode blank (a) before and (b) after plating. Note the edge strips on the permanent cathodes. Copper sheets are (c) stripped from the blank by an (d) automatic or (e) robotic stripping machine, (f) corrugated or (g) clinched, packed, and (h) strapped for shipping. The corrugations or clinches are introduced so that the cathode stack melts more easily when the cathodes are melted and cast (see Chapter 21). *(Photographs (c) Courtesy of M. Nicol, Murdoch University; (d), (e), and (g) Courtesy of Glencore Technology.)*

copper is plated in the EW process is known as *permanent cathode technology* (Fig. 16.2). The selection of cathode material is based on a trade-off between cost, lifetime, conductivity (a higher conductivity provides lower resistance), better weldability to copper header bars, and the need for high corrosion resistance (such as in tankhouses that operate with a high-chloride electrolyte) (Glencore Technology, 2019, 2020a; Pranowo, Aslin, Eriksson, & Heferen, 2019). A few older EW tankhouses use copper starter sheets as the cathodes. The starter sheets are obtained from an electrorefinery or are made in a separate section of the EW tankhouse itself using stainless steel or titanium as starter cathode blanks.

The size of the cathodes is determined by industry standards (to fit efficiently into shipping containers) and by the openings in the furnaces in which the copper will be remelted (see Chapter 21). The standard size is 1.2 m × 1 m; jumbo cathodes with a length up to 1.4 m are sometimes produced, but these can be more difficult to handle.

Plastic edge strips are used on the sides of the blanks to avoid plating around the sides of the cathode. The bottoms of the blanks are either wax dipped or have a V groove to allow for easy removal of the plated copper.

Cathode blanks are attached to a hanger bar that provides the electrical contact between the cathode and the electrical circuit. The material of the two points of contact is copper, owing to its high conductivity. The method of attachment of the cathode blank to the hanger bar is important in ensuring that the acidic electrolyte or generated acid mist (see Section 16.5.4) does not damage the cathode.

Automated cathode stripping machines are widely used today, particularly using robots. These enable stripping rates of up to 600 cathodes/h (Aslin, Eriksson, Heferen, & Sue Yek, 2010; Larinkari, Nord, & Schmidt, 2016, 2019). Automation enables rapid and repeatable removal of the copper from the blank, while maximizing the physical quality of the copper and minimizing damage to the blanks.

16.4.2 Anodes

16.4.2.1 Lead alloy anodes

Most modern copper EW tankhouses use anodes comprising cold-rolled lead (Pb) alloyed with 1.35% Sn and 0.07%–0.08% Ca. Sn provides strength, corrosion resistance, and corrosion layer conductivity; Ca improves the mechanical properties and decreases the anode potential; cold rolling adds strength (Mirza et al., 2016; Prengaman, Ellis, & Mirza, 2010). A few older plants still use Pb–Sn or Pb–Sb alloys, but these are being gradually replaced with Pb–Sn–Ca anodes.

A Pb–Sn–Ca alloy slab is soldered onto a slotted copper hanger bars for support in the EW cells (Fig. 16.3A). Lead is then electrodeposited around the joints to protect them from corrosion. The anode size is slightly smaller than the cathode size (about 30 mm on

(a)

(b)

Figure 16.3 (a) Pb alloy anodes. (b) Coated titanium anode installed at Kristiansand, Norway. *((a) Courtesy of Castle Lead Works and RSR Technologies. (b) Courtesy of Metso Outotec and Kristiansand.)*

the width and height) to modify the electric field lines to ensure that Cu plating does not occur around the sides of the cathode.

Anodes are fitted with several polymer spacer insulators, buttons, or straps that prevent contact with the cathode to avoid short circuiting and minimize lead corrosion product carryover from anodes to electrodepositing copper.

The life expectancy of a lead anode is typically 6−9 years (Beukes & Badenhorst, 2009; Sole, Moats, Sandoval, Robinson, & Davenport, 2019). The anodes slowly corrode in the sulfuric acid electrolyte, forming an adherent, inert, insoluble PbO_2 layer on the surface (Cifuentes et al., 2005; Mirza et al., 2016; Nikoloski, Nicol, & Stuart, 2010). This minimizes Pb contamination of the cathode copper and extends anode life. The components of the anode either dissolve in the electrolyte or fall to the bottom of the EW cell as insoluble lead sludge. This is periodically removed from the cells and sent to a recycling facility for the recovery of lead.

16.4.2.2 Coated titanium anodes

Titanium-based anodes are now increasingly installing in new tankhouses (Åkre & Rosseland, 2019; Zhang et al., 2017, Zhang, 2018). These comprise a titanium mesh or sheet with a precious metal coating, usually an oxide of Ir or Ru (see Fig. 16.3B) (Morimitsu, Oshiumi, & Wada, 2010). These anodes are therefore also known as *mixed metal oxide* or *dimensionally stable anodes*.

Use of these anodes minimizes Pb contamination, there is no need for cell cleaning, and operator exposure to heavy metals is reduced due to the elimination of lead sludge (Sandoval, Clayton, Dominguez, Unger, & Robinson, 2010; Sandoval, Garcia, Neff, & Schnebly, 2013). There is no need for Co^{2+} addition to the electrolyte (see Section 16.5.3.2). These anodes offer a 0.3−0.4 V decrease in oxygen overpotential (see Section 16.3), which lowers energy consumption by about 15%. The oxygen bubble size generated by these anodes is also smaller, so acid mist control requirements are reduced (see Section 16.5.4). These anodes are expensive owing to their precious metals content and titanium structure. They require careful handling to avoid damaging the oxide surface layer. They are not suitable for operation in electrolytes containing significant amounts of Mn.

16.4.3 Cell design

A typical cell design is shown in Fig. 16.4. EW cells, anodes, and cathodes are somewhat larger than those used for electrorefining:

cells: up to 9.4 m long × 1.2 m wide × 1.5 m deep
anodes: 1.1 m long × 0.9 m wide × 0.006 m thick
cathodes: 1.2 m long × 1.0 m wide × 0.003 m thick.

Most modern cells are constructed of polymer concrete (Currie & Miller, 2018; Moats & Free, 2019; Robinson et al., 2012). Some designs use reinforced concrete with a chemically resistant lining, such as welded or spray-coated polyvinyl chloride. Older tankhouses have cell linings of rubber, lead, or an epoxy resin coating. The length

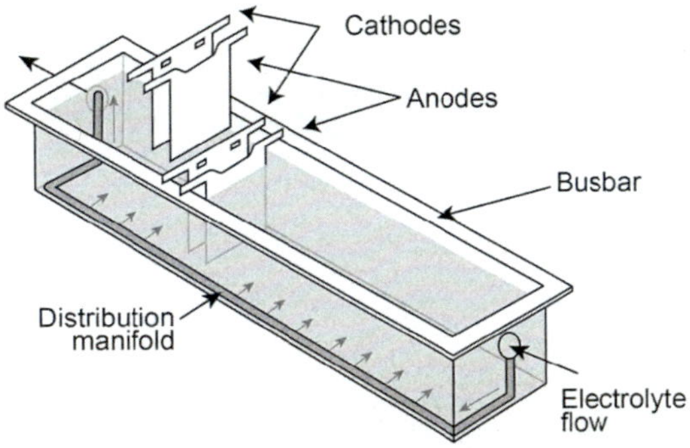

Figure 16.4 Schematic view of an electrowinning cell. Anodes and cathodes are alternately arranged. Electrical contact is via busbars. An electrolyte input manifold for uniformly distributing electrolyte over all the cathode faces is shown. Single multiholed horizontal pipes on the cell floor are also used.

Figure 16.5 Anode—cathode arrangement showing different contact systems that minimize the occurrence of poor contacts of electrodes with the busbar. The plastic beads on the surface of the electrolyte minimize the effects of acid mist (see Section 16.5.4). *(Photographs courtesy of Multotec Outotec, M. S. Moats (Missouri University of Science and Technology), and F. K. Crundwell (CM Solutions).)*

of the cell depends on the spacing between the electrodes. Modern designs tend to reduce the anode—cathode spacing to accommodate more plating capacity for a given floor area, with up to 90 cathodes per cell (Forner et al., 2018). Typical anode—cathode spacing is 45—50 mm.

The anodes and cathodes are interleaved as in electrorefining (Fig. 16.5). All anodes in a cell are held at a single potential; the cathodes are at a lower potential. The electrodes within a cell are connected in parallel, but the cells of a tankhouse are connected in series. Current is carried between electrodes and between cells using copper busbars. To minimize heat losses through the busbars, the current is limited to 1—1.2 A/m^2. Where parallel busbars are run, a gap between them allows for cooling.

16.5 Operational practice

16.5.1 Current density

Cathode current density in modern plants varies from 200 to above 450 A/m^2 (Flores, Muñoz, & Zárate, 2010; Sole, Moats, Sandoval, Robinson, & Davenport, 2019). This gives a plating rate of 0.4—1.0 kg/h of copper on each cathode.

Figure 16.6 Harvesting of plated cathodes by crane. Note that only every third cathode is removed from the cell each time. This allows the cell to continue plating without having to stop the current supply to that cell. *(Photograph courtesy of K. Voogt, Anglo American.)*

Cathodes are plated to a thickness of 5–8 mm (typically 50–55 kg Cu on each side of a stainless-steel blank), which means that the cathodes are harvested every 4–7 days, depending on the applied current density. In a modern tankhouse using an overhead crane, every second or third cathode is harvested from a particular cell at a time, leaving the remaining cathodes available to continue plating (Fig. 16.6). The cathodes remaining in the cell will temporarily experience a considerably higher current density, so it is important to minimize the time required for cathode change-outs. This procedure maintains an adherent corrosion product layer on the anodes and minimizes lead contamination of the electrodepositing copper. The longer plating is continued, the greater is the likelihood of shorts occurring due to the growth of nodules or dendrites on the cathodes.

If starter sheets are used, these are often pulled out after 2 days of plating, straightened in a press, and then reinstalled to continue the plating cycle. This minimizes growth of nodules or dendrites due to bent cathodes and avoids short circuiting.

The rate of copper plating increases with increasing current density (see Eq. 16.5); however, excessive current density can promote rough, nodular cathode deposits and decrease copper purity if not properly managed. The use of air sparging of the electrolyte can help to minimize inhomogeneity in the electrolyte composition throughout each cell and permits high-quality copper to be plated at increasingly higher current densities.

16.5.2 Electrolyte

To produce high-quality copper at all cathode positions in a cell, the electrolyte composition should be essentially the same throughout the cell. The *advance electrolyte* from SX typically contains 45 g/L Cu^{2+} and 170 g/L H_2SO_4. This is combined with a bleed of *spent electrolyte* leaving the EW cells to give the *recirculating electrolyte* that is the feed to the EW cells. The spent electrolyte contains 2–5 g/L Cu^{2+} less than the recirculating

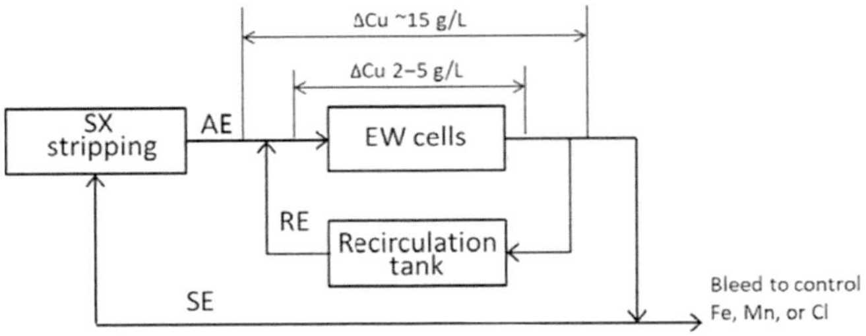

Figure 16.7 Recirculating electrolyte flow configuration that is employed to maintain an approximately constant electrolyte composition throughout the cells. *AE*, advance electrolyte; *RE*, recirculating electrolyte; *SE*, spent electrolyte.

electrolyte, *i.e.*, a single pass through an EW cell takes a "bite" or ΔCu of 2–5 g/L Cu^{2+} from the electrolyte.

Depleted spent electrolyte is replenished with Cu^{2+} by sending it back to SX where the acid that it contains is used to strip Cu^{2+} from the loaded organic extractant (Chapter 15). About 25%–50% of the spent electrolyte is sent to SX. The remainder is sent to an active storage tank where it is mixed with Cu-enriched advance electrolyte returning from SX. A typical flow configuration is shown in Fig. 16.7.

Electrolyte flows continuously from an active storage tank into one end of each cell through a manifold that is usually situated at the bottom of the cell (Fig. 16.4). Electrolyte overflows at the other end of the cell into a collection system. Electrolyte is added at ~15 m^3/h into each cell. This gives constant availability of high Cu^{2+} electrolyte over the surfaces of all the cathodes. The specific flowrate of electrolyte past the face of the cathode (the *face velocity*) is typically designed at 0.08 m^3/h per m^2 of cathode area (Beukes & Badenhorst, 2009).

Efficient circulation of the electrolyte to all cathodes in a cell is critical to good cathode quality. As electrode-spacing distances decrease and current densities increase, this becomes even more important. Modern cell designs may introduce the electrolyte at several places into the cell, including at multiple points from a manifold at the base of the cell or just beneath the level of the electrolyte, or from feed points at the side or end of a cell.

16.5.2.1 Organic contamination

The advance electrolyte from SX often contains small amounts of organic phase that is carried over by entrainment or incomplete phase separation (see Section 15.8.2). The presence of organics is detrimental to copper EW because the cathodes may become discolored at the electrolyte–air interface (so-called *organic burn*), which decreases the value of the copper on the commercial market. The presence of organic on the surface of the electrolyte also represents a fire risk owing to the high currents employed in the tankhouse. Organic components of the advance electrolyte are filtered by passing it through activated carbon columns and/or flotation cells before it reenters the EW tankhouse.

In certain operations, particularly where a high extractant concentration is employed, organic carryover can be considerable, even with the installation of mitigating equipment. In such tankhouses, the electrolyte first enters so-called *scavenger cells*. The copper plated in these cells is below LME standard and receives a lower value on sale; however, these cells serve to "clean up" the electrolyte and most of the organic component is retained within these cells. The treated electrolyte then goes forward to the main tankhouse for production of high-quality cathode.

16.5.2.2 Conductivity

Conductivity of the electrolyte is an important consideration in minimizing power consumption (see Section 16.3) (Tozawa et al., 2003). A temperature of 45—55°C is optimum for maximizing electrolyte conductivity; however, the spent electrolyte returning to SX needs to be kept below 40°C to avoid degradation of the organic phase. The incoming advance electrolyte is heated to 45—50°C by heat exchange using outgoing spent electrolyte that has been resistively heated in the EW cell.

Higher acidity can improve electrolyte conductivity, but too high an acidity can cause hydrolytic degradation of the organic phase in SX (see Section 15.9.3). For this reason, the spent electrolyte acid concentration is generally limited to 190—200 g/L. Higher acidities also increase the corrosion effects of acid mist in the tankhouse and decrease anode lifetime. A higher Cu concentration in the electrolyte will improve cathode quality, but can lead to $CuSO_4$ crystallization. The effects of Cu, H_2SO_4, and temperature variation on electrolyte conductivity, σ, are given empirically by Eq. (16.6) (Price & Davenport, 1980):

$$\sigma(\mathrm{S}) = 13.4 - 0.356[\mathrm{Cu}] + 0.249[\mathrm{H_2SO_4}] + 0.426T, \tag{16.6}$$

where the concentrations are in g/L and temperature is in °C.

16.5.2.3 Effect of iron

Iron is often present in electrolytes. This is attributed to transfer of iron from the leach liquor through the SX circuit by both chemical complexation with the extractant and physical entrainment (see Chapter 15). Iron in the electrolyte offers both positive and negative effects.

Reduction of Fe^{3+} at the cathode and the corresponding oxidation of Fe^{2+} at the anode

$$\mathrm{Fe^{3+}} + \mathrm{e} \leftrightarrow \mathrm{Fe^{2+}} \tag{16.7}$$

compete with the Cu EW reactions (Eq. 16.3). Consumption of current by these cycling reactions lowers the current efficiency for plating of copper by ~2.5% for each 1 g/L Fe. For example, 5 g/L Fe^{3+} in the electrolyte can drop the current efficiency to as low as

80% (Khouraibchia & Moats, 2009). However, the presence of 1−2 g/L Fe in the electrolyte improves the physical quality of the copper cathodes.

16.5.2.4 Effect of manganese

Manganese is often present in electrolytes, particularly in South America and the African Copper belt. This contributes to the formation of MnO_2 on the anode, which can redissolve when power is lost, contributing to entrapment of Pb and Mn in the cathode and sludge formation at the bottom of the cell (Tjandrawan & Nicol, 2010). Anodes need to be periodically cleaned with high-pressure water to remove this layer.

Mn(II) is also problematic in SX−EW circuits because it can be oxidized to Mn(III), Mn(IV), or Mn(VII) at the anodes (Cheng et al., 2000). To avoid the formation of the permanganate ion (MnO_4^-) at the anode—a highly oxidizing species that can have very detrimental effects on degradation of the SX organic-phase components—a ratio of Fe: Mn in the electrolyte of approximately 10:1 is recommended (Miller, 2010, 2011).

16.5.2.5 Effects of other metal contaminants

The behavior of other metal species present in the electrolyte can be understood with reference to their position in the electrochemical series (Table 16.1). Those species with a more positive reduction potential than Cu can plate at the cathode, while those with a more negative potential will remain in solution.

Table 16.1 Standard reduction potentials of half-cell reactions relevant to Cu electrowinning.

Reaction	Standard potential (V)
$Mn^{2+} + 2e \rightarrow Mn$	−1.18
$Fe^{2+} + 2e \rightarrow Fe$	−0.44
$PbSO_4 + 2e \rightarrow Pb + SO_4^{2-}$	−0.35
$Co^{2+} + 2e \rightarrow Co$	−0.28
$Ni^{2+} + 2e \rightarrow Ni$	−0.26
$Pb^{2+} + 2e \rightarrow Pb$	−0.13
$2H^+ + 2e \rightarrow H_2$	**0.00**
$AgCl + e \rightarrow Ag + Cl^-$	0.22
$Cu^{2+} + 2e \rightarrow Cu$	**0.34**
$Fe^{3+} + e \rightarrow Fe^{2+}$	0.77
$Ag^+ + e \rightarrow Ag$	0.80
$O_2 + 4H^+ + 4e \rightarrow 2H_2O$	**1.23**
$MnO_2 + 4H^+ + 2e \rightarrow Mn^{2+} + 2H_2O$	1.33
$Cl_2 + 2e \rightarrow 2Cl^-$	1.36
$Mn^{3+} + e \rightarrow Mn^{2+}$	1.49
$MnO_4^- + 8H^+ + 5e \rightarrow Mn^{2+} + 4H_2O$	1.51
$PbO_2 + SO_4^{2-} + 4H^+ + 2e \rightarrow PbSO_4 + 2H_2O$	1.70

16.5.2.6 Effect of chloride

The presence of chloride ion in the electrolyte can have both positive or negative effects, depending on its concentration. Chloride ions are naturally present in the electrolyte (particularly in Chilean operations that treat ores with chloride-containing minerals) or can be added as HCl. Low levels of chloride ions promote the growth of dense, fine-grained, low-impurity copper deposits on the cathode; higher concentrations negatively affect the growth of copper crystals, and can lead to a powdery deposit in extreme cases. Chloride is necessary to achieve polarization when polyacrylamides or glue are used as the smoothing agent (see Section 16.5.3.1). When using permanent cathode technology (stainless-steel cathode blanks), the chloride concentration must be kept below 30 mg/L to avoid *pitting corrosion* of the steel blade at the electrolyte—air interface, which causes the depositing copper to stick to the blank and resist detachment.

16.5.2.7 Control of impurities

The buildup of impurities in the electrolyte is controlled by bleeding a small stream from the tankhouse (see Fig. 16.7). In addition to impurities that need to be removed from the circuit, the electrolyte also contains significant quantities of copper and acid, so this is typically returned to the leach, where the acid is used for leaching, or to the SX circuit for recovery of the copper. In some operations, the bleed is treated by precipitation or ion exchange to remove contaminants (Shaw et al., 2019; Sole et al., 2017).

16.5.3 Electrolyte additives

16.5.3.1 Smoothing agents and grain refiners

All copper EW plants add smoothing agents to their electrolytes to promote the plating of dense, smooth copper deposits with minimum entrapment of electrolyte impurities. In the past, guar gum (150—400 g/t of cathode), a natural product, was widely used (Fig. 16.8). This behaves much as glue in electrorefining (Chapter 13), but guar is

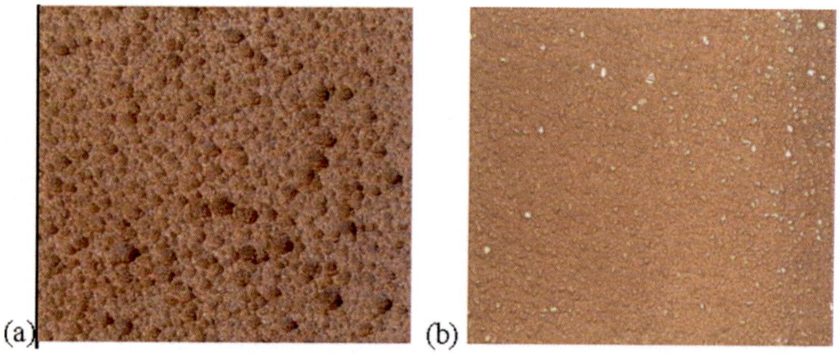

Figure 16.8 The effect on the morphology of copper cathode plated in the (a) absence and (b) presence of guar. Guar, added at concentrations of 150—400 g/t cathode, acts as a smoothing agent, allowing refinement of the copper crystal grains as they are plated. This gives the cathode a smooth morphology and avoids the growth of dendrites and nodules that could short circuit the current between anode and cathode. *(Courtesy of M. Nicol, Murdoch University.)*

compatible with SX organics. Smoothing agents originating from natural products tend to break down above current densities of about 300 A/m^2, so the modern trend is toward the use of synthetic products that have more consistent and predictable compositions and performance (Flores, Muñoz, & Zárate, 2010). Modified starches (polysaccharides) and polyacrylamides (typically added at 200–700 g/t cathode) are widely used (Chibwe & Tadie, 2021; Luyima et al., 2016; Moats et al., 2016). These products are compatible with the sensitivity of SX organics to surface-active additives. The oligosaccharide product, DXG-F7, is in widespread use in Chile and various other global operations, and has particular benefits in reducing sulfur contamination of the cathode (Cifuentes, Bravo, & Schwarz, 2015; Maldonado, 2010). Recent advances in the development of synthetic smoothing agents allow high-quality plating at current densities well above 400 A/m^2.

16.5.3.2 Cobalt sulfate

When using lead-based anodes, it is necessary to maintain a cobalt concentration in the electrolyte of 100–200 mg/L Co^{2+} to promote the O_2 evolution reaction at the anode (Eq. 16.2), rather than Pb oxidation (Nikolosi & Nicol, 2008). This stabilizes the PbO_2 layer on the anode surface, minimizes Pb contamination of the depositing copper, and extends anode life (Mirza et al., 2016). The addition of 350 mg/L Co halves the rate of anode corrosion. As the current density increases, the required Co concentration typically increases (Sole, Moats, Sandoval, Robinson, & Davenport, 2019).

The presence of cobalt also decreases the anode overpotential (see Fig. 16.1), which results in a lower cell voltage (V) and improves power consumption. The empirical relationship is given by the following (Miller, 2011):

$$V\ (\mathrm{V}) = 2.15 - \left(0.35/6.4[\mathrm{Co}]^{-0.21}\right), \tag{16.8}$$

where the cobalt concentration is given in g/L. A more comprehensive relationship for the anode overpotential, which includes the effects of Mn and Fe, is given by the following (Abbey & Moats, 2018):

$$V\ (\mathrm{V}) = 1.961 - 0.0962[\mathrm{Co}]^{0.32} + 0.0524[\mathrm{Mn}] + 0.0114[\mathrm{Fe}]^{0.25}. \tag{16.9}$$

Most operations add cobalt to the electrolyte as the cobalt sulfate salt. Many mines on the African Copper belt contain significant cobalt mineralization in association with the copper ores, so cobalt deports to the leach liquor with copper and adequate concentrations are transferred to the advance electrolyte by entrainment through the SX circuit.

A significant advantage of coated titanium anodes (see Section 16.4.2.2) is that no cobalt addition to the electrolyte is required.

16.5.4 Acid mist suppression

Oxygen bubbles that are produced at the anodes (Eq. 16.2) rise to the top of the electrolyte where they burst into the atmosphere, carrying tiny droplets of sulfuric acid with them. This *acid mist* can create a corrosive, unpleasant, and unhealthy environment in the tankhouse. Acid mist is a known human carcinogen and may cause health problems for operators if adequate personal protection equipment is not employed. To minimize tankhouse footprint and capital costs, modern tankhouse design is tending toward the use of higher current densities (Sole, Moats, Sandoval, Robinson, & Davenport, 2019). As indicated by Eq. (16.3), the more copper that is plated per unit time, the more O_2 is produced, which gives rise to increased acid mist production. The intensity and detrimental effects of acid mist increase with increasing current density. Combinations of various measures are employed to minimize its detrimental effects (McGinnity & Nicol, 2014; Mohanarangam, Yang, Rajasingam, Tadesse, & Robinson, 2019; Robinson, White, & Grassi, 2013).

Good ventilation of the tankhouse is essential, particularly incorporating cross-flow or draft effects. In some plants, the mist is cleared from the workplace by large blowers or extraction fans on the sides of the tankhouse building or, where the climate permits, the sides of the building may be open. Modern tankhouse designs include hoods over the cells. The acid mist is captured and sent to a scrubber; an added advantage is a reduction in evaporation of the electrolyte (Fig. 16.9) (Vainio & Weatherseed, 2009).

Hollow polyethylene balls (~2 cm diameter) are usually floated, 5–10 cm deep, on the electrolyte (Fig. 16.5). Polypropylene beads (3 mm diameter) are sometimes used. It is essential that at least three layers of balls or beads are used on the electrolyte surface to

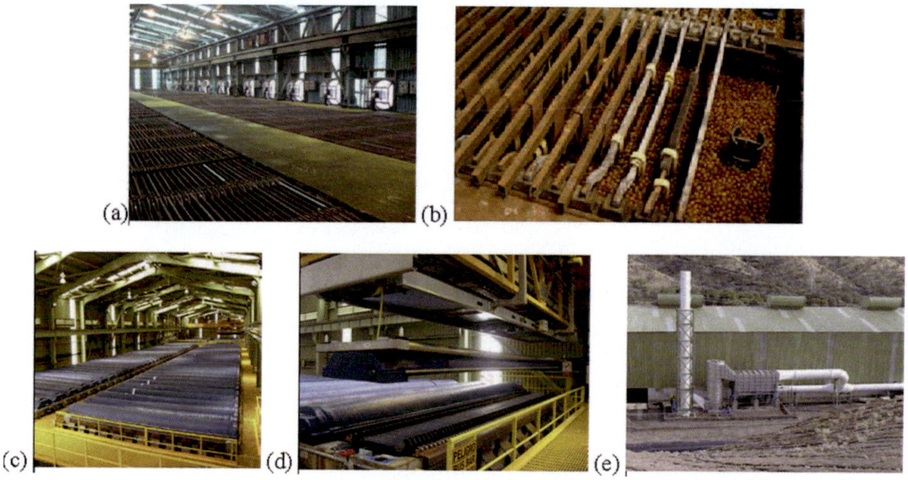

Figure 16.9 Tankhouse equipment to avoid and mitigate the effects of acid mist. (a) Older tankhouses use extraction fans and ventilation across the cells, and (b) polypropylene balls on the surface of the electrolyte. (c) Modern tankhouses use hoods to cover the cells, which are lifted for harvesting of cathodes (d), requiring very accurate positioning, so computer control of the crane movements is essential. (e) Acid mist is captured and scrubbed to avoid releasing corrosive gases to the environment. *(Photograph courtesy of Metso Outotec.)*

provide effective acid mist mitigation. Owing to their mechanical and operating difficulties, balls and beads are now being replaced by anode brushes for acid mist control (Åkre & Rosseland, 2019).

In addition to physical mitigation approaches, chemical acid mist suppression agents are employed. These are typically a foaming agent or surfactant, which modifies the surface characteristics of the electrolyte such that the energy of the bursting bubbles is reduced on their release (Bender, 2010).

Natural products, such as *Quillaja saponaria*, can be used at low current density (San Martin et al., 2005). Other synthetic surfactants, added at levels of 10–20 mg/L of electrolyte, are effective at higher current densities. It is extremely important that mist suppressants that have surfactant character do not detrimentally affect phase separation when returned in the spent electrolyte to the SX circuit.

16.6 Maximizing copper quality

The demands of a modern EW tankhouse are to produce copper of increasingly higher purity, using higher current densities, while minimizing power consumption (Marsden, 2008; Nieminen, Virtanen, Ekman, Palmu, & Tuuppa, 2010). Copper quality is determined by both the purity and physical appearance of the cathode product.

16.6.1 Copper purity

From a thermodynamic point of view, only metals that have a higher electrochemical potential will codeposit with Cu (Table 16.1). Because Cu is relatively noble, it is easily plated with high purity. Impurities typically arise from poor tankhouse housekeeping and poor deposit morphology.

The three main impurities in electrowon cathode copper are as follows:

- Pb (1–2 ppm), resulting from physical entrapment of anode corrosion products (if lead alloy anodes are used);
- S (~5 ppm), resulting from incorporation of $PbSO_4$ anode corrosion product (when using lead-based anodes) and electrolyte in the deposit;
- Fe (1–2 ppm), resulting from electrolyte entrapment.

Cathode purity is therefore enhanced by minimizing (or containing) anode corrosion and avoiding entrapment of the electrolyte within a growing copper deposit.

Factors that enhance cathode purity include the following:

- maintaining straight, vertical, equi-spaced cathodes and anodes with no anode–cathode contact, which promotes a uniform current distribution throughout the tankhouse;
- immediate thorough washing of harvested copper cathodes with water at 70–80°C, to remove electrolyte from the surface and avoiding surface S contamination;
- using rolled Pb–Sn–Ca anodes, which give adherent corrosion products (Mirza et al., 2016; Nikoloski, Nicol, & Stuart, 2010);

- frequent removal of anode corrosion products (lead sludge) from the bottom of the EW cells, to avoid suspension of fine particles in the circulating electrolyte and their incorporation into growing deposits;
- addition of Co^{2+} to the electrolyte to minimize Pb corrosion;
- minimizing the Mn^{2+} concentration in the electrolyte to avoid rapid Pb corrosion and the formation of flaky corrosion products (Miller, 2010);
- maintaining a steady electrical current at all times to maintain an adherent anode corrosion layer;
- controlled resumption of current after power outages to minimize anode corrosion (Castle Lead Works, 2020; Glencore Technology, 2020b);
- keeping the electrolyte free of solids and SX organics;
- ensuring a steady uniform flow of purified Cu^{2+} electrolyte at appropriate Cu and acid concentrations across all cathode faces;
- an electrolyte temperature of 45−50°C to give pure fine-grained copper deposits;
- use of smoothing agents and chloride addition agents to give dense, smooth copper deposits (but with $Cl^- < 30$ mg/L in the electrolyte) (Section 16.5.2.6);
- maintaining Fe in electrolyte to < 2 g/L to minimize Fe deporting to cathode copper and reduce losses in current efficiency (Section 16.5.2.3).

16.6.2 Physical appearance

The physical appearance of the cathodes is also important in determining their commercial value.

Cathodes should be of uniform thickness. This can be achieved by the following:

- ensuring vertical and horizontal alignment in cells;
- even spacing between electrodes;
- uniform solution flow in the cells;
- consistent current density and plating time throughout the tankhouse and plating cycles.

Cathodes should have a smooth surface appearance and good color. This is promoted by the following:

- preventing organic from SX from entering the cells;
- ensuring appropriate conditions for ordered crystal growth;
- use of correct concentration of smoothing agents;
- matching the current density to the electrolyte composition, flowrate, and cell agitation;
- avoiding shorts and dendrite formation;
- maintaining a good surface finish of the cathode blanks (Larinkari, Nord, & Schmidt, 2016);
- avoiding a plating cycle that is too long and leads to dendrite formation.

16.7 Maximizing energy efficiency

Energy consumption in EW is determined by the following:

(i) the cell voltage;
(ii) the applied current (or current density); and
(iii) the current efficiency.

As shown in Fig. 16.1, cell voltage depends largely on the cell design, electrode materials, and the reactions occurring at the anode and cathode. The applied current is usually determined by the production rate required. Tankhouse operators can, however, influence and manage energy consumption by understanding how to maximize current efficiency (Moats, 2012, 2018).

The *theoretical* mass of Cu that can be plated for a given current is given by Faraday's Law (Eq. 16.5). The ratio of the *actual* mass of Cu plated, compared with this theoretical value, is defined as the *current efficiency*, ξ:

$$\xi(\%) = \frac{\text{measured mass of Cu plated} \times 100}{\text{theoretical mass of Cu plated}} \tag{16.10}$$

Current efficiencies in modern Cu EW plants range from 85% to 95% (Sole, Moats, Sandoval, Robinson, & Davenport, 2019). Current that is not available for plating of copper is consumed by the following (Moats, 2012; Nicol, 2006):

- anode-to-cathode short circuits;
- stray currents to ground;
- reduction of Fe^{3+} or dissolved oxygen at the cathode;
- oxidation of Fe^{2+} at the anode;
- redissolution/oxidation of the deposited metal (particularly in the case of power failures);
- poor-quality deposited metal that falls off the cathode and cannot be harvested.

High current efficiency is important because it maximizes copper plating rate and minimizes electrical energy consumption per tonne of copper produced.

Strict attention to good housekeeping practices is an important criterion in maximizing cathode purity and energy efficiency (Joy et al., 2010). Electrolyte spills should be avoided and cleaned up immediately if they occur. Anode–cathode short circuits are detected by frequent infrared and gauss meter inspections and by individual voltage measurements on cells. Shorts are minimized by ensuring that the electrodes are evenly spaced, straight, and vertical (Aylwin, Lagos, & Melani, 2013; Blackett & Nicol, 2010). Bending of stainless-steel blanks during copper stripping must be avoided. Regular polishing and buffing of cathode blanks assists in minimizing difficulties in removing deposited cathode plates (Larinkari, Nord, & Schmidt, 2016). Automation and advanced instrumentation in modern tankhouses can assist greatly in maintaining good electrode arrangements, uniformity of current distribution, and rapid short detection (Fraser et al., 2014).

Control of electrolyte purity is essential. Current consumption by cycling of the $Fe^{3+} \leftrightarrow Fe^{2+}$ redox reaction is minimized by controlling the Fe concentration in the electrolyte to <2 g/L (Section 16.5.2.3). This is done by preventing leach solution carryover to electrolyte from the SX circuit and by regularly bleeding a small portion of Fe-bearing electrolyte from the tankhouse.

Current efficiency generally increases with increasing Cu and acid concentrations in the electrolyte and with current density (j), but decreases with increasing Fe^{3+} concentration and temperature (Zhang et al., 2018). A useful practical correlation that can be used to optimize current efficiency on a daily basis in a tankhouse has been provided by Khouraibchia & Moats (2010) for operation at 40°C:

$$\begin{aligned}\xi(\%) &= 88.19 - 4.91\left[Fe^{3+}\right] + 0.52\left[Cu^{2+}\right] + 1.81 \times 10^{-3} j - 6.83 \times 10^{-3}\left[Cu^{2+}\right]^2 \\ &\quad + 0.028\left[Fe^{3+}\right]\left[Cu^{2+}\right] + 4.015 \times 10^{-3}\left[Fe^{3+}\right] j,\end{aligned} \tag{16.11}$$

where all concentrations are in g/L and current density is in A/m^2.

16.8 Modern industrial electrowinning plants

Differences between EW and electrorefining are summarized in Table 16.2. EW differs from electrorefining mainly in the following:

(a) the use of inert Pb alloy or coated titanium anodes instead of soluble copper anodes;
(b) the higher applied voltage.

The use of inert anodes means that (i) the anodes remain in the cells for 5+ years, (ii) there is little anode scrap, and (iii) no precious metal slimes (although lead sludge accumulates at the bottom of the EW cells and needs to be periodically removed). In the case of coated titanium anodes, there is no lead sludge, but residual precious metal oxides from the coatings can accumulate at the base of the cells.

Table 16.3 shows the operating conditions and equipment in five EW tankhouses that process electrolyte from an SX circuit (Sole, Moats, Sandoval, Robinson, & Davenport, 2019). San Francisco currently operates with the highest current density (498 A/m^2) in the world; Mantoverde and San Francisco have among the highest current efficiencies (>92%). Morenci is one of the largest tankhouses in the world (350 kt/a). Tenke Fungurume represents typical operating conditions in the African Copper belt, while Sepon is a typical Asian operation. Further information regarding these and other EW plants is available from T. Robinson, White, & Grassi, 2013; T.G. Robinson et al., 2013b and Sole, Moats, Sandoval, Robinson, & Davenport, 2019.

Table 16.2 Summary of similarities and differences between electrowinning and electrorefining.

Parameter	Electrorefining	Electrowinning
Anode	Cast copper anode, replaced every 3 weeks	Lead–alloy "permanent" or coated titanium anode Oxygen evolution at the anode
Cathode	Stainless-steel permanent cathode blank	Stainless-steel permanent cathode blank
Copper deposition	All cathodes plated for the same time	Three different plating cycles in a single cell
Electrolyte	Common electrolyte Greater variety of impurities 45 g/L Cu, 170 g/L H_2SO_4, 45 mg/L Cl 25–35 L/min/cell, 62–64°C	Advance and spent electrolytes Smaller variety of impurities 45 g/L Cu, 170 g/L H_2SO_4, <30 mg/L Cl 250–350 L/min/cell, 45–50°C
Current density	200–250 A/m^2	290–460 A/m^2
Cell voltage	~0.2 V	2.0–2.2 V
Power consumption	0.25 MWh/t	2.0–2.3 MWh/t
Current efficiency	97%–98%	88%–92%
Plating time	21 days	7.5 days
Cell bottom	Anode slimes	Lead sludge
Current interruption	No battery effect, reverse polarity	Battery effect from galvanic reaction

Adapted from Glencore Technology. (2020b). *Tankhouse shutdown and restart.* Webinar; Houlachi, G., & Bas, A. D. (2019). Copper electrowinning and electrorefining: Fundamentals and practical aspects. In Copper hydrometallurgy: Principles and practice. Wassink, B. and Asselin, E. (Eds.) (pp. 148–188). Montreal: Canadian Institute of Mining, Metallurgy and Petroleum; Sole, K C., Moats, M. S., Sandoval, S., Robinson, T. G., & Davenport, W. E. (2019). Copper electrowinning: 2018 global survey of tankhouse operating practice and performance. *Copper–cobre 2019 (Paper 581092).* Montreal: Canadian Institute of Mining, Metallurgy and Petroleum.

16.9 Direct electrowinning from agitated leach solutions

16.9.1 From ore leach solutions

A few operations, mainly in African and Asian countries, treat leach solutions directly by EW for Cu recovery. The liquors, obtained by agitation leaching of high-grade oxides and roasted sulfide concentrates or mattes, contain 30–40 g/L Cu. These leach liquors can be purified or upgraded by SX, but are also processed by direct EW to produce Cu cathode.

Direct EW electrolytes are not as pure as those obtained using SX, so (i) the purity of the copper cathode is lower and (ii) current and energy efficiencies tend to be lower,

Table 16.3 Selected industrial copper electrowinning plant data (2018).

Operation	San Francisco, Chile	Mantoverde, Chile	Morenci Stargo, USA	Sepon, Laos	Tenke Fungurume, DRC
Startup date	2002	1995	2000	2005	2009, 2012
Cathode capacity (kt/a)	41	67	350	80	400
Electrolytic cells					
Number	72	168	324	166	540
Construction material	Polymer concrete	Polymer concrete	Polymer concrete	Polymer concrete	Polymer concrete
Length × width × depth (m)	6.6 × 1.27 × 1.45	6.17 × 1.24 × 1.42	6.71 × 1.22 × 1.57	8.6 × 1.3 × 1.7	7.3 × 1.3 × 1.8
Anodes/cathodes per cell	64/63	49/48	66/65	85/84	67/66
Cathode–cathode spacing (mm)	104.6	95	106	95	100
Acid mist suppression system					
Polypropylene balls	×		×		×
BBs				×	
Chemical suppressant	×	×	×	×	
Hoods	×			×	

Anodes					
Pb (%)	98.44–98.69	98.55	98.5	98	98.425
Sn (%)	1.25–1.50	1.375	1	1.35	1.30–1.45
Ca (%)	0.06–0.07	0.075	0.5	0.07	0.065–0.085
Cast or rolled	Rolled	Rolled	Rolled	Rolled	Rolled
Length × width × thickness (mm)	1095 × 940 × 7	1106 × 940 × 6	1041 × 921 × 6	1135 × 940 × 6	1253 × 872 × 8
Lifetime (years)	5	7	7	2	6
Cathodes					
Material of blanks	St. steel	St. steel	St. steel	St. steel	LDX 2101 duplex
Length × width × thickness (mm)	1210 × 1048 × 3.25	1232 × 1018 × 3.25	1297 × 1007 × 4.25		1403 × 968 × 3
Plating time (d)	3–4	5	5	7	9
Mass of Cu plated each side (kg)	42	35–40	42–50	50–55	45–55
Pb (ppm)	<2	<1	0–5	<1.5	<3
S (ppm)	<20	8	<8	8–15	<8
Fe (ppm)	<1	1		2–5	<0.8
Electrical					
Cathode current density (A/m^2)	498	275	346		
Cathode current efficiency (%)	92.65	92	88	89–90	91.0
Cell voltage (V)	2.2	1.9	1.95		2.0
Power consumption (MWh/t Cu)	2.1	2.1	1.98		1.8

Continued

Table 16.3 Selected industrial copper electrowinning plant data (2018).—cont'd

Operation	San Francisco, Chile	Mantoverde, Chile	Morenci Stargo, USA	Sepon, Laos	Tenke Fungurume, DRC
Electrolyte					
Circulation into each cell (m^3/min)	0.43	0.18	0.21	0.42–0.48	0.33 and 0.66
Cu into cells (g/L)	38	32	36–38	37	42–43
H_2SO_4 into cells (g/L)	191	219	200	185	200–215
Temperature into cells (°C)	55	41	45–48	47.5	44
Fe^{2+} in electrolyte (g/L)	0.2	0.3	0.25	2–2.4 (total)	0.8–1.2
Fe^{3+} in electrolyte (g/L)	0.6	1.6	2.8		0.9
Co^{2+} in electrolyte (mg/L)	360	135	170–180	220	1380
Smoothing agent	DXG-F7	Guar	Modified polysaccharide	Guar	Polyacrylamide/modified polysaccharide
Smoothing agent addition (g/t Cu)	400	180	750	300–350	378
Bleed to control	Fe, Cl	Fe	Fe, Mn	Fe	

Additional details of other major EW plant operations (see Table 2.7) are given by Robinson, Sole, et al. (2013) and Sole, Moats, Sandoval, Robinson, & Davenport (2019).

mainly because of the high Fe content of the electrolytes. There is no recycle of the spent electrolyte to SX in direct EW, so it is necessary to electrowin Cu from the electrolyte to much lower concentrations. This is usually carried out in a two-step EW process: the first EW circuit will reduce the Cu from ~40 to ~20 g/L (a bit lower than SX–EW concentrations) and the second circuit, known as *liberator cells*, will reduce the Cu to as low as 5 g/L. The quality of the cathodes produced in liberator cells is very poor: this material is usually sent for melting and electrorefining. The spent electrolyte is very high in acid and is returned directly to the leach circuit.

16.9.2 From concentrate or matte leach solutions

Leaching of sulfide-containing materials, as is practiced in several base metal refineries, can produce a leach liquor that contains much higher levels of copper than leaching of an ore: up to ~70 g/L Cu. Such solutions can be treated by direct EW and produce LME grade copper. The new Kristiansand copper EW circuit represents current state-of-the-art technology (Holmin et al., 2019; Åkre & Rosseland, 2019). This refinery uses coated titanium anodes, anode brushes, current tracking software, and extensive automation to achieve high-purity cathode copper without the need for an upstream SX circuit.

16.10 Copper electrowinning in EMEW cells

As discussed in Section 16.5.2.7, most plants are required to bleed a portion of the electrolyte to control the levels of Fe, Mn, or Cl. Associated with this bleed is loss of copper. EMEW technology employs a cylindrical cathode with a central anode, and the electrolyte is tangentially injected into the cell at high velocity. This configuration is capable of producing high-purity LME Grade A copper from an electrolyte containing as low as 3 g/L Cu, thereby enabling mitigation of such losses (Dixon et al., 2003). Fig. 16.10 shows a single cathode and a modular arrangement of EMEW cells.

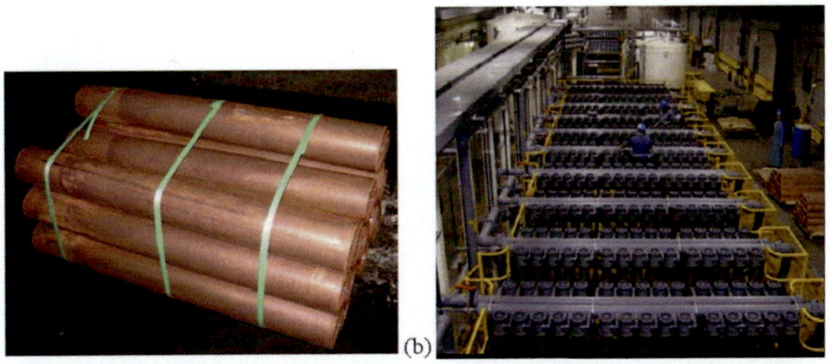

(a) (b)

Figure 16.10 EMEW technology for the electrowinning of copper from dilute solutions. (a) copper cathodes and (b) modular installation of cells. *(Photographs courtesy of EMEW Clean Technologies.)*

16.11 Safety in electrowinning tankhouses

An EW tankhouse operates with high acid concentrations, a conductive electrolyte, the presence of acid mist, high electrical currents, and routine mechanical handling of heavy items. Stray current (see Fig. 16.1) can create conditions for a shock hazard and potential injury to an operator. Recommended housekeeping and electrical practice for safe operation include the following (adapted from Sandoval, 2018 and Domingues, Gebrehiwot, Richardson, & Sandoval, 2019):

- Repair cell leaks timeously to avoid creating a path for current to ground;
- Remove accumulations of crystallized copper sulfate salts around cells and connections by regularly washing down the tankhouse using water;
- Ensure that all spillage is cleaned up and avoid accumulation of electrolyte in walkways and beneath cells;
- Regularly inspect the integrity of concrete foundations;
- Use insulated hand tools and avoid any connections of tools to ground;
- Metal pipe fittings in contact with electrolyte should be insulated;
- Railings and electrical guards can prevent operator contact;
- Careful design and insulation of crane components to avoid shocks;
- Personal protective equipment, such as boots and gloves, should be insulating.

16.12 Future developments

New developments in EW are dominated by (i) improving energy efficiency and (ii) producing high-purity copper at higher current densities (Marsden, 2008; Nieminen, Virtanen, Ekman, Palmu, & Tuuppa, 2010). One of the most significant advances is the increasing adoption of new anode materials that minimize voltage requirements. In particular, titanium mesh and sheet anodes coated with precious metal oxides are becoming more widely used. The commercial application of alternative anodes (nonlead anodes) in the United States (Sandoval, Clayton, Dominguez, Unger, & Robinson, 2010, Sandoval, Garcia, Neff, & Schnebly, 2013), Europe (Åkre & Rosseland, 2019), and South America is a significant advance.

Considerable development work has been carried out using an alternative electrolyte composition in which the anode reaction is the Fe(II) → Fe(III) conversion (Sandoval, Cook, Morales, & Robinson, 2010). This enables a reduction in energy consumption of up to 50% and there is no generation of acid mist. Although this process is not yet commercialized, it is receiving increasing interest as a potential "green" technology (Iacopetti, Calderara, Perrone, Perego, & Furusawa, 2019; Protech, 2020).

Maintaining cleanliness of the electrolyte is essential for the production of high purity copper. Electrolyte filters and the use of anode brushes to remove loose corrosion material from anodes have been introduced.

Developments in hardware, software, automation, and robotic control promote improved tankhouse management and operator safety, reduce labor requirements, and enable consistent production of high quality copper (Larinkari, Nord, & Schmidt, 2019; Åkre & Rosseland, 2019). Crane automation facilitates the use of hoods over cells for acid mist mitigation, allows accurate positioning of electrodes, and minimizes the need for operator activities in the tankhouse. Online monitoring of cell temperatures and voltages and wireless interfaces enable rapid detection of shorts (Fraser et al., 2014). Alternating of direct current with alternating current can improve the smoothness of the deposit and minimize electrolyte entrapment when operating at high current density (Hecker et al., 2010). Advanced process control and the use of artificial intelligence are also being increasingly adopted in many tankhouses. All of these developments improve current efficiency, productivity, and operator health and safety.

16.13 Summary

EW produces pure metallic copper from leach/SX electrolytes. In 2019, about four million tonnes of pure copper were electrowon, which represents 18% of world cathode production.

In the EW process, an electrical potential is applied between an inert Pb alloy or coated Ti anode and stainless-steel cathode in $CuSO_4-H_2SO_4$ electrolyte. Pure copper electroplates on the cathode; O_2 and H_2SO_4 are generated at the anode.

Copper is stripped from the cathode and sold. The O_2 is evolved to the atmosphere and is usually captured to a scrubber to contain the acid mist that it carries. The Cu^{2+}-depleted electrolyte is returned to SX as the spent electrolyte, where the contained acid is used for stripping Cu^{2+} from the loaded organic phase and Cu^{2+} is replenished in the advance electrolyte.

Electrowon copper is as pure as or purer than electrorefined copper. Its only significant impurities are S (~5 ppm) and Pb and Fe (1−2 ppm each). Careful control and attention to detail in housekeeping and tankhouse management can decrease these impurity concentrations to the low ends of these ranges.

Modern trends in copper EW include the following:

(a) increasing current densities above 450 A/m^2;

(b) reducing energy consumption;

(c) improved acid mist management;

(d) increasing automation and robotic control and monitoring of the process.

References

Abbey, C., & Moats, M. (2018). Pb−Ca−Sn anode potential as a function of cobalt, iron and manganese in synthetic sulfuric acid electrolytes. In *Extraction 2018* (pp. 1535−1546). Montreal: Canadian Institute of Mining, Metallurgy and Petroleum.

Åkre, T., & Rosseland, E. (2019). Copper electrowinning developments at Glencore Nikkelverk. In *Copper 2019 (paper 594865)*. Montreal: Canadian Institute of Mining, Metallurgy and Petroleum.

Aslin, N. J., Eriksson, O., Heferen, G. J., & Sue Yek, G. (2010). Developments in cathode stripping machines. An integrated approach for improved efficiency. In *Copper 2010: vol. 4. Electrowinning and -refining* (pp. 1253—1271). Gesellschaft für Bergbau, Metallurgie, Rohstoff- und Umwelttechnik, Clausthal-Zellerfeld: GDMB.

Aylwin, P., Lagos, N., & Melani, P. (2013). SELE® technology: An alternative to boosting current efficiency and cathode quality in EW plants. In *HydroProcess 2013 (CD-ROM)*. Santiago: Gecamin.

Bender, J. T. (2010). Evaluation of mist suppression agents for use in copper electrowinning. In *Electrowinning and —refining: vol. 4. Copper 2010* (pp. 1271—1279). Gesellschaft für Bergbau, Metallurgie, Rohstoff- und Umwelttechnik, Clausthal-Zellerfeld: GDMB.

Beukes, N. T., & Badenhorst, J. (2009). Copper electrowinning: Theoretical and practical design. *Journal of the Southern African Institute of Mining and Metallurgy, 109*(6), 343—356.

Blackett, A., & Nicol, M. (2010). The simulation of current distribution in cells during the electrowinning of copper. In , *Copper 2010: vol. 4. Electrowinning and -refining* (pp. 1291—1292). Clausthal-Zellerfeld: Gesellschaft für Bergbau, Metallurgie, Rohstoff- und Umwelttechnik.

Castle Lead Works. (2020). *Anode care and maintenance*. Retrieved from https://www.castlelead.co.za/anode-technology.html.

Cheng, C. Y., Hughes, C. A., Barnard, K. R., & Larcombe, K. (2000). Manganese in copper solvent extraction and electrowinning. *Hydrometallurgy, 58*, 135—150.

Chibwe, C., & Tadie, M. (2021). An experimental review of the physicochemical properties of copper electrolytes in the presence of polyacrylamide additives. *Mining, Metallurgy and Exploration, 38*, 1225—1237.

Cifuentes, L., Astete, E., Crisostomo, G., Simpson, J., Cifuentes, G., & Pilleux, M. (2005). Corrosion and protection of lead anodes in acidic copper sulphate solutions. *Corrosion Engineering, Science and Technology, 40*(4), 321—327.

Cifuentes, R., Bravo, R., & Schwarz, N. (2015). Effective grain modification in copper electrowinning with DXG-F7. In *Copper cobalt Africa* (pp. 389—395). Johannesburg: Southern African Institute of Mining and Metallurgy.

Currie, B., & Miller, G. (2018). Design and material selection considerations for solvent extraction and electrowinning plants. In *ALTA nickel cobalt copper conference* (pp. 283—308). Melbourne: ALTA Metallurgical Services.

Dixon, R. E., Fester, R., García, C., Contreras, C., Romero, F., & Ewart, I. D. (2003). Adding value to LIX/SX/EW operations by recovering copper from bleed, raffinate and other plant streams. In *Hydrometallurgy 2003*. Montreal: Canadian Institute of Mining, Metallurgy and Petroleum.

Domingues, S., Gebrehiwot, E., Richardson, S., & Sandoval, S. (2019). Worker safety from stray current shock hazards in electrolytic tankhouses. In *Copper 2019 (paper No. 568074)*. Montreal: Canadian Institute of Mining, Metallurgy and Petroleum.

Flores, J., Muñoz, C., & Zárate, G. (2010). Improving cathode quality at high current density at San Francisco EW plant. In *Presentation at XII seminario técnico Cytec, Chile*.

Forner, E. L., Scheepers, J., du Toit, A. J., & Miller, G. M. (2018). Copper electrowinning circuit design: Optimized costing as a function of cell arrangement, productivity, rectiformer size, and throughput. *Journal of the Southern African Institute of Mining and Metallurgy, 118*(11), 1141—1148.

Fraser, R., Boon, C., Jarami, K., Johnston, T., Yesberg, J., Åkre, T., & Rosseland, E. (2014). Cathode current monitoring technology enables current efficiency improvements. In *Automining 2014, 4th International congress on automation in mining*. Santiago: Gecamin.

Glencore Technology. (2019). ISAKidd refining and electrowinning. *Presentation at Copper 2019*. Montreal: Canadian Institute of Mining, Metallurgy and Petroleum.

Glencore Technology. (2020a). *Cathode plates*. Retrieved from https://www.isakidd.com/en/technology-services/Pages/Cathode-Plates.aspx.

Glencore Technology. (2020b). *Tankhouse shutdown and restart*. Webinar.

Hecker, C., Vera, M., Martínez, G., Bustos, J. B., Villavicencio, C., & Beas, E. (2010). Improvement in copper EW and ER processes by using a multi-frequency AC + DC current. In , *Copper 2010: Vol. 4. Electrowinning and -refining* (pp. 1355—1367). Gesellschaft für Bergbau, Metallurgie, Rohstoff- und Umwelttechnik, Clausthal-Zellerfeld: GDMB.

Holmin, S., Afvander, Å., Gustavsson, J., Zimmerman, E., Åkre, T., & Rosseland, E. (2019). Investigation of Permascand coated titanium anodes in copper electrowinning at Glencore Nikkelverk. In *Copper 2019 (paper 594788)*. Montreal: Canadian Institute of Mining, Metallurgy and Petroleum.

Iacopetti, L., Calderara, A., Perrone, P., Perego, M., & Furusawa, T. (2019). Advancements in commercialization of De Nora's "self protected anode – SPA" for metal electrowinning. In *Copper 2019 (paper 592622).* Montreal: Canadian Institute of Mining, Metallurgy and Petroleum.

International Copper Study Group. (2019). *The world copper factbook 2019.* Lisbon: International Copper Study Group.

Joy, S., Staley, A., Moats, M., Perkins, C., Uhrie, J., & Robinson, T. (2010). Understanding and improvement of electrowinning current efficiency at FMI Bagdad. In *Copper 2010, Vol. 4, Electrowinning and – refining* (pp. 1379–1392). Gesellschaft für Bergbau, Metallurgie, Rohstoff- und Umwelttechnik, Clausthal-Zellerfeld: GDMB.

Khouraibchia, Y., & Moats, M. (2009). Effective diffusivity of ferric ions and current efficiency in stagnant synthetic copper electrowinning solutions. *Minerals and Metallurgical Processing, 26,* 176–190.

Khouraibchia, Y., & Moats, M. (2010). Evaluation of copper electrowinning parameters on current efficiency and energy consumption using surface response methodology. In F. M. Doyle, R. Woods, & G. H. Kesall (Eds.), *Electrochemistry in mineral and metal processing VIII ESC transactions* (Vol. 28, No. 6, pp. 295–306). Pennington, NJ: The Electrochemical Society (6).

Larinkari, M., Nord, P., & Schmidt, M. (2016). Cathode condition based maintenance and improved quality control. In *Copper 2016* (pp. 1965–1977). Kobe: The Mining and Materials Processing Institute of Japan and Japan Mining Industry Association.

Larinkari, M., Nord, P., & Schmidt, M. (2019). Latest development in cathode stripping machines. In *Copper 2019 (paper 594055).* Montreal: Canadian Institute of Mining, Metallurgy and Petroleum.

London Metal Exchange. (2020). *Special contract rules for copper grade A.* Retrieved from https://www.lme.com/en-GB/Metals/Non-ferrous/Copper/Physical.

Luyima, A., Moats, M. S., Cui, W., & Heckman, C. (2016). Examination of copper electrowinning smoothing agents. Part II: Fundamental electrochemical examination of DXG-F7. *Mineral Processing and Extractive Metallurgy Review, 33*(1), 8–16.

Maldonado, G. R. (2010). *Evaluación efectos secundarios en SX de utilización de afinador de grano DXG-F7.* Cytec, Chile: Informe técnico.

Marsden, J. O. (2008). Energy efficiency and copper hydrometallurgy. In C. Young, P. R. Taylor, C. G. Anderson, & Y. Choi (Eds.), *Hydrometallurgy 2008* (pp. 29–42). Littleton, CO: Society for Mining, Metallurgy and Exploration.

McGinnity, J., & Nicol, M. J. (2014). Sulphuric acid mist: Generation, suppression, health aspects and analysis. *Mineral Processing and Extractive Metallurgy Review, 35*(3), 149–192.

Miller, G. (2010). Methods of managing manganese effects on copper SX-EW operations. In *ALTA nickel cobalt copper 2010.* Melbourne: ALTA Metallurgical Services.

Miller, G. (2011). Methods of managing manganese effects on copper solvent extraction plant operations. *Solvent Extraction and Ion Exchange, 29*(5–6), 837–853.

Mirza, A., Burr, M., Ellis, T., Evans, D., Kakengela, D., Webb, L., Gagnon, J., Leclercq, F., & Johnston, A. (2016). Corrosion of lead anodes in base metals electrowinning. *Journal of the Southern African Institute of Mining and Metallurgy, 116*(6), 533–538.

Moats, M. (2012). How to evaluate current efficiency in copper electrowinning. In *Separation technologies for minerals, coal and earth resources* (pp. 333–340). Littleton, CO: Society for Mining, Metallurgy and Exploration.

Moats, M. S. (2018). Energy efficiency of electrowinning. *Energy efficiency in the minerals industry: Best practices and research directions* (pp. 213–232). Berlin: Springer Verlag.

Moats, M. S., & Free, M. L. (2019). Electrowinning. In R. C. Dunne, S. K. Kowatra, & C. A. Young (Eds.), *SME minerals processing and extractive metallurgy handbook* (pp. 1369–1390). Littleton, CO: Society for Mining, Metallurgy and Exploration.

Moats, M. S., Luyima, A., & Cui, W. (2016). Examination of copper electrowinning smoothing agents. Part I: A review. *Minerals and Metallurgical Processing, 33*(1), 1–7.

Mohanarangam, K., Yang, W., Rajasingam, R., Tadesse, B., & Robinson, D. J. (2019). Total acid mist quantification within a full-scale copper electrowinning cell. In *Copper 2019 (paper 593974).* Montreal: Canadian Institute of Mining, Metallurgy and Petroleum.

Morimitsu, M., Oshiumi, N., & Wada, N. (2010). Smart anodes for electrochemical processing of copper production. In , *Electrowinning and −refining: vol. 4. Copper 2010* (pp. 1511−1520). Gesellschaft für Bergbau, Metallurgie, Rohstoff- und Umwelttechnik, Clausthal-Zellerfeld: GDMB.

Nicol, M. (2006). *Electrowinning and electrorefining of metals. Short course.* Perth: Murdoch University.

Nieminen, V., Virtanen, H., Ekman, E., Palmu, L., & Tuuppa, E. (2010). Copper electrowinning with high current density. In , *Electrowinning and −refining: vol. 4. Copper 2010* (pp. 1545−1558). Gesellschaft für Bergbau, Metallurgie, Rohstoff- und Umwelttechnik, Clausthal-Zellerfeld: GDMB.

Nikolosi, A. N., & Nicol, M. J. (2008). Effect of cobalt ions on the performance of lead anodes used for the electrowinning of copper—a literature review. *Mineral Processing and Extractive Metallurgy Review, 29*, 143−172.

Nikoloski, A., Nicol, M., & Stuart, A. (2010). Managing the passivation layer on lead alloy anodes in copper electrowinning. In , *Electrowinning and −refining: Vol. 4. Copper 2010* (pp. 1559−1568). Gesellschaft für Bergbau, Metallurgie, Rohstoff- und Umwelttechnik, Clausthal-Zellerfeld: GDMB.

Pranowo, A., Aslin, N. J., Eriksson, P. O., & Heferen, G. J. (2019). Stainless steel "permanent" cathode plates—fixed assets or consumables?. In *Copper 2019 (paper 590965).* Montreal: Canadian Institute of Mining, Metallurgy and Petroleum.

Prengaman, R. D., Ellis, T., & Mirza, A. (2010). 10 years' experience with rolled lead-calcium-silver anodes. In A. Siegmund, I. Centomo, C. Geenen, N. Piret, G. Richards, & R. Stephens (Eds.), *Lead-zinc 2010* (pp. 819−837). Warrendale, PA: The Minerals, Metals and Materials Society.

Price, D. C., & Davenport, W. G. (1980). Densities, electrical conductivities and viscosities of $CuSO_4$/ H_2SO_4 solutions in the range of modern electrorefining and electrowinning electrolytes. *Metallurgical Transactions B, 11*(1), 159−163.

Protech. (2020). *Proceso de Ecowinning®.* Retrieved from https://www.protech.cl/ecowinning/.

Robinson, T., White, D., & Grassi, R. (2013). Acid mist abatement in base metal electrowinning. In *Nickel cobalt 2013* (pp. 141−153). Warrendale, PA: The Minerals, Metals & Materials Society.

Robinson, T. G., Sole, K. C., Sandoval, S., Moats, M., Siegmund, A., & Davenport, W. E. (2013). Copper electrowinning − 2013 world operating tankhouse data. In *Copper 2013* (Vol. V, pp. 3−14). Santiago: Gecamin.

Robinson, T., Sole, K. C., Moats, M., Crundwell, F., Morimitsu, M., & Palmu, L. (2012). Developments in base metal electrowinning cell house design. In *Electrometallurgy 2012 (CD-ROM).* Warrendale, PA: The Minerals, Metals and Materials Society.

San Martin, R. M., Otero, A. F., & Cruz, A. (2005). Use of quillaja saponins (*Quillaja saponaria* Molina) to control acid mist in copper electrowinning process. Part 2: Pilot plant and industrial scale evaluation. *Hydrometallurgy, 77*, 171−181.

Sandoval, S. (2018). Potential hazards of stray current in an electrolytic tankhouse. In *HydroProcess 2018*. Santiago: Gecamin.

Sandoval, S., Clayton, C., Dominguez, S., Unger, C., & Robinson, T. (2010). Development and commercialization of an alternative anode for copper electrowinning. In , *Copper 2010: vol. 4. Electrowinning and -refining* (pp. 1635−1648). Gesellschaft für Bergbau, Metallurgie, Rohstoff- und Umwelttechnik, Clausthal-Zellerfeld.

Sandoval, S., Cook, P., Morales, C., & Robinson, T. (2010). Demonstration of the ferrous/ferric anode reaction for copper electrowinning. In , *Copper 2010: vol. 4. Electrowinning and -refining* (pp. 1617−1634). Clausthal-Zellerfeld: Gesellschaft für Bergbau, Metallurgie, Rohstoff- und Umwelttechnik.

Sandoval, S., Garcia, R., Neff, T., & Schnebly, N. (2013). Operation of alternative anodes at Chino SXEW. In *Copper 2013* (Vol. 5, pp. 145−152). Santiago: Gecamin.

Shaw, D. R. (2019). Ion exchange in hydrometallurgy. In B. Wassink, & E Asselin (Eds.), *Copper hydrometallurgy: Principles and Practice* (pp. 189−212). Montreal: Canadian Institute of Mining, Metallurgy and Petroleum.

Sole, K. C., Moats, M. S., Sandoval, S., Robinson, T. G., & Davenport, W. E. (2019). Copper electrowinning: 2018 global survey of tankhouse operating practice and performance. In *Copper 2019 (paper 581092).* Montreal: Canadian Institute of Mining, Metallurgy and Petroleum.

Sole, K. C., Mooiman, M. B., & Hardwick, E. (2017). Ion exchange in hydrometallurgical processing: An overview and selected applications. *Separation and Purification Reviews, 47*(2), 159−178.

Tjandrawan, V., & Nicol, M. J. (2010). The oxidation of manganese ions on lead alloys during the electrowinning of copper. In *Copper 2010: vol. 4. Electrowinning and -refining* (pp. 1699–1712). Clausthal-Zellerfeld: Gesellschaft für Bergbau, Metallurgie, Rohstoff- und Umwelttechnik.

Tozawa, K., Umetsu, Y., Su, Q., & Li, Z. (2003). Constitution of electrical conductivity of acidic copper sulfate electrolytes acidified with sulfuric acid for copper electrowinning and -refining processes. In F. Kongoli, K. Itagaki, C. Yamauchi, & H. Y. Sohn (Eds.), *Yusawa international symposium. Metallurgical and materials processing: Principles and technologies, vol. III, Aqueous and electrochemical processing* (pp. 105–119). Warrendale, PA: The Minerals, Metals and Materials Society.

Vainio, T., & Weatherseed, M. (2009). Acid mist capture and recycling for copper electrowinning tankhouses. In *ALTA copper 2009*. Melbourne: ALTA Metallurgical Services.

Zhang, T. (2018). Outotec coated titanium anodes technology. *Journal of the Mining and Materials Processing Institute of Japan, 5*(2). Paper No. 3101-06-02.

Zhang, W., Ghali, E., & Houlachi, G. (2017). Review of oxide coated catalytic anode performance for metal electrowinning. *Hydrometallurgy, 169*, 456–467.

Zhang, Z., Werner, J., & Free, M. (2018). A current efficiency prediction model based on electrode kinetics for iron and copper during copper electrowinning. In G. Lambotte, J. Lee, A. Allanore, & S. Wagstaff (Eds.), *Materials processing fundamentals* (pp. 111–113). Warrendale, PA: The Minerals, Metals and Materials Society.

Suggested reading

Hopkins, W. (2009). Electrowinning – some recent developments. In *ALTA copper 2009*. Melbourne: ALTA Metallurgical Services.

Lillo, A. R. (2010). Intensification of copper electrowinning. In *Copper: vol. 4. Electrowinning and - refining* (pp. 1449–1462). Gesellschaft für Bergbau, Metallurgie, Rohstoff- und Umwelttechnik, Clausthal-Zellerfeld: GDMB.

Minakshi, M., & Nicol, M. (2009). The mechanism of the incorporation of lead into copper cathodes. In E. Domic, & J. Casas (Eds.), *V hydrometallurgy workshop, HydroCopper 2009* (pp. 382–391). Santiago: Gecamin.

Pagliero, A. (2010). Evolution of the behaviour of anodes for copper electrowinning. In J. Casas, P. Ibáñez, & M. Jo (Eds.), *Third international workshop on process hydrometallurgy, HydroProcess 2010* (pp. 76–77). Santiago: Gecamin.

Salas, J., Weishaupt, R. A., & Guzman, J. A. (2019). Chiquicamata: Quality challenges throughout 100 years of copper electrorefining and electrowinning. In *Copper 2019 (paper 595038)*. Montreal: Canadian Institute of Mining, Metallurgy and Petroleum.

CHAPTER 17

Collection and processing of recycled copper

Previous chapters describe production of *primary* copper—extraction of copper from ore. This chapter and the next describe production of *secondary* copper—recovery of copper from scrap. About half the copper reaching the marketplace has been scrap at least once, so scrap recycle is of the utmost importance.

This chapter describes scrap recycling in general, major sources and types of scrap, and physical beneficiation techniques for isolating copper from its coatings and other contaminants.

Chapter 18 describes the chemical aspects of secondary copper production and refining.

17.1 The materials cycle

Fig. 17.1 shows the materials cycle flowsheet. It is valid for any material not consumed during use. Its key components are as follows:

(a) raw materials—ores from which primary copper is produced
(b) primary production —processes described in previous chapters of this book
(c) engineering materials—the final products of smelting/refining, mainly cast copper and predraw copper rod, ready for manufacturing
(d) manufacturing—production of goods to be sold to consumers
(e) obsolete products—products that have been discarded or otherwise taken out of use
(f) discard—sending of obsolete products to a discard site, usually a landfill.

Obsolete copper products are increasingly being recycled rather than sent to landfills. This is encouraged by the value of their copper and the increasing cost and decreasing availability of landfill sites.

17.1.1 Home scrap

The arrow marked (1) in Fig. 17.1 shows the first category of recycled copper, known as *home* or *run-around* scrap. This is copper that primary producers cannot further process or sell. Off-specification anodes, cathodes, bar, and rod are examples of this type of scrap. Anode scrap is another example.

The arrow shows that this material is reprocessed directly by the primary producer, usually by running it through a previous step in the process. Off-specification copper is usually put back into a converter or anode furnace, then electrorefined. Defective rod and bar are remelted in the vertical shaft furnace and recast.

Extractive Metallurgy of Copper, Sixth Edition
ISBN 978-0-12-821875-4, https://doi.org/10.1016/B978-0-12-821875-4.00021-3

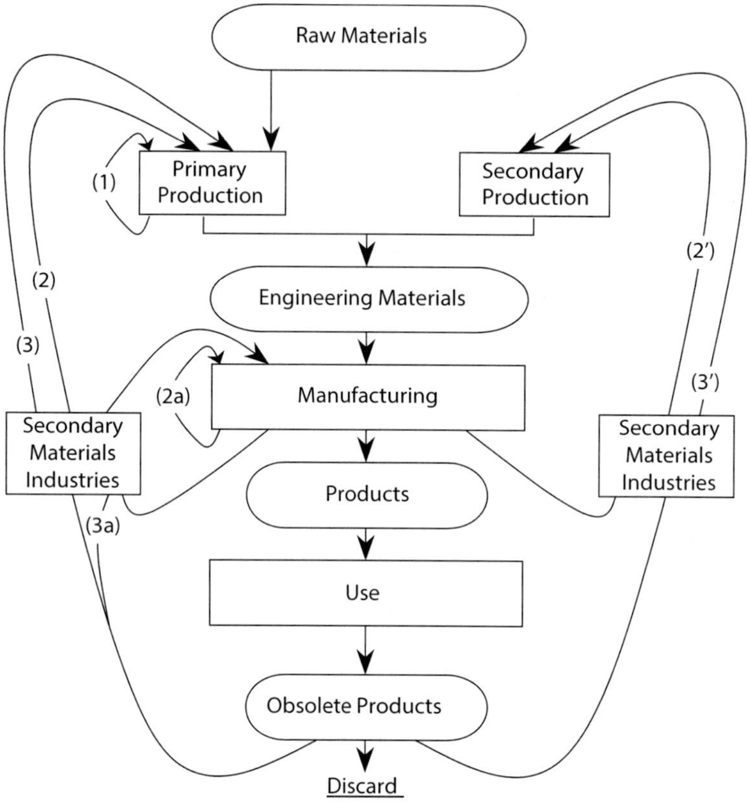

Figure 17.1 Flowsheet of "materials cycle." This is valid for any material not consumed during use. The arrow marked (1) shows *home* or *run-around* scrap. The arrows marked (2), (2a), and (2′) denote *new*, *prompt industrial*, or *internal arising* scrap. The paths marked (3), (3a), and (3′) show *old*, *obsolete*, *post-consumer*, or *external arising* scrap.

The annual amount of home scrap production is not known because it is not reported. However, industrial producers try to minimize its production to avoid recycle expense.

17.1.2 New scrap

The arrows marked (2), (2a), and (2′) in Fig. 17.1 denote *new*, *prompt industrial*, or *internal arising* scrap. This is scrap that is generated during manufacturing. The primary difference between this and home scrap is that it may have been adulterated during processing by alloying or by applying coatings and coverings. Examples of *new* scrap are as numerous as the products made with copper, since no manufacturing process is 100% efficient.

The pathway taken by new scrap depends on its chemical composition and the degree to which it has become entwined with other materials. The simplest approach is to recycle it internally (2a). This is common practice with gatings and risers from castings. They are simply remelted and cast again. Direct recycling has the advantages of (a) retaining the value of added alloying elements such as zinc or tin which would be lost if the alloy were sent to a smelter and (b) eliminating the cost of removing the alloying elements, which would be required if the metal were reprocessed at a smelter.

Similar reprocessing is done for scrap copper tube and uncoated copper wire. Path (2a) is the most common recycling route for new scrap. As much as 90% of new U.S. copper scrap is recycled along this path (Edelstein, 1999).

If the new scrap has coatings or attachments that cannot easily be removed, or if the manufacturing facility cannot directly reuse its new scrap (e.g., a wire-drawing plant without its own melting facilities), then paths (2) and (2′) are followed.

The secondary materials industries described in Fig. 17.1 fill the role that mining and ore beneficiation facilities fill for primary copper production. In many cases, they simply remove the coatings, insulation, or attachments from the scrap to make it suitable for reuse by the manufacturing facility. If purification or refining is needed, the cleaned-up scrap is sent to a primary or secondary smelter/refinery (Forsén et al., 2017). Since these facilities produce cathode grade copper, alloying elements present in the scrap are lost.

Specific activities of secondary materials industries are described later in this chapter.

17.1.3 Old scrap

The final category of copper scrap (paths (3), (3a), and (3′)) is termed *old*, *obsolete*, *postconsumer*, or *external arising* scrap. It is obtained from products that have ended their useful life. Old scrap is a huge potential source of recyclable copper. It is also more difficult to process. The challenges for processing old scrap include the following:

(a) low Cu grades—old copper scrap is often mixed with other materials and must be separated from this waste

(b) unpredictability—deliveries of materials and objects vary from day to day, making processing difficult

(c) location—old scrap is scattered about the landscape rather than being concentrated in a specific location like primary ore or new scrap.

The sources of old scrap have been divided into six classes (Bertram et al., 2002; Ruhrberg, 2006):

End-of-life vehicles (ELVs) include cars, trucks, and buses, but not planes, ships, or trains. The copper in ELV consists primarily of wiring, in particular electric motors and the wiring harness that connects the lights and motors (McGlothlin, 2018). The coverall copper content of ELV is about 0.8%.

Construction and demolition (C&D) waste includes copper from the wiring of buildings, pipes and fittings from the plumbing, and brass from door handles and frames, among other items (Wildnauer, 2019). C&D debris as a whole contains about 0.30% copper, or 0.05−0.24 kg/m^3 (Koutamanis, 2018). The current high price of copper encourages copper recovery from buildings; unfortunately, this often occurs as theft from job sites or vandalism.

Waste from electrical and electronic equipment (WEEE) is the largest source of old copper scrap (Glöser et al., 2013). WEEE has three subcategories (European Copper Institute, 2007): *white goods* (domestic electrical appliances such as washing machines, refrigerators, dishwashers, etc.), *brown goods* (audiovisual appliances such as televisions and DVD players), and *gray goods* (computers and telecommunication appliances). The estimated copper content of WEEE ranges from 3% to 21% (Barnwal et al., 2018).

Industrial electrical equipment waste (IEW) includes power cables (underground and surface), transformers, and other electrical equipment. The relative amount of this material is small, but its copper content is high (5%−80%).

Industrial nonelectrical equipment waste (INEW, sometimes grouped with IEW) includes large transportation equipment (planes, ships, and trains), spent ammunition and ordinance, and other machinery. Although no formal estimates of its copper content exist, the percentage is likely to be small (Ruhrberg, 2006).

Municipal solid waste (MSW) is the hardest waste stream to process, and contains very little copper (0.05%−0.20%). Loose coins in MSW contain much of its copper, and small appliances contain much of the rest.

Recovery of old copper scrap from waste streams depends on the type of waste, the location of the waste stream, and the price of copper (Vexler et al., 2004). High-grade streams such as IEW are profitable to process and recycle, so most of the copper from this waste is recovered (Glöser et al., 2013). Although few statistics are kept, the same is likely true for INEW. The presence of infrastructure for dismantling and processing ELV means that recovery of copper from this stream is higher in the developed world than in less-developed areas. Recovery of copper from MSW is always low. Recovery of copper from WEEE has also been low, but is increasing as a result of government mandates to process it.

17.2 Secondary copper grades and definitions

The Institute of Scrap Recycling Industries (ISRI, 2019) currently recognizes 45 grades of copper-base scrap. However, most of these are for alloy scrap, which is much less available than "pure" copper scrap. Alloy scrap is also more likely to be directly recycled than copper scrap. As a result, ISRI designations are not always important to copper recyclers. Table 17.1 describes the most significant grades of nonalloy copper scrap, including ISRI and trade names (Fig. 17.2).

Table 17.1 Common scrap grades (ISRI, 2009).

Scrap type (ISRI name)	Description
№ 1 copper wire (Barley)	Shall consist of no. 1 bare, uncoated, unalloyed copper wire, commonly known as Bare Bright copper wire. Wire gauge subject to agreement between buyer and seller. Green copper wire and hydraulically compacted material to be subject to agreement between buyer and seller.
№ 1 copper wire (Berry)	Shall consist of clean, untinned, uncoated, unalloyed copper wire and cable, and free of brittle burnt wire. Wire gauge subject to agreement between buyer and seller. Free of copper tubing. Hydraulically briquetted copper subject to agreement.
№ 2 copper wire (Birch)	Shall consist of miscellaneous, unalloyed copper wire having a nominal 96% copper content (minimum 94%) as determined by electrolytic assay. Should be free of the following: Excessively leaded, tinned, soldered copper wire; brass and bronze wire; excessive oil content, iron, and nonmetallics; copper wire from burning; insulation; hair wire; brittle burnt wire; and should be reasonably free of ash. Hydraulically briquetted copper subject to agreement.
№ 1 heavy copper (Candy)	Shall consist of clean, unalloyed, uncoated copper clippings, punchings, bus bars, commutator segments, and clean copper tubing. Hydraulically briquetted copper subject to agreement.
№ 2 copper (Cliff)	Shall consist of miscellaneous, unalloyed copper scrap having a nominal 96% copper content (minimum 94%) as determined by electrolytic assay. Should be free of the following: Excessively leaded, tinned, soldered copper scrap; brasses and bronzes; excessive oil content, iron and nonmetallics; copper tubing with other than copper connections or with sediment; copper wire from burning; insulation; hair wire; brittle burnt wire; and should be reasonably free of ash. Hydraulically briquetted copper subject to agreement.
№ 1 copper wire nodules (Clove)	Shall consist of no. 1 bare, uncoated, unalloyed copper wire scrap nodules, chopped or shredded, free of tin, lead, zinc, aluminum, iron, other metallic impurities, insulation, and other foreign contamination. Minimum copper 99%. Gauge smaller than no. 16 B and S wire and hydraulically compacted material subject to agreement between buyer and seller.

Continued

Table 17.1 Common scrap grades (ISRI, 2009).—cont'd

Scrap type (ISRI name)	Description
№ 2 copper wire nodules (Cobra)	Shall consist of no. 2 unalloyed copper wire scrap nodules, chopped, or shredded, minimum 97% copper. Maximum metal impurities not to exceed 0.50% aluminum and 1% each of other metals or insulation. Hydraulically compacted material subject to agreement between buyer and seller.
Copper wire nodules (Cocoa)	Shall consist of unalloyed copper wire scrap nodules, chopped, or shredded, minimum 99% copper. Shall be free of excessive insulation and other nonmetallics. Maximum metal impurities are as follows: Al 0.05% Sn 0.25% Ni 0.05% Sb 0.01% Fe 0.05%. Hydraulically compacted material subject to agreement between buyer and seller.
Electronic scrap metals EM3	Shall consist of whole or shredded copper/precious metal bearing populated or unpopulated circuit boards from the manual dismantling of electronic products. May also consist of shredded circuit boards from end-of-life electronic product processing systems with a maximum piece size of 2 inches. Maximum acceptable metal contamination: Aluminum, 5%; ferrous, 2%; zinc, 2%; magnesium, 1%; and beryllium, 200 ppm. Other elements subject to agreement between buyer and seller. Maximum plastic content: 40%. Typically sold on an assay basis and classified into different categories denominated by the gold levels contained in the material. Major classifications are as follows: <50 g/mt, <200 g/mt, and >200 g/mt.
Light copper (Dream)	Shall consist of miscellaneous, unalloyed copper scrap having a nominal 92% copper content (minimum 88%) as determined by electrolytic assay and shall consist of sheet copper, gutters, downspouts, kettles, boilers, and similar scrap. Should be free of the following: Burnt hair wire; copper clad; plating racks; grindings; copper wire from burning, containing insulation; radiators and fire extinguishers; refrigerator units; electrotype shells; screening; excessively leaded, tinned, soldered scrap; brasses and bronzes; excessive oil, iron and nonmetallics; and should be reasonably free of ash. Hydraulically briquetted copper subject to agreement. Any items excluded in this grade are also excluded in the higher grades above.

Table 17.1 Common scrap grades (ISRI, 2009).—cont'd

Scrap type (ISRI name)	Description
Insulated copper wire scrap (Druid)	Shall consist of no. 1 bare, uncoated, unalloyed copper wire (see Barley), not smaller than no. 16 B and S wire gauge (unless smaller wire gauge is mutually agreed upon), with various types of insulation. To be sold on a sample or recovery basis, subject to agreement between buyer and seller.
Mixed electric motors (Elmo)	Shall consist of whole electric motors and/or dismantled electric motor parts that are primarily copper wound. May contain some aluminum-wound material, subject to agreement between buyer and seller. No excessive steel attachments such as gear reducers, iron bases, and pumps, or loose free iron allowed. Specification not to include sealed units or cast iron compressors.
Shredded electric motors (Shelmo)	Shall consist of mixed copper-bearing material from ferrous shredding, comprised of motors without cases. May contain up to 5% aluminum-wound material and may contain insulated copper harness wire, subject to agreement between buyer and seller. Trace percentages of other contaminants and fines may be present. No free iron or sealed units.
High-density shredded nonferrous (Zebra)	Shall consist of high-density nonferrous metals produced by media separation technology containing brass, copper, zinc, nonmagnetic stainless steel, and copper wire. Material to be dry and free from excess oxidation. The percentage and types of metals other than these, as well as the percentage and types of nonmetallic contamination, are to be agreed upon between the buyer and seller.
Shredded insulated copper wire (Zeyda)	Shall consist predominantly of recovered ICW (insulated copper wire) that has been derived by either mechanical or physical separation. This material is likely to have other metals, in small percentages (should be less than 5%) that were not successfully separated. The percentage of each metal within the overall product shall be subject to agreement between buyer and seller. Shall have passed one or more magnets to reduce or eliminate free iron and/or large iron attachments. Shall be free of radioactive material, dross, or ash. Material to be bought/sold under this guideline shall be identified as "Zeyda" with two numbers to follow, indicating

Continued

Table 17.1 Common scrap grades (ISRI, 2009).—cont'd

Scrap type (ISRI name)	Description
	the estimated percentage recoverable copper wire and the second indicating the maximum amount of other metals or contaminants expected (e.g., "Zeyda 45/3" means the material contains approximately 45% copper and up to 3% other material). May also be screened to permit description by specific size ranges.

The most significant recent changes have been replacement of electronic scrap grades "Dallas" and "Depot" with grade EM3, which describes printed circuit boards (PCBs) recovered from e-waste. In addition, Grade "Zeyde" (shredded insulated copper wire) has been added.

Copper recycling also includes the treatment of *process wastes*. The definition of this word is a matter of debate in industrialized countries, because the sale and transportation of materials designated as waste is more heavily regulated than that of materials designated as scrap. In fact, material graded as copper-bearing scrap is defined in many countries as waste, despite the fact that it can be recycled profitably. Wastes generally have a low copper content, low economic value, and a high processing cost per kg of contained copper. As a result, recyclers sometimes charge a fee for processing these materials (Hagelüken, 2006). Examples of process wastes include the sludge generated from copper plating, refining slags from foundries, and the residue from leaching of anode slimes.

17.3 Scrap processing and beneficiation

17.3.1 Wire and cable processing

Wire and cable are by far the most common forms of old scrap. It is these forms for which the most advanced reprocessing technology exists. Nijkerk and Dalmijn (1998) divide scrap wire and cable into three types:

(a) *Above-ground*, mostly high-tension power cable. These cables are high grade (mainly copper, little insulation) and fairly consistent in construction. They are easy to recycle.

(b) *On-the-ground*, with a variety of coverings and sizes. These are usually thin wires, so the cost of processing per kg of recovered copper is higher than that for cable. Wire is also more likely to be mixed with other waste, requiring additional separation. Automotive harnesses and appliance wire are examples.

(c) *Below-ground*, which feature complex construction and many coverings. These cables often contain lead sheathing, bitumen, grease, and mastic. This means that fairly complex processing schemes are required to recover their copper without creating safety and environmental hazards.

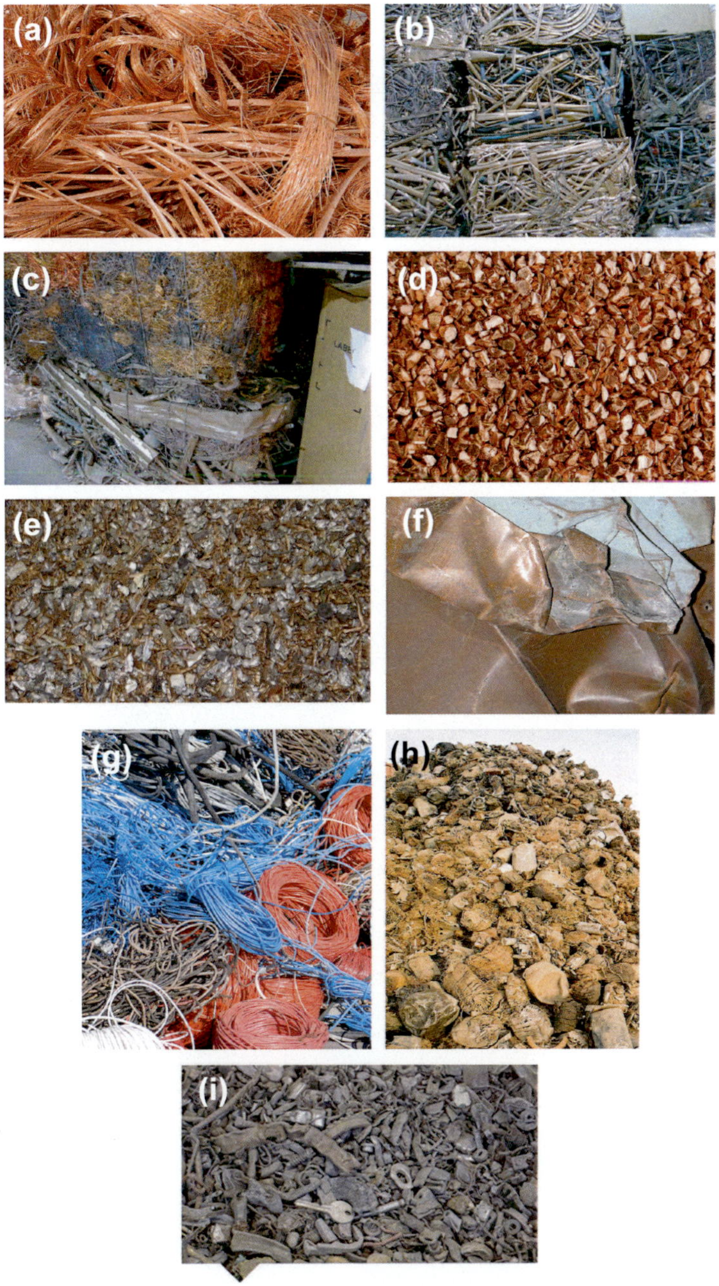

Figure 17.2 Grades of copper scrap: (A) No. 1 copper wire (Barley/Berry), (B) No. 1 heavy copper (Candy), (C) No. 2 heavy copper (Cliff), (D) Chopped copper wire (Clove/Cobra/Cocoa), (E) Upgraded shredded electronic scrap (Dallas), *(Credit: Used with permission from Brian Taylor and Recycling Today, https://www.recyclingtoday.com/article/cohen-coppper-scrap-recycling-cathodes-project-ohio/.)* (F) Light copper scrap (Dream), (G) Insulated wire scrap (Druid), (H) Shredded electric motors (Shelmo), (I) Upgraded auto shredder residue (Zebra). *(Credit: Used with permission from REDWAVE, https://redwave.com/loesungen/recycling/metall.)*

Copper recovery from scrap cable by shredding (also known as chopping or granulating) has its origins in World War II when it was developed to recover rubber coatings (Sullivan, 1985). Shredding has since become the dominant technology for scrap wire and cable processing (Nijkerk & Dalmijn, 1998).

Fig. 17.3 shows a typical cable-chopping flowsheet. Before going to the first *granulator*, the scrap cable is sheared into lengths of 36 inches or less (Nijkerk & Dalmijn, 1998; Sullivan, 1985). This is especially important for larger cables. The first granulator, or "prechopper," is typically a rotary knife shear with one rotating shaft. The knives on this shaft cut against a second set of stationary knives. Rotation speed is about 120 rpm, and a screen is provided to return oversize product to the feed stream. Its primary task is size reduction rather than separation of the wire from its insulation. Depending on the type of material fed to the prechopper, the length of the product pieces is 10−100 mm (Reid, 2003).

The primary granulator also liberates any pieces of steel that may be attached to the scrap cable. These are removed from the product by a magnetic separator (Reid, 2003).

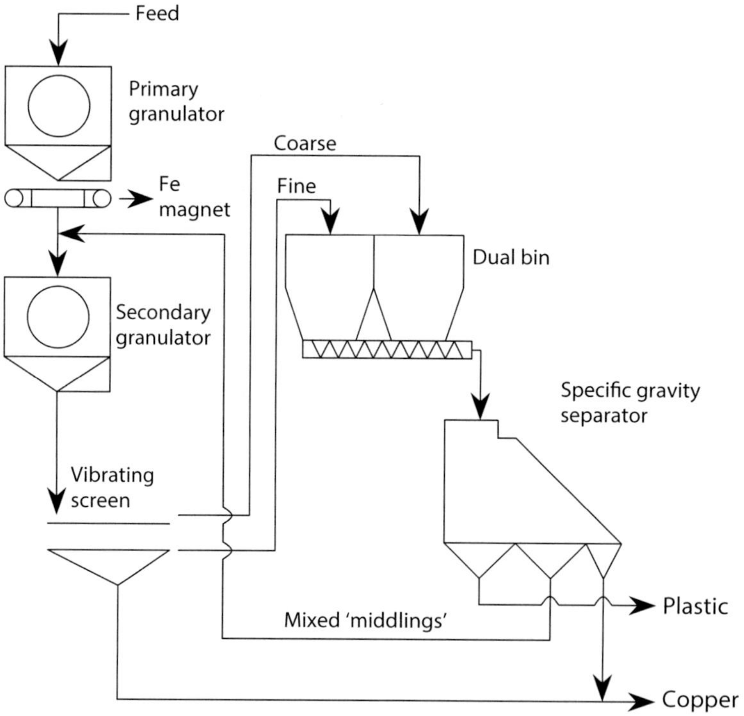

Figure 17.3 Recycle cable "chopping" flowsheet for separating copper from insulation and Fe. The specific gravity separator is an air table which blows insulation upwards while allowing copper to "sink." *(After Sullivan, J.F. (1985). Recycling scrap wire and cable: The state of the art.* Wire Journal International, 18*(11), 36−50.)*

The partially chopped cable is then fed to a second granulator (Sullivan, 1985). The second granulator is similar to the first, but operates at much higher speeds (400 rpm) and has more knives and smaller blade clearances (as small as 0.05 mm). It chops the cable to lengths of 6 mm or smaller, mostly liberating the copper from its insulation. Again, a screen is used to return oversize material. The recent trend is to operate the second granulator at lower speeds, which reduces metal losses to the plastics fraction (Reid, 2003).

The final unit process in scrap cable and wire processing is separation of copper from insulation. This is normally accomplished using the difference between the specific gravity of the copper (8.96) and that of the insulating plastic and rubber (1.3—1.4). Fig. 17.3 shows a *specific gravity separator*, which typically produces three fractions:

(a) a "pure" plastic fraction
(b) copper *chops* meeting Number 1 or Number 2 scrap purity specifications (Table 17.1).
(c) a middlings fraction which is returned to the second granulator for reprocessing.

The separator is most commonly an air table (Sullivan, 1985) which ultimately recovers 80%—90% of the input copper. Compact units that include the granulator and air separator in one unit are increasingly popular for smaller processors (Fowler, 2008).

In larger chopping operations, elutriation or electrostatic separation are sometimes used as "cleaner" steps to recover the last bit of copper from the plastic fraction (Pita and Castilho, 2018; Reid, 2003). A significant difference between copper chops and plastic is particle shape—the copper chops are longer than the plastic. Efforts have been made to develop separation processes based on this characteristic (Koyanaka et al., 1997). Results from Pita and Castillo suggest that elutriation is the most effective of these techniques.

Underground cable processing is complicated by the complexity of its construction, the flammability of its coverings, and the presence of aluminum or lead in the shredding product. Nijkerk and Dalmijn (1998) describe the use of cable stripping for larger cables. This follows shearing and involves slicing open the cable and removing the copper wire by hand. Smaller underground cables (<5 cm) can be successfully chopped without prior stripping (Reid, 2003). Attempts have been made to introduce cryogenic shredding to reduce the flammability hazard. Eddy-current separators can be used following shredding to separate lead and aluminum from copper (Steinert, 2020).

17.3.2 Automotive copper recovery (ELV)

Fig. 17.4 shows a flowsheet for recovering materials from junked automobiles (Suzuki et al., 1995). About 90% of the copper in an ELV is found in the various wire harnesses (McGlothlin, 2018; Sprovieri, 2014); lesser amounts are found in the entertainment system, motors, alternator, sensors, and transmission. As a result, the dismantling step produces very little copper for recycle.

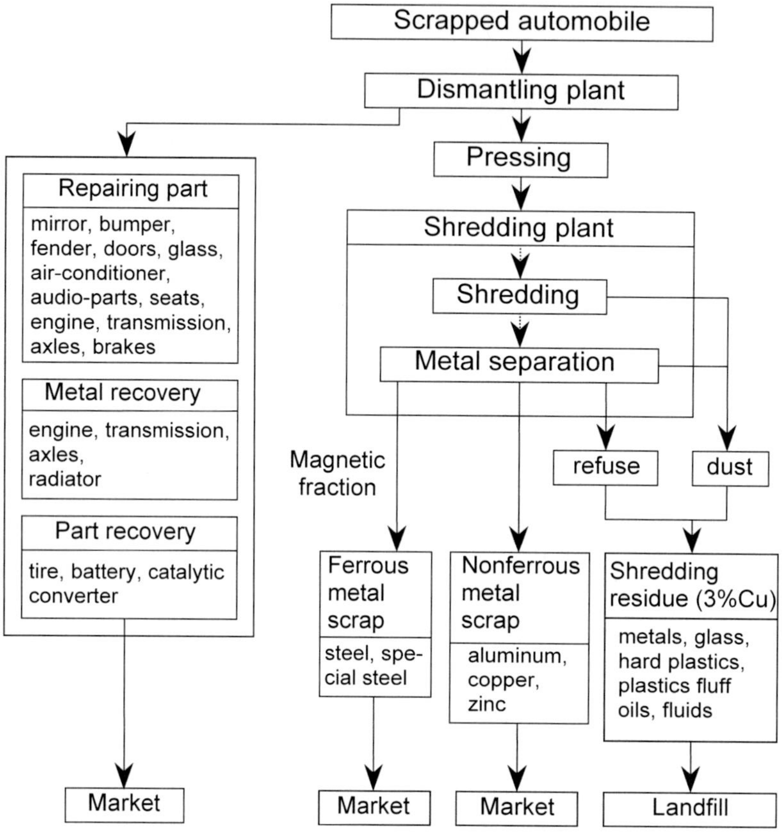

Figure 17.4 Flowsheet for recovering metals from scrap automobiles. Procedures for recovering copper are detailed in Section 17.3.2.

Most of the copper from ELVs is in the nonferrous metal scrap stream remaining after the vehicle has been shredded and its iron and steel have been magnetically removed. Three metals dominate this stream: aluminum, copper, and zinc. The copper consists mostly of wire from the electrical circuits (McGlothlin, 2018). Several techniques are used to separate the metals in this fraction (Dalmijn & De Jong, 2007; Steinert, 2020). Eddy-current separation (ECS) uses the different electrical conductivity of the metallic and nonmetallic particles in the shredded material to remove most of the nonmetallics. Heavy-media separation removes the lower-density aluminum and magnesium from the metallic fraction. What remains is Zebra (see Table 17.1), a mixture of mostly zinc, brass, and copper.

Two methods can be used to upgrade Zebra, both based on the difference in color of the copper, brass, zinc, and other metals in it (Pretz & Mutz, 2006). The first is hand sorting, which is widely practiced in China and India but may become less viable as labor costs there increase over time. The second is automated sorting, which has been commercially practiced for over 10 years (Steinert, 2020).

Figure 17.5 *Meatballs* are created when electric motors are shredded, resulting in a commingled mix of copper wire and iron motor core. *(Credit: Used with permission from Stokkermill Recycling Machinery https://www.stokkermill.com/applications/meatball-processing-copper-recycling.)*

Better efforts have been made over the past 10 years to recover copper from the ferrous scrap stream produced by ELV shredding. This stream contains copper in the form of *meatballs* (Fig. 17.5), the intermingled remains of electric motors (Kiser, 2007; McGlothlin, 2018). The copper is a significant contaminant in the production of steel, and can be smelted to produce secondary copper if the meatballs are collected (see next chapter). Induction sorters have been developed that can sense the differing electrical conductivity of copper-containing scrap, and separate this from the rest of the ferrous material (Sicon, 2019; Steinert, 2020).

The final potential source of copper in Fig. 17.4 is the auto shredder residue (ASR) which remains after the metals have been removed. This residue consists primarily of dust and organic matter—plastic from the dashboard and steering wheel, fluff from the seat cushions, and pieces of carpet and fabric. ASR also contains about 1% copper and has some fuel value (Khodier et al., 2018). Current practice is to landfill this material, but this practice is increasingly unacceptable. As a result, extensive research efforts have been made over the past decade to find a means of processing it to recover valuables. Commercial facilities now operate specifically for processing ASR, including copper recovery (Jody et al., 2010). ASR can also be directly charged to smelting furnaces (Nishiwaki & Hayashi, 2005).

17.3.3 Electronic scrap treatment

Although it consists of a variety of items, the overall composition of electronic scrap can be divided into three categories: (a) plastic (~30% in 1991); (b) refractory oxides (~30%); and (c) metals (~40%). About half of the metal content is copper. It also contains significant amounts of gold and silver. Copper smelting/refining is already set up to recover gold and silver, so it is a logical destination for treating electronic scrap.

A potential problem with smelting electronic scrap is incomplete combustion of its plastic fraction and consequent evolution of organic compounds. However, high temperature oxygen smelting completely avoids this problem.

A more serious problem is the declining metal content of electronic scrap. The producers of circuit boards and other assemblies have learned over time to reduce the amount of metal needed in their products (Zhang & Forssberg, 1999). This makes the scrap increasingly difficult to profitably recycle. The result has been development of "minerals processing" strategies for isolating the metals in electronic scrap. The approach is similar to that used for automobiles (Chen, 2009; Cui & Forssberg, 2003; Kaya, 2016):

(a) disassembly to recover large items
(b) shredding to reduce the size of the remaining material and liberate the metal
(c) separation of metals based on differences in density and electrical conductivity.

Shredding is typically done using low-speed, high-torque knife shredders (Anon, 2004). A primary shredding step generates relatively large particles (4 cm × 4 cm), from which copper is recovered by air tabling, screening, or hand-picking (Burat & Özer, 2018). Secondary shredding generates a finer material with better liberation. Copper recovery from this stream is accomplished using ECS or electrostatic separation (Cui & Forssberg, 2003; Huang et al., 2009; Kaya, 2016).

However, shredding does not liberate metal that has been plated onto surfaces, such as PCBs. In addition to copper, precious metals such as gold and silver also remain behind in this fraction, and this makes it worth recovering. As a result, the plastic residue from PCB processing is also frequently treated as a raw material. There are two general approaches for processing it, the *formal* and *nonformal.*

The nonformal approach for metals recovery from PCB board surfaces is a hydrometallurgical process commonly used in the Third World, particularly China and India (Chatterjee & Kumar, 2009; Huang et al., 2009; Kaya, 2016). This process replaces mechanical processing with burning in open air to liberate pieces of copper. Two acid treatments follow, the first to dissolve solder and the second to dissolve copper. The solution is then boiled away to recover copper sulfate. This process is simple to operate, especially on a small scale. It is also environmentally destructive, and increasingly unacceptable to local governments and international bodies.

The formal approach uses mechanical processing, and sends the residues to secondary smelters which use it as feed in their process (Kaya, 2016). The gold and silver report to the impure copper metal (along with any copper) and are recovered using the same flowsheet used for gold and silver recovery from primary concentrates (Hagelüken, 2006). This will be described in more detail in the next chapter.

17.4 Summary

About half the copper reaching the market today has been scrap at least once. Scrap is generated at all stages in the life span of a copper product, including production (home scrap), manufacturing (new scrap), and postconsumer disposal (old scrap).

The purest copper scrap is simply remelted and recast in preparation for manufacture and use. Less pure copper scrap is resmelted and rerefined. Alloy scrap is usually recycled directly to make new alloy.

Considerable scrap must be physically treated to isolate its copper from its other components. An important example of this is recovery of copper from wire and cable. It is done by the following:

(a) *chopping* the wire and cable into small pieces to liberate its copper

(b) physically isolating its copper by means of a specific gravity separation (air table).

Copper recovery from used automobiles and electronic devices follows a similar pattern, i.e.,

(a) liberation by size reduction (*shredding*)

(b) isolation of copper by magnetic, specific gravity, and eddy-current separation.

The copper from these processes is then resmelted and rerefined.

Old (obsolete) scrap is often discarded in landfills. There is, however, an increasing tendency to recycle this material due mainly to the increased cost and decreased availability of landfill sites.

References

Anon. (2004). *Equipment spotlight: Electronic shredders* Accessed 26.04.04 www.americanrecycler.com/0104spotlight.html.

Barnwal, A., Vishvakarma, S., & Dhawan, N. (2018). Comparison of different routes for recovery of metals from electronic scrap. *Materials Today: Proceedings, 5*, 17046–17054.

Bertram, M., Graedel, T. E., Rechberger, H., & Spatari, S. (2002). The contemporary European cycle: Waste management subsystem. *Ecological Economics, 42*, 43–57.

Burat, F., & Özer, M. (2018). Physical separation route for printed circuit boards. *Physicochemical Problems of Mineral Processing, 54*, 554–566.

Chatterjee, S., & Kumar, K. (2009). Effective electronic waste management and recycling process involving formal and non-formal sectors. *International Journal of the Physical Sciences, 4*, 893–905.

Chen, B. X. (2009). *Where gadgets go to die: Facility, strips, rips and recycles* Accessed 21.01.20 https://www.wired.com/2009/03/where-gadgets-go-to-die-facility-strips-rips-and-recycles/?currentPage=all.

Cui, J., & Forssberg, E. (2003). Mechanical recycling of waste electric and electronic equipment: A review. *Journal of Hazardous Materials, B99*, 243–263.

Dalmijn, W. L., & De Jong, T. P. R. (2007). The development of vehicle recycling in Europe: Sorting, shredding, and separation. *JOM, 59*(11), 52–56.

Edelstein, D. L. (1999). Copper. In *Recycling – metals*. Washington, DC: U.S. Geol. Survey Accessed 21.01.20 www.minerals.usgs.gov/minerals/pubs/commodity/recycle/870499.pdf.

European Copper Institute. (2007). *Recycling of copper waste and WEEE: A response to growing demand for raw materials* Accessed 12.07.10 www.eurocopper.org/files/presskit/pk_en_recycling_31.05.07_final.pdf.

Forsén, O., Aromaa, J., & Lundström, M. (2017). Primary copper smelter and refinery as a recycling plant — a system integrated approach to estimate secondary raw material tolerance. *Recycling, 2*, 19—29.

Fowler, J. (2008). Equipment focus: Compact wire processors. *Scrap, 65*(2), 231—242.

Glöser, G., Soulier, M., & Tercero Espinoza, L. A. (2013). Dynamic analysis of global copper flows. Global stocks, postconsumer material flows, recycling indicators, and uncertainty evaluation. *Environmental Science and Technology, 476*, 6564—6572.

Hagelüken, C. (2006). Recycling of electronic scrap at Umicore's integrated metals smelter and refinery. *World of Metallurgy — Erzmetall, 59*, 152—161.

Huang, K., Guo, J., & Xu, Z. (2009). Recycling of waste printed circuit boards: A review of current technologies and treatment status in China. *Journal of Hazardous Materials, 164*, 399—408.

ISRI. (2019). *Scrap specifications circular 2019* Accessed 21.01.20 www.isri.org/recycling-commodities/scrap-specifications-circular.

Jody, B. J., Daniels, E. J., Duranceau, C. M., Pomykala, J. A., & Spangenberger, J. S. (2010). *End—of—life vehicle recycling: The state of the art of resource recovery from shredder residue* Accessed 15.07.15 www.ipd.anl.gov/anlpubs/2011/02/69114.pdf.

Kaya, M. (2016). Recovery of metals and nonmetals from electronic waste by physical and chemical recycling processes. *Waste Management, 57*, 64—90.

Khodier, A., Williams, K., & Dallison, N. (2018). Challenges around automotive shredder residue production and disposal. *Waste Management, 73*, 566—573.

Kiser, K. (2007). Motors & meatballs: From trash to treasure. *Scrap, 64*(3), 50—60.

Koutamanis, A., van Reijn, B., & van Bueren, E. (2018). Urban mining and buildings: A review of possibilities and limitations. *Resources, Conservation and Recycling, 138*, 32—39.

Koyanaka, S., Ohya, H., Endoh, S., Iwata, H., & Ditl, P. (1997). Recovering copper from electric cable wastes using a particle shape separation technique. *Advanced Powder Technology, 8*, 103—111.

McGlothlin, S. (2018). *Copper recycling process technology in end of life (ELV) shredder plants* Accessed 21.0120 www.metalbulletin.com/events/download.ashx/document/speaker/6539/a0ID000000X0jUWMAZ/Presentation.

Nijkerk, A. A., & Dalmijn, W. L. (1998). *Handbook of recycling techniques*. Amsterdam: Nijkerk Consultancy.

Nishiwaki, M., & Hayashi, S. (2005). Stabilization of recycling technology for shredder residue in copper smelting process. *Shigen-to-Sozai, 121*, 357—362.

Pita, F., & Castilho, A. (2018). Separation of copper from electric cable waste based on mineral processing methods: A case study. *Minerals, 8*, 517—528.

Pretz, T., & Mutz, S. (2006). Metal recovery in automobile recycling. *Acta Metallurgica Slovaca, 12*, 456—462.

Reid, R. L. (2003). Equipment focus: Wire processing. *Scrap, 60*(5), 63—69.

Ruhrberg, M. (2006). Assessing the recycling efficiency of copper from end—of—life products in Western Europe. *Resources, Conservation and Recycling, 48*, 141—165.

Sicon. (2019). *Shredding & ferrous downstream* Accessed 21.01.20 https://sicontechnology.com/wp-content/uploads/shredding-and-ferrous-downstream-2019-1.pdf.

Sprovieri, J. (2014). *Wire harness recycling* Accessed 21.01.20 https://www.assemblymag.com/articles/92263-wire-harness-recycling.

Steinert. (2020). *Non-ferrous metals* Accessed 28.01.20 https://steinertglobal.com/us/metal-recycling/non-ferrous-metal-recycling/#.

Sullivan, J. F. (1985). Recycling scrap wire and cable: The state of the art. *Wire Journal International, 18*(11), 36—50.

Suzuki, M., Nakajima, A., & Taya, S. (1995). Recycling scheme for scrapped automobiles in Japan. In P. B. Queneau, & R. D. Peterson (Eds.), *Third International Symposium on Recycling of Metals and Engineered Materials.* (pp. 729—737). Warrendale, PA: TMS.

Vexler, D., Bertram, M., Kapur, A., Spatari, S., & Graedel, T. E. (2004). The contemporary Latin American and Caribbean copper cycle: 1 year stocks and flows. *Resources, Conservation and Recycling, 41*, 23—46.

Wildnauer, M. (2019). *Life cycle assessment of copper tube and sheet* Accessed 28.01.20 www.copper.org/environment/sustainability/pdfs/copper_life_cycle_assessment_tube_and_sheet.pdf.

Zhang, S., & Forssberg, E. (1999). Intelligent liberation and classification of electronic scrap. *Powder Technology, 105*, 295—301.

CHAPTER 18

Chemical metallurgy of copper recycling

About one-third of the copper currently produced in the world is derived from *secondary* materials—the copper scrap and waste described in the previous chapter. Secondary material is recycled in numerous ways. New scrap is often recycled directly back to the melting furnace where it was produced in the first place (Wood et al., 2011). Old scrap and waste streams (and some new scrap) travel a more complex path. They can either be added to one of the furnaces used to produce primary copper—the smelters, converters, and anode furnace described in previous chapters. They can also be reprocessed by secondary smelters specifically designed to handle such material. This chapter looks at the different methods used to turn old scrap and waste back into new copper metal.

18.1 Characteristics of secondary copper

Several factors determine what happens to copper-containing scrap and waste. These include the following:

- *The condition of the scrap*. The different types of № 1 scrap (see Table 17.1) contain very small levels of oxidation or impurities, and need to only be remelted in a nonoxidizing environment. As a result, this scrap commands a higher price, and is more likely to be charged to furnaces requiring a purer grade of input material. № 2 scrap is more oxidized, and will require refining. Heavily oxidized waste materials like plating sludges require remelting in a reducing environment to avoid copper loss.
- *Location*. Copper scrap is often collected at locations which are very far from primary smelters. The cost of transporting it determines where it winds up.
- *Alloying elements*. Impurity elements in secondary copper often have sufficient value to be recovered in their own right. This includes the gold and silver content of electronic scrap, the lead in copper drosses, and the lead, tin, and zinc in brass and bronze scrap. Recovering these elements often dictates recycling strategy.
- *Government regulations*. Some governments place export tariffs on copper-containing scrap to encourage recycling at home. Restrictions on the exporting of some electronic scrap have also encouraged local reprocessing efforts. Restrictions on air emissions from processing plants have had an impact on processing strategy (Tang, 2016). The recent decision by the Chinese government to restrict imports of outside scrap material (including copper) has created ongoing turmoil that impacts processing decisions (Home, 2019).

Extractive Metallurgy of Copper, Sixth Edition
ISBN 978-0-12-821875-4, https://doi.org/10.1016/B978-0-12-821875-4.00001-8

18.2 Scrap processing in primary copper smelters

Secondary copper can be added at three locations in the primary copper making process. The most common is the converting furnace, but additions are also made in the smelter and anode furnace.

18.2.1 Scrap use in smelting furnaces

Because secondary copper has no sulfur in it, adding it to a furnace is a net energy consumer. This makes it difficult to add to matte smelting furnaces, which already require some fuel use. Making the size of scrap or waste particles small enough to use in flash furnace concentrate burners is also difficult. As a result, the extensive use of waste materials in the flash furnace at Kosaka (Watanabe & Nakagawara, 2003) is rare. However, small-size shredded scrap can be added in limited quantities to flash furnaces (Maeda et al., 2000). In fact, the smelting furnace is preferred to the converter for feeding electronic scrap, due to its plastic content. There are two reasons for this:

(a) the plastic has fuel value which provides heat for smelting (Tang, 2016)

(b) when burned intermittently, plastic often gives off smoke and other particulates which might escape through the mouth of a Peirce—Smith converter, adversely affecting workplace hygiene (Yamashita et al., 2016). Burned in a sealed flash furnace, these particulates are efficiently captured by dust collection devices.

Feeding scrap and waste to other types of primary smelting furnace is easier, and both the Noranda and Mitsubishi smelting processes have been adapted to include scrap in the feed (Huitu, 2003). Fig. 18.1 illustrates the flowsheet for scrap usage in the Mitsubishi process at Naoshima (Ariizumi et al., 2016; Oshima et al., 1998). Particulate scrap is mixed with concentrate and blown into the smelting furnace through its rotating lances. Larger scrap pieces are charged to the smelting and converting furnaces through roof and wall chutes. Mitsubishi converting is particularly exothermic, allowing large amounts of scrap to be melted in the converting furnace. The Onahama smelter in Japan has for some time used reverberatory furnaces to add automotive shredder residue (ASR) to sulfide matte. The enthalpy of combustion of the organic matter in the residue helps offset the extra fuel required for this (Kikumoto et al., 2000).

18.2.2 Scrap additions to converters and anode furnaces

In contrast to smelting furnaces, scrap is usually needed in copper converters, to use the excess heat created by the converting reaction. Depending on matte grades, oxygen enrichment, and scrap quality, the fraction of scrap used in converter charges can range from 0% to 35% or higher.

A higher quality of scrap is needed for primary converters: low-alloy scrap, № 1 and № 2 scrap if available, along with compressed turnings and anode scrap. Low-grade material and plant reverts may also be fed if their plastic content is not too large (Oshima et al., 1998).

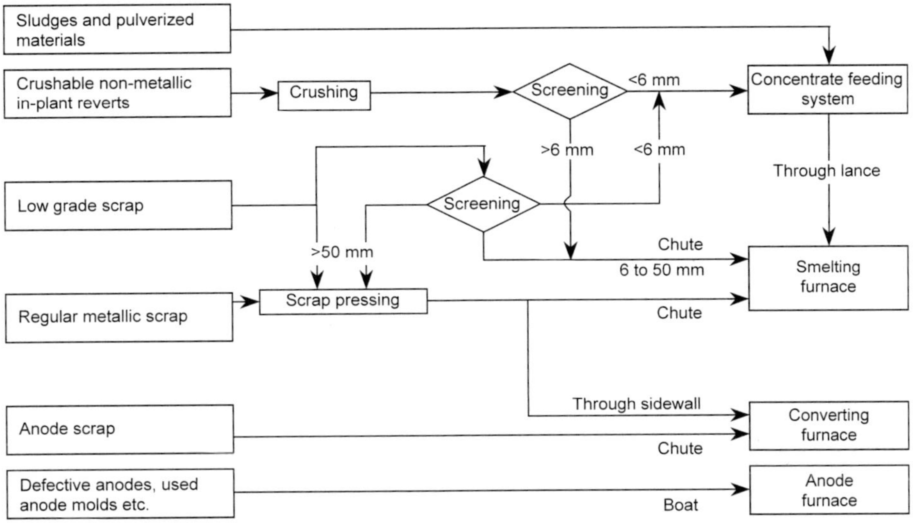

Figure 18.1 Scrap smelting in the Mitsubishi smelter, Naoshima (Oshima et al., 1998). Low Cu-grade particulate scrap is fed into the smelting furnace through its rotating lances. Larger scrap is fed through roof and sidewall chutes. Large high-grade scrap is fed into the anode furnace. *(Redrawn from Oshima, E., Igarashi, T., Hasegawa, N., & Kumada, H. (1998) Recent operation for treatment of secondary materials at Mitsubishi Process. In J.A. Asteljöki and R.L. Stephens (Ed.), Sulfide smelting '98 (pp. 597–606). Warrendale, PA: TMS.)*

Scrap can also be added to anode furnaces, although the most common addition is directly recycled scrap anodes, rather than externally purchased scrap. Because anode furnaces were designed to handle molten blister copper, scrap melting can be slow (Potesser et al., 2007; Suping & Qinghua, 2013). New furnace and burner designs can make anode furnaces better for melting scrap (Deneys & Enriquez, 2009).

18.3 The secondary copper smelter

Although each secondary copper smelter is unique, they can be divided into two groups. The first type are *metal* smelters, which treat only higher-grade metallic scrap. Many of these smelters are located in China (Risopatron, 2007; Tang, 2016). The second are *black copper* smelters, who process low-grade scrap and waste along with higher-value scrap (Shuva et al., 2017). Hoboken in Belgium and Brixlegg in Austria are examples of this type of smelter (Hageluken, 2006; Messner et al., 2005).

18.3.1 High-grade secondary smelting

Rinnhofer and Zulehner (2005) described the typical process for scrap remelting in tiltable reverberatory furnaces. The furnaces typically have a small capacity (5–30 tonnes) and a fairly long cycle time (up to 24 h tap-to-tap; Risopatron, 2007). They have the advantage of flexibility, and can be used for both oxidation of impurities and deoxidation afterward. Side tuyeres are used both for injection of air during oxidation and natural gas during reduction (Tang, 2016).

While reverbs are effective furnaces for refining operations, their efficiency as scrap melters is low (Potesser et al., 2007), and this has created incentive for more effective furnace design. Recent designs have added a reverb to a shaft furnace (see next chapter), which delivers molten copper to the reverb at a much quicker rate (Air Products, 2007; Properzi & Djukic, 2008; Tang, 2016). The refining capability of the reverb allows the use of more scrap in casting facilities than normal.

18.3.2 Smelting to black copper

Fig. 18.2 is a flowsheet for pyrometallurgical processing of low-grade scrap in a secondary black copper smelter. The "Melting/Reducing furnace" at the top accepts the "copper bearing scrap" described in Section 18.2. This scrap includes the following:

(a) automobile shredder product from which the copper and iron cannot be separated, along with motors, switches, and relays (Elmo and Shelmo from the previous chapter)
(b) dross from decopperizing lead bullion
(c) dusts from copper melting and alloying facilities
(d) sludges from copper electroplating operations
(e) auto shredder residue

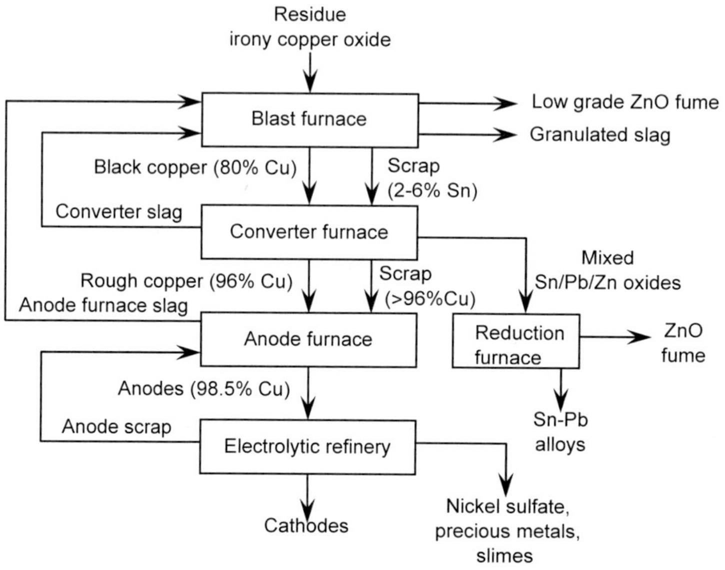

Figure 18.2 Flowsheet for Cu scrap treatment.

The feed to these furnaces is low grade and highly oxidized. It requires reduction to metallic copper. Major metallic impurities are lead and tin (from bronze scrap, solder, and decopperizing dross), zinc (from scrap brass), iron (from automotive scrap), and nickel (from scrap monel and other alloys). These elements are often present as mixtures of metal and oxide.

Heat and CO reductant are supplied to this type of furnace by combusting metallurgical coke or natural gas, e.g.,

$$C\ (\text{coke}) + {}^{1}/_{2}\,O_2 \rightarrow CO + \text{heat} \tag{18.1}$$

The carbon monoxide in turn reduces the oxides of the feed to metal or a lower oxide, i.e.,

$$CO + Cu_2O \rightarrow CO_2 + 2Cu(l) \tag{18.2}$$

$$CO + ZnO \rightarrow CO_2 + Zn(g) \tag{18.3}$$

$$CO + PbO \rightarrow CO_2 + Pb(l,g) \tag{18.4}$$

$$CO + NiO \rightarrow CO_2 + Ni(l) \tag{18.5}$$

$$CO + SnO_2 \rightarrow CO_2 + SnO(l,g) \tag{18.6}$$

$$CO + SnO \rightarrow CO_2 + Sn(l) \tag{18.7}$$

Metallic iron in the scrap also performs some reduction, especially of easily reduced oxides like Cu_2O:

$$Fe + Cu_2O \rightarrow FeO(l) + Cu(l) \tag{18.8}$$

As a result of these reactions, secondary copper reducing furnaces generate three products. They are as follows:

(a) molten *black copper*, 74%–80% Cu, 6%–8% Sn, 5%–6% Pb, 1%–3% Zn, 1%–3% Ni, and 5%–8% Fe (Ayhan, 2000)

(b) molten slag containing FeO, CaO, Al_2O_3, and SiO_2 along with 0.6%–1.0% Cu (as Cu_2O), 0.5%–0.8% Sn (as SnO), 3.5%–4.5% Zn (as ZnO), and small amounts of PbO and NiO

(c) off-gas containing CO, CO_2, H_2O, and N_2, plus metal and metal oxide vapors.

Cooling and filtering of off-gas (c) recovers oxide dust containing 1%–2% Cu, 1%–3% Sn, 20%–30% Pb, and 30%–45% Zn (Suping & Qinghua, 2013). The dust also contains chlorine from chlorinated plastics in the feed. It is always reprocessed to recover its metal content (Hanusch & Bussmann, 1995).

Traditionally, a blast furnace was used as the vessel for initial melting and reducing secondary copper. However, this furnace has all but disappeared in recent years

(Tang, 2016), due to its thermal inefficiency and the need for expensive metallurgical coke. The blast furnaces at Brixlegg in Austria and Kovohuty in Slovakia are among the few remaining examples (Pesl, 2010). Two other types of furnace have become more prominent for this purpose.

The first is a Kaldo furnace (Outotec, 2018; Tang, 2016). This is an upgraded version of the well-known top-blown rotary converter (TBRC). The Kaldo furnace is a batch processor, which burns natural gas or oil. The use of oxygen or enriched air reduces heat losses, and the oxygen/fuel ratio can be adjusted to produce a reducing or oxidizing atmosphere. The furnace rotates during operation, which improves kinetics, and the overall fuel usage is less than that of a blast furnace. Because they rotate, they are generally preferred for smaller operations. Kaldo furnaces are currently in use for secondary copper processing in Belgium, China, India, and Sweden (Pesl, 2010; Tang, 2016; Zhu & Liu, 2000). A Chinese variation of the technology has also been put into service.

The second type of commonly used furnace for secondary copper melting and reduction is the top submerged lance furnace that is increasingly prevalent for sulfide matte smelting and converting (Edwards & Alvear, 2007; Hughes, Jåfs, Johto, Stål, & Karonen, 2019; Wood et al., 2011). Like the Kaldo furnace, Ausmelt and Isasmelt furnaces accept a wide variety of feeds, and can process almost anything with copper in it (Ayhan, 2000; Alvear, Nikolic & Mackey, 2014). As stationary furnaces, they can burn coal or coke fines as a fuel, along with fuel oil and natural gas. They can also use the plastic in electronic scrap and auto shredder residue as a fuel (Brusselaers et al., 2006). They can change from a reducing to an oxidizing environment like a Kaldo furnace, which allows them to remove the impurities from black copper without needing another furnace. Isasmelt furnaces are being used for secondary copper production in Belgium and Germany, and three Ausmelt furnaces are in service in Asia.

Tang (2016) lists other types of furnace developed in recent years in China. These include a horizontal rotary furnace, which operates using a pure O_2 environment (Suping & Qinghua, 2013); a tilting stationary furnace; and a top-blown furnace, used at one Chinese smelter (Jinchuan).

18.3.3 Converting black copper

The impurities in black copper can be divided into two groups—those that are more easily oxidized than copper (Fe, Pb, Sn, and Zn) and those that are difficult or impossible to remove by oxidation (Ni, Ag, Au, and platinum group metals). These impurities are removed sequentially by a strategy similar to that for purifying primary copper.

The first step in refining black copper is oxidation. The black copper produced by blast furnaces or Kaldo units is typically oxidized in a Peirce–Smith converter (Lennartson et al., 2018). Black copper produced in a TSL furnace is refined as a second stage of operation in that furnace (Wood et al., 2011). In either case, air is blown into the molten black copper, oxidizing Fe, Pb, Sn, and Zn along with some Ni and Cu. Alloyed copper scrap (the light copper described in Section 18.2) is also added to the converter. Most of its impurities are also oxidized.

This oxidation generates slag containing 10%–30% Cu, 5%–15% Sn, 5%–15% Pb, 3%–6% Zn, and 1%–5% Ni, depending on the composition of the converter feed (Ayhan, 2000). The slag can be returned to the blast furnace to recover the copper and nickel (Messner et al., 2005; Pesl, 2010), or processed first to recover the lead and tin (Hageluken, 2006). Lead, tin, and zinc will ultimately be recovered from slag or dust generated by one of the furnaces, but several trips through the circuit may be required before these elements wind up in a separate phase.

An off-gas is also generated which, when cooled and filtered, yields dust analyzing 0.5%–1.5% Cu (as Cu_2O), 0.5%–1.5% Sn (as SnO), 10%–15% Pb (as Pb and PbO), and 45%–55% Zn (as ZnO). This dust is usually reduced to recover its Pb and Sn as solder (Ayhan, 2000).

Oxidation of black copper provides little heat to the converter (unlike oxidation of matte, Chapter 9). Heat for the converting process must, therefore, be provided by burning hydrocarbon fuel.

18.3.4 Fire refining and electrorefining

The main product of the converter is molten *rough copper*, 95%–97% Cu. It is added to an anode furnace for final, controlled oxidation before casting it as anodes. № 1 scrap (Table 17.1) is also often added to the anode furnace for melting and casting as anodes. Plant practice is similar to that for fire refining of primary copper (Chapter 12 and Chapter 13).

The rough copper in Fig. 18.2 usually contains nickel or tin, which is not completely removed by oxidation converting. It may also contain appreciable amounts of gold, silver, and platinum group metals from the original scrap. Recovery of these metals is important to the profitability of a recycling facility. As a result, the anode furnace product is almost always cast as anodes for electrorefining. The impurity level in secondary anodes is higher than that in most anodes from primary smelting operations. As a result, electrolyte purification facilities need to be larger. Otherwise, plant practice is similar to that described in Chapter 14.

The principal electrorefining products are high-purity cathode copper, nickel sulfate from electrolyte purification, and anode slimes. Processing of anode slimes is described in Chapter 21.

18.4 Summary

Copper scrap is smelted in primary (concentrate) and secondary (scrap) smelters. Primary smelters mainly smelt concentrate. Some, however, are well adapted to smelting all grades of scrap. Smelters with Isasmelt, Mitsubishi, Noranda, reverberatory, and TBRC smelting furnaces are particularly effective.

Scrap is also extensively recycled to the converters in primary smelters. The heat from the converter's exothermic Fe and S oxidation reactions is particularly useful for melting scrap, especially if considerable oxygen is used for the oxidation reactions.

Secondary scrap smelters primarily use TSL furnaces and TBRCs for smelting low-Cu grade scrap. The main smelting product is molten black copper (80% Cu), which is converted to rough copper (96% Cu) then fire refined and cast into anodes (98.5% Cu).

These processes do not completely remove Ni and Sn from Cu, so the refining furnace product must be electrorefined. Electrorefining also recovers Ag, Au, and Pt group metals.

Secondary copper refining is similar to primary copper refining. However, scrap may contain more impurities than concentrates so larger electrolyte purification and slimes treatment facilities may be required.

References

Air Products. (2007). *Melting and refining technology for copper produced from scrap charge* Accessed 21.01.20 http://www.airproducts.com/media/files/PDF/industries/metals-melting-refining-technology-copper-scrap-charge-pdf&usg=AOvVaw0vDvfTfp9VJOZQkZ8Pza7I9.

Alvear Flores, G. R. F., Nikolic, S., & Mackey, P. J. (2014). ISASMELT™ for the recycling of E-scrap and copper in the U.S. case study of a new compact recycling plant. *JOM, 66*(5), 823–832.

Ariizumi, M., Takagi, M., Inoue, O., & Oguma, N. (2016). Integrated processing of E-scrap at Naoshima smelter and refinery. In *Proceedings of copper 2016* (pp. 2587–2597). Tokyo: MMIJ.

Ayhan, M. (2000). Das neue HK—verfahren für die verarbeitung von kupfer—sekundärmateralien. *Intensivierung Metallurgische Prozesse, 27*, 197–207.

Brusselaers, J., Mark, F. E., & Tange, L. (2006). *Using metal-rich WEEE plastics as feedstock/fuel substitute for an integrated metals smelter* Accessed 24.08.10 http://www.plasticseurope.org/Documents/Document/20100312155603-Using_metal-rich_WEEE_plastics_as_feedstock_fuel_substitute_for_an_integrated_metals_smelter.pdf.

Deneys, A., & Enriquez, A. (2009). Scrap melting in the anode furnace and the development of coherent jet technology in copper refining. In J. Kapusta, & T. Warner (Eds.), *International Peirce—Smith converting centennial* (pp. 321–338). Warrendale, PA: TMS.

Edwards, J. S., & Alvear, F. G. R. (2007). *Converting using Isasmelt™ technology* Accessed 21.01.20 https://www.yumpu.com/en/document/read/27705578/converting-using-isasmelt-technology.

Hageluken, C. (2006). Recycling of electronic scrap at Umicore's integrated metals smelter and refinery. *World of Metallurgy — Erzmetall, 59*, 152–161.

Hanusch, K., & Bussmann, H. (1995). Behavior and removal of associated metals in the secondary metallurgy of copper. In P. B. Queneau, & R. D. Peterson (Eds.), *Third international symposium on recycling of metals and engineered materials* (pp. 171–188). Warrendale, PA: TMS.

Home, A. (2019). *China is redrawing the global copper scrap map* Accessed 04.07.20 https://www.reuters.com/article/metals-copper-ahome/column-china-is-redrawing-the-global-copper-scrap-map-andy-home-idUSL8N28L4GB.

Hughes, S., Jåfs, M., Johto, H., Stål, J., & Karonen, J. (2019). Outotec solutions for processing E-scrap. In G. Gaustad, C. Fleriault, M. Gökelma, J. A. Howarter, R. Kirchain, K. Ma, … M. Zhang, et al. (Eds.), *REWAS 2019* (pp. 283–294). Warrendale, PA: TMS.

Huitu, K. (2003). Using of secondary raw materials in copper processing. *Helsinki University of Technology Publications. Metallurgy and Materials Science, TKK-MK, 152*, 43–51.

Kikumoto, N., Abe, K., Nishiwaki, M., & Sato, T. (2000). Treatment of industrial waste material in reverberatory furnace at Onahama smelter. In P. R. Taylor (Ed.), *EPD congress 2000* (pp. 19–27). Warrendale, PA: TMS.

Lennartsson, A., Engström, F., Samuelsson, C., Björkman, B., & Pettersson, J. (2018). Large–scale WEEE recycling integrated in an ore–based Cu–extraction system. *Journal of Sustainable Metallurgy, 4*, 222–232.

Maeda, Y., Inoue, H., Kawamura, S., & Ohike, H. (2000). Metal recycling at Kosaka smelter. In D. L. Stewart, R. Stephens, & J. C. Daley (Eds.), *Fourth International Symposium on recycling of metals and engineered materials* (pp. 691–700). Warrendale, PA: TMS.

Messner, T., Antrekowitsch, H., Pesl, J., & Hofer, M. (2005). Prozessoptimierung durch stoffstromanalyse am schachtofen der Montanwerke Brixlegg AG. *BHM, 150*, 251–257.

Oshima, E., Igarashi, T., Hasegawa, N., & Kumada, H. (1998). Recent operation for treatment of secondary materials at Mitsubishi Process. In J. A. Asteljöki, & R. L. Stephens (Eds.), *Sulfide smelting'98* (pp. 597–606). Warrendale, PA: TMS.

Outotec. (2018). *Outotec® PM Kaldo furnace* Accessed 21.01.20 https://www.outotec.com/products-and-services/technologies/smelting-and-converting/pm-kaldo-furnace/.

Pesl, J. (2010). State of slag cleaning in secondary copper smelting. In *Presented at short course on slag cleaning at copper 2010, Hamburg, Germany, 05.06.10.*

Potesser, M., Antrekowitsch, H., & Hollets, B. (2007). The elliptical anode furnace — a metallurgical comparison — part one. In A. E. M. Warner, C. J. Newman, A. Vahed, D. B. George, P. J. Mackey, & A. Warczok (Eds.), *(Book 1): The Carlos Díaz symposium on pyrometallurgy: vol. III. Cu2007* (pp. 667–680). Montreal: CIM.

Properzi, G., & Djukic, V. (2008). Fire refined copper rod production in a clean environment. *Wire Journal International, 41*(8), 88–92.

Rinnhofer, H., & Zulehner, U. (2005). Gas-fired furnaces for copper and copper alloys. In A. von Stark, A. Mühlbauer, & C. Kramer (Eds.), *Handbook of thermoprocessing technologies: Fundamentals, processes, components, safety* (pp. 358–364). Essen: Vulkan–Verlag.

Risopatron, C. (2007). *Copper scrap recycling project* Accessed 16.11.07 http://www.economia-dgm.gob.mx/dgpm/doctos/cedoc/Tres.pdf.

Shuva, M. A. H., Rhamdhani, M. A., Brooks, G. A., Masood, S., Reuter, M. A., & Firdaus, M. (2017). Analysis for optimum conditions for recovery of valuable metals from E-waste through black copper smelting. In J. Y. Hwang, T. Jiang, M. W. Kennedy, O. Yücel, P. Pistorius, V. Seshadri, … E. Keskinkilic, et al. (Eds.), *8th international symposium on high–temperature metallurgical processing* (pp. 419–427). Warrendale, PA: TMS.

Suping, Y., & Qinghua, L. (2013). Process features and operation practices of NGL furnace in secondary copper smelter. In *Proceedings of copper 2013* (pp. 323–335). Santiago: IIMCH.

Tang, X. (2016). *The technological developments of scrap copper smelters in China* Accessed 21 January 2020 www.metalbulletin.com/events/download.ashx/document/speaker/8955/a0ID000000ZP0tpMAD/Presentation.

Watanabe, K., & Nakagawara, S. (2003). The behavior of impurities at Kosaka smelter. In F. Kongoli, K. Itagaki, C. Yamauchi, & H. Y. Sohn (Eds.), *High-temperature metals production: vol. II. Yazawa international symposium* (pp. 521–531). Warrendale, PA: TMS.

Wood, J., Creedy, S., Matusewicz, R., & Reuter, M. (2011). Secondary copper processing using Outotec Ausmelt TSL technology. In *Proceedings of MetPlant 2011* (pp. 460–467). Melbourne: AusIMM.

Yamashita, M., Hoshi, M., & Suzuki, Y. (2016). JX Nippon Mining & Metals Group's comprehensive copper recycling process. In *Proceedings of copper 2016* (pp. 2575–2586). Tokyo: MMIJ.

Zhu, Y., & Liu, W. (2000). Kaldo technology for copper scrap treatment. *Youse Jinshu, 52*, 72–74.

CHAPTER 19

Melting and casting

About 95% of the copper currently produced in the United States has existed as cathode copper at some time during its processing (Edelstein, 2000). The cathodes are produced by electrorefining pyrometallurgical anodes (from ore and scrap) and by electrowinning copper leached from oxide and chalcocite ores. To make it useful, this copper must be melted, alloyed as needed, cast, and fabricated.

Much of the fabrication process for copper and its alloys is beyond the scope of this book; see Joseph (1999) and Davis (2001, p. 15) for more information. However, melting and casting are often the last steps in a copper smelter or refinery. A discussion of these processes is therefore in order.

19.1 Product grades and quality

The choice of melting and casting technology is defined by (a) the quality of the input copper, (b) the required chemistry of the desired product, and (c) the type of final product, e.g., wire or tube. Table 19.1 lists the *copper cathode* impurity limits specified by various national standards (ASTM, 2010; Camurri et al., 2012; London Metals Exchange, 2018). Customers usually require purer copper than these specifications (Nairn, 2001). The adoption of stainless steel cathodes for electrorefining and electrowinning has improved cathode purity to match these customer requirements.

The tightest impurity limits in copper cathode are for selenium, tellurium, and bismuth. All three of these elements are nearly insoluble in solid copper. They form distinct grain boundary phases upon casting and solidification. Selenium and tellurium form Cu_2Se and Cu_2Te, while bismuth exists as pure Bi (Zaheer, 1995). These phases are brittle and cause rod cracking and poor drawability (Nairn, 2001). Fig. 20.1 in the previous edition of this book illustrates the effect of impurities on the resistivity of solid copper. Several of the impurities in Table 19.1 have a significant impact (Camurri et al., 2012), which is why they are included in the specifications.

The Unified Numbering System currently recognizes 45 grades of wrought coppers (99.3% Cu or better) and 6 grades of cast coppers (Davis, 2001, p. 15). Several of these coppers are alloyed with small amounts of phosphorus to combine with oxygen when they are being welded.

Unalloyed coppers can be divided into two general classes. The first is *tough pitch* copper (ETP), which purposefully contains 100–650 ppm dissolved oxygen (Table 19.2; Davis, 2001, p. 15; Nairn, 2001). Dissolving oxygen in molten copper accomplishes

Extractive Metallurgy of Copper, Sixth Edition
ISBN 978-0-12-821875-4, https://doi.org/10.1016/B978-0-12-821875-4.00018-3

Table 19.1 Upper impurity limits for copper cathodes (ppm = weight parts per million) as specified in the United States (ASTM), Chile (Codelco), Great Britain (BS EN), and China (GB).

	ASTM B115–10				
Element	**Grade 1**	**Grade 2**	**BS EN 1978:1998**	**GB/T467–2010**	**Codelco**
% Cu + Ag		99.95			
Se (ppm)	2	10	2	2	0.01
Te (ppm)	2	5	2	2	0.05
Bi (ppm)	1	3	2	2	0.005
Sb (ppm)	4	15	4	4	0.02
Pb (ppm)	5	40	5	5	0.01
As (ppm)	5	15	5	5	0.005
Fe (ppm)	10	25	10	10	3.4
Ni (ppm)	10	20		20	0.19
P (ppm)				10	
Sn (ppm)	5	10			1.8
S (ppm)	15	25	5	15	3.4
Ag (ppm)	25	70	25	25	0.22
Zn (ppm)				30	
Bi + Se + Te (ppm)	3		3	3	
As + Cd + Cr + Mn + P + Sb (ppm)			15	15	
Co + Fe + Ni + Si + Sn + Zn (ppm)			20	20	
Total (ppm)	65		65	65	

EN 1978 is equivalent to LME Grade A (Cu-CATH-1).

two goals. The first is removal of inadvertently adsorbed hydrogen during melting by the reaction (Cochrane & Yeoman, 2006; Pops, 2008):

$$[O] + 2\,[H] \rightarrow H_2O\ (g) \tag{19.1}$$

This reduces the amount of porosity created by water vapor formation during casting and welding.

The second is reaction of the oxygen with metallic impurities, precipitating them as oxides at grain boundaries during solidification (Cochrane & Yeoman, 2006; Pops, 2008). These oxide precipitates have a smaller adverse effect on drawability than the compounds that would form if oxygen were not present. Removing the impurities from solution also reduces their impact on conductivity.

Most copper is cast and fabricated as *electrolytic tough pitch* (ETP), *chemically refined tough pitch*, or *fire-refined high conductivity*. The chemical specifications for these grades (which are identical) are shown in Table 19.2 (ASTM, 2003; ASTM International, 1998, ASTM International, 2010; Davis, 2001, p. 15; Davis, 2001; Nairn, 2001; Ningbo, 2010).

Table 19.2 Upper impurity limit specifications for tough pitch copper.

Element	C11000 (ETP) C11020 (FRHP) C11030 (CRTP) CW004A	C11040	ASTM B216−16	GB T2
%Cu (min.)	99.90	99.90	99.88	99.90
Ag (ppm)		25		
As (ppm)	0.5	5	120	20
Sb (ppm)		4	30	20
Bi (ppm)	0.2	1	30	10
Fe (ppm)		10		50
Pb (ppm)	5	5	40	50
Ni (ppm)		10	500	50
O (ppm)	200−400	100−650		600
Se (ppm)	0.2	2		
S (ppm)		15		50
Sn (ppm)		5		20
Te (ppm)	0.2	2		
Se + Te (ppm)			250	
Se + Te + Bi (ppm)		3		
Total (ppm)		65		

ASTM, American Society for Testing and Materials (ASTM International, 1998); *GB*, Chinese Standards; *ppm*, parts per million.

The second class of pure coppers are the *oxygen-free* (oxygen-free copper [OFC] or oxygen-free high conductivity copper [OFHC]) grades. The amount of oxygen in these grades is so low that no visible amount of Cu_2O is present in the solid copper microstructure. The maximum permissible oxygen level in OFC Grade 2 is 10 ppm. In the best grade (oxygen-free electronic, OFE Grade 1), it is only 5 ppm (ASTM, 2010; Davis, 2001, p. 15).

Because no Cu_2O is generated in the grain boundaries, the electrical conductivity of OFC is higher than that of tough pitch copper. As a result, OFC is primarily used for demanding electrical applications, such as bus tube and wave guides (Joseph, 1999). The lack of Cu_2O precipitates also reduces the chance of wire breakage during fine wire drawing. As a result, magnet wire is often produced with OFC (Nairn, 2001), although ETP copper can be used if the impurity levels in the melted cathode are low enough (Cochrane & Yeoman, 2006; Southwire, 2010). The fraction of wire rod copper produced as OFC appears to be growing; data presented by Keung (2009) suggest that it may be as high as 18%.

19.2 Melting technology

19.2.1 Furnace types

According to CRU International (Keung, 2009), about 50% of world copper production in 2006–07 was used for wire rod manufacture. Another 25% is also unalloyed copper. It is mostly fabricated into pipe and tube. As a result, most current melting and casting technology produces (a) copper rod for drawing into wire or (b) billets for extrusion to pipe and tube. The vast majority of this copper is tough pitch.

Most tough pitch copper is produced from cathode in shaft furnaces (Hugens, 2013), shown in Fig. 19.1. The furnace operates counter-currently, with rising hot hydrocarbon combustion gas heating and melting descending copper cathodes. High-quality scrap can also be melted. Natural gas is the usual fuel; propane, butane, naphtha, and LPG are also used (Kamath, 2003; Montgomery, 2008). Avoiding high-sulfur fuels prevents sulfur pickup in the metal (Meseha & Meseha, 2005). Oxygen enrichment is also practiced in some of these furnaces. The melting process is continuous.

An important feature of the furnace is its burners. Each burner uses a high-velocity premix flame in a burner tile. This eliminates the need for an external manifold (Amato, 2009; Ware, 2007). This design reduces accretions, shortens downtime for cleaning, and allows individual control of each burner. The burners are frequently mounted in two or more rows. Firing the upper row at a higher rate may improve fuel efficiency (Spence & Hugens, 2005).

Automatic burner control using CO or hydrogen analysis of the off-gas is a common feature of these furnaces (Kamath, 2003; Montgomery, 2008; Ware, 2007; Takazawa et al., 2016). A flame with a moderately reducing atmosphere is produced, resulting in molten metal with about 60–90 ppm oxygen and 0.3–0.4 ppm hydrogen (Amato, 2009). Other impurity concentrations are largely unaffected. The reducing conditions allow the use of silicon carbide refractories, usually silicon nitride bonded (Hugens & Spadaccia, 2016; Yurkov, Malakho & Avdeev, 2017). These refractories are more wear resistant than oxide refractories, reducing the number of inclusions in the product metal. The recent use of β Si_3N_4 as the bonding phase instead of the α modification has improved the life of these refractories (Hugens, 2013).

Three types of shaft furnace are currently in use. The most common is the Asarco furnace, of which more than 100 are currently operating (Montgomery, 2008). Approximately 70 Southwire shaft furnaces are in operation (Ware, 2007). The third type is a Properzi/Luxi furnace, which is less common. The basic principle of the three furnaces is similar. Asarco furnaces have individual fuel and air mixing points for each burner, while Southwire furnaces use multiple premix manifolds. As a result, Asarco furnaces can be adjusted more easily for changing conditions, while Southwire furnaces are simpler to operate.

Efforts to improve the shaft furnace center around increasing thermal efficiency, which is currently about 50% (Spence & Hugens, 2005). Heat transfer from the hot

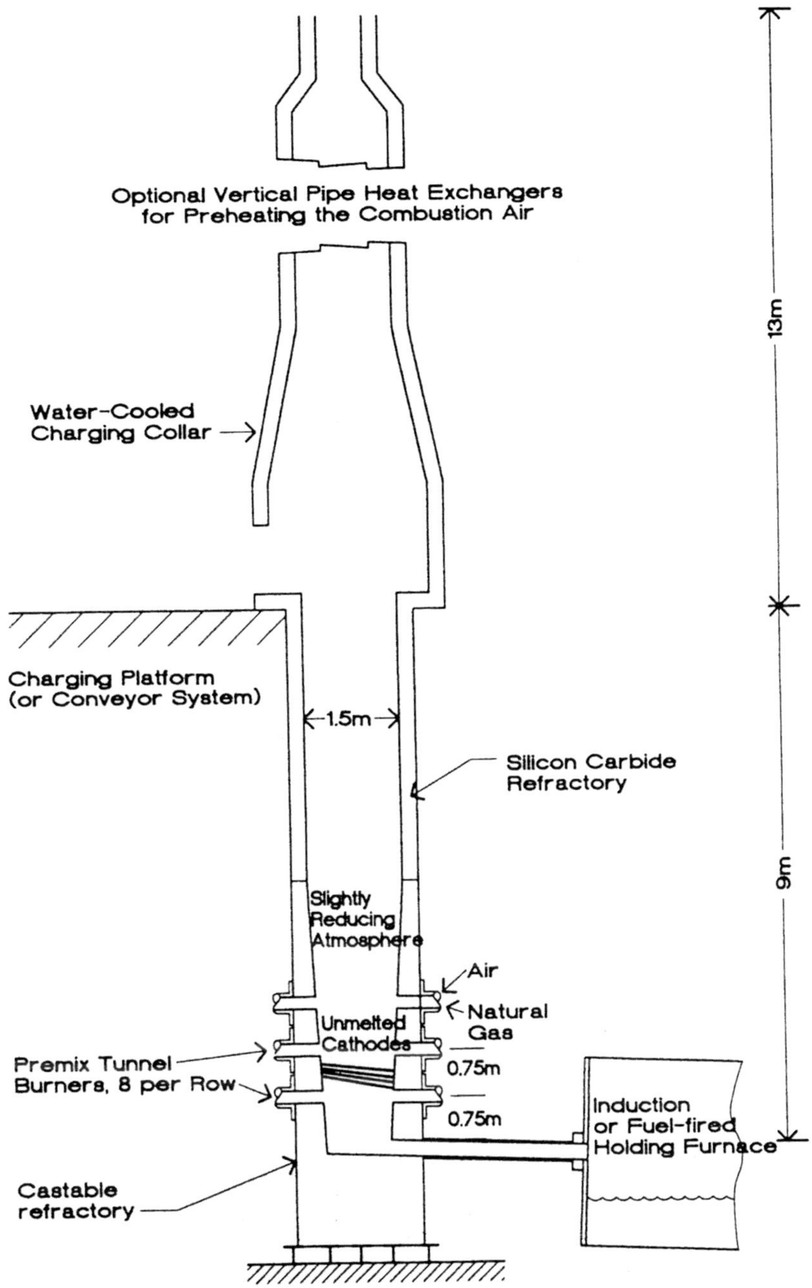

Figure 19.1 The Asarco-type shaft furnace for melting cathode copper.

combustion gases to the charge is a current limitation of the furnace. Recent developments include increasing the height of the furnace, which increases contact time (Kamath, 2003). A new cyclonic burner has been developed, which enhances circulation in the middle of the furnace (Amato, 2009). Spacing the cathodes and keeping them in a vertical position during charging also seems to help. With sufficient improvement, thermal efficiencies of 80% might be possible.

Lower-quality scrap is less suitable for shaft furnaces, which have no refining ability. As a result, some producers have combined reverberatory furnaces with a shaft furnace (Air Products, 2007; Arderiu & Properzi, 1996). Metal charged to these furnaces can be fire refined. This allows the furnaces to melt lower grade cathode and better grades of scrap.

Another melting option is the induction furnace, either the channel or coreless type (Bebber & Phillipps, 2001; Eklin, 2005; Induga, 2011; Nairn, 2008). Induction furnaces are usually used to melt oxygen-free or phosphorus-deoxidized grades, since the absence of a combustion atmosphere prevents oxygen and hydrogen from being absorbed into the molten copper. Feed to induction furnaces which produce OFC is limited to high-quality cathode and scrap. Melting capacities are small, typically 10,000—20,000 tonnes per year. Electric resistance heating using graphite elements is also used (Cochrane & Yeoman, 2006; Nairn, 2015). The use of graphite crucibles rather than oxide refractory linings in resistance furnaces reduces the number of inclusions that end up in the metal.

Molten copper tapped from melting furnaces flows into a holding furnace before being directed to continuous casting. This ensures a steady supply of molten copper to the casting machines. It also creates more consistent metal temperature and composition. Holding furnaces vary considerably in size and type, but are usually induction heated to minimize hydrogen pickup from combustion gases. The copper may also be covered with charcoal to minimize oxygen pickup (Bebber & Phillipps, 2001; Eklin, 2005; Nairn, 2008). Automation of the holding furnace to produce a steady flow of metal at constant temperature is an important part of casting operations (Meseha & Meseha, 2005; Ware, 2007).

Although slag production during melting is minimal, the desire to eliminate inclusions from the cast rod has made its removal increasingly important. As a result, a deslagging vessel is now placed between the shaft furnace and the holding furnace (Bebber & Phillipps, 2001; Ware, 2007). Ceramic filters are also increasingly used to remove inclusions caused by erosion of the furnace system refractories or precipitation of solid impurities from the molten copper (Kamath, 2003; Zaheer, 1995). These inclusions are a leading cause of wire breaks (Joseph, 1999; Meseha & Meseha, 2005).

Multichamber induction furnaces are increasingly popular (Brey, 2005). The 'storage' chambers in these furnaces eliminate the need for multiple holding furnaces.

19.2.2 Hydrogen and oxygen measurement/control

As previously mentioned, control of hydrogen and oxygen in molten copper is critical. Oxygen is monitored in one of two ways. The first is Leco infrared absorbance, which measures the amount of CO_2 generated when the oxygen in a heated sample of copper reacts with admixed carbon black. This method requires external sample preparation, and so does not offer an immediate turnaround.

The second approach is an oxygen sensor, which is applied directly to the molten copper (Richerson, 2005). The electrode potential of the dissolved oxygen in the copper is measured against a reference electrode in the sensor. This relative potential is converted to an equivalent oxygen content in the metal at the measurement temperature. Dion et al. (1995) have shown that the two methods yield similar results. The amount of oxygen in the molten copper is controlled by adjusting burner flames and by injecting compressed air into the copper.

Hydrogen is more difficult to monitor and control. Analysis of solid samples is the usual practice (Strand et al., 1994), but efforts have been made to adapt aluminum industry technology to on-line measurement of hydrogen in molten copper (Hugens, 1994). Japanese investigators have had success using α—alumina as a solid electrolyte for a hydrogen sensor (Okuyama et al., 2009).

19.3 Casting machines

Casting machines are usually one of three main types:

(a) billet ("log") casting, for extrusion and drawing to tube (Fig. 19.2)

(b) bar casting, for rolling to rod and drawing to wire (Figs. 19.3 and 19.4, Table 19.3)

(c) strip casting, for rolling to sheet and forming of welded tube.

19.3.1 Billet casting

Billet casting is usually performed in vertical direct-chill casters, such as that shown in Fig. 19.2 (SMS Meer, 2015). Graphite-lined copper or graphite—ceramic molds are used. Diameters up to 30 cm are cast (Hugens & DeBord, 1995). Oscillation of the water-cooled molds (60—360 min^{-1}) improves surface quality and prevents sticking in the mold.

Billets are often cast from phosphorus-deoxidized grades of copper (DHP), rather than tough pitch (Induga, 2007). The first reason is that tough pitch often performs poorly when extruded from billet. The second reason is the use of graphite molds. The dissolved oxygen in tough pitch copper reacts with the graphite to form carbon monoxide, which corrodes the mold.

Figure 19.2 Semicontinuous vertical chill caster for extrusion. *(Courtesy SMS Meer (2008).)*

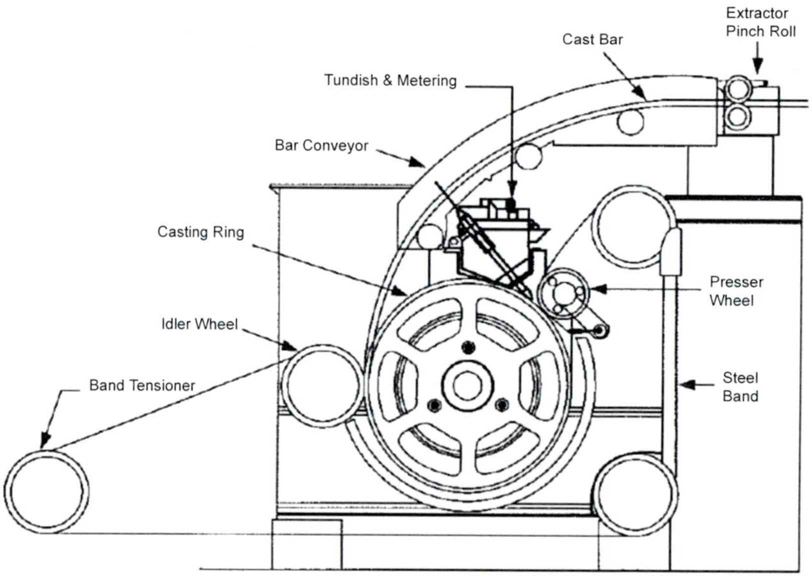

Figure 19.3 Southwire casting machine for continuously casting copper bar (Adams & Sinha, 1990).

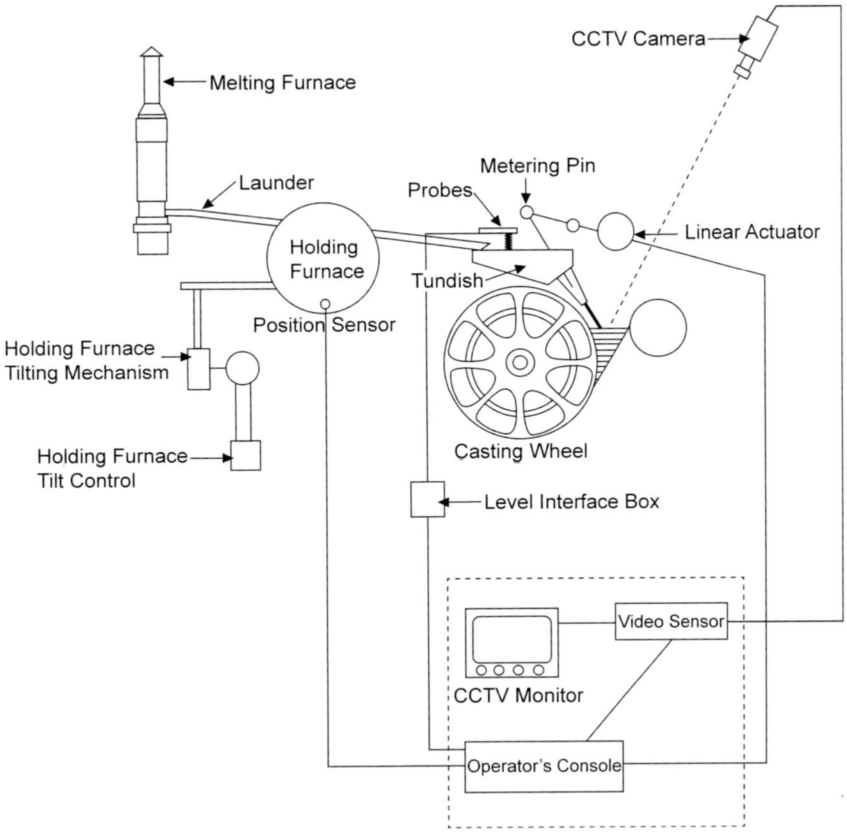

Figure 19.4 System for controlling molten copper level in Southwire continuous casting machine (Adams & Sinha, 1990). *(Reprinted courtesy TMS.)*

Over the past decade, horizontal casters have begun to replace vertical billet casters, due to their lower cost (Müller & Schneider, 2005). Horizontal continuous casting of hollow billets has become an attractive alternative for smaller operations (Induga, 2007; Schwarze, 2007). These billets are rolled directly to tube, eliminating the need for extrusion and piercing.

19.3.2 Bar and rod casting

Copper bar is mostly cast in continuous wheel-and-band and twin-band casting machines (Table 19.3; Figs. 19.2 and 13.6). Fig. 19.3 shows a Southwire wheel-and-band caster. Its key features are (a) a rotating copper—zirconium alloy rimmed wheel with a mold shape machined into its circumference and (b) a cold-rolled steel band which moves in the same direction and at the same speed as the wheel circumference.

Table 19.3 Operating details of Asarco cathode melting shaft furnaces.

Melting plant	Nexans Canada, Montreal	FMI refinery El Paso, USA	Halcor S.A., Greece	Encore wire USA
Inputs	Cathodes and "runaround" scrap rod	Cathodes	Cathodes and internal scrap	Cathodes and recycled scrap
Molten copper destination	Hazelett caster and rod mill	Hazelett caster and rod mill	Hazelett caster and rod mill	Southwire caster and rod mill
Melting rate, tonnes of copper per hour	48	58	10.7	22
Feed system	skip hoist	Elevator with automatic trip	Forklift truck and skip hoist	Forklift truck and skip hoist
Furnace details, m				
Height, taphole to charge floor	13	12.2	7.5	10
Inside diameter at charge floor	1.7	1.75	1.68	1.68
Inside diameter at taphole	1.3	1.37	1.37	1.37
Burner details				
Number of burners	23	27	10	15
Rows of burners	3	4	2	2
Fuel	Natural gas	Natural gas	Natural gas	Natural gas
Combustion rate, Nm^3/hour		2400	400	
Nm^3 of natural gas burnt per tonne of copper melted	1.9 GJ	1.8 GJ 1.706 Dth	53 (including holding furnace and launders)	1,435,000 kJ
Refractory life, tonnes of copper				
Above burners	500,000	500,000	300,000	400,000
Below burners	250,000	300,000	200,000	400,000

Molten copper is poured from a *pour pot* into the mold just as the steel band joins the wheel to form the fourth side of the mold (Ware, 2007). The wheel and band move together through water sprays as the copper solidifies. After 180−250 degrees of rotation, the band moves off to an idler wheel and the solidified copper bar is drawn away (under minimum tension) to a rolling mill. Pouring to bar separation takes about 15 s. The cast bar is removed at about 0.25 m/s. The Properzi casting machine is similar, but uses individually controlled zones for spray cooling the solidifying bar (Meseha & Meseha, 2005).

The Hazelett twin-band caster is shown in Fig. 13.6 in its role as an anode-casting machine. Molten copper is fed from a pour pot into the space between two sloped moving steel bands (von Gal, 2007). The bands are held apart by moving alloyed copper dam blocks on each side, creating a mold cavity ranging between 6 and 13 cm in width and 3.5−7.5 cm in thickness (for bar casting). Both separations are adjustable, allowing variable product size. Solidification times are similar to those of the Southwire and Properzi machines (Strand et al., 1994).

The three types of moving band casting devices have several features in common. All require lubrication of the bands and mold wheel or dam blocks, using silicone oil or acetylene soot (Kamath, 2003; Meseha & Meseha, 2005). Leftover soot is removed from the bands after each revolution, and then reapplied. This ensures an even lubricant thickness and a constant heat transfer rate. The casters all use similar input metal temperatures, 1110−1130 °C (Table 19.3). All require smooth, low-turbulence metal feed into the mold cavity, to reduce defects in the solidified cast bar. Lastly, all require steady metal levels in the pour pot and mold.

Control of mold metal level is done automatically (Fig. 19.4). Metal level in the mold cavity is measured with a laser pool level sensor (Hazelett Corp, 2020a, 2020b) or with a television camera (Southwire). It is controlled with a stainless-steel or ceramic metering pin in the pour pot. Metal level in the pour pot is determined using a bubble tube or radar sensor. It is controlled by changing the tilt of the holding furnace which feeds it (Ware, 2007). The temperature of the solidified copper departing the machine is controlled to 940−1015 °C by varying casting machine cooling water flow rate.

Copper cast in the Hazelett, Properzi, and Southwire machines is usually fed directly to a rolling machine for continuous production of copper rod. Southwire Continuous Rod and Hazelett Contirod are prominent (Bebber & Phillipps, 2001; Hazelett, 2020a, 2020b; Ware, 2007; Zaheer, 1995). Both systems produce up to 60 tonnes of 8−22 mm diameter rod per hour (Table 19.3) (Table 19.4).

19.3.3 Oxygen-free copper casting

The low oxygen and hydrogen content of OFC minimizes porosity when these grades are cast. As a result, the rolling step which is used to turn tough pitch copper bar into rod is not necessary. This has led to the development of processes for direct casting of OFC

Table 19.4 Operating details of Hazelett and Southwire continuous casting machines.

Casting plant	Nexans Canada	FMI refinery	Halcor S.A.	Encore wire
Casting machine	Hazelett twin band	Hazelett twin band	Hazelett	Southwire wheel and band
Bar size, cm × cm	7 × 13	7 × 13.0	3.5 × 6	5.4 × 7.9
Casting rate of this bar, tonnes/hour	48	63	10.7	20
Molten copper level control in caster	Electromagnetic pool level measurement	Electromagnetic pool level measurement	Electromagnetic pool level control	Infrared scanner
Casting temp., °C	1125	1130	1135–1140	1100
Bar temperature leaving caster, °C	~950	1015	930–950	930
Target O in copper, ppm	250	250	180–320	270
Measurement technique	Electronite cell in launder; Tempolab in holding furnace; Leco on rod	Leco on rod	Ströhlein on finished rod	Leco on rod
Control system	Manual	Compressed air injection into molten Cu	Controlled natural vent on transfer launder	Compressed air injection into molten Cu
Wheel and band details				
Wheel diameter, m				2.01
Rotation speed, rpm				1.8
Rim materials				Cu–Cr–Zr
Rim life, tonnes of cast, Copper				28,000
Band material				Cold-rolled steel
Band life				2200 t
Lubrication				Acetylene soot

Table 19.4 Operating details of Hazelett and Southwire continuous casting machines.—cont'd

Casting plant	Nexans Canada	FMI refinery	Halcor S.A.	Encore wire
Twin band details				
Caster length, m	3.7	3.7	3	
Band material	Low carbon steel	Titanium steel	Titanium steel	
Life	24 h	1300 tonnes Cu	36–48 h	
Lubrication	Oil	Union carbide Lb-300× oil	Oil Breox	
Dam block material	Si bronze	Cu with 1.7%–2% Ni and 0.5%–0.9% Si	CuNiSiCr	
Dam block life	100,000 tonnes cast copper	~1000 h	20,000 tonnes	

copper rod. Nearly, all of these are vertical casting machines. Upcast OY (formerly Outokumpu, now spun off as a separate company) is the best known OFC caster type; others include Rautomead and Conticast (Fig. 19.5; Cochrane & Yeoman, 2006; Eklin, 2005; Nairn, 2015). Numerous Chinese vertical casting machines are also in use (Keung, 2009).

Upward vertical casting machines use a vacuum to draw metal into water-cooled graphite-lined dies partially submerged in the molten copper. As it freezes, the rod is

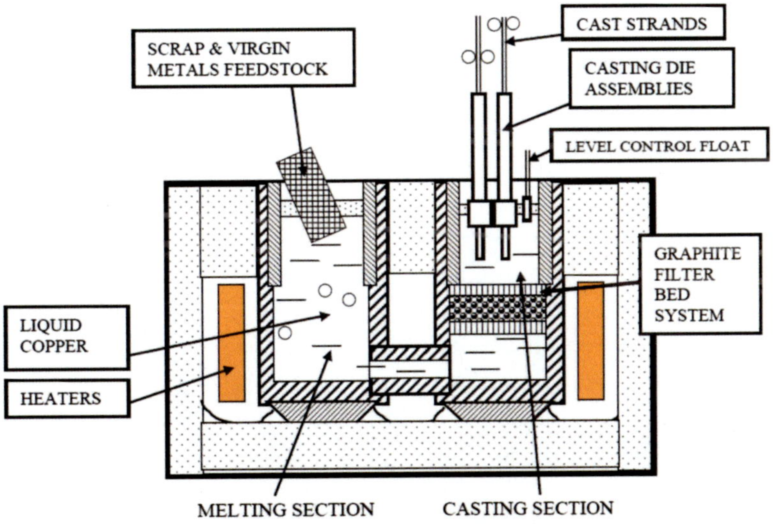

Figure 19.5 Conticast vertical casting machine for OFC. *(Cochrane, S., & Yeoman, G.* Cu-OF or Cu-ETP: Some general comparisons of small diameter copper wire rod products. *(2006). http://www.conticast.com/documents/CuOF%20or%20CuETP%20-%20GENERAL%20COMPARISONS.pdf. Accessed 15.01.20.)*

mechanically drawn upward and coiled (Nairn, 2015). It is about the same size as rolled rod. Vertical casting machines produce 3–12 rod strands simultaneously, and have a capacity of 500–700 kg/h. These are integrated with a single electrically heated furnace which is used for both melting and casting (Cochrane & Yeoman, 2006). Newer machines have separate furnaces for melting and casting, and produce up to 56 strands. Their production capacity is as high as 40,000 tonnes per year (Eklin, 2005).

19.3.4 Strip casting

The development of strip casting for copper and copper alloys parallels developments in the steel industry, in that continuous processes are favored. The newer the technology, the less rolling is required. One approach taken by small-volume producers is to roll strip from the bar produced by a Hazelett caster (Roller et al., 1999). This can be combined with continuous tube rolling/welding to make optimum use of the casting machine for a mix of products.

However, direct strip casting which avoids rolling is the goal. Current horizontal casters can produce *thick strip* (15–20 mm), which requires some rolling (Roller & Reichelt, 1994). Development efforts are being made to develop *thin-strip* (5–12 mm) casting to avoid rolling completely. Hazelett Strip-Casting (2010) recently put into service the first continuous strip caster/in-line rolling facility in Hettstedt, Germany.

19.4 Summary

The last step in copper extraction is melting and casting of electrorefined and electrowon cathodes. The main products of this melting and casting are as follows:

(a) continuous rectangular bar for rolling to rod and drawing to wire
(b) round billets ("logs") for extrusion and drawing to tube
(c) flat strip for rolling to sheet and forming into welded tube.

The copper in these products is mostly "tough pitch" copper, i.e., cathode copper into which 100–500 ppm oxygen has been dissolved during melting/casting. The remainder of unalloyed copper production is in the form of OFHC with 5–10 ppm dissolved oxygen. It accounts for up to 18% of wire rod production.

These pure copper products account for about 70% of copper use. The remainder is used in the form of copper alloy, mainly brass and bronze.

The principal melting tool for cathodes is the shaft furnace. It is thermally efficient and provides good oxygen-in-copper control. Its molten copper is mainly cast as rectangular bar in continuous wheel-and-band and twin-band casters, or as round billets ("logs") in horizontal and vertical direct chill casters.

The quality of cathode copper is tested severely by its performance during casting, rolling, and drawing to fine wire. Copper for this use must have high electrical conductivity, good drawability, and good annealability. These properties are all favored by maximum cathode purity.

References

Adams, R., & Sinha, U. (1990). Improving the quality of continuous copper rod. *Journal of Metals, 42*(5), 31–34.

Air Products. (2007). *Melting and refining technology for copper produced from scrap charge* Accessed 09.02.20 http://www.airproducts.com/nr/rdonlyres/906a3970-3a98-461c-89b2-.498549239ff0/0/refining_scrap_copper_33407004us.pdf.

Amato, K. (2009). *SFIP — using cyclonic burner technology* Accessed 09.02.20 http://www.https://combustion.fivesgroup.com/literature/north-american/technical-sheets.html.

Arderiu, G., & Properzi, G. (1996). Continuous copper rod production from 100 percent scrap. *Wire Journal International, 29*(3), 60–67.

ASTM International. (1998). ASTM B 49–98: Standard specification for copper rod drawing stock for electrical purposes. In *Annual book of ASTM standards, section 2, nonferrous metal products*. West Conshohocken, PA: American Society for Testing and Materials.

ASTM International. (2003). ASTM B 216–97: Standard specification for tough pitch fire–refined copper — refinery shapes. In *Annual book of ASTM standards, section 2, nonferrous metal products*. West Conshohocken, PA: American Society for Testing and Materials.

ASTM International. (2010). ASTM B115–10: Standard specification for electrolytic cathode copper. In *Annual book of standards, section 2, nonferrous metal products*. West Conshohocken, PA: American Society for Testing and Materials.

Bebber, H., & Phillipps, G. (2001). The CONTIROD furnace concept at MKM. *Metall, 55*, 433–435.

Brey, M. (2005). Horizontal continuous casting of copper alloy billets. In K. Ehrke, & W. Schneider (Eds.), *Continuous casting* (pp. 325–332). Weinheim, Germany: Wiley–VCH.

Camurri, C., Carrasco, C., Leite, R., Mangalaraja, R., & Dille, J. (2012). Influence of Impurities in cathodic copper on the ductility of copper wires. *Journal of Materials Engineering and Performance, 21*, 1474–1478.

Cochrane, S., & Yeoman, G. (2006). *Cu-OF or Cu-ETP: Some general comparisons of small diameter copper wire rod products* Accessed 15.01.20 http://www.conticast.com/documents/CuOF%20or%20CuETP%20-%20GENERAL%20COMPARISONS.pdf.

Davis, J. R. (Ed.). (2001). *ASM specialty handbook: Copper and copper alloys materials*. Park, OH: ASM International.

Dion, J. L., Sastri, V. S., & Sahoo, M. (1995). Critical studies on determination of oxygen in copper anodes. *Trans. Am. Foundryman's Soc., 103*, 47–53.

Edelstein, D. E. (2000). *Copper. In 1999 minerals yearbook*. United States Geological Survey Accessed 15.01.20 https://s3-us-west-2.amazonaws.com/prd-wret/assets/palladium/production/mineral-pubs/copper/240499.pdf.

Eklin, L. (2005). The Outokumpu UPCAST® system. In K. Ehrke, & W. Schneider (Eds.), *Continuous casting* (pp. 333–340). Weinheim, Germany: Wiley–VCH.

Hazelett Corp. (2020a). *Copper bar casters* Accessed 09.02.20 https://www.hazelett.com/copper-bar.

Hazelett Corp. (2020b). *Copper strip casting technology* Accessed 09.02.20 https://www.hazelett.com/copper-strip-technology.

Hugens, J. R. (1994). An apparatus for monitoring dissolved hydrogen in liquid copper. In G. W. Warren (Ed.), *EPD congress 1994* (pp. 657–667). Warrendale, PA: TMS.

Hugens, J. R. (2013). Future directions for scrap melting in copper shaft furnaces. In *Proceedings of copper 2013* (pp. 309–322). Santiago: IIMCH.

Hugens, J. R., & DeBord, M. (1995). Asarco shaft melting and casting technologies '95. In W. J. Chen, C. Díaz, A. Luraschi, & P. J. Mackey (Eds.), *Pyrometallurgy of copper: vol. IV. Copper 95–cobre 95* (pp. 133–146). Montreal: CIM.

Hazelett Strip-Casting. http://www.hazelett.com/casting_machines/strip_casting_machines/copper_strip_casting_machines/copper_strip_casting_machines/php.

Hugens, J., & Spadaccia, S. (2016). Third generation silicon carbide linings for copper melting shaft furnaces. In *Proceedings of copper 2016* (pp. 553–560). Tokyo: MMIJ.

Induga. (2007). *Induga copper tube furnaces.* http://www.furnacesinternational.com/download/IndugaCopperTubeFurnaces.pdf.

Induga. (2011). *Channel type melting furnace for non-ferrous metals* Accessed 09.02.20 https://www.induga.de/com/pdf/Channel-metals.pdf.

Joseph, G. (1999). *Copper: Its trade, manufacture, use and environmental status.* Materials Park, OH: ASM International.

Kamath, B. P., Adiga, R., Sharma, L. K., & Gupta, S. (2003). Productivity improvements at continuous casting rod plant of Sterlite Copper at Silvassa. In G. E. Lagos, M. Sahoo, & J. Camus (Eds.), *Plenary lectures, economics and applications of copper: vol. 1. Copper 2003—cobre 2003* (pp. 720—727). Montreal: CIM.

Keung, S. (2009). *Recent trends in copper wirerod markets.* http://www.icsg.org/index.php?option=com_docman&task=doc_download&gid=201&Itemid=62.

London metals Exchange special contract rules for copper Grade A.(2018) Accessed 15.01.19 https://www.lme.com/-/media/files/Trading/Brands/Chemical-composition/Non-ferrous/Chemical-composition-Copper.pdf?la=en-GB.

Meseha, J., & Meseha, G. (2005). Comparison of competing continuous casting processes. *Wire Journal International, 38*(3), 185—193.

Montgomery, M. C. (2008). *Shaft melting furnace.* http://www.asarco.com/Asarco_Shaft_Melting_Furnace.html.

Müller, W., & Schneider, P. (2005). Horizontal casting technology for copper products. In H. R. Müller (Ed.), *Continuous Casting* (pp. 329—335). Wertheim: Wiley-VCH.

Nairn, M. (2001). *Graphite crucible furnace technology as a basis of highest quality copper redraw rod* Accessed 07.08.10 http://www.rautomead.com/files/WireSing_Reprint.pdf.

Nairn, M. (2008). *Adding value and growing markets through wire rod production* Accessed 09.02.20 www.rautomead.com/uploads/files/1515577532/addingvalueandgrowingmarkets.pdf.

Nairn, M. (2015). *Technology for continuous casting of oxygen—free copper, copper—magnesium & other copper alloys* Accessed 09.02.20 www.bluetoad.com/publication/?i=251043&article_id=1962060&view=articleBrowser.

Ningbo Jintian Copper (Group) Co., Ltd. (2010). *China (mainland) high—precision copper strip exhibition.* http://www.hisupplier.com/sell-148954-High-precision-Copper-Strip.

Okuyama, Y., Kurita, N., Yamada, A., Takami, H., Ohshima, T., Katahira, K., & Fukatsu, N. (2009). A new type of hydrogen sensor for molten metals useable up to 1600 K. *Electrochimca Acta, 55*, 470—474.

Pops, H. (2008). *The metallurgy of copper wire* Accessed 07.08.10 http://www.litz-wire.com/pdf%20files/Metallurgy_Copper_Wire.pdf.

Richerson, David W. (2005). *Modern ceramic engineering: Properties, processing, and use in design* (3rd ed.). Boca Raton, FL: CRC Press.

Roller, E., Kalkenings, P., & Hausler, K. H. (1999). Continuous narrow strip production line for welded copper tubes. *Tube International, 18*, 28—31.

Roller, E., & Reichelt, W. (1994). Strip casting of copper and copper alloys. In *In proceedings of METEC congress 94* (vol. 1, pp. 480—486). Düsseldorf, Germany: Verein Deutscher Eisenhüttenleute.

Schwarze, M. (2007). Innovations in the field of non-ferrous metal processing. In J. Hugens, K. Sadayappan, J. Spooner, Smith, L.D., C. Twigge-Molecey, J. Kapusta, , … N. L. Piret, et al. (Eds.), *Plenary/copper and alloy casting and fabrication/copper — economics and markets: vol. I. Copper 2007* (pp. 151—172). Montreal: CIM.

SMS Meer. (2015). *Continuous casting plants for copper* Accessed 09.02.20 http://www.polymet-solutions.com/press-media/downloads/download-detail/21392/.

Southwire. http://www.southwire.com/scr/scr-copper-rod-systems.htm.

Spence, G., & Hugens, J. (2005). Copper shaft furnace melting and charging: A new approach. *Wire Journal International, 38*, 110—115.

Strand, C. I., Breitling, D., & DeBord, M. (1994). Quality control system for the manufacture of copper rod. In *1994 conference on proceedings of wire association international* (pp. 147—151). Guilford, CT: Wire Association International.

Takazawa, T., Teshigawara, S., Ode, K., & Segawa, H. (2016). Manufacturing technology for oxygen—free copper for electronic devices using shaft furnace. In *Proceedings of copper 2016* (pp. 2405—2414). Tokyo: MMIJ.

Von Gal, R. (2007). Innovations in twin–belt compliant mold casting technology. In J. Hugens, K. Sadayappan, J. Spooner, J. Kapusta, A. Fuwa, L. D. Smith, , … N. L. Piret, et al. (Eds.), *Plenary/copper and alloy casting and fabrication/copper — economics and markets: vol. I. Copper 2007* (pp. 189–202). Montreal: CIM.

Ware, P. W. (2007). The SCR® process produces high quality copper rod around the world. In K. Sadayappan, J. Spooner, J. Kapusta, A. Fuwa, L.D. Smith, C. Twigge-Molecey, N.L. Piret (Series Eds.) & J. Hugens, et al. (Vol. Eds.), *Plenary/copper and alloy casting and fabrication/copper — economics and markets: vol. I. Copper 2007* (pp. 175–188). Montreal: CIM.

Yurkov, A., Malakho, A., & Avdeev, V. (2017). Corrosion behavior of silicon nitride bonded silicon carbide refractory material by molten copper and copper slag. *Ceramics International, 43*, 4241–4245.

Zaheer, T. (1995). Reduction of impurities in copper. *Wire Industry, 62*(742), 551–553.

Suggested reading

Tyler, D. E., & Vierod, R. P. (2008). Copper continuous casting. In S. Viswanathan, et al. (Eds.), *Casting: vol. 15. ASM handbook* (pp. 1019–1025). Materials Park, OH: ASM International.

CHAPTER 20

Byproduct and waste streams

20.1 Molybdenite recovery and processing

Many porphyry Cu deposits contain economic concentrations of Mo, virtually all of it as molybdenite, MoS_2. Table 20.1 shows the Cu and Mo contents of several porphyry Cu—Mo ores.

Molybdenite floats with chalcopyrite, chalcocite, and other Cu—S and Cu—Fe—S minerals. It can then be separated from the resulting Cu—Mo bulk concentrate by depressing the Cu minerals with sodium hydrosulfide (NaHS), Nokes reagent, or sodium cyanide, while refloating the MoS_2 in a Mo/Cu separator plant (Amelunxen et al., 2019; Yin et al., 2017). Mo/Cu separator flotation plants are much smaller than their preceding bulk flotation plant because their feed (bulk Cu—Mo concentrate) mass is only 2% or 3% of the initial ore feed.

Typical Cu—Mo bulk concentrates and their equivalent Mo/Cu separator molybdenite and Cu concentrates are given in Table 20.2. A representative flowsheet is presented in Fig. 20.1.

20.1.1 Flotation reagents

The principal collector for molybdenite is diesel fuel or similar oil, ~0.01 kg per tonne of ore (Amelunxen et al., 2019; Lin et al., 2017). It is added to the bulk concentrator ore slurry feed to maximize molybdenite recovery from ore. Its effect carries over to flotation of molybdenite in the Mo/Cu separator flotation plant. Some Mo/Cu separation plants add more in the Mo/Cu separator plant.

The most common Cu—Fe—S mineral depressant is NaHS (Table 20.2). It is added to the Mo/Cu separator plant bulk concentrate slurry feed, 4—10 kg per tonne of bulk concentrate. However, toxicity concerns with NaHS and other current depressants have spurred extensive research in the past decade for new options. Amelunxen et al. (2019) review a long list of candidates. Thermal pretreatment to oxidize chalcopyrite surfaces and make them hydrophilic is another possibility (Tang et al., 2019).

Molybdenite flotation frothers are conventional mixed alcohols. In recent years, MIBC and X-133 have become popular (Amelunxen et al., 2019). The pH is raised with CaO and lowered with sulfuric acid, occasionally CO_2. The pH requirement varies from ore to ore.

Extractive Metallurgy of Copper, Sixth Edition
ISBN 978-0-12-821875-4, https://doi.org/10.1016/B978-0-12-821875-4.00009-2

Table 20.1 Ore, tailings, and concentrate compositions for four Cu, Mo concentrate producers.

Ore		Ore concentrator tailing		Ore concentrator bulk concentrate (Mo/Cu separator flotation plant feed)		Final Cu concentrate		Final Mo concentrate	
%Cu	%Mo	%Cu	%Mo	%Cu	%Mo	%Cu	%Mo	%Cu	%Mo
0.3	00.01	00.035	0.003	34	00.9	34.6	0.0	3	48
0.4	0.025	0.05	0.006	30	1.67	30.9	0.1	04.3	48
0.6	0.015	0.07	0.004	26	0.53	26.2	0.1	02.5	50
0.8	0.025	0.09	0.009	30	0.68	30.3	0.1	5	45

Data courtesy Peter Amelunxen, Amelunxen Mineral Engineering Ltda., Santiago, Chile.

Table 20.2 Reagents, cells, and pH used in floating molybdenite from bulk Cu–Mo concentrate (Amelunxen and Amelunxen, 2009).

Plant	Cu–S and Cu–Fe–S mineral depressants	NaHS addition amount, kg per tonne of bulk concentrate feed	Flotation gas	Cells enclosed?	Rougher feed (Fig. M.1) pH
Andina, Chile	NaHS + H_2SO_4	4	Recycled air	Yes	7–8
Bagdad, U.S.A.	NaHS	4.5	N_2 + air	No	11.5
Cerro Verde, Peru	NaHS + CO_2	12	recycled air	Yes	8–9.5
Chuquicamata, Chile	NaHS + H_2SO_4	6	N_2	No	7–8
Cuajone 87, Peru	NaHS	4	N_2	No	10–11
El Salvador, Chile	NaHS + As_2O_3	7.4	N_2	No	
Gibraltar 84, Canada	NaHS + COOH	5	N_2	No	7–8
Las Tortolas, Chile	NaHS + H_2SO_4	4	Recycled air	Yes	7–8
Pelambres, Chile	NaHS + H_2SO_4	4.6	N_2	No	7–8
Sierrita, U.S.A.	NaHS	5–10	N_2	No	11–12
Teniente, Chile	$Na_3PS_2O_2$ (Nokes reagent) + NaHS	4	N_2	No	9

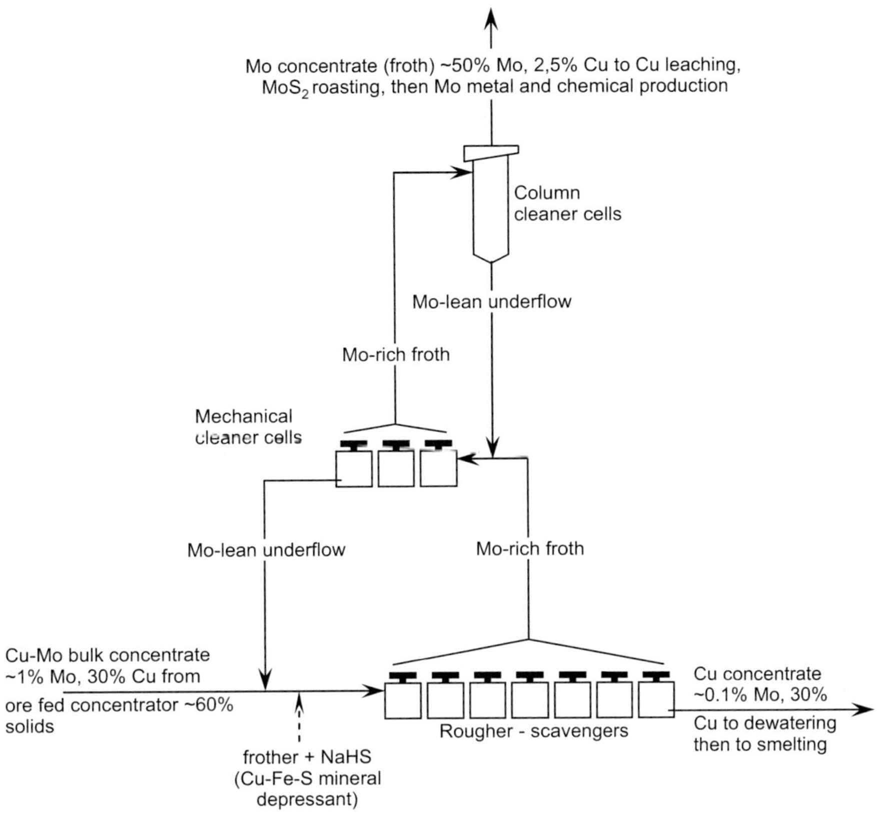

Figure 20.1 Simplified molybdenite flotation (Mo/Cu separator) flowsheet.

20.1.2 Operation

Flow through the Mo/Cu separator plant is continuous. Depressant and frother (sometimes oily collector) are continuously added where needed. Product Mo froth and Cu underflow continuously depart the plant for dewatering and subsequent transport to production of metal and chemical products. Flows and Mo and Cu concentrations are continuously measured and applied to process optimization.

Mo/Cu separator flotation plants often contain regrind mills (Lin et al., 2017). Their objective is to liberate molybdenite grains from Cu—Fe—S mineral particles (Amelunxen et al., 2019). However, too small a particle size can be a depressant for MoS_2.

20.1.3 Optimization

The objectives of a Mo—Cu bulk concentrator are (a) maximum Cu and Mo recovery to the bulk concentrate and (b) maximum rejection of non-Cu, non-Mo minerals to the tailings. The objectives of a Mo/Cu separation flotation plant are (a) maximum recovery of molybdenite to the Mo concentrate and (b) minimum recovery of Cu—Fe—S minerals to the Mo concentrate. The bulk concentrator objectives are met by adequate grinding, efficient rougher scavenger flotation, and quiescent concentrate cleaning in column cells.

The Mo/Cu separator plant objectives are obtained by

(a) efficient collection of molybdenite and efficient depression of Cu—Fe—S minerals in the rougher (mechanical) cells with adequate concentrations of oily collector and depressant

(b) efficient cleaning of molybdenite concentrate in ~5 column flotation cells in series (Amelunxen et al., 2019)

(c) use of N_2 in the flotation cells

(d) adequate washing of cleaner froth with fresh water.

Molybdenite concentrates often contain 3%—4% Cu. This presents a problem, since iron and steel production consumes much of the molybdenum produced, and copper is not a desired input to steelmaking processes. To remove the copper, a ferric chloride leaching step is commonly performed (Amelunxen et al., 2019). The ferric ion oxidizes the copper, and the chloride content dissolves any lead present in the concentrate. However, the solution is corrosive, and this has led to research on alternatives (Padilla, Letelier & Ruiz, 2013).

20.2 Anode slimes

20.2.1 Anode slime composition

Table 20.3, taken largely from the review by Moats et al. (2007), tabulates the composition of several anode slimes from electrolytic refineries around the world. These compositions reflect the composition of the raw materials input to the copper making process. Concentrates with high levels of gold and silver will produce anode slimes with high levels of these elements; the same is true for other impurities. Secondary smelters use little or no concentrate in their feed, and so produce slimes with little or no sulfur, selenium, or tellurium. However, anode slimes from secondary smelters often contain alloying elements from the scrap and waste streams fed to the furnace. This means higher levels of lead than those in anode slimes from primary smelters, and significant levels of elements such as tin and nickel. The production of higher matte grades in smelters has led to increased levels of arsenic, antimony, and bismuth in anode slimes over recent decades (Hoffmann, 2008). The reason for this is the reduced amount of converter slag produced when higher matte grades are used.

The mineralogy of anodes slimes varies with content, but is essentially a mixture of oxide and chalcogenide solid solutions (Chen & Dutrizac, 2005; Chen et al., 2015; Hait et al., 2009). Lead is found mostly as insoluble $PbSO_4$, and silver and gold occur mostly as selenide and telluride phases. Antimony, arsenic, and bismuth are found as a series of multiple oxides. Copper occurs in many phases.

Table 20.3 Composition of anode slimes (wt.-%).

	Cu	Ag	Au	Se	Te	As	Sb	Bi	Pb	Fe	Ni	Sn	S	Zn
Primary smelters														
La Caridad (Mexico)	12.7	15.4	0.036	4.2	0.37	5.0	5.0	1.2	14.8	2.1	0			0.81
BGMK (Kazakhstan)	3.0	20	1.2	6.6	1.2	5.3	19		11.9					
Olympic Dam (Australia)	30	9.9	1.19	14.9	3.5	3.9		2.5	12.3					
Port Kembla (Australia)	29	3.1	1.0	8	0.9	1.6	0.5	2.5	22					
Townsville (Australia)	21	6.1	0.63	3	0.5	3.9	0.6	0.7	9.8	0.04	0.3		2.5	
Bahia (Brazil)	4.38	7.59	0.40	11.5	2.6	3.8	4.1	0.7	6.7		0.1			
Med (Bulgaria)	25.27	4.86	0.49	9.25	0.83	2.96	1.16	0.18	10.3	0.06	0.136		10	
Timmins (Canada)	17	23	0.26	8.4	0.7	2.3	0.9	0.8	23		0.9			
Montreal (Canada)	15	29	1.4	5	2	1.5	4.5	1.5	16		3.8			
Potrerillos (Chile)	7.80	15.42	0.47	8.65	0.66	9.22	10.45	0.41	1.16	0.19	0.02			
Las Ventanas (Chile)	24.27	14.77	5.4	7.9	0.8	6.2	5.5	0.3	8.1	0.1	0.1			
Gresik (Indonesia)	0.6	4.9	2.1	11.1	0.2	1.2	0.3	2.8	51.4	0.1	0			
Saganoseki (Japan)	25.2	19	2.1	12.1	3.9	3.3	1.8	0.7	4.0	0.21	0.4		3.2	
Nishibara (Japan)	21.4	10.8	1.4	6.9	1.7	3.3	1.7	1.4	14		0.7		8.1	
Almalyk (Uzbekistan)	29			9	11	4	2	0.2	24	0.1			6	
Amarillo (USA)	19.6	18.6	0.12	14.1	1.45	2.91	2.86	0.44	0.15		0.74			
El Paso (USA)	1	22	0.2	20	0.4	2	4	0.7	5	0.04	0.05		12	
Kennecott (USA)	30	5	0.5	5	1	5	1	3	30	0.25	0.05			
Kayseri (Turkey)	18	1.5	0.08	1	0.4	0.3	0.4	0.04	16.9		0.2			
Secondary smelters														
Lünen (Germany)	1	8	0/16	0.7	0.7	2.5	9	0.4	30		1	9.5	8.5	
Beerse (Belgium)	19	5.5	0.06	0.5	0.5	2.5	16	1.5	4					
Boliden (Sweden)	15.69	22.95	0.578	5.17	0.94	3.21	4.15	0.777	9.69		5.09			
Montanwerke Brixlegg (Austria)	7.00	8.751	0.125	0.30	0.59		3.80	0.41	24.56		10.86	12.14		

The treatment of slimes is an important aspect of a refinery's operation. A significant fraction of the world's supply of gold, silver, selenium, and tellurium are obtained from these slimes, and the value of the recovered gold and silver is a significant part of a copper producer's profitability. In fact, there have been times in the past when the price of copper was so low that only the production of gold and silver justified continued operation of a smelting and refining plant.

20.2.2 The slime treatment flowsheet

There are a number of commercial processing schemes for processing anode slimes, depending on the composition and workplace hygiene concerns. Fig. 20.2 illustrates the slimes treatment process used by CCR in Canada (Hait et al., 2009), which is typical of several current facilities. The first step is *decoppering*, which was previously done by

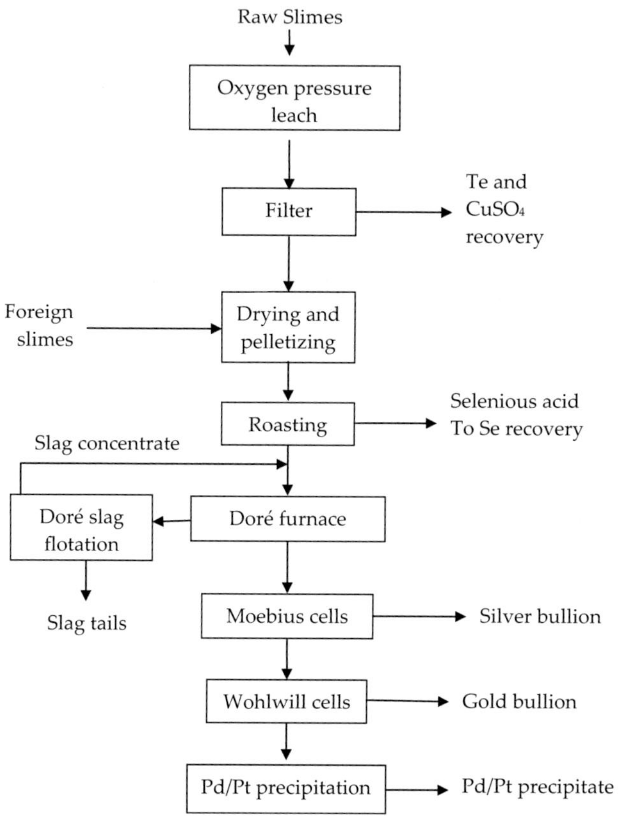

Figure 20.2 The slimes treatment process at CCR in Montreal, Canada.

leaching under atmospheric conditions in return electrolyte from the electrorefining cells (Hoffmann, 2008). The use of oxygen pressure leaching at 125−150°C improves both process kinetics and overall copper recovery from the slime. However, it also dissolves much of the tellurium in the slimes. This is followed by roasting with sulfuric acid at ~300°C, which converts most of the selenium in the slime to a vapor:

$$Se + 2\ H_2SO_4 \rightarrow SeO_2\ (g) + 2\ SO_2 + 2\ H_2O \qquad (20.1)$$

An alternative is to eliminate the leaching step and roast at 600−800°C (Chen et al., 2015), which converts to copper to soluble copper sulfate while oxidizing all of the selenium as well.

In Fig. 20.2 the remaining slime is smelted to produce a slag containing the arsenic, antimony, bismuth, and tellurium, along with a metal phase called *Doré* bullion containing the gold, silver, lead, and remaining copper (Chen et al., 2015; Hait et al., 2009; Hoffmann, 2008; Tanisaki et al., 2016). Some of the precious metal content winds up in the slag, which is treated using flotation to recover the silver and gold or returned to the smelter. Doré smelting has traditionally been performed in a small reverberatory furnace, but top-blown rotary converters such as the Kaldo furnace are more popular (Vartiainen, 2015). The Doré bullion is refined electrolytically.

A change in recent decades has been the development of hydrometallurgical processes for anode slime. Several different processes are in use; the best known (Fig. 20.3) is the wet chlorination process (Chen et al., 2015; Nexhip, 2015; Vartiainen, 2015), which leaches the slime with chlorine gas bubbled into a pulp of as-received slime. This produces leach liquor containing gold, selenium, tellurium, and some of the copper, and a residue containing silver chloride and the rest of the copper. Solvent extraction recovers the gold from the leach liquor, and precipitation with SO_2 sequentially recovers the Se and Te. The slime is sometimes decopperized before leaching, and hydrogen peroxide and hydrochloric acid are a less expensive alternative to chlorine gas (Hait et al., 2009; Furuzono et al., 2018).

20.3 Dust treatment

Chapter 7 briefly described the formation of dust during matte converting operations. Dusts are generated during all stages of pyrometallurgical processing of copper concentrates, and the mass of dust can be as much as 10% of the mass of input concentrate (Xu et al., 2020). These dusts are collected by electrostatic precipitator or baghouse (see Chapter 11). Common practice has been to recycle these dusts to the smelter to recover the copper content. However, this approach also recycles impurities such as arsenic, antimony, bismuth, and lead. As the levels of these impurities build up, their

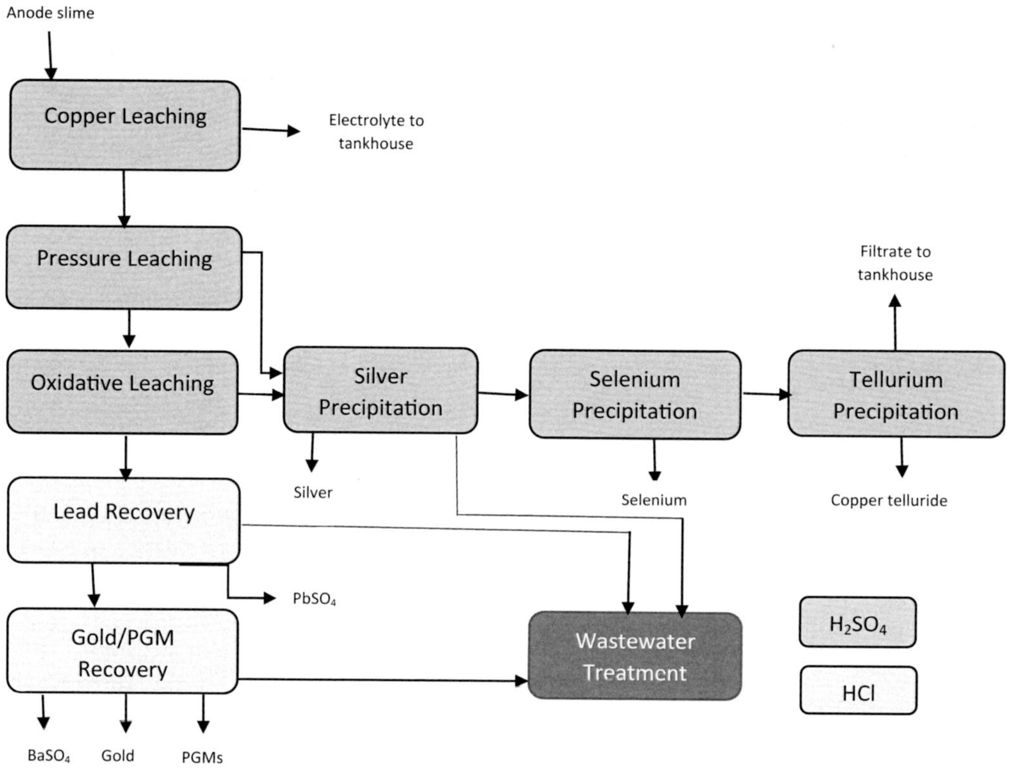

Figure 20.3 Flowsheet for hydrometallurgical processing of anode slimes. *(After Vartiainen, A. (2015) By-products recovery technologies in primary non-ferrous metals production. http://ec.europe.eu/DocsRoom/documents/14054/Attachments/1/translations/en/renditions/native. Accessed 9 August 2020.)*

concentration in the matte increases as well, and this eventually means high impurity levels in the blister copper (Montenegro et al., 2010). This in turn makes electrolytic refining more difficult. The recycle of dust also decreases furnace capacity, which hurts productivity (Parra & Parada, 2008). As impurity levels in concentrates increase, the need to find a way of treating smelter dust to separate the copper from the impurities has increased.

Table 20.4, taken from data published by Samuelsson and Carlsson (2001), provides typical compositions of dust from various locations in the copper making process at the Rönnskär smelter of Boliden. Dusts from concentrate roasting and smelting feature high copper and sulfur levels, the result of concentrate particles reporting directly to the dust stream. Dusts from converting and slag cleaning have lower copper levels and higher levels of lead, antimony, and arsenic. As a result, direct recycle of the roaster and smelter dusts is more feasible, and the incentive to treat converter and slag cleaning dusts to remove impurities is higher.

Table 20.4 Composition of dusts from the Rönnskär smelter of Boliden (Samuelsson & Carlsson, 2001).

Dust source	Collection device	% Cu	% Fe	% Zn	% Pb	% Sb	% As	% S	% Sn
Roaster	ESP	20.0	30.0	5.7	2.5	0.26	1.40	11.5	<0.01
Smelter	ESP	20.2	25.6	12.3	3.0	0.05	1.88	7.5	0.02
Converter	Baghouse	8.3	1.56	15.1	20.2	0.21	4.51	8.6	23.6
Slag cleaning	Baghouse	1.96	1.22	33.3	16.6	0.92	11.5	1.8	20.0

Numerous hydrometallurgical and pyrometallurgical processes have been developed to treat copper making dusts (Xu et al., 2020). The biggest concern with these processes is recovering the arsenic in a stable form for disposal. An example is the process for treating Teniente furnace dust described by Ichimura et al. (2007). The flowsheet (Fig. 20.4) is designed to generate a copper electrolyte solution that can be sent to the electrorefining circuit, a copper sulfide concentrate that is returned to the smelter, purified zinc sulfate for sale, and crystalline ferric arsenate suitable for disposal.

20.4 Use or disposal of slag (Gorai et al., 2003)

Chapter 10 described the processing of smelting slags to recover the copper content. These processes leave behind a substantial amount of material to be disposed of or otherwise used. About 25 million tonnes of slag are generated per year worldwide (Gorai et al., 2003). Over the years, several applications for smelting slag have been found. These include the following:

- *Cement and concrete* (Pavez, Rojas, Palacios, & Sanchez, 2004; Shi et al., 2008). The iron content of copper slag makes it a useful replacement for iron powder in cement clinker production. Copper slag has also been used as a partial substitute for Portland cement in concrete, with mixed effects on properties. The most beneficial aspect of using copper slag as a replacement is the reduction in energy use required to produce the concrete.
- *Abrasive grit* (Flynn & Susi, 2004; Pavez, Rojas, Palacios, & Sanchez, 2004). Copper slag is a common abrasive for blasting operations. Its advantages include higher hardness and density than other abrasives, and a lack of free silica (a workplace hygiene hazard). The biggest disadvantage is the presence of hazardous metal content, which can potentially contaminate job sites.
- *Land reclamation* (Lim & Chu, 2006). Copper slag is denser than the sand normally used for fill, and has a better particle-size distribution. The friction angle is higher as well, enhancing slope and foundation stability.

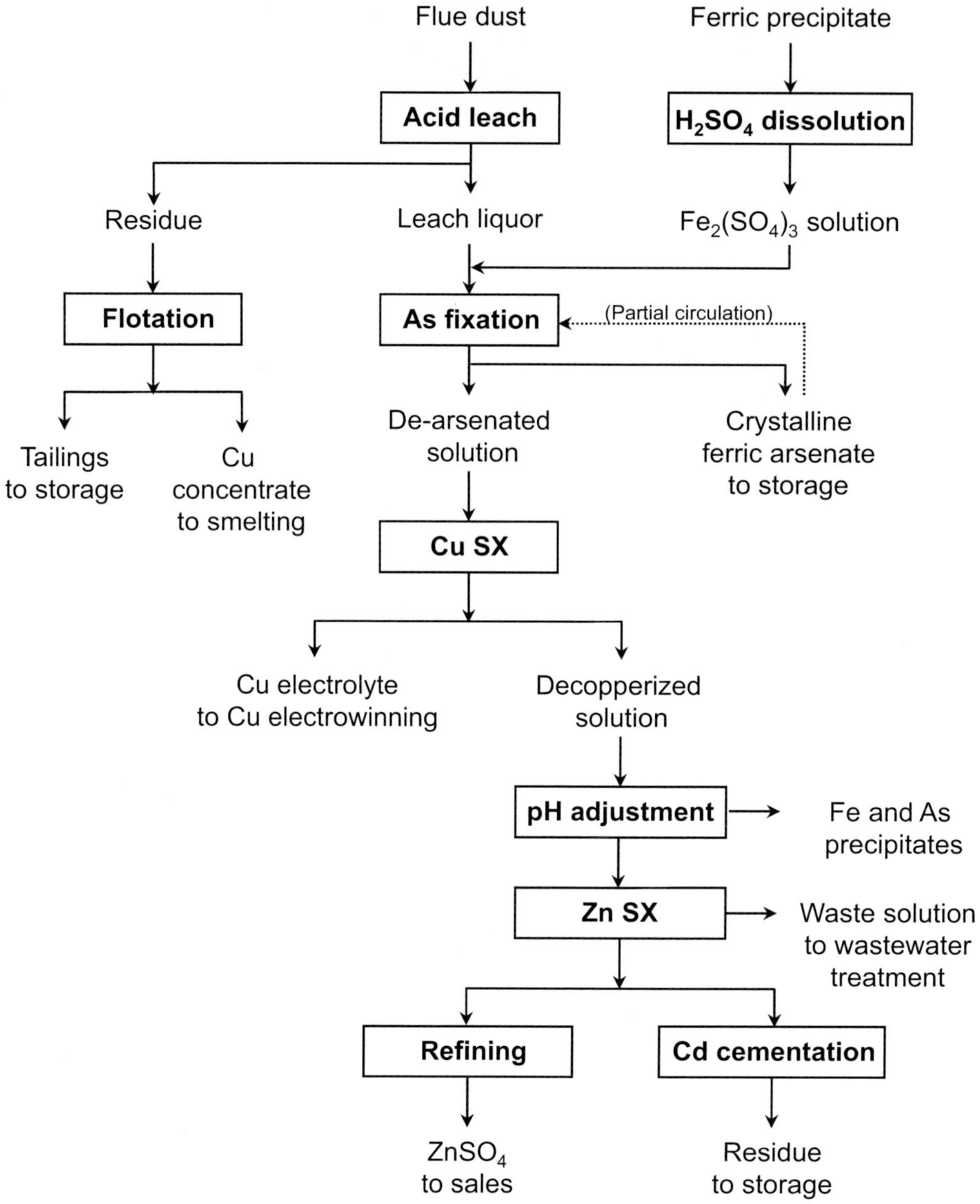

Figure 20.4 Process flowsheet for treatment for Teniente furnace dust.

- *Iron ore* (Cao et al., 2009). The iron content of smelter slag is 40%–60% (Shi et al., 2008), which is higher than that of many iron ores. However, the iron is largely present as fayalite, which is difficult to concentrate. Experiments with reverse magnetic flotation have upgraded the iron content to a level which might be suitable for some ironmaking processes. González et al. (2004) smelted copper slag with carbon to generate a molten Fe–1.9 Cu–5 C alloy, but this has too much copper to be of value to steelmakers.
- *Ceramic tile* (Marghussian & Maghsoodipoor, 1999). The use of copper slag in tile production improves the product strength and acid resistance. However, the sulfur content results in SO_2 emission and bloating during firing, limiting the amount that can be used.

Other potential uses of copper slag include roofing granules, hot-mix asphalt, and cutting tools (Gorai et al., 2003). However, all of the applications described here are cost sensitive, and the cost of transporting slag from a remote location to a manufacturing facility often makes its use uncompetitive. As a result, most slag is still landfilled after processing.

20.5 Summary

MoS_2 occurs in economic quantities in many porphyry Cu deposits. It is recovered to MoS_2 flotation concentrates by (a) floating MoS_2 and Cu–Fe–S minerals together in a bulk Cu–Mo concentrate, then (b) depressing Cu–Fe–S minerals and floating MoS_2 in a Mo–Cu flotation separator plant. Typical Mo recovery to final MoS_2 concentrate is ~70% (80% to the bulk concentrate and 90% of that to the final MoS_2 concentrate). Other byproducts from copper production include the slime from electrorefining, dust recovered from baghouses and electrostatic precipitators, and smelting slag. All of these can be treated to recover their copper content, and the remaining material is also valuable. The slime in particular contains significant levels of gold, silver, selenium, and tellurium. The recovery of these metals has a significant impact on the profitability of copper production facilities.

References

Amelunxen, P., & Amelunxen, R. (2009). Moly plant design considerations. In *, 09-136. Proceddings of the SME Annual Meeting. Preprints* (pp. 1–5). Littleton, CO: SME.

Amelunxen, P., Schmitz, C., Hill, L., Goodweiler, N., & Andres, J. (2019). Molybdenum. In R. C. Dunne, S. K. Kawatra, & C. Y. Young (Eds.), *SME minerals processing and extractive metallurgy handbook* (pp. 1891–1916). Littleton, CO: SME.

Cao, H., Zhang, L., Wang, C., Fu, N., Sui, Z., & Feng, N. (2009). Study of selectively separating iron constituents in copper smelting slags. In *EPD congress 2009* (pp. 939–941). Warrendale, PA: TMS.

Chen, T. T., & Dutrizac, J. E. (2005). Mineralogical characterization of a copper anode and the anode slimes from the La Caridad Copper Refinery of Mexicana de Cobre. *Metallurgical and Materials Transactions B, 36B*, 229–240.

Chen, A., Peng, Z., Hwang, J.-Y., Ma, Y., Lu, X., & Chen, X. (2015). Recovery of gold and silver from copper anode slimes. *JOM, 67*, 493–502.

Flynn, M. R., & Susi, P. (2004). A review of engineering control technology for exposure generated during abrasive blasting operation. *Journal of Occupational and Environmental Hygiene, 1*, 680–687.

Furuzono, T., Fujimoto, A., Takeuchi, T., & Takebayashi, K. (2018). Unique hydrometallurgical process for copper–anode slime treatment at Saganoseki Smelter and Refinery. In B. Davisa, et al. (Eds.), *Extraction 2018* (pp. 2075–2083). Warrendale, PA: TMS.

González, C., Parra, R., Klenovcanova, A., Imris, I., & Sánchez, M. (2004). Reduction of Chilean copper slags: A case of waste management project. *Scandinavian Journal of Metallurgy, 34*, 143–149.

Gorai, B., Jana, R. K., & Premchand, M. (2003). Characteristics and utilization of copper slag — a review. *Resources, Conservation and Recycling, 39*, 299–313.

Hait, J., Jana, R. K., & Sanyal, S. K. (2009). Processing of copper electrorefining anode slime: A review. *Transactions of the Institution of Mining Metallurgy Section C, 118*, 240–252.

Hoffmann, J. E. (2008). The world's most complex metallurgy revisited. *World of Metallurgy — Erzmetall, 61*, 6–13.

Ichimura, R., Tateiwa, H., Almendares, C., & Sanchez, G. (2007). Arsenic immobilization of Teniente furnace dust. In D. Rodier, & W. Adams (Eds.), *Sustainable development, HS&E, and recycling: vol. VI. Copper 2007* (pp. 275–288). Montreal: CIM.

Lim, T.-T., & Chu, J. (2006). Assessment of the use of spent copper slag for land reclamation. *Waste Management and Research, 24*, 67–73.

Lin, Q., Gu, G., Wang, H., Liu, Y., Wang, C., Fu, J., Zhao, J., & Huang, L. (2017). Recovery of molybdenum and copper from porphyry ore via iso-flotability flotation. *Transactions of Non-Ferrous Metals Society of China, 27*, 2260–2271.

Marghussian, V. K., & Maghsoodipoor, A. (1999). Fabrication of unglazed floor tiles containing Iranian copper slag. *Ceramics International, 25*, 717–722.

Moats, M., Robinson, T., Davenport, W., Karcas, G., & Demetrio, S. (2007). Electrolytic copper refining — 2007 world tankhouse operating data. In G. E. Houlachi, J. D. Edwards, & T. G. Robinson (Eds.), *Copper electrorefining and electrowinning: vol. V. Cu2007* (pp. 195–241). Montreal: CIM.

Montenegro, V., Sano, H., & Fujisawa, T. (2010). Recirculation of Chilean copper smelting dust with high arsenic content to the smelting process. *Materials Transactions, 49*, 2112–2118.

Nexhip, C. (2015). *By-products recovery via integrated copper operations at Rio Tinto Kennecott*. http://ec.europe.eu/DocsRoom/documents/14046/Attachments/1/translations/en/renditions/native. (Accessed 8 August 2020).

Padillla, R., Letelier, H., & Ruiz, M. C. (2013). Kinetics of copper dissolution in the purification of molybdenite concentrates by sulfidation and leaching. *Hydrometallurgy, 137*, 78–83.

Parra, R., & Parada, R. (2008). Minor element control by flue dust treatment in copper smelting. In B. Mishra, C. Ludwig, & S. Das (Eds.), *REWAS 2008* (pp. 1027–1038). Warrendale, PA: TMS.

Pavez, O., Rojas, F., Palacios, J., & Sanchez, M. (2004). A review of copper pyrometallurgical slags utilization. In S. R. Rao, F. W. Harrison, J. A. Kozinski, L. M. Amaratunga, T. C. Cheng, G. G. Richards, et al. (Eds.), *Waste processing and recycling in mineral and metallurgical industries V* (pp. 291–298). Montreal: CIM.

Samuelsson, C., & Carlsson, G. (2001). Characterization of copper smelter dusts. *CIM Bulletin*, (1051), 111–115.

Shi, C., Meyer, C., & Behnood, A. (2008). Utilization of copper slag in cement and concrete. *Resources, Conservation and Recycling, 52*, 1115–1120.

Tang, X., Chen, Y., Liu, K., Zeng, G., Peng, Q., & Li, Z. (2019). Selective flotation separation of molybdenite and chalcopyrite by thermal pretreatment under air atmosphere. *Colloids and Surfaces A, 583*, 123958.

Tanisaki, K., Matsubaru, S., Manabe, Y., Nagai, H., Asano, S., & Hattori, Y. (2016). Development of the through hydrometallurgical process for precious metals recovery from copper anode slime. In *Copper 2016* (pp. 1806–1817).

Vartiainen, A. (2015). *By-products recovery technologies in primary non-ferrous metals production*. http://ec.europe.eu/DocsRoom/documents/14054/Attachments/1/translations/en/renditions/native. (Accessed 9 August 2020).

Xu, B., Ma, Y., Gao, W., Yang, J., Yang, Y., & Jiang, T. (2020). A review of the comprehensive recovery of valuable elements from copper smelting open – circuit dust and arsenic treatment. *JOM, 72*. https://doi.org/10.1007/s11837-020-04242-0. . (Accessed 9 August 2020)

Yin, Z., Sun, W., Hu, Y., Zhai, J., & Qingjun, J. (2017). Evaluation of the replacement of NaCN with depressant mixtures in the separation of copper–molybdenum sulphide ore. *Seperation and Purification Technology, 173*, 9–16.

CHAPTER 21

Costs of copper production

This chapter

(a) describes the investment and production costs of producing copper metal from ore

(b) discusses how these costs are affected by such factors as ore grade, process route, and inflation

(c) indicates where cost savings might be made in the future.

The discussion centers on mine, concentrator, smelter and refinery costs. Costs of producing copper by leach/solvent extraction/electrowinning and from scrap are also discussed.

The cost data have been obtained from published information and personal contacts in the copper industry. They were obtained during 2019 and are expressed in 2019 U.S. dollars. The data are directly applicable to plants in South and North America. Costs elsewhere may vary considerably, due to fluctuations in exchange rates and raw materials costs (Boulamanti & Moya, 2016).

Investment and operating costs are significantly affected by inflation. Fortunately, inflation has been small since the previous edition of this book. Inflation in the chemical processing industry is represented by the *Chemical Engineering* Plant Cost Index (Fig. 21.1). The basic equation for using this index for identical equipment is the following:

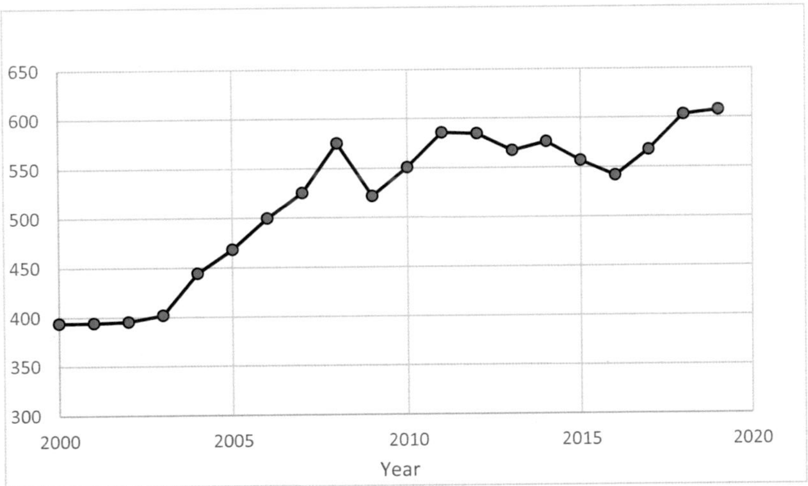

Figure 21.1 Chemical Engineering Plant Cost Index from 2000 to 2019 (1958 value = 100). Steep inflation from 2003 to 2008 and flatter inflation from 2009 to 2020 are notable. Data is obtained from July issues.

Extractive Metallurgy of Copper, Sixth Edition
ISBN 978-0-12-821875-4, https://doi.org/10.1016/B978-0-12-821875-4.00019-5

$$\frac{\text{Cost}_{\text{(year A)}}}{\text{Cost}_{\text{(year B)}}} = \frac{\text{Index}_{\text{(year A)}}}{\text{Index}_{\text{(year B)}}} \tag{21.1}$$

The investment and operating costs in this chapter are at the *study estimate* level, which is equivalent to an accuracy of ±30%. Data with this accuracy can be used to examine the economic feasibility of a project before spending significant funds for piloting, market studies, land surveys, and property acquisitions (Christensen & Dysert, 2005).

21.1 Overall investment costs: mine through refinery

Table 21.1 lists study estimate investment costs for a mine/concentrator/smelter/refinery complex designed to produce electrorefined cathodes from 0.5% Cu ore. These costs are for a *greenfield* (new) complex starting on a virgin site with construction beginning July 1, 2019. It should be noted that construction of completely integrated complexes is now rare. Typically, new mines and concentrators are built in South America and new smelters/refineries are built in coastal Asia. The investment costs are expressed in terms of investment cost per annual tonne of product copper. This is defined by the following equation:

$$\begin{aligned}\text{Plant cost} &= \text{investment cost per annual tonne of copper} \\ &\quad \times \text{plant capacity, tonnes of copper per year}\end{aligned} \tag{21.2}$$

This equation shows, for example, that the investment in a concentrator which (a) costs \$10,000 per annual tonne of copper and (b) produces 200,000 tonnes of Cu–in–concentrate per year will be $\$2000 \times 10^6$ (\$2 billion).

Table 21.1 indicates that the fixed capital investment for a complex which produces electrorefined copper from 0.5% Cu ore is ~\$30,000 per annual tonne of copper. To this must be added working capital to cover the initial operating expenses of the complex

Table 21.1 Copper extraction investment costs.

Facility	Fixed investment cost ($U.S. per annual tonne of Cu)
Mine (open pit)	11,000
Concentrator (including water acquisition and recycle)	11,000
Smelter (Outotec flash furnace smelting/converting), including sulfuric acid plant	10,000
Electrolytic refinery (excluding precious metals refinery)	1,000
Total	**33,000**

Fixed investment costs for a copper extraction complex, starting with 0.5% Cu ore. The costs are at the study estimate level of accuracy. Cost effects of underground mining and ore grade are discussed in Section 21.1.1.

(about 10% of fixed capital investment; Peters & Timmerhaus, 1968). It means that a new mine/mill/smelter/refinery complex which is to produce 200,000 tonnes of copper per year will cost ~ \$7.2 billion (\$36,300 × 200,000).

21.1.1 Variation in investment costs

Mine investment costs vary considerably between mining operations. This is due to (a) differences in ore grades, mine sizes, mining method, topography, and ground condition and (b) infrastructure requirements for electric power and water. Underground mine development costs considerably more than open pit mine development, per annual tonne of mined ore. This and the high cost of operating underground explain why underground ore bodies must contain (a) higher ore grade than open pit ore bodies and/or (b) valuable byproducts such as Ag, Au, Co, Mo, and Ni.

Ore grade has a direct effect on mine investment costs. Consider (for example) two identical ore bodies, one containing 0.5% Cu ore and the other 1% Cu ore. Achievement of an identical annual production of Cu requires that the 0.5% Cu ore be mined at twice the rate of the 1% Cu ore. This, in turn, requires about (a) twice as much plant and equipment (e.g., trucks) and (b) twice as much investment. The same is true for concentrators and heap leach pads; they will have to treat 0.5% Cu ore twice as fast as 1% Cu ore to achieve the same annual production of Cu. This will require about twice the amount of equipment and twice the investment.

A significant investment cost for new concentrators is electric energy and water production/distribution. This is particularly true in South America, where virtually all new Cu mines are in high, dry desert locations.

Smelter investment costs, per annual tonne of copper production, are influenced by concentrate grade rather than by ore grade. The higher the % Cu in the concentrate, the smaller the smelter (and smelter investment) for a given annual production of copper. High-grade concentrates also minimize smelter operating costs (such as materials handling costs, fuel consumption costs, and gas handling costs) per tonne of copper.

Refinery investment costs are not affected by mine/concentrator/smelter characteristics. This is because copper refineries treat 99.5% Cu anodes, irrespective of the preceding processes.

21.1.2 Economic sizes of plants

Mines can be economic at any size, depending upon the Cu grade of their ore. Thus, copper mines operate at production rates between 10,000 tonnes of ore per day (a high-grade operation) and 250,000 tonnes per day (a large open pit operation; ICSG, 2018).

Concentrators vary similarly. A new large concentrator unit typically consists of a semiautogenous grinding mill, high pressure grinding rolls, two ball mills, and a flotation circuit. It can treat up to 170,000 tonnes of ore per day (Sartain, 2010). Larger concentrators consist of multiples of this basic concentrating unit.

Smelters are almost always large because their minimum economic output is that of a single, fully used high intensity smelting furnace. These furnaces typically smelt up to 4000 tonnes of concentrate per day.

Copper refineries are usually sized to match the anode output of an adjacent smelter. The advantage of having a smelter and refinery at the same site is the presence of shared site facilities, particularly for anode casting and anode scrap remelting. A few refineries treat the anodes from several smelters. The largest electrorefineries produce up to 2500 tonnes of copper cathodes per day (ICSG, 2018).

21.2 Overall direct operating costs: mine through refinery

Direct operating (cash) costs (excluding depreciation, capital repayment, and income taxes) for mining/concentrating/smelting/electrorefining are given in Table 21.2. The table shows that direct operating costs for the major steps are, in descending order, concentration, open pit mining, smelting, and electrorefining. Overall direct operating costs for extraction are ~ $3.9 per kg of copper.

21.2.1 Variations in direct operating costs

The operating costs which vary most are those for mining and concentrating. The amounts of ore which must be handled by these operations, per tonne of Cu, vary directly with ore grade—and this significantly affects operating costs. Underground mining can also cost up to five times as much as open pit mining, which must be offset by (a) higher ore grade and (b) the presence of byproduct metals in underground ore.

Table 21.2 Copper extraction operating costs.

Activity	Direct operating cost ($U.S. per kg of Cu)
Open pit mining, 0.5% Cu ore at $3.9/tonne of ore	0.78
Concentration from 0.5% Cu ore to 30% Cu concentrate at shipping point, including tailings disposal at $5/tonne of ore	1.1
Smelting at $80/tonne of 30% Cu concentrate including sulfuric acid production	0.27
Electrolytic refining, excluding precious metals recovery	0.15
Local services and overhead	0.2
Total direct operating cost	**2.50**

Direct operating costs for producing electrorefined copper cathodes from a 0.5% Cu ore (assuming 90% Cu recovery). Maintenance is included. The costs are at the study estimate level. Factors affecting these costs are discussed in Section 21.2.1.

21.3 Total production costs, selling prices, profitability

The total cost of producing copper from ore is made up of (a) direct operating costs and (b) finance (indirect) costs, including interest and capital recovery. A reasonable estimate for (b) is 12% of the total capital investment per year. Based on a fixed capital investment of $33,000 (including 10% working capital) per annual tonne of copper, this is equivalent to $4000 per tonne of copper ($4/kg). Thus, the direct ($3.9) plus indirect ($4.3) operating costs of producing electrorefined copper in a new operation total approximately $8.2 per kg. For a new operation to be profitable, the selling price of copper must exceed these costs.

Mines and plants that have been in operation for many years may have repaid much of their original capital investment. In this case, direct operating costs (plus refurbishing) are the main cost component. This type of operation will be profitable at selling prices of ~$3 per kg of copper.

In summary, the price—profit situation is

(a) At copper selling prices above $6 per kg, copper extraction is profitable and expansion of the industry is encouraged. Underground ore bodies containing about 1.5% Cu are viable, as are open pit ore bodies containing about 0.5% Cu.

(b) At selling prices below about ~$3 per kg, some mines and plants are unprofitable. Some operations begin to shut down.

These costs and prices all refer to July 1, 2019. They will increase at about the same rate as the cost index in Fig. 21.1. The 2019 selling price of copper has been about $6 per kg so that operating + finance costs were met in most cases.

21.3.1 Byproduct credits

Many Cu ore bodies contain Au and Ag (Chadwick, 2010; Kendrick et al., 2003; Supomo et al., 2008). These metals follow Cu during concentration, smelting, and refining. They are recovered during electrorefining (with some additional treatment; see Chapter 20) and sold. Other ore bodies contain MoS_2 (Section 20.1) which is recovered in the concentrator and usually sold. The credits (sales minus extra costs for recovery) for these byproducts should be included in all project evaluations (Freeport—McMoRan Copper and Gold, 2011). Sulfuric acid is also often a valuable byproduct, especially when the smelter is near leach/solvent extraction/electrowinning operations.

21.4 Concentrating costs

The investment costs of constructing a Cu concentrator are of the order of $10,000 per annual tonne of copper (Table 21.1). This means that a concentrator producing 200,000 tonnes of copper-in-concentrate will cost ~$2 billion (Section 21.1).

Table 21.3 Concentrator investment costs.

Section	Percent of total investment cost
Ore handling, storage, conveying equipment	20
Semiautogenous grinding mill, high pressure grinding rolls, ball mills, and size classifiers	30
Flotation cells and associated equipment	15
Water acquisition/distribution system, dewatering/filtering equipment, tailings dam, concentrate loading facilities	35
Total	**100**

Investment costs for a copper concentrator by section, expressed as a percentage of the total investment cost. Control equipment costs are included in each section.

Table 21.3 breaks concentrator investment costs into major cost components, expressed as a percentage of total investment cost. The largest cost for this concentrator is water acquisition, which indicates that it is in a desert region. Grinding is second.

Concentrator direct operating costs (Tables 21.4 and 21.5) are of the order of \$4.8/tonne of ore, which is equivalent to about \$1.1/kg of Cu (assuming 0.5% Cu ore and 90% Cu recovery). Grinding is by far the largest operating cost, followed by flotation.

Grinding and flotation costs vary markedly for different ores. Grinding costs are high for hard primary ores and low for secondary (altered) ores. Flotation costs are low for simple Cu sulfide ores. They increase with increasing ore complexity.

21.5 Smelting costs

The investment cost of a new Outotec flash furnace/flash converter smelter is ~\$9000 per annual tonne of copper (Table 21.1). A smelter designed to produce 200,000 tonnes of new anode copper per year will cost, therefore, about \$1800 $\times$ 10^6—about \$2 billion.

Table 21.4 Concentrator operating costs by activity.

Activity	Cost per tonne of ore (\$U.S.)
Crushing, conveying, storage	0.50
Semiautogenous grinding, ball mill grinding, size classification	2.50
Flotation	0.80
Dewatering, filtering, drying, storage, and loading of concentrate	0.30
Tailings disposal, effluent control, water recycle	0.75
Local services and overhead (accounting, clerical, environmental, human resources, laboratory, management, property taxes, safety)	0.15
Total	**5.00**

Direct operating costs of producing 30% Cu concentrate from 0.5% Cu ore. Ore cost is not included.

Table 21.5 Concentrator operating costs by cost component.

Component	Percent of concentrating cost
Electric energy	14
Water (new + recycled)	10
Operating and maintenance labor	15
Maintenance and operating supplies	20
Reagents and lime	10
Grinding media	21
Local services and overhead (accounting, clerical, environmental, human resources, laboratory, management, safety	10
Total	**100**

Expenditures on energy, manpower, supplies, and overhead are shown. Ore cost is not included.

Table 21.6 Smelter investment costs.

Item	Percent of smelter cost
Concentrate handling and drying, including delivery of dry concentrate to smelting furnace	10
Oxygen plant	10
Flash furnace	15
Flash converter, including matte granulation and crushing	15
Cu-from-slag recovery equipment (electric furnace or flotation) including barren slag disposal	10
Anode furnaces and anode casting equipment	10
Gas handling system including waste heat boilers, electrostatic precipitators, and sulfuric acid plant	30
Total	**100**

Investment costs of a flash furnace/flash converter smelter by section, expressed as a percentage of total fixed investment cost. The costs include installation and housing of the units. Control systems are also included.

Table 21.6 breaks this investment cost into its major components. About 70% of the investment goes into concentrate handling/smelting/converting/anode casting and about 30% into gas handling/sulfuric acid manufacture.

In 2020, there are three major intensive smelting processes available for installing in new smelters or for modernizing old smelters—Outotec flash, Isasmelt, and Ausmelt. Each has been installed extensively since 2005. Each appears to be competitive for new and replacement smelting units.

Table 21.7 shows the direct costs of operating an oxygen-enriched Outotec flash smelting/flash converting smelter. The total is about $80 per tonne of concentrate after sulfuric acid and energy credits. For a 30% Cu concentrate, this is equivalent to about $0.3 per kg of new copper anodes. Table 21.8 breaks down these direct operating costs

Table 21.7 Smelter operating costs by activity.

Activity	Cost, $U.S. per tonne of concentrate
Concentrate reception, storage, and delivery to dryer	5
Flash furnace smelting including concentrate drying, gas handling, and delivery of 70% Cu crushed matte granules to flash converting	20
Flash converting including delivery of molten copper to anode furnaces	20
Cu recovery from smelting slag	15
Anode-making including desulfurization and deoxidation of molten copper, anode casting, and loading for transport to electrorefinery	5
Sulfuric acid plant including acid storage and loading of rail cars and trucks. Costs of treating acid plant blowdown and credit for sulfuric acid are included	10
Local overhead (accounting, clerical, environmental, human resources, laboratory, management, property taxes, safety)	5
Total	**80**

Direct operating costs for producing anodes from 30% Cu concentrate in a flash smelting/flash converting smelter, including maintenance. Concentrate cost is not included.

Table 21.8 Smelter operating costs by cost component.

Component	Percent of smelting cost
Operating and maintenance manpower, including supervision	20
Electricity (including electricity used for making oxygen and sulfuric acid)	30
Hydrocarbon fuel	20
Flux and refractories	10
Maintenance supplies	15
Local services and overhead (accounting, clerical, environmental, human resources, laboratory, management, property taxes, safety)	5
Total	**100**

Expenditures on manpower, utilities, and supplies in a flash smelting/flash converting smelter, by percentage. Concentrate cost is not included.

into labor, fuel, oxygen, and supplies. Energy is shown to be the largest cost, including the energy used in oxygen production (Bunjaku et al., 2019). Flash furnaces appear to have slightly lower operating costs than bath smelters; cost per kilogram is highly dependent on smelter capacity (Castillo & Cantallopts, 2016).

21.6 Electrorefining costs

The investment cost of a new electrorefinery using stainless steel cathode technology is ~\$1000 per annual tonne of electrorefined cathodes. This means that a refinery producing 200,000 tonnes per year of cathodes will cost of the order of $\$200 \times 10^6$.

The relative investment costs of various sections of a refinery are shown in Table 21.9. The production electrorefining section (including stainless steel blanks) is by far the largest investment cost component of the refinery.

The direct cost of producing electrorefined cathodes in an electrolytic refinery is ~\$0.12 per kg of cathode copper, Table 21.10. The largest component of that cost is energy, Table 21.11.

21.7 Production of copper from scrap

Chapter 18 shows that copper scrap varies in grade from 99.5+% Cu (manufacturing wastes) to 5% Cu (recycled mixed-metal scrap). The high-grade manufacturing wastes require only reclamation, melting, casting, and marketing, which costs about \$0.2/kg of copper. Low-grade scrap, on the other hand, requires reclamation, sorting, smelting, refining, and marketing, which costs about \$0.6 per kg of copper. Intermediate grade scrap treatment lies between these two extremes.

Table 21.9 Electrorefinery investment costs.

Component	Percent of total fixed investment cost
Anode reception, weighing, straightening, lug milling, sampling equipment	17
Production electrorefining equipment including stainless steel blanks, polymer concrete cells, transformers, rectifiers, electrical distribution system	39
Electrolyte circulation and purification equipment including filters, heaters, pumps, storage tanks, reagent addition equipment, electrowinning cells	18
Cathode handling equipment including stripping, washing, weighing, sampling and bundling equipment	11
Anode (and purchased) scrap melting and anode casting equipment—including Asarco shaft furnace, holding furnace, pouring equipment (including mass control system), and mold-on-wheel caster	14
Total	**100**

Investment costs of components in an electrolytic copper refinery expressed as a percentage of the total fixed investment cost.

Table 21.10 Electrorefinery operating costs by activity.

Activity	Cost, $U.S. per kg of cathode Cu
Anode reception, weighing, straightening, lug milling, delivery to tankhouse	0.01
Production electrorefining, including cell cleaning, electrolyte purification and reagent addition, delivery of cathodes to washing, and delivery of slimes to Cu/precious metal recovery plant	0.06
Cathode handling including stripping, washing, weighing, quality control, and delivery to loading docks	0.01
Anode scrap washing and melting, purchased scrap melting, anode casting, and anode delivery to tankhouse	0.03
Local services and overhead (accounting, clerical, environmental, human resources, laboratory, management, property taxes, safety)	0.01
Total	**0.12**

Direct operating costs including maintenance, for producing electrorefined cathode plates from 99.5% Cu anodes in a stainless steel mother blank electrorefinery. Anode cost is not included.

Table 21.11 Electrorefinery operating costs by cost component.

Component	Percentage of electrorefining cost
Operating and maintenance manpower, including supervision	28
Electricity	35
Maintenance, excluding labor	14
Natural gas	12
Local services and overhead (accounting, clerical, environmental, human resources, laboratory, management, property taxes, safety)	11
Total	**100**

Expenditures on manpower, electricity, and supplies (excluding anodes and scrap), by percentage. Anode cost is not included.

For scrap recovery to be profitable, the difference between refined copper sales price and scrap purchase price must exceed these treatment charges. When this is not so, scrap is held off the market. In 2011, many smelters are actively seeking scrap and scrap collection is very profitable.

21.8 Leach/solvent extraction/electrowinning costs

The investment and operating costs of heap leach/solvent extraction/electrowinning plants are listed in Tables 21.12 and 21.13. The costs are shown to be considerably lower

Table 21.12 Fixed investment costs for a heap leach/solvent extraction/electrowinning plant.

Component	$U.S. per annual tonne of copper
Heap leach system including leach pad, crusher, agglomerating drum, on—off heap building and removal equipment, piping, pumps, solution collection ponds, etc.	1450
Solvent extraction plant including mixer-settlers, pumps, piping, storage tanks, and initial first fill extractant and diluent	1040
Electrowinning plant including electrical equipment, polymer concrete cells,, Ir—Ta oxide—coated Ti anodes, 316L stainless steel cathodes, cranes, cathode stripping, washing and handling equipment	1060
Total	**3550**

The plant produces copper cathode plates ready for shipment from 0.5% Cu oxide ore. Ir—Ta oxide—coated Ti anodes, 316L stainless steel cathodes, and polymer concrete cells are used. Utilities and infrastructure are included in the plant capital costs. Mine investment cost is not included. Sulfur-burning sulfuric acid plant investment is not included.

Table 21.13 Direct operating costs of a heap leach/solvent extraction/electrowinning system (Mokmeli, 2020).

Item	$/kg of copper
Heap leach operation including crushing, acid curing, agglomeration, on—off heap construction/removal, solution delivery, and collection, excluding sulfuric acid	0.53
Sulfuric acid	0.45
Solvent extraction plant operation, including maintenance	0.05
Reagent make-up: extractant, diluent, guar, and $CoSO_4 \cdot 7H_2O$	0.18
Electrowinning tankhouse operation, delivering cathode plates to loadout platform	0.05
Local services and overhead (accounting, clerical, environmental, human resources, laboratory, management, property taxes, safety)	0.09
Total	**1.35**

The plant produces copper cathode plates ready for shipment from 0.5% Cu oxide ore. Ir—Ta oxide—coated Ti anodes, 316L stainless steel cathodes, and polymer concrete cells are used. Ore cost is not included.

than those for conventional concentration/smelting/refining complexes (Mokmeli, 2020). This accounts for the rapid adoption of leaching around the world, especially in Chile.

Unfortunately, chalcopyrite ore (the world's largest source of copper) cannot be processed by heap leach/solvent extraction/electrowinning (Chapter 13). Chalcopyrite ores must be treated by conventional concentration/smelting/refining/refining, irrespective of cost.

The small investment requirement of leach/solvent extraction/electrowinning plants is due to the small equipment and infrastructure requirements of these processes. Specifically, leaching and solvent extraction require much less equipment than concentrating, smelting, converting, and anode casting.

An interesting aspect of pyrometallurgical and hydrometallurgical copper extraction is sulfuric acid production and use. Hydrometallurgical copper extraction requires sulfuric acid (Chapter 14), and pyrometallurgical copper extraction produces it (Chapter 11). Companies with both processes benefit significantly from this synergistic effect, especially if the operations are close together.

21.9 Profitability

The key to a profitable mine-to-market copper operation is, of course, a large, high Cu-grade ore body. Such an ore body maximizes copper production per tonne of ore mined, moved, and processed. Optimal use of an ore body requires that each part of the ore body be processed by its most efficient method—leaching or concentrating/smelting. Separation of the ore body into milling ore, leaching ore, leaching waste, and unleachable waste is crucial for optimal utilization of the resource.

21.10 Summary

The total direct plus indirect cost of producing electrorefined copper from ore by conventional mining/concentration/smelting/refining is in the range of \$3–\$6 per kg of copper. The total direct plus indirect cost of producing electrowon copper cathodes from oxide and chalcocite ores (including mining) is \$1–\$2 per kg of copper.

Copper extraction is distinctly profitable when the selling price of copper is >\$6 per kg. It is unprofitable for some operations when the selling price falls below \$3 per kg. At the former price, the industry tends to expand. At the latter, it begins to contract.

The selling price of copper in mid-2019 is ~\$6 per kg, and has been rising since.

References

Boulamanti, A., & Moya, J. A. (2016). Production costs of the non-ferrous metals in the EU and other countries: Copper and zinc. *Resources Policy, 49*, 112–118.

Bunjaku, A., Johto, H., & Pesonen, L. (2019). *Detailed OPEX comparison of modern copper smelting technologies using HSC–SIM modeling*. https://www.mogroup.com/insights/blog/mining-and-metals/detailed-opex-comparison-of-modern-copper-smelting-technologies-using-hsc-sim/. (Accessed 22 October 2020).

Castillo, E., & Cantallopts, J. (2016). Copper smelting technologies: Chilean competitiveness analysis. In *Proceedings of copper 2016* (pp. 2977–2986). Tokyo: MMIJ.

Chadwick, J. (2010). Grasberg concentrator. *International Mining, 27*(5), 8–20.

Christensen, P., & Dysert, L. R. (2005). Cost estimate classification system – as applied in engineering, procurement and construction for the process industries. *AACE International Recommended Practice*,

18R—97R. http://www.costengineering.eu/Downloads/articles/AACE_CLASSIFICATION_SYSTEM.pdf. (Accessed 28 January 2021).

Freeport—McMoRan Copper and Gold. (2011). *Freeport—McMoRan copper and gold 4th quarter 2010 earnings conference call, Jan 20, 2011*. http://www.fcx.com/ir/2011present/FCX_4Q10CC.pdf. (Accessed 1 February 2011).

International Copper Study Group. (2018). *The world copper factbook 2018* Accessed 01.11.18 www.icsg.org.

Kendrick, M., Baum, W., Thompson, P., Wilkie, G., & Gottlieb, P. (2003). The use of the QemSCAN automated mineral analyzer at the Candelaria concentrator. In C. O. Gomez, & C. A. Barahona (Eds.), *Mineral processing: Vol. III. Copper/cobre 2003* (pp. 415—430). Montreal: CIM.

Mokmeli, M. (2020). Pre feasibility study in hydrometallurgical treatment of low—grade chalcopyrite ores from Sarcheshmeh copper mine. *Hydrometallurgy, 191*. https://doi.org/10.1016/j.hydromet.2019.105215. . (Accessed 18 December 2020)

Peters, M. S., & Timmerhaus, K. D. (1968). *Plant Design and economics for chemical Engineers* (2nd ed.). New York, NY: McGraw—Hill Book Company.

Sartain, C. (2010). *Xstrata copper: Driving the organic growth strategy*. http://www.xstrata.com/assets/pdf/xcu_speech_201004071_world_copper_conference.pdf. (Accessed 10 May 2010).

Supomo, A., Yap, E., Zheng, X., Banini, G., Mosher, J., & Partanen, A. (2008). PTFreeport Indonesia's mass—pull control strategy for rougher flotation. *Minerals Engineering, 21*, 808—816.

CHAPTER 22

Toward a sustainable copper processing

Since the first edition of *Extractive Metallurgy of Copper*, the processing of primary and secondary copper containing materials has experimented dramatic changes. Efforts have been made in the last 40 years to

- Improve energy efficiency by reducing the specific energy consumption.
- Increase sulfur capture.
- Adapt existing processes to declining copper grades.
- Develop new copper processing technologies.
- Optimize recovery of valuable elements
- Develop environmentally friendly technologies to stabilize and safely storage metallurgical wastes
- Develop technical solutions to improve equipment reliability
- Increase productivity
- Develop process synergies to recover valuable metals from secondary materials and industrial wastes as part of existing flowsheets or by coupling new processes to existing flowsheets.

Many of the above-mentioned aspects have been at the core of flowsheet transformations required to meet the increasing copper demand. This explosive copper demand has been driven by the urbanization needs of emerging economies and a growing middle class with more expectation for comfort and goods all over the world (Alvear Flores, Löbbus, Springub, Fallah−Mehrjardi, & Tappe, 2018). Although we are living unprecedented times in 2020 due to the CV19 pandemic, there is a consensus that the future will require even more copper to meet the expectations of a more sustainable planet.

22.1 Resource complexity and flowsheet solutions

The primary and secondary copper processing industry will continue facing challenges associated with resource complexity, regulatory legislation, lower ore grades and higher impurity levels, flowsheet optimization, energy utilization, water access, and pressure to reduce operating costs and remain competitive.

A global copper mine review reported by the ICSG found that the average global copper ore grade was 0.45% in reported reserves, and 0.65% copper in copper mine production in 2015 (Alvear Flores, Risopatron, & Pease, 2020). This reduction in grade can be seen in the declining copper concentrates grades processed by the main custom

Extractive Metallurgy of Copper, Sixth Edition
ISBN 978-0-12-821875-4, https://doi.org/10.1016/B978-0-12-821875-4.00014-6

smelters (Table 6.1, also see fifth and sixth editions of *Extractive Metallurgy of Copper*). Copper smelters have responded by increasing their processing capacity to achieve at least the same copper output, which at the same time increases acid production and slag generation, and in some cases operating costs.

In addition, the content in copper concentrates of deleterious impurity elements has also increased. Arsenic is a clear example of this trend. According to Metals Economics Institute of Japan, the arsenic content in copper concentrates processed in Japanese smelters has increased since 1991 from 400 ppm to over 1000 ppm in 2016 (Japan Oil, Gas and Metals National Corporation, 2020). When combining this figure with the reduction of copper concentrate grade, this represents a threefold increase in the units of As per unit of Cu in the feed.

Complexity in secondary processing is also a relevant aspect to consider. With nonferrous metals recycling intended to recover copper, nickel, tin, and precious metals, electronic equipment have consistently reduced the valuable metal content in their products, and equipment design have become more complex, making it more difficult to enrich the metallic content in these end-of-life materials before processing. As a result, pretreatment facilities are required to eliminate before the main smelting process elements that might negatively impact the direct or indirect processing of these materials. Off-gas handling also requires special attention to manage postcombustion and heat release, proper handling of halides, and avoidance of formation of dioxins and furans during cooling of the metallurgical off-gas streams.

Chapter 5 showed how copper smelters have gradually increased their processing capacities, up to 400,000 tonnes of copper per year, aiming to meet increasing demand for copper and remain competitive at the same time. From this perspective, a continuous growth of processing capacity to reduce unitary costs seems to be reaching slowly and steady reaching a plateau, from which it will become more difficult to improve profitability. Smelters have to be creative and look for alternative feedstocks that could help them to increase their margins without upsetting their metallurgical operations. Intake of diverse input materials, not only from primary sources but also from urban mining resources, with a multimetal recovery approach, will be one of the differentiating factors.

22.2 Multimetal flowsheet integration

Processing of complex copper-bearing materials requires a multimetal recovery approach to optimize metals recovery, minimize waste, and maximize commercial benefits. This approach involves a combination of pyrometallurgical and hydrometallurgical unit operations, and ideally synergies between base metals operations. Flowsheet interactions involving copper, lead, nickel, and zinc operations can allow the maximization of metallurgical recoveries at an optimum cost—benefit balance, taking advantage of the collector capacity of base metals, and more efficient and integrated minor metals and waste processing.

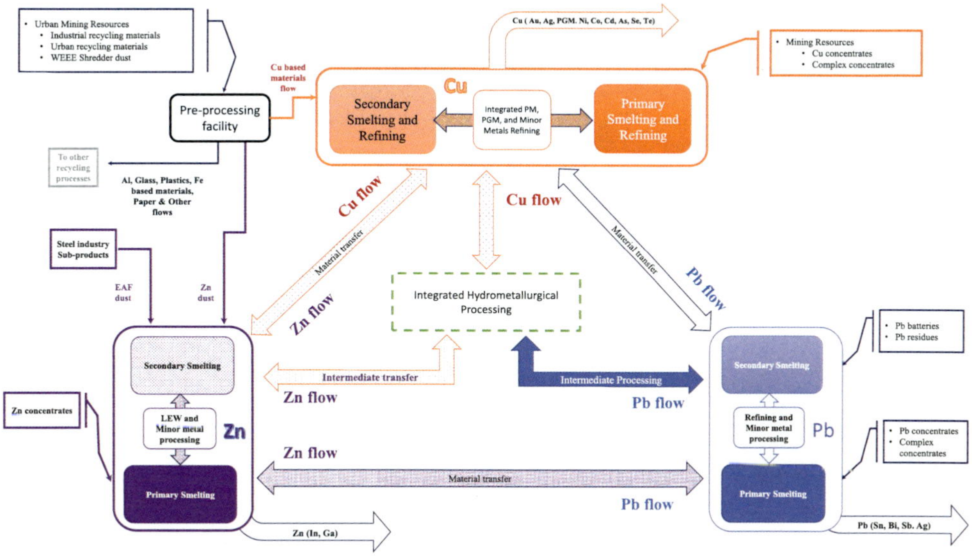

Figure 22.1 Integrated multimetal recovery concept with synergies between copper, lead and zinc processing facilities, processing primary and secondary materials. Option for an integrated hydrometallurgical plant in dotted green line highlight the to optimize minor, and hazardous metal processing.

The above-mentioned approach is illustrated by Fig. 22.1. Taking advantage of the base metal carrier properties, materials and metal flows can be optimized by processing materials according to their relative physical and chemical properties in a copper, lead, and/or zinc circuits. Primary and secondary resources can be pretreated as required to optimize recovery of other containing materials such as plastic, paper, aluminum base materials, and steel. For instance, waste incineration can help to remove zinc as a valuable subproduct prior to copper recycling and generate a more suitable product for copper processing. Halogen, dioxins, and furans removal can be also accomplished prior to incorporating the materials into more traditional base metal flowsheets.

Examples of the above mentioned multimetal processing principle have been successfully applied in processing of primary and secondary resources. The following examples can be mentioned:

- Kazzinc Ust-Kamenogorsk metallurgical complex: Integration of copper, lead, and zinc operations (Burrows & Azekenov, 2018)
- Boliden Rönnskär Smelter: Integration of primary copper and lead, and secondary copper (Testaye et al., 2017)
- JX Nippon mining multisite integration: Integration of primary and secondary copper, end-of-life Li Ion batteries, and industrial wastes (JX Metals, 2020)

- Dowa multisite integration: Combination of copper, lead, and zinc primary and secondary operations with the addition of industrial waste and recycle raw materials (Tokumoto, 2018)
- Mitsubishi Naoshima and Onahama smelters: Integration of primary and secondary copper sources with pretreatment facilities (Takagi, 2018)
- Horne smelter: Integration of primary and secondary copper sources (Glencore Fonderie Horne, 2017)
- Korea Zinc operations: Lead, zinc, and copper operations from primary and secondary sources (Korea Zinc, 2019)
- Umicore: Hoboken integrated base metal flowsheet for processing of complex primary and secondary materials (Alvear Flores, Löbbus, Springub, Fallah—Mehrjardi, & Tappe, 2018)
- Aurubis multisite operations: Integration of primary and secondary copper operations and a lead primary operation (Zschiesche et al., 2018)
- Codelco Chuquicamata: Integration of complex processing into a primary copper operation by addition of a roaster facility and a dust processing plant (Holmström et al., 2013).
- La Caridad smelter: standard and complex concentrates with dust leaching operations (Lopez Garcia, 2009)

Some of these examples are described in the following sections and summarised in Table 22.1.

22.2.1 Primary base metal integration: the Ust-Kamenogorsk metallurgical complex

The Kazzinc Ust-Kamenogorsk Metallurgical Complex (UKMC) is an example of a multimetal recovery plant combining copper, lead, and zinc (Fig. 22.2). The UKMC flowsheet is shown in Fig. 22.3. The copper plant involves processing of complex copper concentrates and other valuable materials in a process configuration involving a TSL ISASMELT™ furnace, electric furnace, Peirce—Smith converters, anode refining, and copper electrorefining. The lead circuit involves processing of lead concentrates in a TSL lead ISASMELT™ furnace, followed by the production of lead bullion in a blast furnace. A pyrometallurgical lead refining plant allows further lead refining, and recovery of copper intermediates. The zinc plant involves roasting, leaching, and electrowinning processes and a Waelz kilns plant to process zinc residues and recover valuable and precious metals. The anode slimes from the copper circuit and the Parkes crust from the lead circuit are refined to recover precious metals. Lead—silver alloys produced from Parkes crust and dried anode slimes from the copper refinery are processed in cupels. The doré metal produced in the cupels is spooned manually for casting into molds, creating anodes ready for refining (Burrows & Azekenov, 2018). A hydrometallurgical

Table 22.1 Some metallurgical integrated plants with multimetal recovery.

Plant and company	Base metals	Furnace type	Processing materials	Synergy with	Products
La Caridad, Grupo Mexico	Cu	Teniente converter Flash furnace Peirce–Smith converter	Standard and complex Cu concentrates	Dust leaching plant	Cu, Au, Ag, Se
Kosaka, Dowa	Cu, Pb	**CuAusmelt®** TSL furnace **Pb** Electric furnace	WEEE, Residues from Zn plant	Pb and ZNn Zn plant	Cu, Pb, Sn, precious metals
Ust-Kamenogorsk, Kazzinc	Cu, Pb, Zn	**Cu** ISASMELT™ – Pierce–Smith converter Anode furnace **Pb** ISASMELT™ Blast furnace **Zn** Roaster EW Waelz kiln	Polymetallic complex concentrates Residues	Pb and Zn plant	Cu, Pb, Zn, Cd, Tl, Se, Hg, Te, $ZnSO_4$, Na_2SbO_4 precious metals
Hamburg Pb smelter, Aurubis AG	Cu, Pb	**Cu** Flash FurnaceElectric FurnacePeirce-Smith Converters Anode Furnace **Cu–Pb** Electric furnace Peirce–Smith converter	Complex Cu–Pb concentrates	Copper smelter	Cu matte Pb bullion
Hoboken, Umicore precious metals	Cu, Pb, Zn, Ni	**Cu** ISASMELT™ Leaching-EW **Pb** ISASMELT Blast furnace	Complex Cu–Pb concentrates WEEE, Recycling materials Residues	Base metals and precious metals circuits	Cu, Pb bullion Ni, precious metals

Continued

Table 22.1 Some metallurgical integrated plants with multimetal recovery.—cont'd

Plant and company	Base metals	Furnace type	Processing materials	Synergy with	Products
JX Nippon mining	Cu, Pb	**Cu** Flash furnace Peirce–Smith converters Anode furnace	Standard copper concentrates Industrial residues WEEE Batteries	Intersite synergies with primary and secondary copper, Ni recovery plant	Cu, Ni, Au, Ag, PGM, se, Te, Pb, Bi, Sb, Ni, Co, Mn, Li.
Korea zinc	Cu, Pb, Zn, Ni.	**Pb** TSLAusmelt® TSL QSL Reactor **Zn** TSLAusmelt® TSL **Cu**	Complex Pb–Zn concentrates and residues	Cu, Pb, Zn, and PM plants	Cu, Pb, Zn, Ag, Au, In, Bi, Sb, Cd, Te, Se, Co
Ronnskar smelter, Boliden	Cu, Pb, Zn, Ni.	**Cu – Process line 1** Dryer Flash furnace Peirce–Smith converters Anode furnace Refinery **Cu – Process line 2** Fluidized bed roaster Electric furnace Settling furnace **Pb** Dryer Kaldo plant Lead refinery **Zn** Fuming furnace Clinker plant	Cu blended concentrates WEEE Recycling materials Lead concentrates Complex concentrates Secondary materials concentrates Lead concentrate Electronic scrap	Cu, Pb, and Zn plants	Cu, Cu and Ni sulfates, Au, Ag, Se Copper matte to process line 1 Slag to Zn fuming Pb, Zn-clinker
Naoshima smelter, Mitsubishi materials	Cu	**Cu** Industrial waste incinerator Mitsubishi process	Copper concentrates WEEE Recycling materials	Multisite integration (Cu, Ni, Pb base metals)	Cu, Ni, PM, Se, Te, Pb, Bi, Sb, PGM,

Modified from Flores, GA., Risopatron, C., & Pease, J. (2020) Processing of complex materials in the copper industry: Challenges and opportunities ahead. *JOM 72*, 3447–3461.

Figure 22.2 Overall view of the Ust-Kamenogorsk Metallurgical Complex in Kazakhstan. *(Photo courtesy of Nick Popovic, Glencore.)*

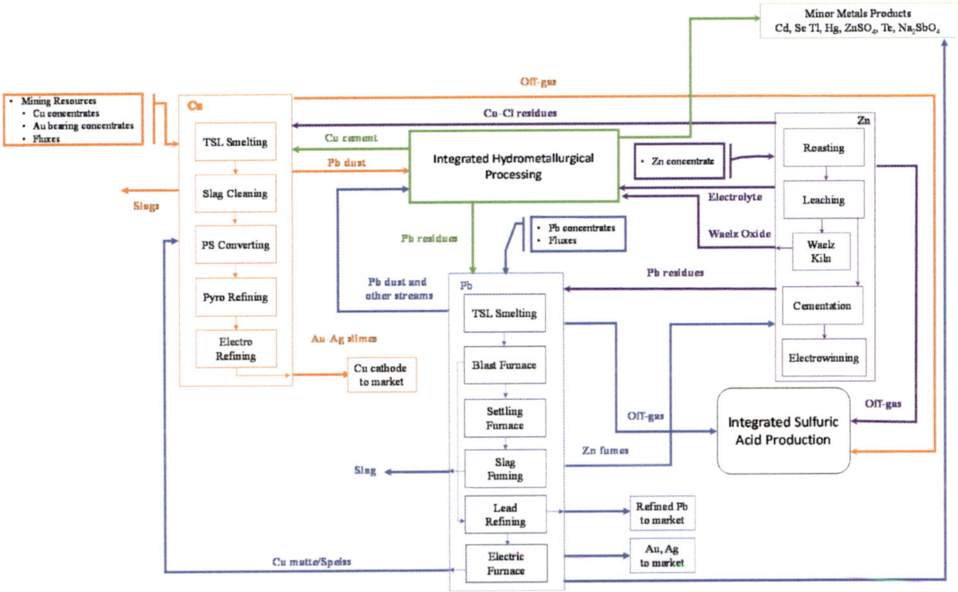

Figure 22.3 Integrated primary multimetal processing and recovery concept developed by Kazzinc at Ust-Kamenogorsk Metallurgical Complex (Burrows & Azekenov, 2018).

plant acts a key element to treat intermediate products generated in the base metal circuits, recovering valuable minor metals, enriching base metal streams, and stabilizing residues.

The integration of the three base metals metallurgical circuits is the key to optimizing metallurgical flows, increasing concentration of elements in different streams, minimizing metallurgical wastes, and optimizing metal recoveries. Some examples of these synergies illustrated in the UKMC flowsheet are as follows:

- Integrated hydrometallurgical plant to recover minor metals
- Integrated precious metal plant
- Integrated zinc residue processing plant in synergy with hydrometallurgical plant.

The integrated approach of the UKMC using, copper, lead, and zinc base metals as the main collectors allows the recovery of precious metals, platinum group metals, selenium, tellurium, cadmium, thallium, mercury, zinc sulfate, and sodium antimonate.

22.2.2 Primary and urban mining integration: Japanese industrial examples

The Japanese nonferrous industry has developed a multimetal recovery approach processing primary raw materials in conjunction with recyclable materials, allowing at the same time the processing and detoxification of industrial wastes.

Industrial wastes and recyclable materials are preprocessed and detoxified in dedicated plants, generating metal containing streams that are then directed to the most suitable processing plant, depending on their base metal content. This approach allows to maximize synergies with existing base metals processing plants. Industrial examples of this approach are as follows:

- JX Nippon Mining multisite integration for processing on industrial waste, recycling materials, and primary materials (JX Nippon Mining, 2020)
- Dowa group Eco-recycling Network (Kosaka, 2019)
- Mitsubishi Onahama and Naoshima smelters primary and recycling materials processing (Takagi, 2018).

Fig. 22.4 shows the integrated approach developed by JX group (JX Nippon Mining & Metals, 2010). In contrast with UKMC, processing plants are in different geographic locations requiring transportation of intermediate and preprocessed materials to the processing plants. The JX integrated approach involves direct processing and/or preprocessing of recycling materials, detoxification of industrial residues, and subsequent direction of subproducts to copper and lead processing plants or battery materials processing facilities.

This eco-processing approach offer several advantages such as the following:

- Incorporate back into the metal flow base metals contained in WEEE (waste electrical and electronic, equipment)
- Minimize waste generation
- Detoxification and minimization of industrial waste prior to landfill

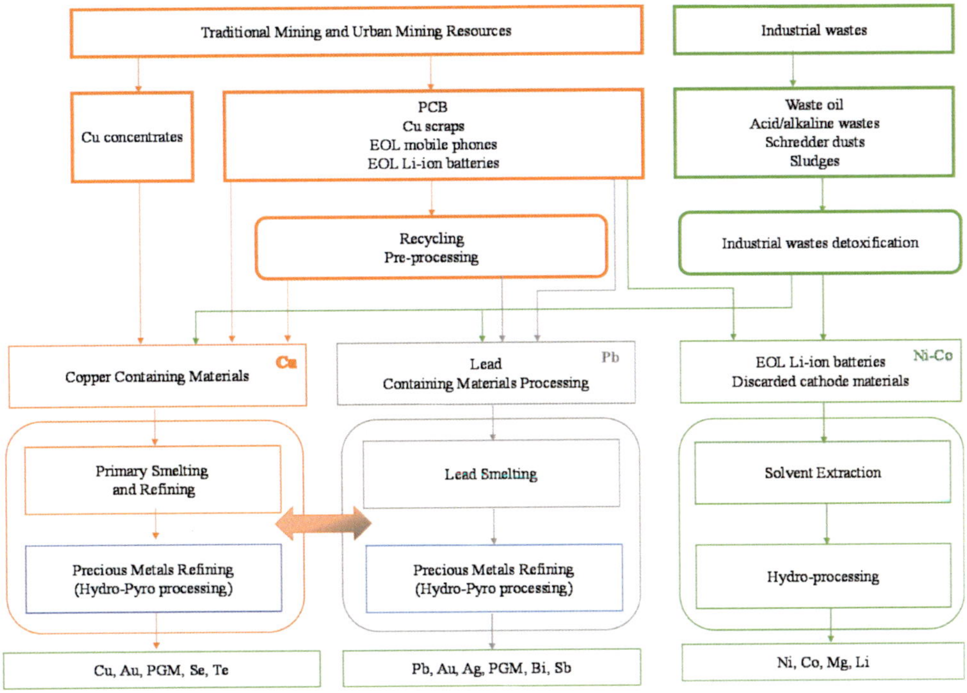

Figure 22.4 Base metal smelters as part of an integrated eco-processing system processing traditional and urban mining resources. Japan JX Group approach (JX Nippon Mining and Metals, 2016).

- Stabilization of residues
- Preprocessing of WEEE and removal of harmful materials prior to processing in primary smelters (i.e., halogens that could chemically attack off-gas equipment)

The main copper recovery hub for JX approach is the Saganoseki copper smelter where urban mining resources are processed with copper concentrates allowing recovery of precious metals from WEEE. Lead-containing materials are processed at the Hitachi lead smelter. A dedicated battery recycling processing plant allows the recovery of cobalt and nickel.

A regional eco-system approach was also developed by Dowa in the northern region of Akita, Japan. Taking advantage of the location of copper, lead, and zinc processing plants, Dowa developed an eco-recycling network to facilitate recovery and recycling of metals from traditional urban mining resources, allowing at the same time the processing of general and industrial wastes. Fig. 22.5 shows the integrated approach developed by Dowa Group. The integrated eco-system allows the collection and preprocessing of end-of-life electric and electronic equipment, industrial wastes, and contaminated soils. These materials become raw materials for processing in zinc, copper, and lead processing lines located in two different locations.

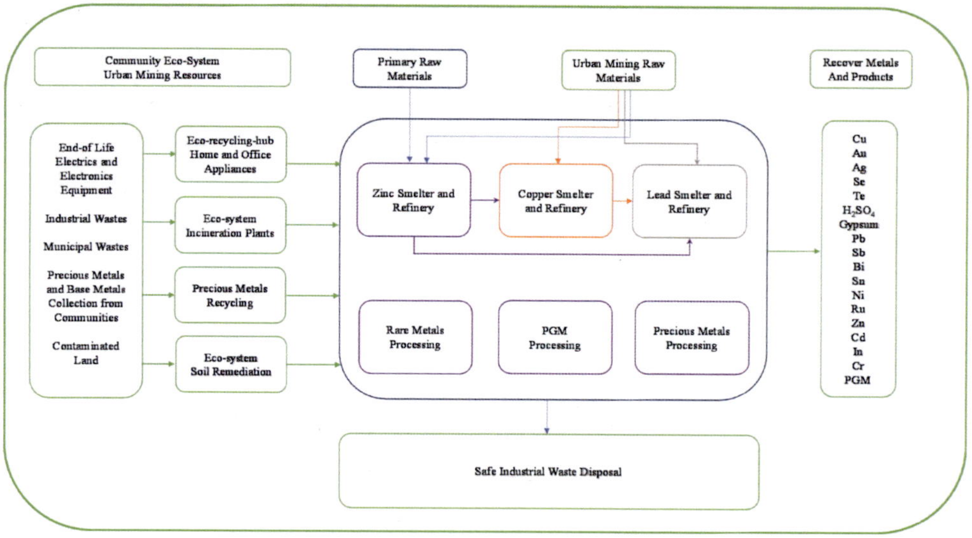

Figure 22.5 Example of integrated industrial metal production eco-system processing traditional and urban mining resources with base smelters as core element to allow efficient metal recovery (Dowa Kosaka, 2019).

22.2.3 European examples of integration

Examples of synergies to optimize processing of primary and secondary raw materials can be found in the integrated flowsheet of the following plants:

- Umicore Hoboken plant: Flowsheet developed to process complex primary and secondary raw materials. The flowsheet combines copper, lead, precious metals refining, and special metals refining operations (Vanbellen, 2010)
- Aurubis: Combination of primary copper and lead operations to optimize processing of complex materials and recovery of precious metals (Aurubis AG, 2020)
- Boliden: Integration of copper, lead, and zinc operations with intake of primary and secondary materials (Testaye, 2017).

Fig. 22.6 shows the integrated processing flowsheet developed by Aurubis to process complex concentrates. The overall flowsheet combination allows the integration and transfer of materials from the lead to the copper smelter. The lead process circuit consists of an electric furnace that processes in different campaign copper—lead containing materials and a converter to produce copper matte and blister copper. Products from the copper campaign are transferred to the copper smelter. Lead generated in the process is pyrorefined. The lead plant is integrated with the anode slimes plant and the precious metal refining plant.

On the secondary copper side, Aurubis has developed the Kayser Recycling System. This flowsheet, shown in Fig. 22.7, and developed to recover copper, lead, tin, and precious metals has been in operation since 2002 in Lünen, Germany. The process

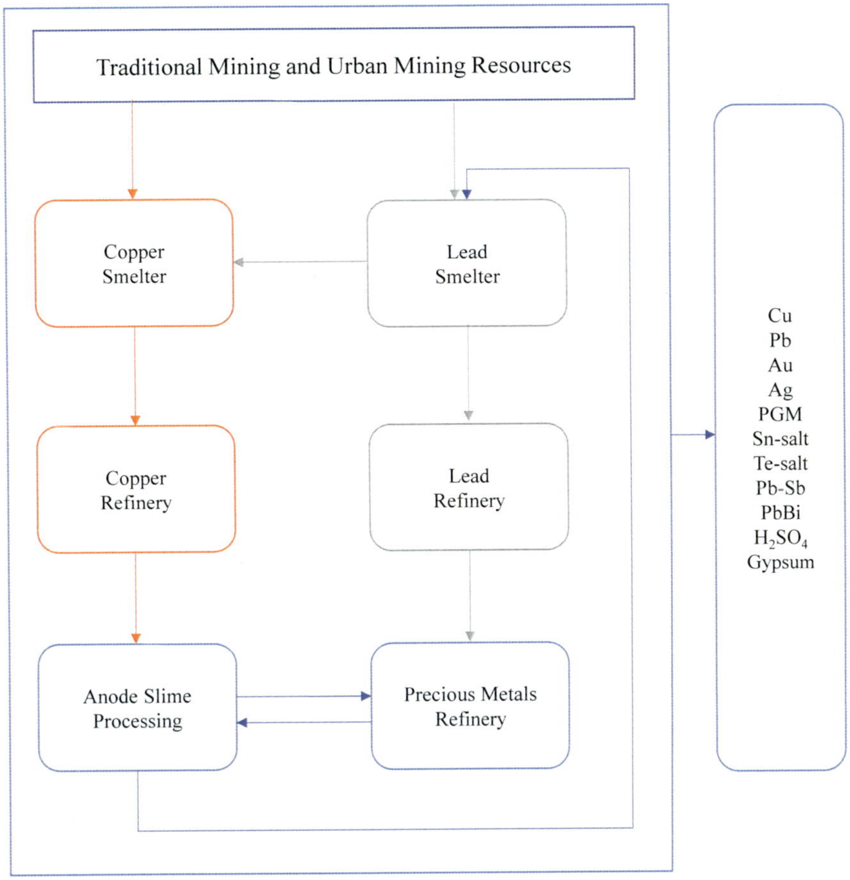

Figure 22.6 Example of integrated copper and lead processing at Aurubis AG. Modified from Zschiesche et al., 2018.

involves the reductive smelting of secondary copper sources in a TSL ISASMELT™ furnace, converting in a TBRC furnace and subsequent anode refining. A lead–tin furnace allows the recovery of a lead–tin alloy. The smelting process conducted at a reductive environment allows the fuming of zinc, which is recovered as a complex zinc–bearing residue. A partnership with a third party company allows the processing of the zinc containing material, recover of copper and lead, safe disposal of residual elements, and valorization of zinc for further use (Aurubis Homepage, 2020).

These examples show how synergies can create opportunities to recover metals, minimize waste streams, optimize recovery, and finally most importantly achieve a sustainable processing of materials. In most of the above-mentioned cases, integration of hydrometallurgical and pyrometallurgical plants allows the optimization of multimetal recovery and minimization of wastes.

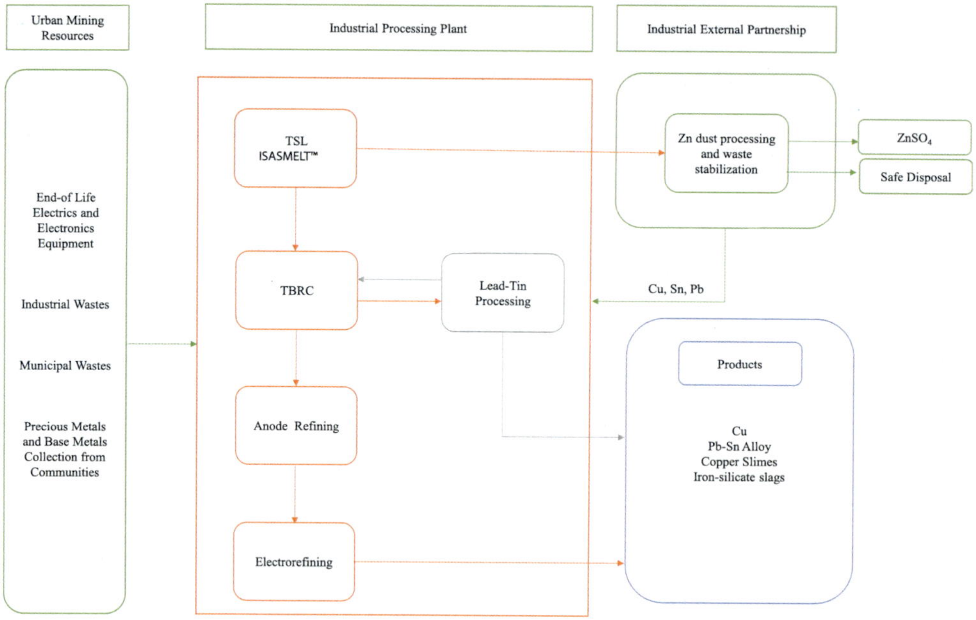

Figure 22.7 Example of integrated copper, zinc, and lead processing from secondary sources at Aurubis AG. *(Modified from Alvear Flores, GA, Risopatron, C., & Pease, J. (2020) Processing of complex materials in the copper industry: Challenges and opportunities ahead.* JOM 72, *3447–3461.; Aurubis AG.* Closing the loop *(2020). www.aurubis.com/de/verantwortung/ressourceneffizienz-und-recycling/closing-the-loop. Accessed 01.09.20.)*

Integration of complexity by processing of primary and urban mining resources, using base metals metallurgy as a core, will become even more relevant for the future of smelter operations as they become essential players of a circular economy enhancing recovery, recycle, and reuse of metals.

Fig. 22.8 shows the flowsheet developed by Umicore Precious Metals for the processing of complex materials. The main processing unit is a TSL ISASMELT™ furnace. Complex lead/copper materials, containing valuable minor elements, are treated in this furnace in a two-step smelting and converting process. The smelting stage involves the oxidation of the feed to form a copper matte and a smelting slag (see Chapter 9 for TSL operating principles). The silica-based lead-rich slag produced during smelting is tapped and slow cooled for further reduction and lead bullion production in a blast furnace. The copper matte is converted in the TSL ISASMELT™ furnace after the tapping of the smelting slag. The flowsheet allows maximization of precious metals recovery. The flowsheet is complemented with a precious and minor metals processing plant (Vanbellen et al., 2006).

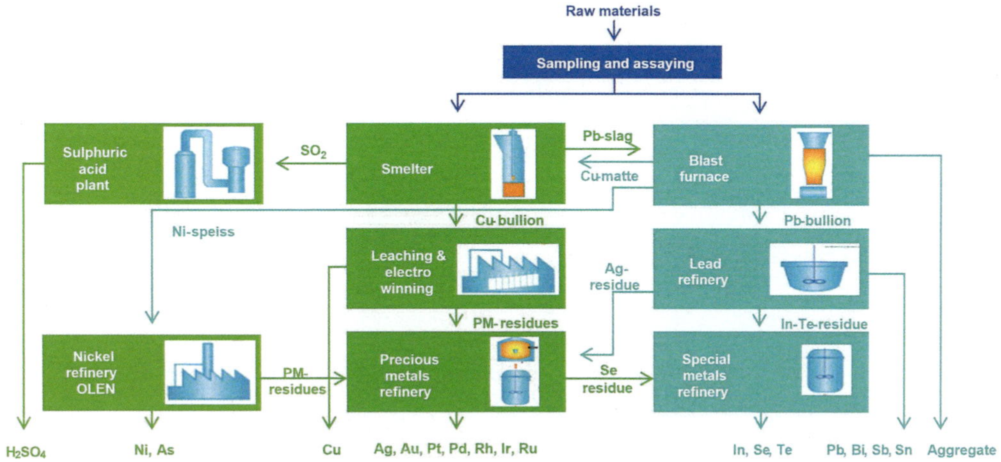

Figure 22.8 Example of integrated metallurgical circuit for processing of complex materials at Umicore Precious Metals. *(Figure reproduced with permission of Umicore Precious Metals.)*

22.3 Concluding remarks

This book has described through many decades the evolution of copper processing, explaining in detail technologies involved in the processing of copper concentrates from mine to cathode, and providing relevant metallurgical information to support the education and training of students and technical personnel. In the last 40 years, the industry has evolved to become more energy efficient (Coursol et al., 2010), environmentally sustainable, and competitive. Pyrometallurgical production has also shifted to China and resources have become more complex. Recycling has also become more relevant as society has transformed its core to become a high intensity metal consuming one.

These changes have also been at the core to evaluate the required technological transformation in metallurgical flowsheet to adequate them to integrate processing of primary and urban mining resources.

As described in this chapter, copper smelters have therefore become industrial assets—part of integrated processing networks enabling the sustainable production of metals from complex sources. As our modern society moves towards a more digital society, with intense use of metals to enhance sustainable living, process engineers and technologists will be required to constantly evaluate how existing metallurgical processes will handle more complex materials recover from urban and traditional mining sources. Eco-processing systems integrating the processing of industrial waste, WEEE and other recycling materials, traditional copper ores, and other potential sources will be required to a larger extent.

In addition, complexity derived from the increment of deleterious elements such as arsenic will require the industry to revisit the current strategy of blending. Preprocessing of complex concentrates, involving hydrometallurgical processing or roasting, might be required.

Eco-processing systems involve an efficient and sustainable recovery of metals. From this perspective, the following elements will continue to be relevant to ensure sustainable eco-friendly copper smelters:

- Base metal integration to allow efficient recovery of minor and valuable metals and safe disposal of metallurgical wastes
- Valorization of copper slags as subproducts
- Integration with industrial waste processing facilities to minimize overall waste generation and maximize recovery of valuable elements
- Maximization of impurity capacity usage in copper smelters
- Development of integrated hydrometallurgical solutions to complement smelting processing

References

Alvear Flores, G., Risopatron, C., & Pease, J. (2020). Processing of complex materials in the copper industry: Challenges and opportunities ahead. *JOM, 72*, 3447—3461.

Alvear Flores, G. R. F., Löbbus, M., Springub, B., Fallah—Mehrjardi, A., & Tappe, A. (2018). Sulfide smelting: Thirty—five years of continuous efforts to find new value adding solutions. In B. Davis, et al. (Eds.), *Extraction 2018* (pp. 39—56). Warrendale, PA: TMS.

Aurubis, A. G. (2020). *Closing the loop* Accessed 01.09.20 www.aurubis.com/de/verantwortung/ressourceneffizienz-und-recycling/closing-the-loop.

Burrows, A., & Azekenov, T. (2018). Ust—kamenogorsk metallurgical complex: A silent achiever. In B. Davis, et al. (Eds.), *Extraction 2018* (pp. 365—378). Warrendale, PA: TMS.

Dowa Kosaka. (2019). *Creating the world's leading combined—recycling smelter*. https://dowa-kosaka.co.jp/sys/wp-content/themes/dowa-kosaka/file/kosakasr_brochure.pdf. (Accessed 28 September 2020).

Glencore Fonderie Horne. (2017). *Nous sommes des alliés*. www.fonderiehorne.ca/fr/Pages/home.aspx. (Accessed 28 September 2020).

Holström, A., Berg, G., Andersson, M., Carrasco, S., & Deutsch, D. (2013). Development of partial roasting technology for arsenic containing copper concentrates. In *Proceedings of copper 2013* (pp. 965—978). Santiago: IIMCH.

Japan Oil, Gas and Metals National Corporation. (2020). *Metals: Technological development*. http://www.jogmec.go.jp/english/stockpiling/metal_10_000002.html. (Accessed 28 September 2020).

JX Nippon Mining & Metals. (2010). *Business segments: Recycling and environmental services*. www.nmm.jx-group.co.jp/english/industry/recycling/. (Accessed 28 September 2020).

JX Nippon Mining & Metals. (2016). *Sustainability report*. https://www.nmm.jxgroup.co.jp/english/sustainabilityreport/pdf/report2016_e_full.pdf. (Accessed 28 September 2020).

Korea Zinc. (2019). *Half a century of operational know—how*. https://www.koreazinc.co.kr/english/main/page/main.aspx. (Accessed 28 September 2020).

López García, J. (2009). Planta de tratamiento de polvos y efluentes Mexicana de Cobre S.A. de C. V. In *XXVIII Convención Minera Internacional* (pp. 498—503). Veracruz: AIMMGM AC.

Takagi, N. (2018). Current status and future developments of E-scrap processing at Mitsubishi Materials Corporation. In *E-scrap symposium 2018*. Tokyo: University of Tokyo.

Testaye, F., Lindberg, D., & Hamuyuni, J. (2017). Valuable metals and energy recovery from electronic waste streams. In L. Zhang, et al. (Eds.), *Energy technology 2017* (pp. 103–116). Warrendale, PA: TMS.

Tokumoto, T. (2018). Recycling initiatives at Kosaka smelter. In *E-scrap symposium 2018*. Institute of Industrial Science University of Tokyo, Non-ferrous Resources Recycling Research Division, 30th November 2018.

Vanbellen, F. (2010). *The Hoboken recycling plant*. www.umicore.com/storage/migrate/2010CMD_Hoboken.pdf. (Accessed 28 September 2020).

Vanbellen, F., & Chintinne, M. (2006). The precious art of metals recycling. In F. Kongoli, & R. G. Reddy (Eds.), *Sohn international symposium* (vol. 1, pp. 43–52). Warrendale, PA: TMS.

Zschiesche, C., Ayhan, M., & Antrekowitsch, J. (2018). Challenges and opportunities of a lead smelting process for complex feed mixture. In B. Davis, et al. (Eds.), *Extraction 2018* (pp. 325–338). Warrendale, PA: TMS.

Suggested readings

Reuter, M. A., & van Schaik, A. (2013). Resource efficient metal and material recycling. In A. Kvithyld, et al. (Eds.), *REWAS 2013* (pp. 332–340). Warrendale, PA: TMS.

Index

Note: 'Page numbers followed by "*f*" indicate figures and "*t*" indicate tables.'

D

E

F

G

N

O

P

Q

R

S

T

U

V

W

X

Y